Stochastic Programming

Applications in Finance, Energy, Planning and Logistics

World Scientific Series in Finance

(ISSN: 2010-1082)

Published

Vol. 1 Bridging the GAAP: Recent Advances in Finance and Accounting
edited by Itzhak Venezia & Zvi Wiener (The Hebrew University of Jerusalem, Israel)

Vol. 2 Calendar Anomalies and Arbitrage
by William T. Ziemba (University of British Columbia, Canada & ICMA Centre, University of Reading, UK)

Vol. 3 Social Security's Investment Shortfall: $8 Trillion Plus — and the Way Forward
Plus How the US Goverment's Financial Deficit Reporting = 64 Madoffs
by Nils H. Hakansson (University of California, Berkeley, USA)

Vol. 4 Stochastic Programming: Applications in Finance, Energy, Planning and Logistics
edited by Horand Gassmann (Dalhousie University, Canada) & William T. Ziemba (University of British Columbia, Canada & ICMA Centre, University of Reading, UK)

Forthcoming

Quantitative Methods in Risk Analysis: A Practitioner's Guide
by Michael Foster & Leonard Maclean

World Scientific Series
in FINANCE vol. 4

Stochastic Programming

Applications in Finance, Energy, Planning and Logistics

Horand I. Gassmann
Dalhousie University, Canada

William T. Ziemba
University of British Columbia, Canada (Emeritus)
ICMA Centre, University of Reading

and Visiting Professor
University of Cyprus
Luiss Guido Carli University, Rome
Sabanci University, Istanbul
Korea Advanced Institute of Science and Technology

World Scientific

NEW JERSEY • LONDON • SINGAPORE • BEIJING • SHANGHAI • HONG KONG • TAIPEI • CHENNAI

Published by

World Scientific Publishing Co. Pte. Ltd.
5 Toh Tuck Link, Singapore 596224
USA office: 27 Warren Street, Suite 401-402, Hackensack, NJ 07601
UK office: 57 Shelton Street, Covent Garden, London WC2H 9HE

Library of Congress Cataloging-in-Publication Data
Stochastic programming : applications in finance, energy, planning and logistics / [edited by]
Horand Gassmann, William Ziemba.
p. cm.
ISBN 978-9814407502
1. Mathematical optimization. 2. Mathematical optimization--Industrial applications.
3. Stochastic processes--Econometric models. 4. Stochastic programming. 5. Decision making.
6. Uncertainty. I. Gassmann, Horand. II. Ziemba, W. T.
HB143.7.S76 2012
519.7--dc23

2012027536

British Library Cataloguing-in-Publication Data
A catalogue record for this book is available from the British Library.

In-house Editor: Lum Pui Yee

Typeset by Stallion Press
Email: enquiries@stallionpress.com

Printed in Singapore by World Scientific Printers.

Acknowledgements

Special thanks to all the authors of the papers in this volume, most of whom also attended the XII International Conference on Stochastic Programming held in Halifax, Nova Scotia in August 2010. Thanks also go to those who presented in Halifax and helped make it a successful conference. Ziemba also thanks Horand Gassmann and others who helped him update the conferences and books sections that appear here.

Horand Gassmann organized ICSP 12 with the help of many people. Stein W. Wallace was the Scientific Program Chair, while the Local Organizing Committee included L.C. MacLean, Y. Zhao and G.F. MacLean. The success of the conference was largely due to the tireless efforts of Janet Lord whose assistance in the organization and the running of the seven-day meeting was indispensable and greatly appreciated. Further thanks are due to the staff at Destination Halifax and Dalhousie Conference Services, as well as all levels of governance at Dalhousie University, from the President, Dr. Tom Traves, to the Dean of the Faculty of Management, Dr. Peggy Cunningham, and the Director of the School of Business Administration, Dr. Greg Hebb. Y. Zhao organized the student volunteers, Jing Liu, Ying Ma, Muting Zhang, Peter Lingyun Ye, Wen Wen Pei, and William Cai.

Lu Ping designed the logo and artwork for the conference; Ronald Hochreiter made available the prototype of the conference web site; further web design was done by John Wang and the staff at the Dalhousie Web Authoring Group. Shabbir Ahmed, Maarten van der Vlerk and the other members of COSP lent their support and expertise on numerous occasions. As has become customary at these meetings, the conference was preceded by a workshop aimed mostly at PhD students and newcomers to the field. Organized by Suvrajeet Sen, the workshop featured seven tutorials by Anton Kleywegt, Huseyin Topaloglu, Werner Römisch, Güzin Bayraksan, René Henrion, Maarten van der Vlerk, and R. Tyrrell Rockafellar. It was attended by more than 60 participants. The tutorial presentations are available online at `http://stoprog.org/resources.html\#tutorials`.

We thank the plenary speakers, Georg Pflug, David Morton, Theodore G. Crainic, Asgeir Tomasgard, Hercules Vladimirou, Andy Philpott and William T. Ziemba.

Thanks go to the sponsors and exhibitors, including Frontline Solvers, Lindo Systems, Inc., OptiRisk Systems, AORDA, Maximal Software, AMPL LLC, Pearson International, Nelson Education, Lily Investments Ltd, RCR Hospitality Group and IBM. In addition to being the major sponsor and exhibitor, IBM also provided the conference bags, for which we are especially grateful.

The field of stochastic programming has grown since its beginning in the 1950s and we would like to acknowledge our debt to George B. Dantzig, Martin Beale, Roy Radner, David F. Votaw, Jr., Abraham Charnes, William Cooper, Albert Madansky, Gerhard Tintner, and Richard Bellman for their early papers that posed problems using optimization.

Horand I. Gassmann
William T. Ziemba
March 2012

List of Contributors

Abramov, A, Department of Risks, Bank ZENIT, Moscow, Russia

Ait-Kadi, D, Interuniversity Research Centre on Enterprise Networks, Logistics and Transportation (CIRRELT), Quebec, Canada

Allevi, E, Department of Quantitative Methods, University of Brescia, Italy

Aranburu, L, Departamiento de Economía Aplicada III, Universidad del Pais Vasco, Bilbao (Vizcaya), Spain

Barnes, JW, Graduate Program in Operations Research and Industrial Engineering, The University of Texas at Austin, Austin, TX, USA

Benajam, W, Laboratoire de Recherche en Informatique, Université de Paris Sud, France

Benk, M, Department of Finance, WHU — Otto-Beisheim School of Management, D-56179 Vallendar, Germany

Bertocchi, M, Department of Mathematics, Statistics, Computer Science and Applications, University of Bergamo, Italy

Consigli, G, Department of Mathematics, Statistics, Computer Science and Applications, University of Bergamo, Italy

Davis, M, Mathematics Department, Imperial College, London, UK

Diniz, AL, Energy Optimization and Environment Department, CEPEL — Brazilian Electric Energy Research Center, Rio de Janeiro, Brazil

Edirisinghe, NCP, College of Business, University of Tennessee, Knoxville, TN, USA

Escudero, LF, Departamiento de Estadística e Investigación Operativa, Universidad Rey Juan Carlos, Móstoles (Madrid), Spain

Fleten, S-E, Department of Industrial Economics and Technology Management, Norwegian University of Science and Technology, Trondheim, Norway

Gaivoronski, A, Department of Industrial Economics and Technology Management, Norwegian University of Science and Technology, Trondheim, Norway

Garín, MA, Departamiento de Economía Aplicada III, Universidad del Pais Vasco, Bilbao (Vizcaya), Spain

Golembiovsky, D, Department of Risks, Bank ZENIT, Moscow, Russia

Hæreid, MB, Department for Industrial Economics and Technology Management, Norwegian University of Science and Technology, Trondheim, Norway

Hellemo, L, Department for Industrial Economics and Technology Management, Norwegian University of Science and Technology, Trondheim, Norway

Iaquinta, G, Department of Mathematics, Statistics, Computer Science and Applications, University of Bergamo, Italy

Kazemi Zanjani, M, Department of Mechanical and Industrial Engineering, Concordia University, Montreal, Canada

Kim, M, KAIST, Daejeon, South Korea

Lie, TT, Department of Electrical & Electronic Engineering, Auckland University of Technology, New Zealand

Lisser, A, Laboratoire de Recherche en Informatique, Université de Paris Sud, France

Lleo, S, Finance Department, Reims Management School, Reims, France

Maceira, MEP, Energy Optimization and Environment Department, CEPEL — Brazilian Electric Energy Research Center, Rio de Janeiro, Brazil

Maggioni, F, Department of Mathematics, Statistics, Computer Science and Applications, University of Bergamo, Italy

Michalopoulos, DP, Graduate Program in Operations Research and Industrial Engineering, The University of Texas at Austin, Austin, TX, USA

Midthun, K, Department for Applied Economics and Operations Research, SINTEF Technology and Society, Trondheim, Norway

Moriggia, V, Department of Mathematics, Statistics, Computer Science and Applications, University of Bergamo, Italy

Morton, DP, Graduate Program in Operations Research and Industrial Engineering, The University of Texas at Austin, Austin, TX, USA

Mulvey, JM, Department of Financial Engineering and Operations Research, Princeton University, Princeton, NJ, USA

Nourelfath, M, Department of Mechanical Engineering, Université Laval, Quebec, Canada

Pérez G, Departamiento de Matemática Aplicada, Estadística e Investigación Operativa, Universidad del Pais Vasco, Leioa (Vizcaya), Spain

Philpott, A, Department of Engineering Science, University of Auckland, New Zealand

Potra, FA, Department of Mathematics & Statistics, University of Maryland, College Park, MD, USA

Pritchard, G, Department of Statistics, University of Auckland, New Zealand

Schütz, P, SINTEF Technology & Society, Trondheim, Norway

Simsek, KD, Business School, Sabanci University, Istanbul, Turkey

Tomasgard, A, Department for Industrial Economics and Technology Management, Norwegian University of Science and Technology, Trondheim, Norway

Wallace, SW, Department of Management Science, Lancaster University Management School, Lancaster, UK

Werner, A, Department for Applied Economics and Operations Research, SINTEF Technology and Society, Trondheim, Norway

Zhang, X, College of Business, Austin Peay State University, Clarksville, TN, USA

Ziemba, WT, University of British Columbia (Emeritus), Canada, and ICMA Centre, Reading, UK

Preface

Most practical decision problems involve uncertainty. Stochastic programming is the study of procedures for decision making under uncertainty over time. The uncertainty can be in the model's parameters or in the model itself. Parameters may be uncertain because of lack of reliable data, measurement errors, future and unobservable events, etc. The uncertainty of events, details of the problem structures and constraints and the risk/payoff of decisions are modeled in an optimization framework. High performance workstations and PCs are used to enable exact and approximate algorithms to determine robust decisions that hedge against future uncertainty. Then as the uncertainty becomes known period-by-period, recourse decisions responding to the new information can be made.

Significant applications of stochastic programming have been made in many areas. These include transportation, production planning, energy and environmental policy, telecommunications, forest and fishery harvest management, hydropower and water resources, various areas of engineering and financial modeling including risk management, portfolio theory and management, currency, bond and asset and liability management.

History of Stochastic Programming

The first international conference on stochastic programming was organized by Michael Dempster in 1974 in Oxford, England. The second international conference was organized by András Prékopa and held in Köszeg, Hungary in 1981. The Committee on Stochastic Programming (COSP) was formed at this meeting as a branch of the Mathematical Programming Society. Subsequent international conferences have been the responsibility of this committee. (COSP is currently organized as a Technical Section within the renamed Mathematical Optimization Society.)

The current committee members are:

Güzin Bayraksan University of Arizona, Tucson, USA
Giorgio Consigli, Bergamo University, Italy

Csaba Fábián, Kecskemét College, Hungary
Horand Gassmann, Dalhousie University, Halifax, Canada
Stein-Erik Fleten, Norwegian University of Science and Technology, Trondheim, Norway
Anton Kleywegt, Georgia Institute of Technology, Atlanta, USA
Jeff Linderoth, University of Wisconsin-Madison, USA
Asgeir Tomasgard (Chair), Norwegian University of Science and Technology, and SINTEF, Trondheim, Norway

The intent of COSP was to hold an International Conference on Stochastic Programming at intervals of approximately three years during the year following the Mathematical Programming meeting. The third international conference on stochastic programming was held in 1983 in Laxenburg, Austria at the International Institute of Applied Systems Analysis (IIASA) and was organized by Roger Wets. The fourth conference followed by three years the IIASA meeting and was held in Prague in 1986 and organized by Jitka Dupačová. This firmly established the three-year cycle, which was continued with the fifth conference in 1989 in Ann Arbor at the University of Michigan, organized by John Birge. The sixth conference was in September 1992 in Udine, Italy, organized by Gianni Andreatta and Gabriella Salinetti. The seventh conference, organized by Reuven Rubinstein in 1995, was held in Naharaya, Israel.

The eighth conference organized by William T Ziemba in 1998 was held in Vancouver at the University of British Columbia. The 9th conference was held in Berlin at Humboldt University in 2001 and organized by Werner Römisch and Rüdiger Schultz. The 10th conference was held in Tuscon, Arizona at the University of Arizona in 2004 organized by Julie Higle and Suvrajeet Sen. The 11th conference was held in Vienna at the University of Vienna and was organized by Georg Pflug. The 12th conference was held in Halifax at Dalhousie University in 2010, organized by Horand Gassmann. The 13th conference will be held in Bergamo, Italy at the University of Bergamo in 2013, organized by Marida Bertocchi and Giorgio Consigli.

There are also many conferences such as those organized by INFORMS and IFIP that feature large segments of the program devoted to stochastic programming. There have also been a number of smaller by invitation only or regional workshops on stochastic programming. The listings of these that follow have the location, date, organizer(s) and meeting title. This list has what I know about — there likely are other meetings.

1. Oberwolfach, Germany January 28-February 3, 1979, Peter Kall and András Prékopa
2. IIASA, Laxenburg, Austria, Workshop on stochastic optimization, December 9–13, 1980, Michael Dempster and Yuri Ermoliev
3. Kiev, Ukraine, Stochastic optimization, September 9–16, 1984, V. S. Michalevich
4. Munich, three GAMM/IFIP Workshops: stochastic optimization: numerical methods and technical applications, May 29–31, 1990, June 15–17, 1993, June 17–20, 1996, Kurt Marti
5. University of Essex, Colchester, England, August 12–16, 1991, IMA Meeting on stochastic optimization in economics and finance, Michael Dempster
6. Gosen, near Berlin, GAMM/IFIP Workshop on stochastic programming, stability, numerical methods and applications, March 23–27, Werner Römisch
7. IIASA and University of Vienna, July 11–23, 1993, Approximations of stochastic programming problems, Andrzej Ruszczyński and Georg Pflug
8. Lillehammer, Norway, January 18–20, 1994, IFIP Workshop Stochastic Programming: Algorithms and Models, Stein W. Wallace
9. University of California, Davis, Workshop in stochastic programming, March 10–11, 1995, Roger Wets
10. Tuscon, Arizona, January 17–19, 1996, NSF/IFIP Workshop on stochastic programming and applications, Julia Higle and Suvrajeet Sen
11. Semmering, Austria, January 8–18, 1996, EURO Winter School on Stochastic Optimization, Georg Pflug and Andrzej Ruszczynski
12. Palo Alto, California, Electric Power Research Institute, Workshop on applications of planning under uncertainty, July 12–16, 1999, Hung-Po Chao, George B. Dantzig, Alan S. Manne and William T. Ziemba
13. University of Minnesota, Minneapolis, Institute for Mathematical Analysis, Decision making under uncertainty: energy and environmental Models, July 19–23 1999, François Auzerais, Robert Burridge, Claude Greengard and Roger Wets
14. Ischia, Italy, High-performance computing for financial planning, April 11–13, 1999, Almerico Murli and Stavros Zenios
15. University of Florida, Gainesville, International Conference on stochastic optimization: algorithms and applications, Center for Applied Optimization, February 20–22, 2000, Stanislav Uryasev and Panos Pardalos

16. Department of Mathematics, Statistics, Computing and Applications, University of Bergamo, Spring School, Stochastic programming: theory and applications, April 10–20, 2007, Giorgio Consigli
17. University of California, Davis, J-P Watson, Optimization in an uncertain environment, March 25–26, 2011, Roger Wets and D. Woodruff
18. Department of Mathematics, Statistics, Computing and Applications, University of Bergamo, The impact of population ageing on financial markets, intermediaries and financial stability, May 27 2011, Marida Bertocchi and Costanza Torricelli
19. Department of Mathematics, Statistics, Computing and Applications, University of Bergamo, Agents' behaviour in face of decisions and models complexity, August 29, 2011, Giorgio Consigli

Current, Recent Past and Upcoming events

The 21st International Symposium on Mathematical Programming (ISMP) will take place in Berlin, Germany, August 19–24, 2012. It is planned to have a Stochastic Optimization cluster organized by Shabbir Ahmed and David Morton, `http://ismp2012.mathopt.org/`

Special Workshop of Stochastic Programming Community Stochastic Programming for Implementation and Advanced Applications (StoProg-2012) will be held in Neringa, Lithuania, on July 3–6, 2012, `http://www.mii.lt/STOPROG-2012`

25th European Conference on Operational Research (EURO XXV), July 8–11, 2012 in Vilnius, Lithuania. It is planned to have a stream on Stochastic Programming organized by Jitka Dupacova, Georg Pflug, Andras Prekopa, Alexander Shapiro and Stein W. Wallace, `http://www.euro-2012.lt/`

The Nordic Optimization Symposium, June 7–9, 2012 in Trondheim, Norway. This is the biannual conference of the Nordic section of the Mathematical Optimization Society `http://www.mathprog.org/`. It will have several stochastic programming sessions organized by Asgeir Tomasgard, `http://www.ntnu.no/censes/nos5`

9th International Conference on Computational Management Science, Imperial College London (UK), April 2012. Organized by D. Kuhn, P. Parpas, E.N. Pistikopoulos, B. Rustem, and W. Wiesemann. It is planned to have a stream on Optimization under uncertainty in the power industry organized by L. Escudero and R. Schultz, `http://www.dcs.bbk.ac.uk/cms2011`

APMOD 2012, International Conference on Applied Mathematical Optimization and Modelling. March 28–30, 2012. Paderborn, Germany. Jointly organized by the DS&OR Lab (Decision Support & Operations Research Lab) of the University of Paderborn, and CARISMA (Centre for the Analysis of Risk and Optimisation Modelling Applications) of Brunel University, `http://www.apmod.org/` .

Workshop on Managing Uncertainty in Energy Systems and Markets, March 21–25, 2012. Oppdal, Norway. This workshop is organized by Asgeir Tomasgard, Stein-Erik Fleten, and Mette Bjørndal and the topics include multi-stage stochastic programming and stochastic equilibrium and capacity expansion, `http://www.ntnu.no/c/document\_library/get\_file?uuid=ee612b6e-19c8-458e-9cc5-2aca655c8b07\&groupId=7414984`.

SPXIII in Bergamo in 2013 is announced at `http://www.stoprog.org/`

Working papers in stochastic programming appear on the Stochastic Programming E-Print Series (SPEPS) organized by Werner Römisch and Suvrajeet Sen along with an editorial board. See website `http://dochost.rz.hu-berlin.de/speps/`. To submit a paper for posting, attach a pdf or postscript file to an e-mail addressed to either Werner Römisch <`romisch@mathematik.hu-berlin.de`> or Suvrajeet Sen <`sen@sie.arizona.edu`>, and the body of the e-mail should say that the attached paper is to be posted on the SPEPS web site. It usually takes about a month to get it posted, and the paper will remain on the site until it is accepted for publication and the author transfers copyright to the journal.

A list of books and collections of papers on stochastic programming that I compiled with some help from Horand Gassmann and others, follows. There are books I know about, not a complete list. Another bibliography that is updated periodically by Maarten van der Vlerk, see `http://mally.eco.rug.nl/mally/spbib.thml`. This contains entries from 1996 to 2007.

William T. Ziemba
March, 2012

Books and Collections of Papers on Stochastic Programming

I. Publications with Primary Focus on Stochastic Programming Theory and Applications

(most of these have *Mathematical Reviews* primary classification 90C15)

Abel, P. *Stochastische Optimierung bei partieller Information.* Math. Systems in Economics 96, Königstein/Ts: Athenäum/Hein/Hanstein (*in German*), 1984.

Abel, T. and R. Thiel. *Mehrstufige stochastische Produktionsmodelle-Eine praxisorientierte Darstellung mit programmierten Beispielen.* Frankfurt a.M.: Fischer Verlag. (*in German*), 1981.

Andreatta, G. B., G. Salinetti, and R. J.-B. Wets (eds.). *Stochastic Programming. Annals of Operations Research* 56. Amsterdam: Baltzer, 1995.

Archetti, F., G. Di Pillo, and M. Lucertini (eds.). *Stochastic Programming.* LN in Control and Information Sci. 76. Berlin: Springer Verlag, 1986.

Arnold, K.-P. Stochastische Transportprobleme. Hamburg: Verlag Dr. Kovac. (*in German*), 1987.

Bertocchi, M., G. Consigli and M.A.H. Dempster (eds.). *Stochastic Optimization Methods in Finance and Energy.* New York: Springer Verlag, 2011.

Birge, J. R., C. Edirisinghe, and W. T. Ziemba (eds.). *Research in Stochastic Programming. Annals of Operations Research* 100. Amsterdam: Baltzer, 2001.

Birge, J. R., and F. Louveaux. *Introduction to Stochastic Programming*, second edition. New York: Springer Verlag, 2011.

Birge, J. R., and R. J.-B. Wets (eds.) *Stochastic Programming. Annals of Operations Research* 30 and 31. Amsterdam: Baltzer, 1991.

Böttcher, J. Stochastische lineare Programme mit Kompensation. Frankfurt a.M: Athenäum. (*in German*), 1989.

Bouza, C. (ed.). *Stochastic Programming: The State of the Art.* Revista Investigación Operational, 14 (2-3), 1993.

Dempster, M. A. H. (ed.). *Stochastic Programming.* London: Academic Press, 1980.

Dempster, M. A. H. (ed.). *Proceedings of the IIASA Task Force Meeting in Stochastic Optimization.* Stochastics, 10 (3–4), 1983.

Ermoliev, Yu. M. *Methods of Stochastic Programming.* Moscow: Nauka. (*in Russian*), 1976.

Ermoliev, Yu. M. and S. P. Uryas'ev. Adaptive Algorithms of Stochastic Optimization and Game Theory. Moscow: Nauka. (*in Russian*), 1990.

Ermoliev, Y., and R. J.-B. Wets (eds.). *Numerical Techniques for Stochastic Optimization Problems.* Berlin: Springer Verlag, 1988.

Ermoliev, Yu. M. and A. I. Yastremski. *Stochastic Models and Methods in Economic Planning.* Moscow: Nauka. (*in Russian*), 1979.

Faber, M. M. *Stochastisches Programmieren.* Würzburg-Wien: Physica-Verlag. (*in German*), 1970.

Frauendorfer, K. *Stochastic Two-Stage Programming.* Lecture Notes in Economics and Mathematical Systems 392. Berlin: Springer Verlag, 1992.

Guddat, J., F. Guerra, K. Tammer, and K. Wendler. *Multiobjective and Stochastic Optimization Based on Parametric Optimization.* Berlin: Akademie-Verlag, 1985.

Hellwig, G., P. Kall, and P. Abel (eds.). *Statistical Methods for Decision Processes.* Stuttgart-Möhringen: Daimler Benz AG, 1994.

Higle, J. and S. Sen. *Stochastic Decomposition. A Statistical Method for Large Scale Stochastic Linear Programming.* Dordrecht: Kluwer, 1996.

Infanger, G. *Planning under Uncertainty: Solving Large-Scale Stochastic Linear Programs.* Danvers: Boyd and Fraser, 1994.

Infanger, G. (ed.). *Stochastic Programming: The State of the Art in Honor of George B. Dantzig.* New York: Springer Verlag, 2011.

Kall, P. *Stochastic Linear Programming.* Berlin: Springer Verlag, 1976.

Kall, P. and J. Mayer. *Stochastic Linear Programming: Models, Theory and Computation*, second edition. International Series in Operations Research 80. Berlin: Springer Verlag, 2011.

Kall, P. and A. Prékopa (eds.). *Recent Results in Stochastic Programming.* LN in Economics and Math. Systems 179. Berlin: Springer Verlag, 1980.

Kall, P., and S. W. Wallace. *Stochastic Programming.* Chichester: Wiley, 1994. Available at no cost (out of copyright) at http://www.lancs-initiative.ac.uk/page/153/Resources.htm.

Kaniovski, Yu. M., P. S. Knopov, and Z. V. Nekrylova. *Limit Theorems for Stochastic Programming Processes.* Kiev: Naukova Dumka. (*in Russian*), 1980.

Kibzun, A. I., and Y. S. Kan. *Stochastic Programming Problems with Probability and Quantile Functions.* Interscience Service on Systems and Optimization. Chichester: Wiley, 1996.

King, A. J. (ed.). *Approximation and Computation in Stochastic Programming.* Mathematical Programming B 75, No. 2. Amsterdam: North-Holland, 1996.

Klein Haneveld W. K. *Duality in Stochastic Linear and Dynamic Programming.* LN in Economics and Math. Systems 274. Berlin: Springer Verlag, 1985.

Klein Haneveld, W. K. and M. H. van der Vlerk. *Stochastic Programming, Lecture Notes* (available on request by email), 2000.

Kolbin, V. V. *Stochastic Programming.* Dordrecht: Reidel, 1977.

Kovalenko, I. N. *Probabilistic Calculation and Optimization.* Kiev: Naukova Dumka. (*in Russian*), 1989.

Kuhn, D. *Generalized Bounds for Convex Multistage Stochastic Programs.* Berlin: Springer Verlag, 2005.

Marti, K. *Approximationen stochastischer Optimierungsprobleme.* Math. Systems in Economics 43. Königstein/Ts.: Verlag A. Hain. (*in German*), 1979.

Marti, K. *Descent Directions and Efficient Solutions in Discretely Distributed Stochastic Programs.* LN in Economics and Math. Systems 299. Berlin: Springer Verlag, 1988.

Marti, K. (ed.). *Stochastic Optimization. Numerical Methods and Technical Applications.* LN in Economics and Math. Systems 379. Berlin: Springer Verlag, 1992.

Marti, K. (ed.). *Stochastic Optimization Techniques.* LN in Economics and Math. Systems 513. Berlin: Springer Verlag, 2002.

Marti, K. *Stochastic Optimization Methods*, second edition. Berlin: Springer Verlag, 2008.

Marti, K. and P. Kall (eds.). *Stochastic Optimization: Numerical Techniques and Engineering Applications.* LN in Economics and Math. Systems 423. Berlin: Springer Verlag, 1995.

Marti, K. and P. Kall (eds.). *Stochastic Programming Methods and Technical Applications.* LN in Economics and Math. Systems 458. Berlin: Springer Verlag, 1998.

Mayer, J. *Stochastic Linear Programming Algorithms: A Comparison Based on Model Management Systems.* Amsterdam: Gordon and Breach Science Publishers, 1998.

Mirzoakhmedov, F. *Mathematical Models and Methods for Production Control Taking Random Factors into Account.* Kiev: Naukova Dumka. (*in Russian*), 1991.

Mirozakhmedov, F., and M. V. Mikhalevich. *Applied Aspects of Stochastic Programming.* Dushanbe: Maorif. (*in Russian*), 1989.

Pflug, G. Ch. *Optimization of Stochastic Models. The Interface between Simulation and Optimization.* Dordrecht: Kluwer, 1996.

Pflug, G. Ch. and A. Ruszczyński (eds.). *Thirteenth EURO Summer Institute: Stochastic Optimization.* EJOR 101 (2). Amsterdam: North Holland, 1997.

Powell, W. B. *Approximate Dynamic Programming: Solving the Curses of Dimensionality*, second edition. New York: Wiley, 2011.

Prékopa, A. *Stochastic Programming.* Dordrecht: Kluwer and Budapest: Academiai Kiado, 1995.

Prékopa, A. and A. Ruszczyński (eds.). *Stochastic Programming.* Optimization Methods and Software 17. Reading, UK: Taylor & Francis, 2002.

Prékopa, A. and R. J.-B. Wets (eds.). *Stochastic Programming 84.* Math. Progr. Study 27 and 28. Amsterdam: North-Holland, 1986.

Rachev, S. T. and W. Römisch (eds.). *Stochastic Programming: Stability, Numerical Methods and Applications. J. of Computational and Applied Mathematics* 56(1–2), 1994.

Rembold, J. T. *Stochastische lineare Optimierung. Mathematical Systems in Economics*, No. 31. Meisenheim am Glan: A. Hain. (*in German*), 1977.

Rubinstein, R. Y. and A. Shapiro. *Discrete Event Systems: Sensitivity Analysis and Stochastic Optimization by the Score Function Method.* New York: Wiley, 1993.

Ruszczyński, A. and A. Shapiro. *Stochastic Models in Engineering, Technology and Management.* Handbooks in Operations Research and Management Science 10. Amsterdam: Elsevier, 2003.

Schneeweiss, A. H. *Entscheidungskriterien bei Risiko.* Berlin: Springer Verlag. (*in German*), 1967.

Schürle, M. Zinsmodelle in der stochastischen Optimierung mit Anwendungen im Asset & Liability Management. Series Bank- und finanzwirtschaftliche Forschungen. Band 279. Bern: Paul Haupt Verlag, 1998.

Sengupta, J. K. *Stochastic Programming: Methods and Applications.* Amsterdam: North Holland, 1972.

Sengupta, J. K. *Optimal Decisions under Uncertainty.* LN in Economics and Math. Systems 193. Berlin: Springer Verlag, 1981.

Sengupta, J. K. *Decision Models in Stochastic Programming: Operational Methods of Decision Making under Uncertainty.* North Holland Series in System Science and Engineering, Vol. 7. New York: North Holland, 1982.

Shapiro, A., D. Dentcheva and A. Ruszczyński. *Lectures on Stochastic Programming: Modeling and Theory.* SIAM, Philadelphia, 2009.

Stancu-Minasian, I. M. *Stochastic Programming with Multiple Objective Functions.* Bucureşti: Editura Academiei, and Dordrecht: Reidel, 1984.

Stougie, L. *Design and Analysis of Algorithms for Stochastic Integer Programming.* CWI Tract. 37. Amsterdam: Stichting Math. Centrum, 1987.

Uryasev, S. P. *Adaptive Algorithms of Stochastic Optimization and Game Theory.* Moscow: Nauka. (*in Russian*), 1990.

Uryasev, S. P. (ed.). *Probabilistic Constrained Optimization.* Dordrecht: Kluwer, 2000.

Uryasev, S. and P. M. Pardalos. *Stochastic Optimization: Algorithms and Applications.* Dordrecht: Kluwer, 2001.

van der Vlerk, M. H. *Stochastic Programming with Integer Recourse.* Groningen: Labyrint Publications, 1995.

Vajda, S. *Probabilistic Programming.* New York: Academic Press, 1972.

Vladimirou, H., S. A. Zenios, and R. J.-B. Wets (eds.). *Models for Planning under Uncertainty.* Annals of Oper. Res. 59. Amsterdam: Baltzer, 1995.

Wallace, S. W., J. Higle and S. Sen (eds.). *Stochastic Programming, Algorithms and Models.* Annals of Oper. Res. 64. Amsterdam: Baltzer, 1996.

Wallace, S. W. and W. T. Ziemba. (eds.). *Applications of Stochastic Programming,* SIAM-Mathematical Programming Series on Optimization, 2005.

Wang, Jinde. *Stochastic Programming.* Beijing: Chinese Presses. (*in Chinese*), 1990.

Werner, M. *Stochastische lineare Optimierungsmodelle.* Frankfurt am Main: Akademische Verlagsgesellschaft. (*in German*), 1973.

Wets, R. J.-B. and W. T. Ziemba. *Stochastic Programming: State of the Art, 1998, Annals of Operations Research* 85. Amsterdam: Baltzer, 1999.

Wolf, H. *Entscheidungsfindung bei der stochastischen linearen Optimierung durch Entscheidungsmodelle mit mehrfacher Zielsetzung.* Math. Systems in Economics, 84, Königstein/Ts: Athenäum/Hein/Hanstein (*in German*), 1983.

Yastremski, A. I. *Stochastic Models of Mathematical Economics.* Kiev: Lybid. (*in Russian*), 1983.

Yudin, D. B. *Mathematical Methods of Management under Incomplete Information.* (Problems and Methods of Stochastic Programming). Moscow: Soviet Radio, (*in Russian*), 1974.

Yudin, D. B. *Problems and Methods of Stochastic Programm*ing. Moscow: Soviet Radio, (*in Russian*), 1979.

II. Other Books and Collections Containing Some Material on Stochastic Programming

Arkin, V. I., and I. V. Evstigneev. *Stochastic Models of Control and Economic Dynamics.* Nauka, Moscow (*in Russian*), 1979. English translation by E. A. Medova-Dempster, and M. A. H. Dempster, London: Academic Press, 1987.

Arkin, V. I., A. Shiraev, and R. Wets (eds.) *Stochastic Optimization.* Proceedings of the International Conference, Kiev, 1984. LN in Control and Information Sciences 81. Berlin: Springer Verlag, 1986.

Bäck, T. *Evolutionary Algorithms in Theory and Practice.* New York: Oxford University Press, 1996.

Batukhtin, V. D. and L. A. Maboroda. *Optimization of Discontinuous Functions.* Moscow: Nauka. (in Russian), 1984.

Bensoussan, A. and J.-P. Verjus (eds.). *Future Tendencies in Computer Science, Control and Applied Mathematics.* Lecture Notes in Computer Science 653. Berlin: Springer Verlag, 1992.

Bodner, V. A., N. E. Rodnishchev, and E. P. Yurikov. *Optimization of Terminal Stochastic Systems.* Moscow: Mashinostroenie. (in Russian), 1987.

Breton, M. and G. Zaccour (eds.). *Advances in Operations Research in the Oil and Gas Industry.* Paris: Éditions Technip, 1991.

Cairoli, R. and R. C. Dalang. *Sequential Stochastic Optimization.* New York: Wiley, 1996.

Cheng, M. X., Y. Li, and D.-Z. Du (eds.). *Combinatorial Optimization in Communication Networks.* Combinatorial Optimization 18. Berlin: Springer Verlag, 2006.

Christer, A. H., S. Osaki, and L. C. Thomas (eds.). *Stochastic Modelling in Innovative Manufacturing.* LN in Economics and Math. Systems 445. Berlin: Springer Verlag, 1997.

Dantzig, G. B., M. A. H. Dempster, and M. Kallio (eds.). *Large Scale Linear Programming, Vols. 1 and 2.* Laxenburg: IIASA, 1981.

Davis, M. H. A. *Markov Models and Optimization.* London: Chapman & Hall, 1993.

Davis, M. H. A. and A. Etheridge (translators and eds.). *Louis Bachelier's "Theory of Speculation": the Origins of Modern Finance.* Princeton, NJ: Princeton University Press, 2006.

Dempster, M. A. H. *Risk Management: Value at Risk and Beyond.* Cambridge: Cambridge University Press, 2002.

Dempster, M. A. H., G. Ch. Pflug, and G. Mitra (eds.). *Quantitative Fund Management.* CRC Financial Math Series. Boca Raton: Chapman & Hall, 2008.

Dinkelbach, W. *Entscheidungsmodelle.* Berlin: W. de Gruyter. (*in German*), 1982.

Dritsas, I. (ed.). *Stochastic Optimization: Seeing the Optimal for the Uncertain.* Rijeka: Intech, 2011.

Dupačová, J., J. Hurt and J. Štěpán, *Stochastic Modeling in Economics and Finance.* Applied Optimization 75. Dordrecht: Kluwer, 2002.

Dynkin, E. B. and A. A. Yushkevich. *Controllable Markov Processes and their Applications.* Moscow: Nauka. (*in Russian*), 1975.

Dzemyda, G., V. Šaltenis, and A. Zhilinskas. *Stochastic and Global Optimization.* Berlin: Springer Verlag, 2002.

Greengard, G and A. Ruszczyński. *Decision Making Under Uncertainty: Energy and Power,* IMA Volumes in Mathematics and its Applications, 2000.

Grötschel, M., S. O. Krumke, and J. Rambau (eds.). *Online Optimization of Large Scale Systems.* Berlin: Springer Verlag, 2001.

Gupal, A. M. *Stochastic Methods for Solving Nonsmooth Extremal Problems.* Kiev: Naukova Dumka. (*in Russian*), 1979.

Henry, J. and J.-P. Yvon (eds.). *System Modelling and Optimization. Proceedings of the 16th IFIP-TC7 Conference, Compiegne, France, July 5–9, 1993.* Lecture Notes in Control and Information Sciences 197. Berlin: Springer Verlag, 1994.

Ivashchenko, P. A. *Adaptation in Economics.* Kharkov: Vishcha Shkola. (*in Russian*), 1986.

Jacobs, O. L. R., M. H. A. Davis, M. A. H. Dempster, C. J. Harris, and P. C. Parks (eds.). *Analysis and Optimization of Stochastic Systems.* London: Academic Press, 1980.

Jäger, W. and H.-J. Krebs (eds.). *Mathematics — Key Technology for the Future. Joint Projects between Universities and Industry 2004–2007.* Berlin: Springer Verlag, 2008.

Kall, P. (ed.). *System Modelling and Optimization.* Proceedings of the 15th IFIP Conference, Zurich, Switzerland, 2–6 September, 1991. Lecture Notes in Control and Information Sciences 180. Berlin: Springer Verlag, 1992.

Kallrath, J., P. M. Pardalos, S. Rebennack and M. Scheidt. *Optimization in the Energy Industry.* Berlin: Springer Verlag, 2009.

Katkovnik, V. Ya. *Numerical Methods for Solving Deterministic and Stochastic Minimax Problems.* Kiev: Naukova Dumka. (*in Russian*), 1979.

Knopov, P. S. and P. M. Pardalos (eds.). *Simulation and Optimization Methods in Risk and Reliability.* New York: Nova Publishers, 2009.

Konno, H., D. Luenberger and J. Mulvey (eds.). *Financial Engineering*. Annals of Operations Research 45. Basel, Switzerland: Baltzer Verlag, 1993.

Konno, H., D. Luenberger, and J. Mulvey (eds.). *Introduction to Financial Optimization*. Special Issue of Mathematical Programming, 89(2), 2001.

Kontoghiorghes, E. J., B. Rüstem and S. Siokos (eds.). *Computational Methods in Decision-Making, Economics and Finance*. Boston: Kluwer Academic Publishers, 2002.

Kovalenko, I. N. and A. A. Nakonechny. *Approximate Calculation and Optimization of Reliability*. Kiev: Naukova Dumka. (*in Russian*), 1989.

Krokhmal, P., R. T. Rockafellar and S. Uryasev. Special issue on risk management and optimization in finance, *Journal of Banking and Finance*, 3(6): pp. 315–796, 2006.

Kulkarni, V. G. *Modeling and Analysis of Stochastic Systems, second edition*. London: Chapman & Hall, 2010.

Lund, D. and B. K. Øksendal (eds.). Stochastic Models and Option Values: Applications to Resources, Environment and Investment Problems. Amsterdam: North-Holland, 1991.

Marti, K. (ed.). *Structural Reliability and Stochastic Structural Optimization. Mathematical Methods of Operations Research*. 46, No. 3. Heidelberg: Physica-Verlag. (*in German*), 1997.

Marti, K. Y. Ermoliev and M. Makowski (eds.). *Coping with Uncertainty: Robust Solutions*. LN in Economics and Mathematical Systems 633. Berlin: Springer Verlag, 2010.

Marti, K. Y. Ermoliev, M. Makowski and G. Pflug (eds.). *Coping with Uncertainty: Modelling and Policy Issues*. LN in Economics and Mathematical Systems 581. Berlin: Springer Verlag, 2006.

Marti, K. Y. Ermoliev and G. Pflug (eds.). *Dynamic Stochastic Optimization*. LN in Economics and Mathematical Systems 532. Berlin: Springer Verlag, 2003.

Mikhalevich, V. S., A. M. Gupal, and V. I. Norkin. *Methods of Nonconvex Optimization*. Moscow: Nauka. (*in Russian*), 1987.

Mockus, J., W. Eddy, A. Mockus, L. Mockus, and G. Reklaitis. *Bayesian Heuristic Approach to Discrete and Global optimization*. Nonconvex Optimization and its Applications, 17. Dordrecht: Kluwer, 1997.

Mockus, J. *Bayesian Approach to Global Optimization. Mathematics and its Applications* (Soviet Series), 37. Dordrecht: Kluwer, 1989.

Mihoc, G. and I. Nadejde. *Mathematical Programming: Parametric, Nonlinear and Stochastic*. Bucureşti: Editura Stiintifica. (*in Romanian*), 1966.

Nurminsky, E. A. *Numerical Methods for Solution of Deterministic and Stochastic Minimax Problems*. Kiev: Naukova Dumka. (*in Russian*), 1979.

Osaki S., D. N. P. Murthy, and R. J. Wilson (eds.). *Stochastic Models in Engineering, Technology and Management*. Papers from the Australian-Japan Workshop held on the Gold Coast, July 14–16, 1994. Math. Comput. Modelling 22, No. 10–12. Exeter: Pergamon Press, 1995.

Penot, J. P. (ed.). *Proceedings of the 2003 MODE-SMAI Conference*. ESAIM: Proceedings, 2003.

Pflug, G. Ch., and W. Römisch. *Modeling, Measuring and Managing Risk.* Singapore: World Scientific Publishing Co, 2007.

Poznyak, A. S. and K. Najim. *Learning Automata and Stochastic Optimization.* LN in Control and Information Sciences 225. London: Springer Verlag, 1997.

Rastrigin, L. A. *Adaptation of Complex Systems.* Riga: Zinatne. (*in Russian*), 1981.

Rebennack, S., P. M. Pardalos, M. V. F. Pereira, and N. A. Iliadis (eds.). *Handbook of Power Systems* (2 volumes). Berlin: Springer Verlag, 2010.

Resende, M. and P. Pardalos (eds.). *Handbook of Optimization in Telecommunications.* Berlin: Springer Verlag, 2006.

Rudolf, M. *Algorithms for Portfolio Optimization and Portfolio Insurance.* Bank- und finanzwirtschaftliche Forschungen 192. Bern: Paul Haupt Verlags AG, 1994.

Słowiński, R. and J. Teghem (eds.). *Stochastic versus Fuzzy Approaches to Multiobjective Mathematical Programming under Uncertainty.* Theory and Decision Library. Series D: System Theory, Knowledge Engineering and Problem Solving, 6. Dordrecht: Kluwer, 1990.

Tarasenko, G. S. *Stochastic Optimization in the Soviet Union. Random Search Algorithms.* Monograph Series on Soviet Union. Falls Church: Delphic Associates, 1985.

Vladimirou, H. (ed.). *Financial Modelling.* Annals of Operations Research 151. Berlin: Springer Verlag, 2007.

Vladimirou, H. (ed.). *Financial Optimization.* Annals of Operations Research 152. Berlin: Springer Verlag, 2007.

Vladimirou, H., I. Maros, and G. Mitra. *Applied Mathematical Programming and Modeling.* Annals of Operations Research 99. Dordrecht: Kluwer, 2000.

Wallace, S. W. (ed.). *Modelling: in Memory of Åsa Hallefjord. Annals of Operations Research 82*, 1998.

Woodruff, D. L. (ed.). *Advances in Computational and Stochastic Optimization, Logic Programming, and Heuristic Search.* Boston: Kluwer, 1998.

Woodruff, D. L. (ed.). *Network Interdiction and Stochastic Programming.* Boston: Kluwer, 2003.

Yan, H., G. Yin and Q. Zhang (eds.). *Stochastic Processes, Optimization, and Control Theory: Applications in Financial Engineering, Queueing Networks, and Manufacturing Systems.* International Series in Operations Research & Management Science 94. Berlin: Springer Verlag, 2006.

Yin, G. and Q. Zhang (eds.). *Recent Advances in Control and Optimization of Manufacturing Systems.* LN in Control and Information Sci. 214. London: Springer Verlag, 1996.

Zavrieva, M. K. *A Combined Penalty and Stochastic Quasigradient Method for the Search for a Connected Maximin.* Moscow: Academiya Nauk SSSR, Vychisleniy Tsentr (Soviet Academy of Sciences, Centre for Computing). (*in Russian*), 1989.

Zenios, S. A. (ed.). *Financial Optimization.* Cambridge: Cambridge University Press, 1993.

Zenios, S. A. and W. T. Ziemba (eds.). *Financial Modeling, Focused Issue.* Management Science 38(11), 1992.

Zenios, S. A. and W. T. Ziemba (eds.). *Handbook of Asset-Liability Management, Volume 1: Theory and Methodology.* Amsterdam: North Holland, 2006.

Zenios, S. A. and W. T. Ziemba (eds.). *Handbook of Asset-Liability Management, Volume II: Applications and Case Studies*, Amsterdam: North Holland, 2007.

Zhigljavsky, A.A. (ed.). *Stochastic Optimization.* Acta Appl. Math., 33(1). Dordrecht: Kluwer, 1993.

Zhigljavsky, A. A. *Theory of Global Random Search. Mathematics and its Applications* (Soviet Series), 65. Dordrecht: Kluwer, 1991.

Zielinsky, R. and P. Neumann. *Stochastic Methods in the Search for the Minimum of a Function.* Mathematical Research 16. Berlin: Akademie-Verlag. (*in German*), 1983.

Ziemba, W. T. and J. M. Mulvey (eds.). *Worldwide Asset and Liability Management.* Cambridge University Press, 1998.

Ziemba, W. T. and R. G. Vickson (eds.). *Stochastic Optimization Models in Finance.* Academic Press, New York, Second revised edition. 2006. Singapore: World Scientific Publishing co., 1975.

Ziemba, W. T. *The Stochastic Programming Approach to Asset-Liability and Wealth Management*, AIMR, Charlottesville, Virginia, 2003. Available for free download at http://www.cfainstitute.org/learning/products/publications/rf/Pages/rf.v2003.n3.3924.aspx. For 72 page appendix, email wtzimi@mac.com for a copy.

Contents

Chapter 1

Introduction and Summary

Summary of papers

Part I has papers on finance. The paper by Kim, Mulvey, Simsek and Kim presents an individual asset liability model for a retired couple where longevity risk is modeled using term life insurance. They present calculations for three cases including using expected lifetimes where the longevity risk is assumed away. The other two cases discuss a shift in expected lifetimes or a decease in retirement income. As expected, the optimal policy depends on these risks and the price of insurance.

Moriggia, Consigli and Iaquinta present a multi-period asset liability planning model for individuals with intermediate and long-term goals. The investments include mutual and pension funds, fixed income instruments, unit-linked annuities as well as complex insurance and retirement protection. Optimal strategies are computed for various levels of risk aversion considering inflation-adjusted income and savings conditions, family consumption and taxes. Various time-distributed investment and payout plans are considered.

Benk presents an intertemporal portfolio choice model with jump risks that can be applied to pension and life insurance funds, and private investors. Following the Rudolf and Ziemba (2004) (the full reference is given in Chapter 4 by Benk) model, these long-term investors maximize the intertemporal expected utility of the surplus of assets net of liabilities. Returns on liabilities are modelled with a pure-diffusion process and the returns on assets are assumed to follow a jump-diffusion process with two jump components. An investor's optimal portfolio consists of three funds: a market portfolio, a liability-hedging portfolio, and a riskless asset. In contrast to the results of Rudolf and Ziemba (2004), a market portfolio not only hedges diffusion risk, but it also hedges systemic risk and it takes into account idiosyncratic jump risk so that the investor is additionally protected against both a systemic risk and an idiosyncratic jump risk.

The paper by Davis and Lleo considers an asset management problem where the assets follow a jump-diffusion process. The objective is to outperform a benchmark as is common in the investment industry for mutual and other funds. They explore two sets of assumptions on the coefficients of the stochastic process that yield unique optimal investment strategies for the stochastic risk-sensitive control problem. The first set of assumptions relate to an affine function of a factor process with constant asset diffusions, and the asset jumps are independent of the factors, which are assumed to be Gaussian diffusion processes. In this case, the associated Bellman equation is a partial differential equation which has a unique classical solution. In the second set of assumptions, both the asset growth rates and volatility depend upon the factors which are jump-diffusion processes. Hence, this model has stochastic volatility. This yields a fully nonlinear controlled jump-diffusion, and the Bellman equation is a partial integer-differential equation [PIDE] for which no analytical solution exists. Proving that the Hamilton-Jacobi-Bellman [HJB] PIDE admits a unique classical (C1,2) solution requires the development of a more sophisticated argument combining viscosity solutions and classical solutions. The argument used in the derivation hinges on only three key points. First, the Lipschitz continuity of the value function provides the ability to rewrite the HJB PIDE as a PDE. Second, viscosity solutions give existence and uniqueness of a weak solution to both of these equations. A proof of existence by Fleming and Rishel based on a policy improvement originally due to Bellman completes the analysis by providing a smooth solution. The robustness of this approach is a clear advantage for control problems: solving benchmarked or ALM investment management problems is not more difficult than addressing an asset-only investment problem.

The paper by Edirisinghe and Zhang presents a dynamic portfolio optimization model for stock portfolio management when there is market uncertainty modeled with regimes. The portfolio of long and short stocks is determined from the regimes and fundamental business strength measured using data envelopment analysis. The data used to calibrate and test the model is for the forty year period 1971–2010. The out of sample test shows that the model is superior to sector based ETF portfolios and the market index for the January to June 2011 period.

Golembiovsky and Abramov present a modification of a standard option portfolio where the goal is to attempt to have excess returns over the risk free rate by taking a small risk. The risk is modeled using chance constraints so is a value at risk concept. Using the Black-Scholes formula, they evaluate European options over a two month horizon. Simulations show that the goal can be achieved.

Aranburu, Escudero, Garín and Pérez present several approaches for the stochastic optimization of immunization strategies. The risk averse measures used are two-stage and multiple stage stochastic dominance and a new multistage VaR stochastic dominance criterion. Scenarios are used and the models consider investing in a market with coupon bonds having different levels of default risk.

The paper by Ziemba demonstrates that racetrack betting is simply an application of portfolio theory. The racetrack offers many bets that involve the results of one to about ten horses. Each race is a special financial market with betting then a race that takes one or a few minutes. Unlike the financial markets, one cannot stop the race when one is ahead or having the market going almost 24/7. There is a well-defined end point. Like standard portfolio theory, the key issues are to get the means right. In this case, it is the probabilities that two, three or four horses finish in the first two, respectively three or four places, in the given order, and to bet well. For the latter, the Kelly capital growth criterion is widely used, which maximises the expected logarithm of final wealth. Transaction and price pressure odds changes fit well into the stochastic programming models. This paper relates the theory, computations and examples of real races and experiences for various bets such as win, place and show, exactas, triactors, superfectas, super hi five, place pick all, double, pick 3, 4, 5 and 6. Many great races can be seen free on the website www.chef-de-race.com.

Part II has papers relating to production planning and logistics. Hellemo, Midthun, Tomasgard and Werner describe a multi-stage model for the design and operation of a network of pipelines and other infrastructure components for the production of natural gas. The model distinguishes two time scales and corresponding uncertainty; the strategic level deals with major investment decisions, such as construction of pipelines and opening of new reservoirs, while the embedded operational level concerns decisions regarding the production, transportation and marketing of the recovered gas. While strategic decisions have implications from one stage to the next, the impact of the operational decisions is assumed to be confined within each strategic stage.

Hæreid, Schütz and Tomasgard build a model that can be used by farmers of salmon in helping to plan their production in the presence of uncertainty about both salmon prices and the development of the fish stock (i.e., growth, mortality, escapes, etc.). The objective is to maximize the farmer's expected profits by considering both the timing and the quantity of the salmon harvest and the introduction of new stock. Important constraints include regulatory limits on the allowable biomass as well as a requirement to keep a particular holding tank empty (fallow) for a certain period after the stock has been harvested.

The paper by Michalopoulos, Morton and Barnes discusses models for interdicting nuclear smugglers. They look at the one-country case where only border-points are potential locations for installations that can detect nuclear material. The original and very general aspect of this model is the fact that not only are the smuggling activities random from the view point of the authorities, but so is the budget for installing detection devises. This is important because the optimal set of check-points for a budget b is normally not a subset of those for a budget b' for $b' > b$. And as is often the case with public investments, decisions of which installations to build are made one by one as budgets are made available. So the output of the model is a priority list of installations which optimally (given a density function for the random budget) trades off the possible budgets.

The paper by Kazemi Zanjani, Nourelfath and Ait-Kadi discusses operational planning in the sawing units of sawmills under two types of uncertainty: Quality uncertainty stemming from the variation in log sizes and qualities and more traditional demand uncertainty. Bad sawing decisions not only lead to more backlog for high quality products but also more inventories of low quality products. Several models are discussed; two- and multi-stage stochastic programming as well as some robust formulations. In some models setup costs are considered, leading to mixed integer formulations. The setup of scenarios is discussed. Since the models are very hard to solve, two approximation schemes are developed, one based on the progressive hedging algorithm, the other on scenario updates.

Part III has papers on energy. Philpott and Pritchard present many different clever modelling ideas to help a resource intensive industry decide on the appropriate manufacturing schedule to take advantage of lower electricity costs while satisfying a known demand schedule. There are many sources of uncertainty in this model, including the price of electricity, which depends on both hydrological processes (for the production of hydro-energy) and the total system demand for electricity. A novel energy surcharge levied on the 100 highest demand periods (30-minute time intervals) during the previous year (collected ex-post) is also considered. The paper is ambitious and far-reaching.

The paper by Fleten and Lie uses a two-stage numerical oligopoly model to study market power in a mixed hydro and thermal system, with example data taken from the Nordic market. Uncertainty is related to water inflows to reservoirs. The point of the model is to understand if market power is exercised and to what extent prices differ from those that would have occurred in a fully competitive market. For the case at hand they find that market prices are about 7% above the competitive prices — a rather small difference. They also show that the largest producer in the region has incentives to reduce its thermal output in order to increase the spot prices. This type of model can be used by regulators to observe the large producers to prevent the exercise of potential market power.

The paper by Diniz and Maceira discusses a classical dispatch problem for a mixed hydro and thermal energy system with one additional difficulty. When thermal units are fueled by liquid natural gas (LNG) the dispatch decision can no longer be made in the same period as the other units (in a medium to long term model), but must be made many periods earlier. This is caused by the way logistics decisions about the LNG supply are made. This change creates algorithmic challenges in both Benders (L-shaped) decomposition and in dual dynamic programming. The paper illustrates that this difficulty can be overcome, and shows how to approach the difficulties. Illustrations are based on the Brazilian energy system.

Part IV has two theoretical papers that deal with telecommunication. The paper by Maggioni, Bertocchi, Allevi, Potra and Wallace uses second-order cone programming to analyze a problem from mobile communication networks. Uncertainty is primarily associated with the movements of the destination node of a message sent from a non-moving sender node. Scenarios are in the form of ellipses describing where the destination node might be based on where it was last observed. The main goal of the paper is very general: To understand when deterministic models are useful, and if they are, in what sense, and when they are not. In particular, a bad deterministic solution (in terms of a large Value of the Stochastic Solution) might have valuable information embedded in it even if it is very bad in its own right. This points to the fact that deterministic solutions are not just good or bad: There is a range of positions in between.

The paper by Benajam, Gaivoronski and Lisser uses stochastic programming and semidefinite approximations to model and analyze the frequency assignment problem from telecommunications. While the basic resource for mobile communications remains basically unchanged, the demand is growing rapidly. At the same time, the services face challenging randomness in demand and operational conditions. The main contribution of this paper is its positive answer to the question: Can adding randomness to the already challenging deterministic versions of the model, lead to models that are solvable? The conclusions are backed up with numerical illustrations of moderately sized problems.

PART I

Papers in Finance

Chapter 2

Longevity Risk Management for Individual Investors

Woo Chang Kim*, John M. Mulvey†,
Koray D. Simsek‡ and Min Jeong Kim§

Summary

We model and numerically solve the optimal asset allocation problem of a retired couple with uncertain lifetime, in the presence of a life insurance policy. The couple maximizes expected utility over their joint lifetime by dynamically adjusting their asset allocation and purchasing term-life insurance. We conduct three numerical analyses: In the base case, we find optimal policies assuming the expected lifetimes are correct. The other two examples introduce longevity risk through either a shift in the expected lifetimes or an unexpected cut in retirement income. We find that optimal asset allocation policy depends on the presence and the type of these risks as well as the relative price of insurance. Furthermore, we show that a generalized linear policy is not likely to help under such circumstances.

1 Introduction

The objective of this study is to determine the optimal portfolio of the traditional asset classes (stocks and bonds) along with the life insurances and the pension plans in the retirement planning framework. The extended life span thanks to the advances in the medical sciences is good news for the mankind. However, it could affect the retirement planning for the individuals, and the pension plan management in a negative fashion. Importantly, the distribution of the life expectancy is not fixed. Thus, one

*KAIST, Daejeon, South Korea, wkim@kaist.ac.kr.
†Princeton University, Princeton, NJ, USA, mulvey@princeton.edu.
‡Sabanci University, Istanbul, Turkey, ksimsek@sabanciuniv.edu.
§KAIST, Daejeon, South Korea, lisaoran@kaist.ac.kr.

should address the longevity risk, while it has not been actively discussed so far.

The optimal portfolio and consumption plan of the individuals along whole life have been studied in the life cycle model. The life cycle model assumed a fixed time horizon when the model was first developed. However the lifetime is random not fixed actually, many researchers have begun to consider a random lifetime in the life cycle model from the 1960's. Yaari (1965) discussed about the optimal consumption plan when a lifetime is random. He introduced the actuarial note for hedging the randomness of a lifetime and enhancing a individual's utility. The optimal asset allocation along the life cycle with uncertain time horizon is investigated by Merton (1969). He studied not only the optimal consumption but also the asset allocation among the several risky assets in the continuous time environment. These two streams of research, Yaari (1965) and Merton (1969), were later combined by Richard (1975).

For hedging of an uncertainty of lifetime, Fischer (1973) and Campbell (1980) focused on an optimal life insurance purchase. In a discrete time environment, Fischer (1973) showed how the optimal insurance investment changes by a simulation study. Campbell (1980) suggested the one time period model and showed a wage earners' random life time can be dealt with a life insurance. Meanwhile, a real data analysis about individual's choice has been done by Hamermesh (1984). He estimated a life expectancy based on the parent's lifetime and studied the effect of that life expectancy on individual's consumption and retirement date.

Recently, there are several attempts to overcome the shortages of previous works and to develop more reasonable life cycle model. Bodie *et al.* (2004) assumed a stochastic wage process and they decided the optimal amount of consumption and labor. Also, Cocco *et al.* (2005) found the optimal consumption and asset allocation when a labor income is uncertain. They estimated parameters by using the calibration and found the optimal solution by simulation.

Bodie *et al.* (2004) and Cocco *et al.* (2005) did not consider a life insurance in their models, but Chen *et al.* (2006), Pliska and Ye (2007), Ye and Pliska (2011), and Huang and Milevsky (2008) used a life insurance as a decision variable. Chen *et al.* (2006) studied the optimal asset allocation and the life insurance purchase under a stochastic labor income. Based on Yaari (1965) model, Pliska and Ye (2007) examined the wage earners' optimal consumption and life insurance demand over the life cycle ended at the earlier time of retirement and death. The result of Pliska and Ye (2007) was extended by Ye and Pliska (2011), they assume the wage earners can invest in the risky assets. Huang and Milevsky (2008) considered a family as a unit of the life cycle model and used HARA utility function to find anoptimal consumption, life insurance purchase, and asset allocation.

The life cycle model has been developed in many different ways, and its results showed an optimal choice under a random lifetime. Recently, longer life span becomes a big problem to individuals, but it seems that the longevity risk does not involve in the life cycle model except recent works by Huang *et al.* (2011) and Cocco and Gomes (2012).

Since Yaari (1965), most of the papers study the wealth management problem *before the retirement* to maximize certain expected utility functions. Our study focuses on the asset allocation problems after the retirement employing the life insurance as the hedging tool along with the traditional asset classes. More importantly, by conducting the sensitivity analysis on the parameters of the life expectancy distribution and retirement income, our work tries to understand the rarely addressed aspects of the longevity risks.

We find that the longevity risk is far too important to be to be simplified to generalized linear rules for asset allocation. A life insurance investment strategy should be couple with an asset allocation policy in order to achieve better risk-adjusted performance. Thus, a systematic approach is required to address the longevity risk management problems.

The rest of the paper is organized as follows. In Section 2, we introduce the model and present the assumptions and the objective functions. In Section 3, we provide numerical examples to illustrate the benefits of our model. We conclude in Section 4.

2 Model

We consider a couple's personal financial planning problem during their retirement years. We assume that the husband, who carries a pension plan, is about to retire. His pension benefits, which are the only source of periodic income for this couple, will be received until his death. In addition, they have a certain amount of savings accumulated prior to the retirement. The couple's objective is to cover their annual living expenses, which are projected for the rest of their lives. They would like to optimally allocate their wealth across a set of asset classes (defined by set $\mathrm{A} = \{1, 2, \ldots, \mathrm{I}\}$, with category 1 representing cash) to achieve this objective. In our numerical analysis, as in standard asset allocation problems, we will assume that they can invest in stocks and bonds. In addition, they can purchase a term life insurance policy for the husband so that his widow can receive a death claim to compensate for the termination of pension benefits. Since the couple doesn't value wealth surplus after their deaths, they buy insurance only for the husband. However, if the wife dies before the husband does, they stop paying the life insurance premium. Having a stochastic model, we represent the uncertainty through a set of scenarios denoted with S and the model's horizon is set to be a maximum lifetime expectancy of T years.

There are several reasons why this study focuses on the retiree's perspective and couple's wealth planning. While the pension funds or the insurance company also face the longevity risk, they have an access to the financial securities such as the longevity bonds or longevity derivatives that can hedge the risks at least partially. However, for individuals, it is difficult, if not impossible, to employ such securities to hedge the longevity risks. Also, the longevity risks are more critical for the retirees than those with the income cash flows. In addition, the previous studies mostly focus on an individual, which leads to a model that terminates upon the retirement or the death of the individual. Such approaches could yield a sub optimal solution for the whole household. In order to overcome these issues, we propose a model considering a couple's wealth management after the retirement. Note that while the suggested framework is relatively simple, one could easily extend it to a variety of directions.

For simplicity, after-tax real dollar terms are employed both in modeling and the numerical analyses; therefore any tax and inflation parameters are omitted.

For each $s \in S$, $i \in A$, and $t \in \{0, 1, \ldots, T\}$, we define the following parameters and decision variables.

Parameters

y^{tg}	Annual living expenses
$x_{\overline{1,0,s}}^{\longmapsto}$	Savings at the initial time
$I_{t,s}^{h(w)}$	$= 1$ for each year t the husband (wife) is alive under scenario, $s = 0$ otherwise
$\tau_s^{h(w)}$	$= \max\{t, \text{s.t.} I_{t,s}^{h(w)} = 1\}$ denotes the last year in which the husband (wife) is alive under scenario s (i.e., the passing occurs in $\tau_s^{h(w)} + 1$)
$\tau_{1,s}$	$= \min\{\tau_s^h, \tau_s^w\}$
$\tau_{2,s}$	$= \max\{\tau_s^h, \tau_s^w\}$
$r_{i,t,s}$	Inflation-adjusted rate of return for asset i in year t under scenario s

$y_{t,s}^{db}$	$= \begin{cases} y^{tg} & \text{if } t \leq \tau_s^h \\ 0 & \text{otherwise} \end{cases}$, annual benefit payments from pension plan in year t under scenario s
$\sigma_{i,t}$	Unit transaction cost for asset i in year t
$cr_{t,s}$	Conversion ratio in year t under scenario s
π_s	Probability of scenario s

Decision Variables

$y_{t,s}^{prem}$	Annual insurance premium for year t under scenario s
$y_{t,s}^{ins}$	$= \begin{cases} cr_{t,s} * y_{t,s}^{prem} & \text{if } t = \tau_s^h \\ 0 & \text{otherwise} \end{cases}$, death claim payment from insurance in year t under scenario s
$x_{i,t,s}^{buy}$	Amount of asset i purchased at the beginning of year t under scenario s
$x_{i,t,s}^{sell}$	Amount of asset i sold at the beginning of year t under scenario s
$x_{i,t,s}$	Amount allocated to asset i after rebalancing at the beginning of year t under scenario s
$x_{t,s}^{ta}$	$= \sum_i x_{i,t,s}$, Total amount in all assets (i.e., wealth) at the beginning of year t under scenario s
$x_{i,t,s}^{\mapsto}$	$= x_{i,t,s} * (1 + r_{i,t,s})$, amount allocated to asset i before rebalancing at the end of period t under scenario s

Given these definitions, the cash-flow balance equations are formulated as follows:

$$x_{i,t,s} = x_{i,t-1,s}^{\longmapsto} + x_{i,t,s}^{buy}(1 - \sigma_{i,t}) - x_{i,t,s}^{sell} \quad \forall s \in S, \quad i \neq 1, \quad t = 1, \ldots, \tau_{2,s}$$

$$x_{1,t,s} = x_{1,t-1,s}^{\longmapsto} + \sum_{i \neq 1} x_{i,t,s}^{sell}(1 - \sigma_{i,t}) - \sum_{i \neq 1} x_{i,t,s}^{buy} + y_{t,s}^{db} - y^{tg} + y_{t,s}^{ins} - y_{t,s}^{prem}$$
$$\forall s \in S, \quad t = 1, \ldots, \tau_{2,s}$$

Please note that the wealth of the couple could be negative, and importantly, once the wealth becomes negative, they will not invest in the risky assets, and only pay the insurance premium and the living expenses. This is a relatively realistic assumption since it is rather hard for the retired couple to borrow money to invest in the risky assets after they declare bankruptcy. Conceptually, once their wealth becomes negative, or equivalently, once they declare bankruptcy, they will be living with some kinds of subsidies (either from the government or from the families) just enough to pay the insurance and the living expense bills, but not more than that. Although the couple doesn't pay the bills after declaring bankruptcy, someone is paying the money for them, thus being a liability for the couple.

Next, we define and formulate two performance measures as well as four downside risk measures. Using these measures, we can analyze the impacts of couple's investment decisions.

Expected final surplus (Z_1) and expected final wealth (Z_2) are the two performance measures, which focus on the horizon date only. These are formulated as follows

$$Z_1 = \sum_{s \in S} \pi_s \max(x^{ta}_{\tau_{2,s},s}, 0)$$

$$Z_2 = \sum_{s \in S} \pi_s x^{ta}_{\tau_{2,s},s}$$

The down-side risk measures are more diverse. Expected final deficit (Z_4) and probability of final deficit (Z_5) are similar measures that focus on the horizon date. Percentage of years in deficit (Z_6) is an attempt to measure the negative impact of decisions across the planning horizon. Expected life time average shortage in consumption (Z_3) is the most comprehensive downside risk measure, which calculates the average deviation from the target consumption (y^{tg}) both across time and over scenarios. These measures are formulated as follows:

$$Z_3 = \sum_{s \in S} \pi_s \sum_{t=1}^{\tau_{2,s}} \frac{\max(y^{tg} - x^{ta}_{t,s}, 0) - \max(-x^{ta}_{t,s}, 0)}{\tau_{2,s}}$$

$$Z_4 = \sum_{s \in S} \pi_s \min(x^{ta}_{\tau_{2,s},s}, 0)$$

$$Z_5 = \sum_{s \in S} \pi_s 1_{x^{ta}_{\tau_{2,s},s} < 0}$$

$$Z_6 = \sum_{s \in S} \pi_s \sum_{t=1}^{\tau_{2,s}} \frac{1_{x^{ta}_{t,s} < 0}}{\tau_{2,s}}$$

The couple's primary objective is to maximize their expected utility function, which is a convex combination of expected final surplus (a performance measure) and the expected life time average shortage in consumption (a risk measure) as follows.

$$\text{Expected Utility} = \alpha Z_1 - (1 - \alpha) Z_3$$

where $\alpha \in [0, 1]$

The value of α indicates the couple's risk preference, so that the large value of α means that the couple is aggressive. Conversely, the conservative couple has small value of α.

We explicitly define several parameters such as stock and bond returns, life insurance premium and death claim payment of life insurance policy. In our simulation study, we assume that stock and bond returns follow a bivariate log-normal distribution. The distribution parameters, however, are based on a regime-switching model with two economic regimes: normal and crash. Under the normal regime, stocks are expected to perform better than bonds and their returns are positively correlated. In the crash regime, stock returns are expected to be negative and are negatively correlated with bond returns. For numerical analysis, the distribution parameters (mean and covariance matrix) of normal/crash regime and the transition probability are needed.

We assume that the insurance policy is the term life insurance. Assume that the term of the insurance policy is T years. Then if the insured person dies within T years after the initial contract, he will receive a designated amount of benefits. If the insured person lives until the end of the coverage period, the insurance policy will be renewed and the premium will be adjusted because of the change of mortality rate. The relationship between the insurance premium and benefit is defined by the conversion ratio. The death claim is constant during the policy period and is formulated as

$$\text{Death claim} = \text{Annual premium}^* \text{ Conversion Ratio}$$

where the annual premium is determined by the insured person and conversion ratio (CR) is a parameter calculated in the base case by setting the expected total premium payment and the expected benefit equal according to the given mortality rates. As an example, in the base case, where the remaining life expectation for the husband is 23 years, expected total premium payment for the first 10 years is 9.68 times the annual premium. Probability of a death claim within 10 years is 0.0898. Thus, $\text{CR} = 9.68/0.0898 = 107.79$. This is equivalent to saying that death claim to be received for a 10-yr policy with \$1 annual premium payments is equal to is \$107.79. Clearly, at every renewal CR decreases since the probability of a death claim goes up. Figure 1 displays this pattern for different assumptions for the expected remaining life.

The procedure described above gives a single ("fair") CR value for every person. However, different population groups tend to get different CRs based on characteristics such as their smoking habits, current wealth,

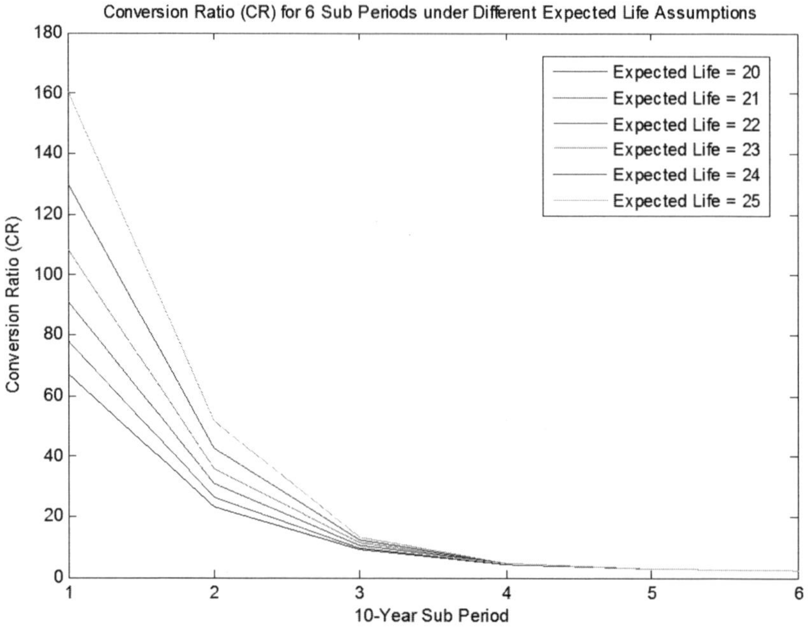

Figure 1 Conversion Ratio for the six 10-yr Insurance Policy Periods for Different Expected Remaining Life Scenarios.

chronic diseases (diabetes, high blood pressure, etc.) and place of residence (car accident rates, crime rates, etc.). Therefore, the "fair" CR should be adjusted to obtain the "actual" CR to reflect the reality. In our numerical tests, we multiply the "fair" CR with an adjustment factor as below and conduct sensitivity analysis on this factor.

$$\text{actual CR} = \text{adjustment factor}^{*} \text{"fair" CR}$$

3 Numerical results

This section reports the results of our optimization-simulation analysis. There are three examples that address different aspects of the longevity risk management. First, we fix the parameters of the life expectancy distribution of the couple, and derive the optimal asset allocation. This will serve as the base case for the following examples. Second, the parameters of the life expectancy distribution are perturbed for the sensitivity analysis, and the losses in the expected utilities caused by employing wrong parameters will be reported. This example directly addresses the impact of the wrong life expectancy parameters, that is, the impact of the longevity risks, to the wealth management problem. For the third example, we consider the uncertainty within the pension plan benefits that can be caused by the extended life spans and the deepened deficits, and examine the asset allocation decisions by the couple.

In our numerical analyses, we assume that both the husband and the wife are currently 65 years old. Their remaining lives are modeled using the mortality table typically employed by the U.S. pension plans (Table 1) assuming that they cannot live beyond 115 years old. For the base case, the average life spans are assumed to be 19 and 21 for the husband and the wife, respectively . For the sensitivity analyses on these average life spans, we employ linear combinations of the mortality rates for 2000 (mean = [17, 20]) and 2010. We also assume that their remaining lives are independent. Figure 2 displays the histogram of remaining life spans of the couple for the base case.

The couple's annual retirement benefits and projected living expenses are assumed to be constant at $40,000. We also assume that living expenses do not change after the death of the spouse. The couple starts with and initial savings of $200,000.

At the beginning of every 10-year period, the couple can purchase (renew) a 10-year term life insurance policy for the husband. If he dies within the 10-year policy period, the wife receives a death claim. For the adjustment factor explained in Section 2, we try values from 20% up to 200% with increments of 20%. A value of 100% implies a fair insurance premium whereas higher (lower) values imply cheap (expensive) insurance policies.

In our simulation study, we assume that stock and bond returns follow a bivariate normal distribution. The distribution parameters, however, are based on a regime-switching model with two economic regimes: normal and crash (Table 2). Under the normal regime, stocks are expected to perform better than bonds and their returns are positively correlated. In the crash regime, stock returns are expected to be negative and are negatively correlated with bond returns.

The assumptions of the transition probability matrix for the regimes are shown in Table 3.

This implies that the stationary probability of a normal regime is 0.75, whereas that of the crash regime is 0.25. Thus, the crash regime comes once every 4 years. Furthermore, the unconditional expected return is 5.5% for stocks and 3.6% for bonds. The histograms for stock and bond returns are displayed in Figure 3.

Table 1 A Sample of Mortality Tables across Time.

1945			1971			1983			2000			2010		
Age	**MR**	**Life Exp**	**Age**	**MR**	**Life Exp**	**Age**	**MR**	**Life Exp**	**Age**	**MR**	**Life Exp**	**Age**	**MR**	**Life Exp**
20	0.06%	53.68	20	0.05%	55.30	20	0.04%	57.90	20	0.03%	59.79	20	0.02%	61.76
	⋮			⋮			⋮			⋮			⋮	
40	0.20%	34.58	40	0.16%	36.06	40	0.12%	38.50	40	0.11%	40.32	40	0.09%	42.21
	⋮			⋮			⋮			⋮			⋮	
60	1.56%	17.77	60	1.31%	18.80	60	0.92%	20.68	60	0.67%	21.78	60	0.49%	23.30
	⋮			⋮			⋮			⋮			⋮	
80	9.97%	6.42	80	8.74%	7.05	80	7.41%	7.68	80	6.44%	7.79	80	5.43%	8.36
	⋮			⋮			⋮			⋮			⋮	
100	36.55%	2.02	100	32.98%	2.25	100	31.92%	2.32	100	34.46%	2.29	100	33.87%	2.31
	⋮			⋮			⋮			⋮			⋮	
120	—	—	120	—	—	120	—	—	120	1	0.5417	120	1	0.5417

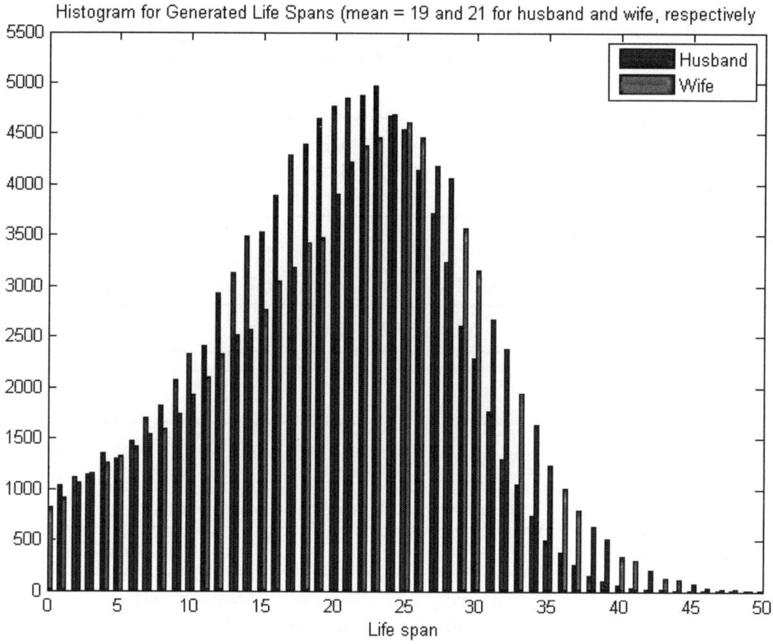

Figure 2 Histogram for Life Expectancy Scenarios of the Couple.

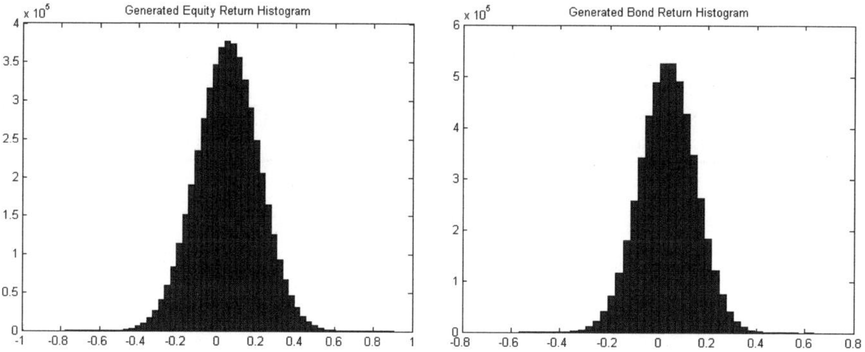

Figure 3 Histogram for the Equity and Bond Returns.

Table 4 is a summary of the input parameters.

We assume that the couple is rational and smart and they will make a decision under the following investment policy. The first investment policy is about the life insurance. We assume that the couple will buy insurance that will provide coverage in proportion to their expected remaining life spans.

Table 2 Parametric Assumptions for the Asset Returns.

	Normal Regime		Crash Regime	
	Stock	Bond	Stock	Bond
Expected Return	9%	3%	−5%	5.5%
Volatility	15%	10%	20%	15%
Correlation	0.40		−0.40	

Table 3 Transition Probability Matrix for the Regime- Switching Model.

	Normal Regime	Crash Regime
Normal Regime	0.9	0.1
Crash Regime	0.3	0.7

Table 4 Summary of Parametric Assumptions.

Parameters	Value
Age of the husband and the wife	65 years old
Average life spans for the husband and the wife	[19, 21]
Maximum life span	115 years old
Annual living expenses	$40,000
Savings at the initial time	$200,000
Annual benefit payments from pension plan	$= \begin{cases} \$40{,}000 & \text{if } t \leq \tau_s^h \\ 0 & \text{otherwise} \end{cases}$
Period of insurance	10 year period
Adjustment factor	20% ~ 200%
Parameters for stock and bond	See Table 2
Transition probability matrix	See Table 3

CRs are calculated so that the coverage is proportionate to the expected life time for the husband given a fixed amount of premium. Thus, we assume that they are paying fixed amount of premium throughout their lives before one of the couple dies. There is a difference between the remaining life spans of the husband and the wife, thus leading to different CR structures required for the coverage proportionate to the expected life of the wife. However, the difference is not large when expected life spans for the couple are set to be similar, so we employ the policy above for simplicity. Therefore, in our analysis we assume that the dollar amount invested in insurance (annual premium) is fixed throughout the couple's life and for sensitivity analysis, we try values ranging from $0 to $40,000.

The other rule is the surplus investment policy. At the beginning of each year, their wealth is reinvested in the portfolio of stock, bond, and insurance. They employ a fixed-mix strategy between stocks and bonds throughout their life span. We do not allow short-selling and for sensitivity analysis, we try mixes ranging from 100–0 (all stocks) to 0–100 (all bonds). If their wealth becomes negative, they simply borrow money from bank at the cash rate with no further investment in stocks or bonds. However, we assume that they still pay insurance premium.

3.1 *First example: Retirement planning without longevity risk consideration*

As a result of these policy rules, the couple has only two parameters to decide upon: Annual constant insurance premium and fixed weight of stocks in their portfolio. Next, we present and discuss the impact of these two decisions on some of the risk and performance measures defined in Section 2 and carry out sensitivity analysis with respect to the CR adjustment factor.[1]

In Figure 4, we show how the expected average life-time shortage in consumption changes, one of the downside risk measure, with respect to the insurance premium and equity allocation decisions for three different values of CR adjustment factor. The red label corresponds to the optimal (minimum) point. Figure 5 displays how the optimal value of this objective changes with different choices for the adjustment factor.

In Figure 6, we show how the probability of a deficit in the final period changes, another downside risk measure, with respect to the insurance premium and equity allocation decisions for three different values of CR adjustment factor. The red label corresponds to the optimal (minimum) point. Figure 7 displays how the optimal value of this objective changes with different choices for the adjustment factor.

We do the same simulation for other downside risk measures, the probability of final deficit and the % years in deficit, and we obtained similar results.

Now we explain the results of simulations for performance measures. In Figure 8, we show how the expected final surplus changes, one of the performance measures, with respect to the insurance premium and equity

[1] We obtained similar results for other objective functions but left them out due to space constraints. These are available from the authors upon request.

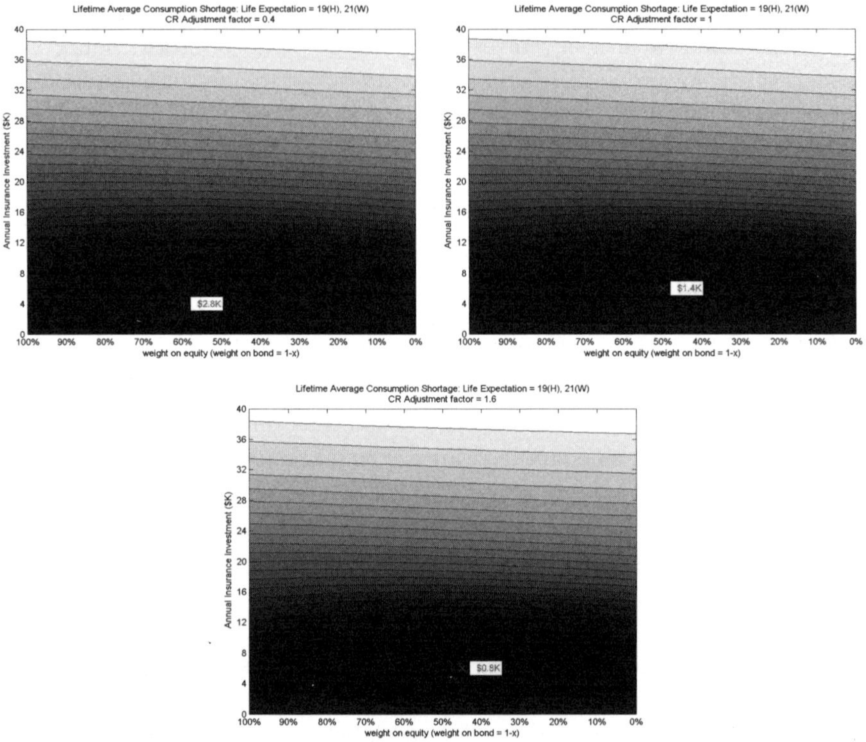

Figure 4 Impact of Equity Allocation and Insurance Premium Decisions on the Average Life-Time Shortage in Consumption (for CR = 0.4, 1.0, 1.6).

allocation decisions for three different values of CR adjustment factor. The red label corresponds to the optimal (maximum) point. Figure 9 displays how the optimal value of this objective changes with different choices for the adjustment factor.

We obtained similar result for the other performance measure, the final wealth.

In Figures. 10, 11, and 12, we present the results for the expected utility maximization by plotting the maximum expected utility, optimal weight of stocks, and optimal insurance investment, respectively, as a function of CR adjustment factor and the degree of risk aversion (α). Due to scaling reasons, the range for α is selected as [0, 0.002].

This first set of results suggests that all risk and performance measures improve as the CR adjustment factor increases (as insurance becomes less expensive). Furthermore, as expected, pure performance maximization is

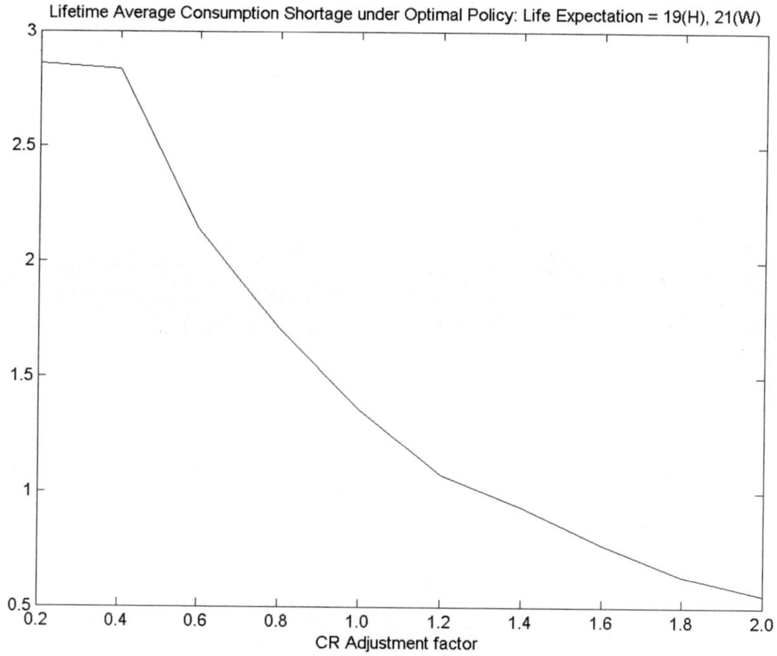

Figure 5 Optimal Value of Life-Time Shortage in Consumption as a Function of CR Adjustment.

achieved through extreme decisions (i.e., no insurance and 100% stock investment) whereas pure risk minimization yields less aggressive optimal investment rules.

Higher CR adjustment factor (i.e., cheaper insurance) implies higher optimal investment in insurance while a lower CR adjustment factor induces more aggressive investment (higher weights in stocks). Yet, beyond a certain value, increasing conversion ratio doesn't affect optimal policies much.

The chosen utility function (linear combination of final surplus and consumption shortage) seems to be representative as the optimal policies are very similar to those for the risk and performance measures. Therefore, in the following analyses we only analyze and optimize this utility function and pick three values for α to represent various risk aversion profiles. For the "conservative" decision makers who minimize downside risk, we choose $\alpha = 0$. For "aggressive" decision makers with very little concern for risk, we pick $\alpha = 0.001$. For a "balanced" strategy, we set $\alpha = 0.0005$.

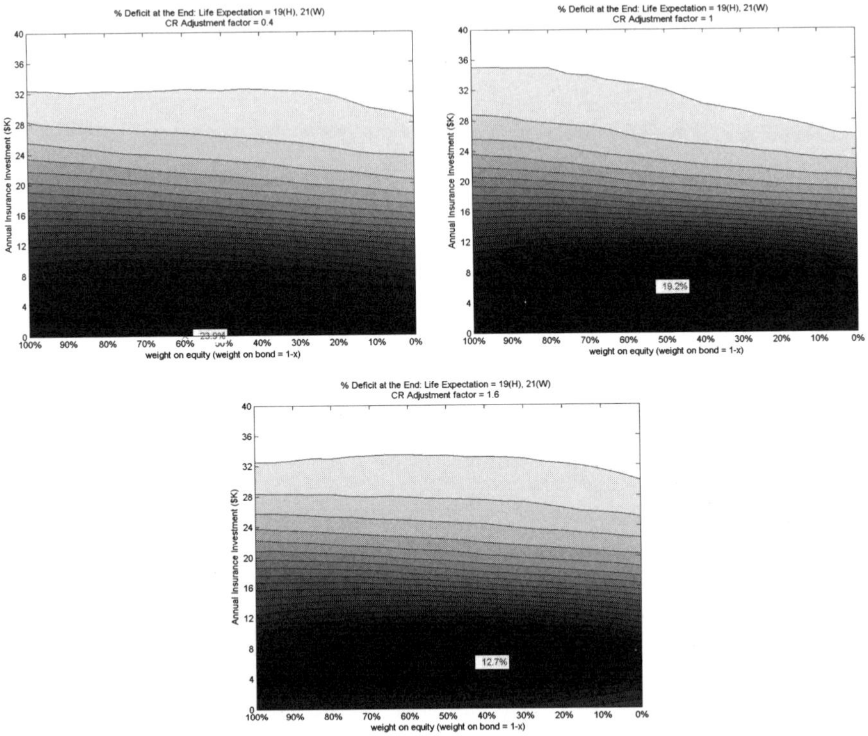

Figure 6 Impact of Equity Allocation and Insurance Premium Decisions on the Probability of Deficit (for CR = 0.4, 1.0, 1.6).

3.2 *Second example: Impact of longevity risk to retirement planning*

In the second set of simulations, we relax the strong assumption that the expected life spans for the couple are correct. The historical data (e.g. the past mortality tables) suggests that the couple is likely to live longer than the original estimates. In reality, the expected life spans change over time. In the first example, we assume that the expected remaining life time of husband and wife is 19 and 21 years. Now we change the husband's expected life span from 14 to 24 years and the wife's expected life span from 16 to 26 years. Therefore, we examine the effects of the changes in the remaining life parameters in various directions and perform expected utility maximization for the three cases — conservative, balanced, and aggressive.

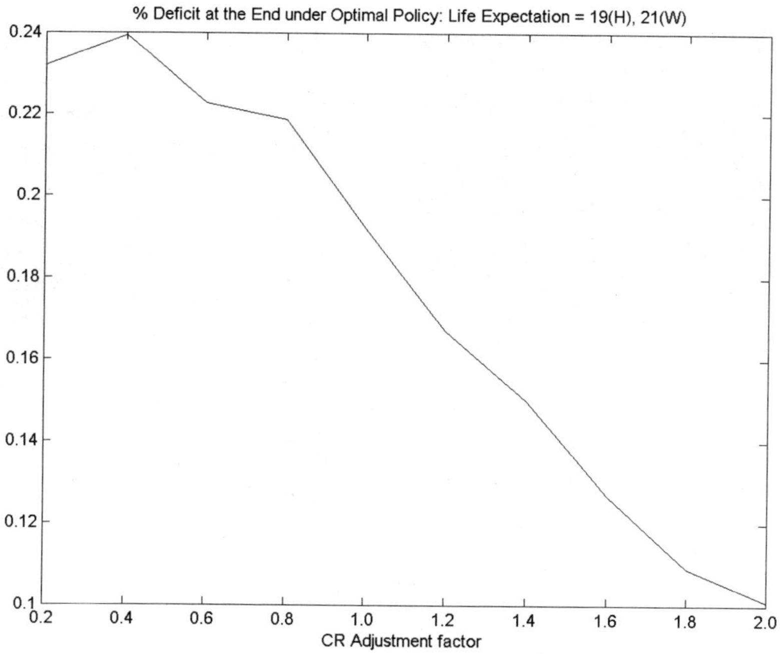

Figure 7 Optimal Value of Probability of Deficit as a Function of CR Adjustment.

We display the optimal expected utility graphs for the "balanced" decision maker in Figure 13. Conservative and aggressive utility maximizing figures have the same pattern. We observe that longevity risk is induced when the husband has a shorter life expectation and the wife has a longer life expectation. Furthermore, under the optimal policy, the expected utility decreases as life expectation for husband decreases or life expectation for wife increases.

Figures 14 and 15 display the optimal stock weights for conservative and balanced utility maximizations, respectively. Aggressive utility maximizing figures have the same pattern of balanced utility maximizing figure. For higher CR adjustment factors, we observe that optimal portfolio becomes less aggressive (lower weight on stocks) in all three utility functions. For the conservative utility, the optimal portfolio becomes more aggressive as the longevity risk increases (i.e., as life expectation for husband decreases and life expectation for wife increases). However, the results from the balanced and aggressive utility functions are somewhat counter-intuitive. For high CR adjustment factors, the optimal strategy becomes more aggressive

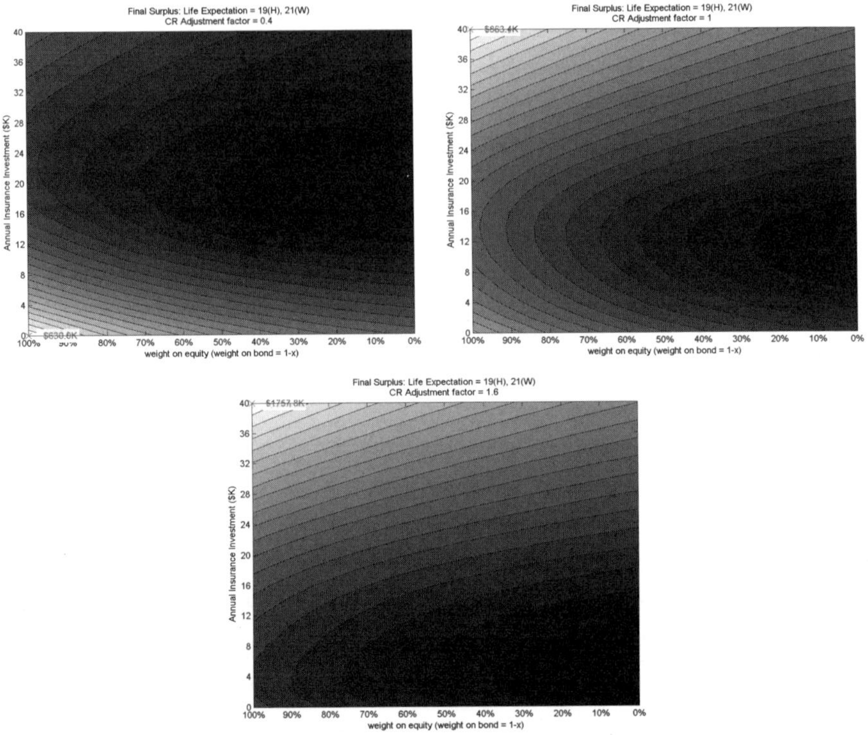

Figure 8 Impact of Equity Allocation and Insurance Premium Decisions on the Final Surplus (for CR = 0.4, 1.0, 1.6).

when longevity risk is lowest (i.e., as life expectation for husband increases and life expectation for wife decreases). For low CR values, the optimal strategy becomes more aggressive when longevity risk is highest and lowest compared to situations in between.

Our explanation for these results is as follows. If the longevity risk could be eliminated with the life insurance at a low cost (high CR), the couple doesn't have an incentive to behave aggressively when their longevity risk increases. However, if the insurance is expensive, the couple should protect themselves from the downside risk by investing heavily on stocks. As their longevity risk diminishes, the couple doesn't need to worry about the downside risk any more. Therefore, they seek for higher surplus to increase their utility by heavily betting on stocks.

We also analyze and display the optimal insurance premium amounts for the conservative utility function in Figure 16. Balanced and aggressive

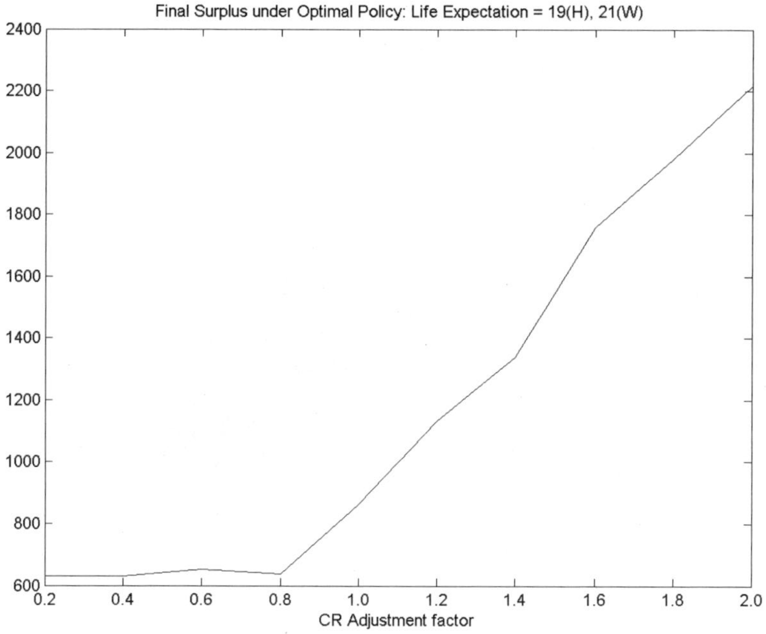

Figure 9 Optimal Value of Final Surplus as a Function of CR Adjustment.

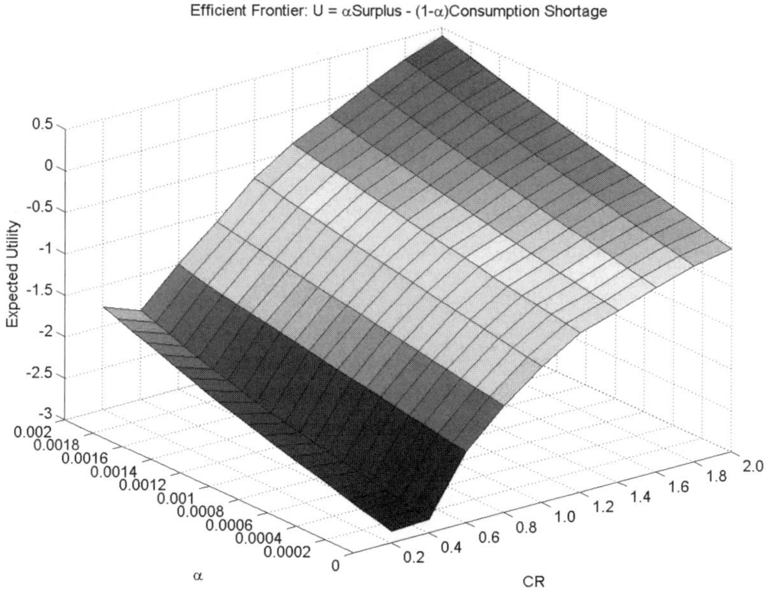

Figure 10 Optimal Expected Utility for Different Risk Aversion Levels and CR Adjustment Factors.

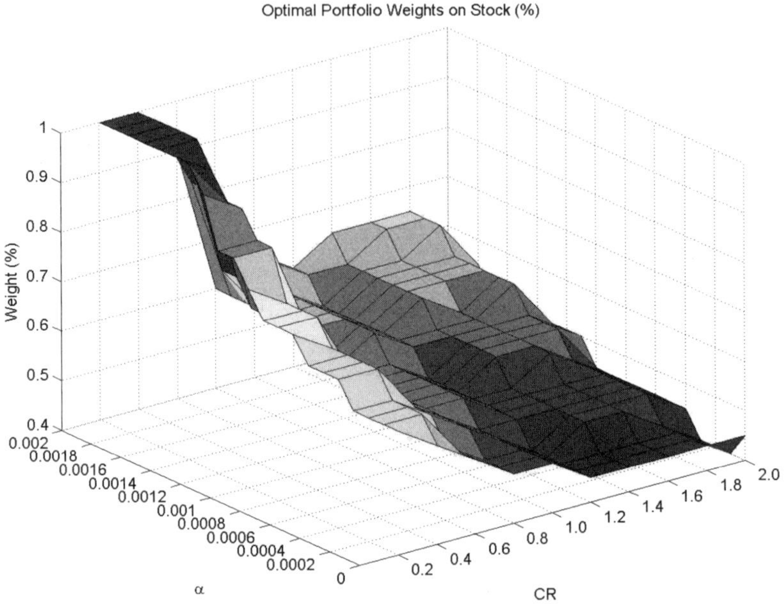

Figure 11 Optimal Stock Allocation for Different Risk Aversion Levels and CR Adjustment Factors.

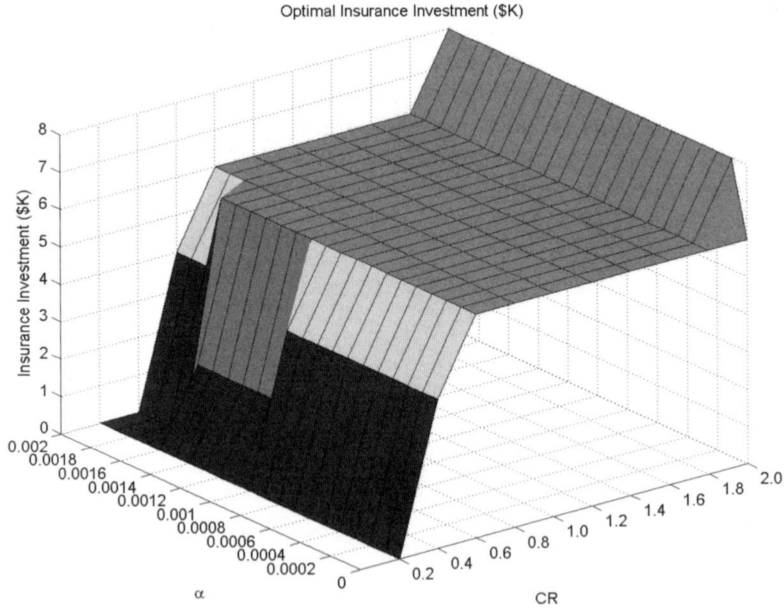

Figure 12 Optimal Insurance Investment for Different Risk Aversion Levels and CR Adjustment Factors.

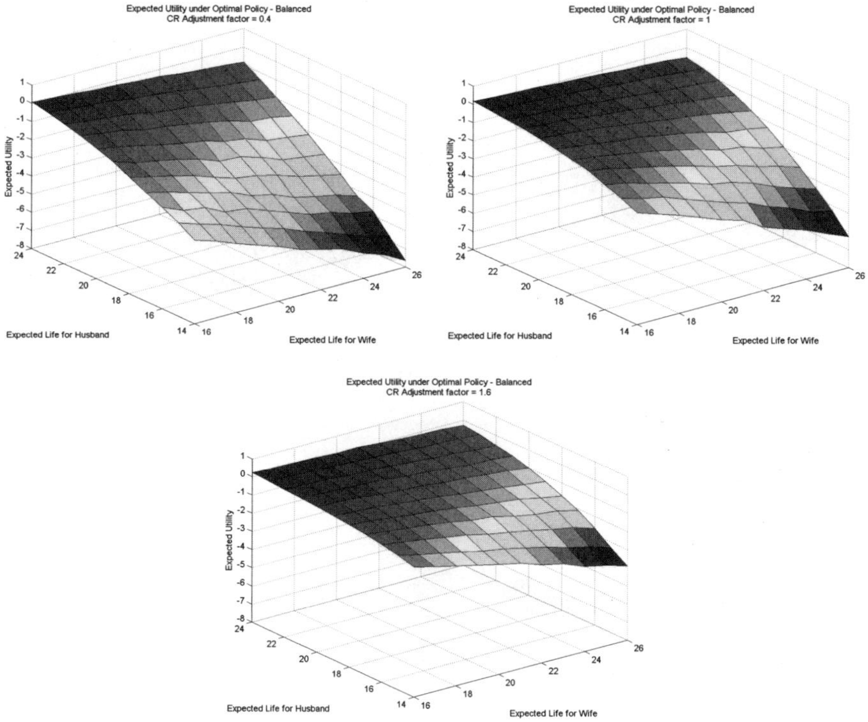

Figure 13 "Balanced" Decision Maker's Optimal Expected Utility for Different Remaining Life Expectancies (for CR = 0.4, 1.0, 1.6).

utility maximizing figures have the same pattern. We observe that the life expectation parameter for the husband is the dominant factor here. If the insurance is expensive (low CR), the couple should not invest in insurance when longevity risk is higher and should moderately invest in insurance as the longevity risk decreases. On the other hand, if the insurance is cheap (i.e., high CR), they should optimally invest more heavily on insurance.

Our interpretation is that high insurance premium makes the couple protect themselves by investing in stocks and bonds as their longevity risk increases. However, as the longevity risk decreases, the couple seeks for higher utility by cutting the downside risk with the insurance, which induces diversification benefits.

We've discussed the changes in the expected utility, optimal weight on stock and insurance investment as the life expectancy parameters of the couple change in order to identify the effect of the longevity risks to the retirement planning. All in all, the results are showing that, although the

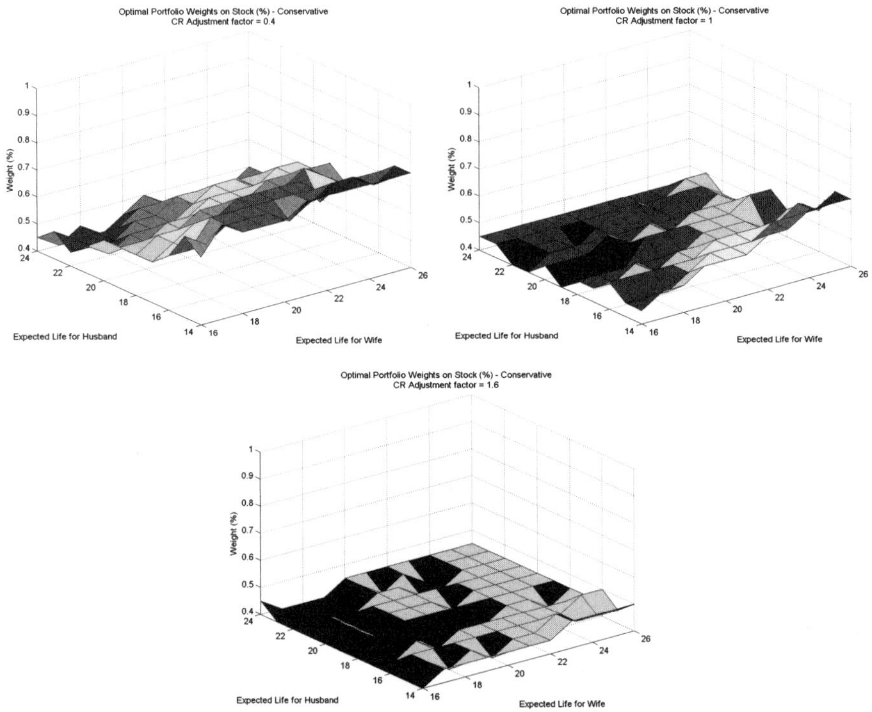

Figure 14 "Conservative" Decision Maker's Optimal Stock Allocation for Different Remaining Life Expectancies (for CR = 0.4, 1.0, 1.6).

impact of the longevity risk is significant, it is hard to find a general pattern for the figures above.

There are 8 possible cases that the life expectancy parameters change as illustrated in the table below. We evaluate the directions of the changes of the expected utility, weight on stock and insurance investment for each case to see if there's any generalizable pattern. Herein, we employ three different utility functions, and vary CR adjustment factor for the analysis.

	Case 1	**Case 2**	**Case 3**	**Case 4**	**Case 5**	**Case 6**	**Case 7**	**Case 8**
Expected life time of husband	—	▲	▲	▲	—	▼	▼	▼
Expected life time of wife	▼	▲	—	▼	▲	▲	—	▼

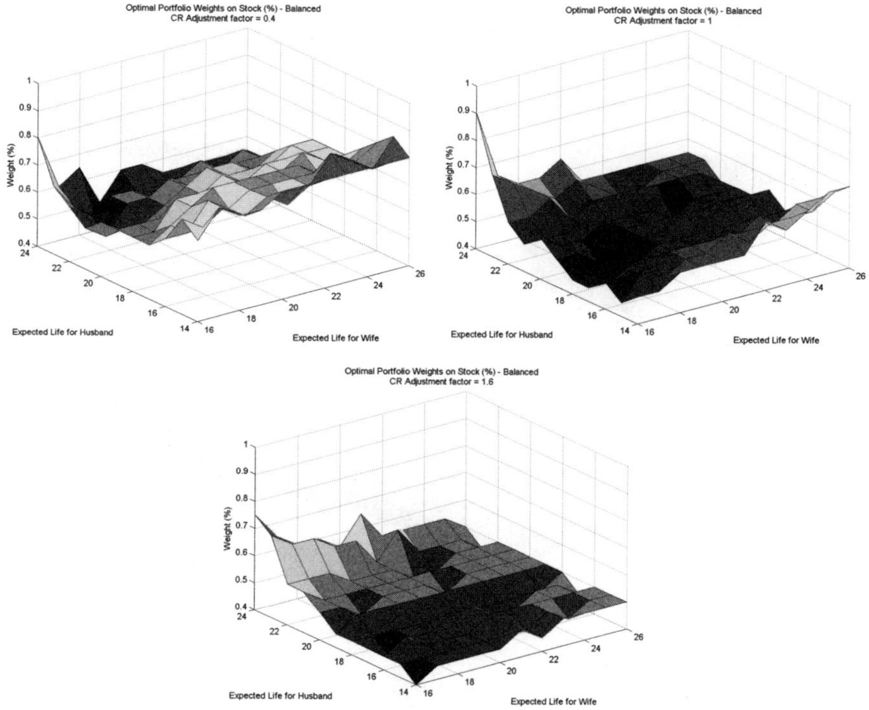

Figure 15 "Balanced" Decision Maker's Optimal Stock Allocation for Different Remaining Life Expectancies (for CR = 0.4, 1.0, 1.6).

The changes of the expected utility function, optimal weight on stock, and insurance investment for each case are shown in the following tables.

	Couple's Utility Function		
	Conservative	**Balanced**	**Aggressive**
Case 1			
Low CR adjustment factor			
Expected utility	▲	▲	▲
Optimal stock weight	—	▼	▼
Insurance investment	▲	▲	▲
High CR adjustment factor			
Expected utility	▲	▲	▲
Optimal stock weight	▼	—	—
Insurance investment	—	—	—

(*Continued*)

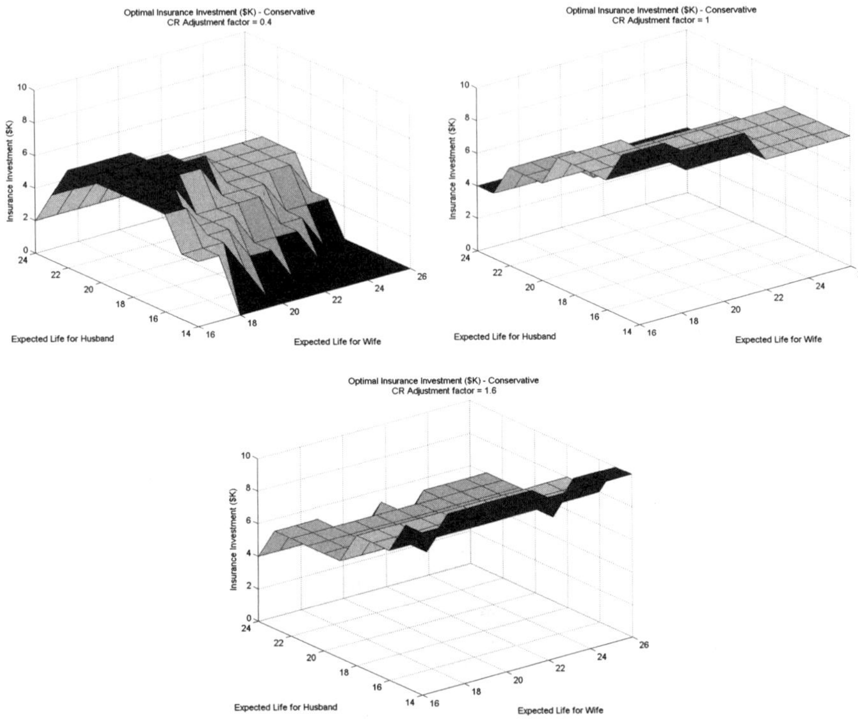

Figure 16 "Conservative" Decision Maker's Optimal Insurance Investment for Different Remaining Life Expectancies (for CR = 0.4, 1.0, 1.6).

(Continued)

	Couple's Utility Function		
	Conservative	**Balanced**	**Aggressive**
Case 2			
Low CR adjustment factor			
Expected utility	▲	▲	▲
Optimal stock weight	▼	▼	▼
Insurance investment	▼	▼	▼
High CR adjustment factor			
Expected utility	▲	▲	▲
Optimal stock weight	—	—	▲
Insurance investment	▼	▼	▼
Case 3			
Low CR adjustment factor			
Expected utility	▲	▲	▲
Optimal stock weight	▼	▼	—
Insurance investment	▼	▼	▼

(Continued)

(*Continued*)

	Couple's Utility Function		
	Conservative	Balanced	Aggressive
Case 3			
High CR adjustment factor			
Expected utility	▲	▲	▲
Optimal stock weight	—	—	▲
Insurance investment	▼	▼	▼
Case 4			
Low CR adjustment factor			
Expected utility	▲	▲	▲
Optimal stock weight	▼	▼	▲
Insurance investment	▼	▼	▼
High CR adjustment factor			
Expected utility	▲	▲	▲
Optimal stock weight	—	▲	▲
Insurance investment	▼	▼	▼

For cases 1–4, the expected utility increases as the life expectancy parameter of the husband increases, and it is of course an intuitive result. However, there is no generalizable pattern in the optimal stock weight and insurance investment.

The following tables illustrate the results for cases 5–8.

	Couple's Utility Function		
	Conservative	Balanced	Aggressive
Case 5			
Low CR adjustment factor			
Expected utility	▼	▼	▼
Optimal stock weight	—	—	—
Insurance investment	▼	▼	▼
High CR adjustment factor			
Expected utility	▼	▼	▼
Optimal stock weight	—	—	—
Insurance investment	—	—	—
Case 6			
Low CR adjustment factor			
Expected utility	▼	▼	▼
Optimal stock weight	▲	▲	▲
Insurance investment	▼	▼	▼
High CR adjustment factor			
Expected utility	▼	▼	▼
Optimal stock weight	—	—	—
Insurance investment	▲	▲	▲

(*Continued*)

(Continued)

	Couple's Utility Function		
	Conservative	**Balanced**	**Aggressive**
Case 7			
Low CR adjustment factor			
Expected utility	▼	▼	▼
Optimal stock weight	▲	▲	▲
Insurance investment	▼	▼	▼
High CR adjustment factor			
Expected utility	▼	▼	▼
Optimal stock weight	—	—	—
Insurance investment	▲	▲	▲
Case 8			
Low CR adjustment factor			
Expected utility	▼	▼	▼
Optimal stock weight	—	—	—
Insurance investment	—	▼	▼
High CR adjustment factor			
Expected utility	▼	▼	▼
Optimal stock weight	▼	▼	▼
Insurance investment	▲	▲	▲

Again, the expected utility is decreasing for these four cases. However, similar to the previous cases, one cannot identify the patterns of changes in the decision variables. This advocates that one should employ a highly systematic approach to incorporate the longevity risk management with the retirement planning, rather relying on the rule-based approach.

The test results up to this point indicate that the investment policies under different parameter sets could differ significantly. Therefore, the couple could suffer from undesirable outcomes if their life expectation parameters are incorrectly chosen. In order to investigate such a case of heterogeneous expectations, we conduct a different type of sensitivity analysis. We adopt the optimal allocation and insurance values obtained according to the original expected remaining life estimates in the first set of results and run simulations based on scenarios generated by varying the couple's expected remaining lives. Thus, we aim to measure the level of sub-optimality by calculating the loss in the expected utility with respect to the optimal utility values found in the second set of results. In Figure 17, we display the losses in the expected utility as a function of the variability in expected remaining life spans of husband and wife for the case of conservative utility.

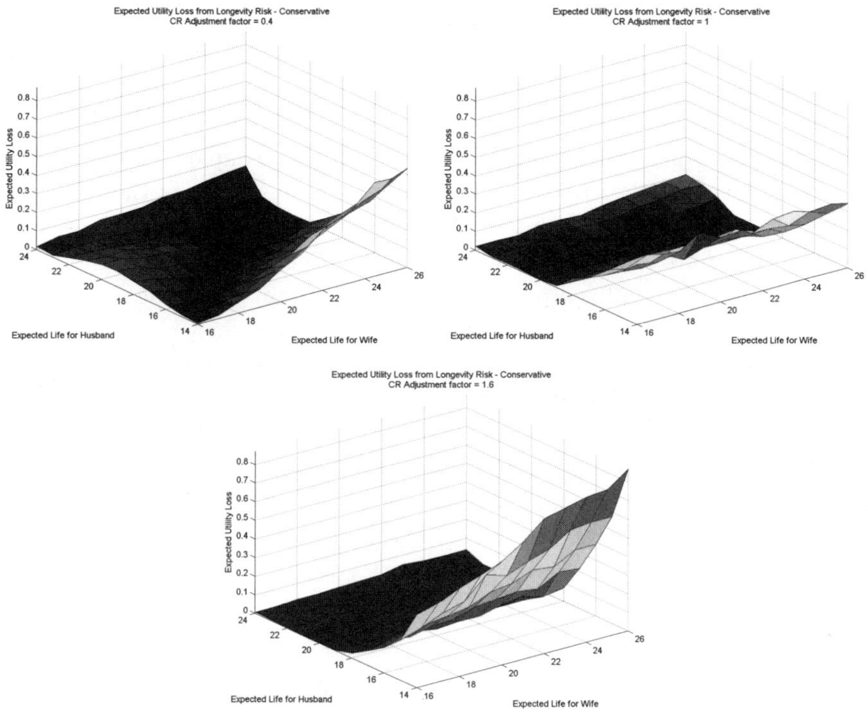

Figure 17 "Conservative" Decision Maker's Loss in Expected Utility Due to Heterogeneous Expectations about Remaining Life Expectancy (for CR = 0.4, 1.0, 1.6).

Figure 17 and other results which are similar but left out due to space constraints indicate that the losses are induced by the misalignment of the life expectation estimates. Error increases if life expectation of husband and that of wife moves away from the original estimate in opposite directions. There is no or little error when these parameters deviate from the original values in the same direction. Moreover, for the high longevity risk domain (i.e., as life expectation for husband lower and life expectation for wife higher), the error increases as the insurance becomes cheaper (high CR adjustment factor). This finding shows that although cheap insurance could protect against the longevity risk efficiently, incorrect estimation of life expectancy leads to inefficient utilization of insurance, causing larger errors.

3.3 *Third example: Longevity risks in pension benefits*

In our final set of simulations, we focus on the impact of the longevity risk on the pension benefits. Long life spans push the value of pension liabilities upwards, thus inflicting larger deficits on pension plans. Defined-benefit pensions plan administrators face two choices in such extreme cases: Increase contributions or decrease benefits. The recent relative popularity for the second choice (Korean national pension plan 2003, State of Illinois public pension plan 2010) could be bad news for the couple in our example. What happens to the couple if their pension plan decreases their benefit payments after a certain point? Once again, we carry out simulation using our base case life expectation scenarios. The difference is that we cut the pension benefits by a fixed percentage after a certain year. The percentage cut is varied from 5% to 25% and the starting point for the cut is varied throughout the planning horizon. As before, we run all simulation for various CR adjustment factors (ranging from 0.5 to 4.0) to see how our results depend on the cost of insurance.

In Figure 18, we show the impact of benefit cuts on the optimal expected utility for the balanced case. The results of other utility functions

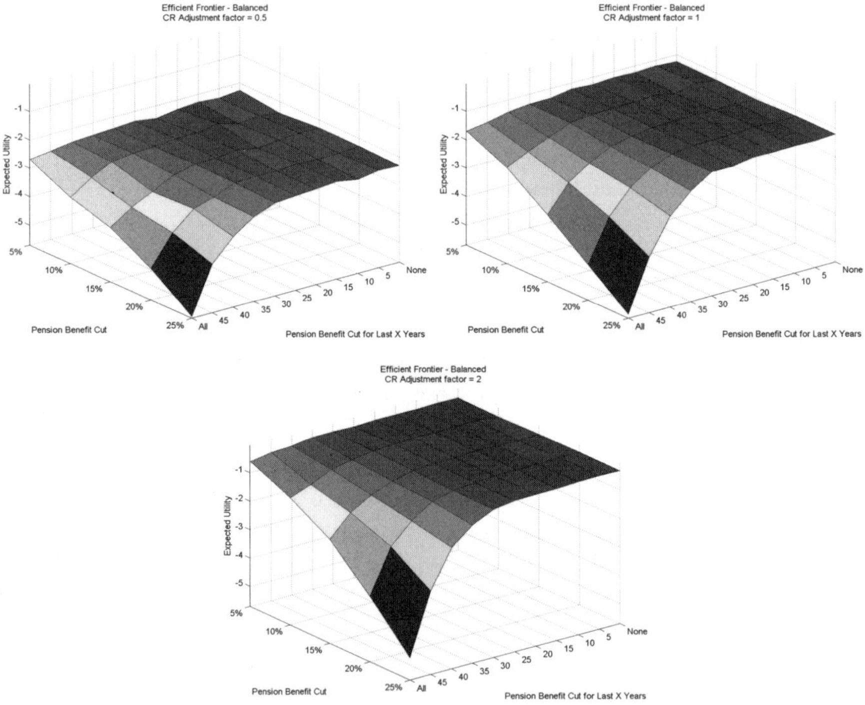

Figure 18 Impact of Pension Benefit Cuts on the Optimal Expected Utility for the "Balanced" Decision Maker (for CR = 0.5, 1.0, 2.0).

have the same pattern. As expected, we observe that the optimal level of expected utility decreases as the pension benefit cut begins earlier or the percentage of cut is higher. Furthermore, the expected utility increases as insurance gets cheaper (higher CR adjustment factor).

Figures 19 and 20 show the impact of benefit cuts on the optimal equity allocation for the balanced and aggressive utilities, respectively. We observe that, regardless of the choice of utility, the portfolio becomes less aggressive (more in bonds) as the insurance premium becomes cheaper (higher CR). When the insurance is most expensive (CR $= 0.5$), the portfolio becomes conservative (more on bonds) as the couple faces higher benefit cuts for a longer period. However, as the insurance becomes less expensive (CR ≥ 1), changes in the pension benefit cuts do not have a noticeable impact on the optimal portfolio choice. Moreover, the portfolio construction becomes less conservative (more in stocks) as the couple employs more aggressive utility.

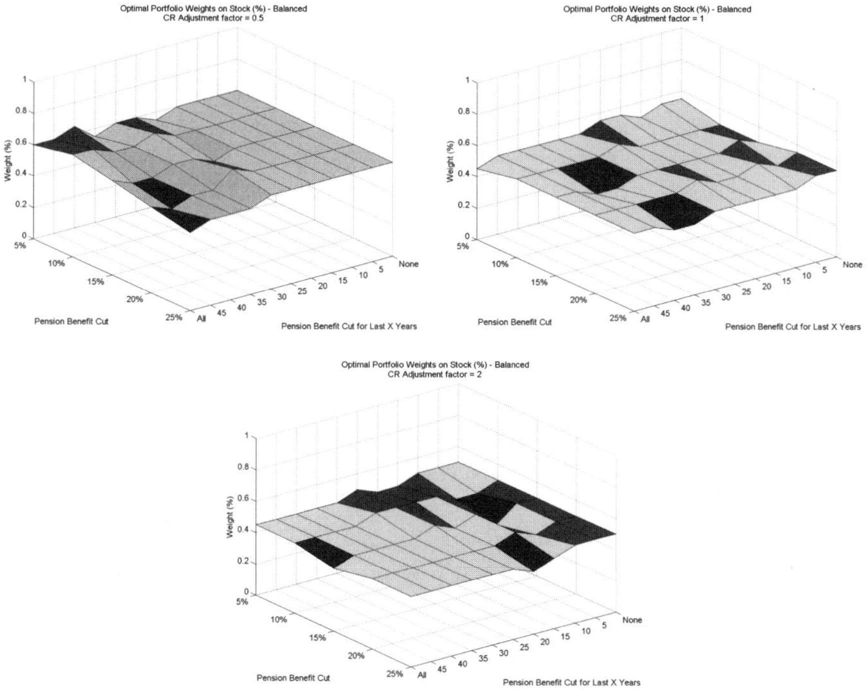

Figure 19 Impact of Pension Benefit Cuts on the Optimal Stock Allocation for the "Balanced" Decision Maker (for CR $= 0.5$, 1.0, 2.0).

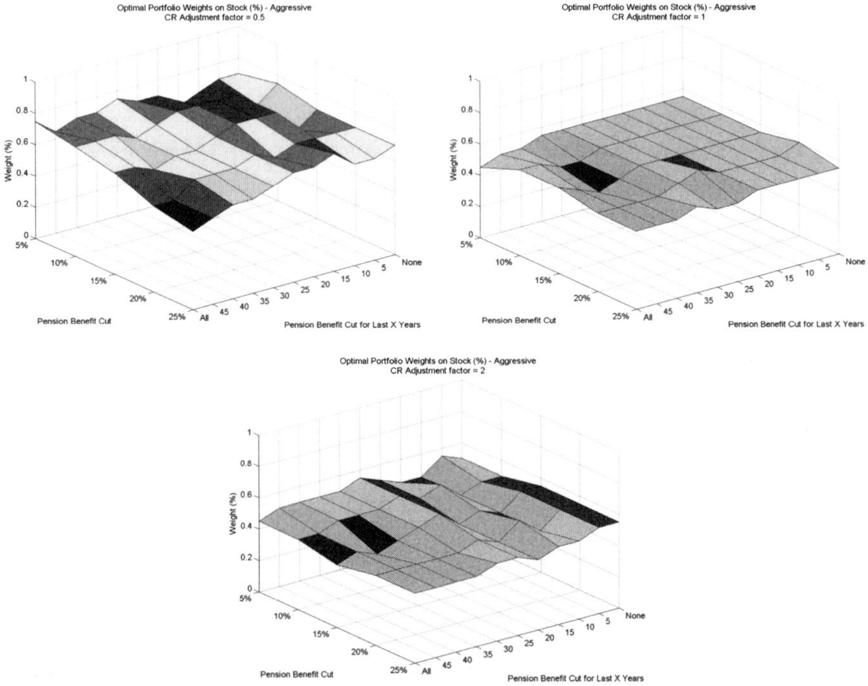

Figure 20 Impact of Pension Benefit Cuts on the Optimal Stock Allocation for the "Aggressive" Decision Maker (for CR = 0.5, 1.0, 2.0).

Figure 21 displays the impact of benefit cuts on the optimal insurance investment for the balanced utility. As expected, as the insurance becomes cheaper, the couple invests more on the insurance. However, as the pension benefits begin to decrease earlier or the percentage of the benefit cut increases, the couple invests less on the insurance. This final result is counter-intuitive as we would expect the couple to invest more in insurance due to higher longevity risk. Since the husband is the only person making income, the couple's financial status stays in a stable condition as long as he is alive. Thus, without the benefit cut, the couple has strong incentives to hedge the risk from earlier death of the husband. However, when the pension benefit cut happens, the amount of income from the benefit does not cover the cost. Therefore, the couple has less incentive to hedge the risk. Therefore, the couple's optimal strategy is to invest less on the life insurance, as opposed to the previous case.

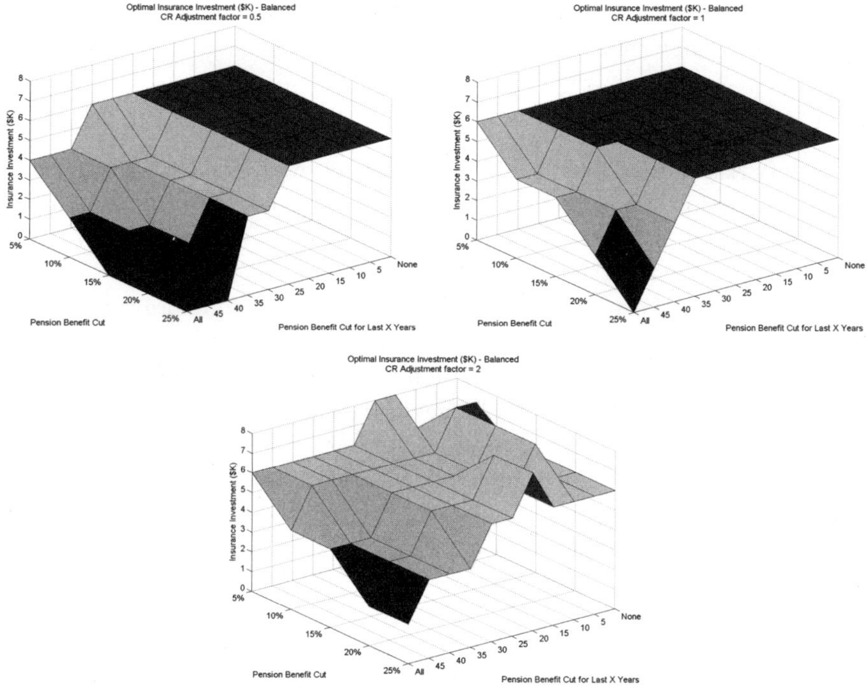

Figure 21 Impact of Pension Benefit Cuts on the Optimal Insurance Investment for the "Balanced" Decision Maker (for CR = 0.5, 1.0, 2.0).

Pension Cut ▲	**Couple's Utility Function**		
Years of Pension Cut ▲	**Conservative**	**Balanced**	**Aggressive**
Low CR adjustment factor			
Expected utility	▼	▼	▼
Optimal stock weight	—	▼	▼
Insurance investment	▼	▼	▼
High CR adjustment factor			
Expected utility	▼	▼	▼
Optimal stock weight	—	—	—
Insurance investment	▼	▼	▼

As we have seen in many of the results above, the problem is highly sensitive on the parameter values, especially the choice of utility function, cost of insurance (hedging cost), and the asset returns (opportunity cost). While one can find reasonable interpretations on the optimal policy, there seems to be no rule to be generalized. This indicates a strong need for a systematic approach to evaluate the situation and an integrated system for longevity risk management.

4 Conclusions

Longevity risk could impose a critical constraint in investment management practice. Portfolio construction policy coupled with insurance investment strategy should be carefully analyzed in order to achieve better risk-adjusted performance.

Our numerical examples show that it is not likely to find a generalized linear rule for optimal investment and insurance policies. Thus, a systematic approach is required to address the longevity risk management problems. Since the parameters related to longevity risk are uncertain, more efforts should be placed on obtaining robust solutions.

While our model and numerical examples focus on the retirement planning problem of a couple, it could be easily extended to full-scaled life-cycle planning problem that also considers pre-retirement investment decisions.

References

Bodie, Z., J.B. Detemple, S. Otruba, S. Walter. Optimal Consumption-Portfolio Choices and Retirement Planning. *Journal of Economic Dynamics and Control*, 28: pp. 1013–1226, 2004.

Campbell, R.A. The Demand for Life Insurance: An Application of the Economics of Uncertainty. *Journal of Finance*, 35: pp. 1155–1172, 1980.

Chen, P., R. Ibbotson, M.A. Milevsky, K. Zhu. Human capital, asset allocation and life insurance. *Financial Analysts Journal*, 62: pp. 97–108, 2006.

Cocco, J.F., F.J. Gomes, P. Maenhout. Consumption and Portfolio Choice over the Life Cycle. *Review of Financial Studies*, 18: pp. 491–533, 2005.

Cocco, J.F., F.J. Gomes. Longevity Risk, Retirement Savings, and Financial Innovation. *Journal of Financial Economics*, 103: pp. 507–529, 2012.

Fischer, S. A Life Cycle Model of Life Insurance Purchase. *International Economics Review*, 14: pp. 132–152, 1973.

Hamermesh, D.S. Life-Cycle Effects on Consumption and Retirement. *Journal of Labor Economics*, 2: pp. 353–370, 1984.

Huang, H., M.A. Milevsky. Portfolio choice and mortality-contingent claims: The general HARA case. *Journal of Banking and Finance*, 32: pp. 2444–2452, 2008.

Huang, H., M.A. Milevsky, T.S. Salisbury. Yaari's Life Cycle Model in the 21st Century: Consumption Under a Stochastic Force of Mortality. Working Paper, 2011.

Merton, R.C. Lifetime Portfolio Selection under Uncertainty: The Continuous-Time Case. *Review of Economics and Statistics*, 51: pp. 247–257, 1969.

Pliska, S.R., J. Ye. Optimal life insurance purchase and consumption/investment under uncertain lifetime. *Journal of Banking and Finance*, 31: pp. 1307–1319, 2007.

Richard, S.F. Optimal consumption, portfolio and life insurance rules for an uncertain lived individual in a continuous time model. *Journal of Financial Economics*, 2: pp. 187–203, 1975.

Yaari, M.E. Uncertain Lifetime, Life Insurance and the Theory of the Consumer. *Review of Economic Studies*, 32: pp. 137–150, 1965.

Ye, J., S.R. Pliska. Optimal Life Insurance Purchase, Consumption and Investment. Working Paper, 2011.

Chapter 3

Optimal Stochastic Programming-Based Personal Financial Planning with Intermediate and Long-Term Goals

Vittorio Moriggia*, Giorgio Consigli† and Gaetano Iaquinta

Summary

We present an asset-liability management (ALM) model for individuals facing a long term allocation problem over a diversified investment universe including mutual and pension funds as well as unit linked contracts with corporate, fixed income or equity exposure. Inflation-adjusted living costs and investment or consumption targets over the planning horizon lead to the definition of a comprehensive decision support tool, whose key building blocks are discussed with reference to a case-study. A structured stochastic model for long-term scenario generation is also briefly analyzed.

1 Introduction

Multistage stochastic programming (MSP) (Zenios and Ziemba, 2007a; Bertocchi *et al.*, 2011; Birge and Louveaux, 2011) is emerging as a natural mathematical framework for asset-liability management problems faced by large institutions as well as individuals (Mulvey and Ziemba, 1998; Zenios and Ziemba, 2007b; Consiglio *et al.*, 2004; Hochreiter *et al.*, 2007; Medova *et al.*, 2008). We describe in this chapter the key elements of a decision support tool for financial management problems relevant for individuals and

*Dept. of Mathematics, Statistics, Informatics and Applications University of Bergamo, via dei Caniana 2 BERGAMO (Italy), vittorio.moriggia@unibg.it.
†Dept. of Mathematics, Statistics, Informatics and Applications University of Bergamo, via dei Caniana 2 BERGAMO (Italy), giorgio.consigli@unibg.it.

households (Consigli, 2007; Consigli *et al.*, 2011) developed in cooperation with a leading insurance company.

The method's key steps include:

1. The accurate definition of an individual planning problem, focusing on evolving inflation-adjusted income and saving conditions, family consumption and income taxes.
2. The specification of an investment universe including financial as well as insurance products spanning a wide range of individual risk attitudes.
3. The definition of the ALM mathematical instance with a risk-reward objective function and an extended set of linear constraints.
4. The implementation and solution of a dynamic optimisation problem leading to strategies based on portfolios of financial and insurance contracts.

The investment universe includes mutual and pension funds as well as unit-linked contracts with insurance features (Consigli *et al.*, 2011). Optimal strategies are analysed for different degrees of risk aversion relying on an underlying stochastic model for equity and fixed income returns with specific long term features.

Relevant application fields of the tool include: long term individual financial planning (Consigli, 2007; Medova *et al.*, 2008); the definition of optimal retirement strategies based on unit-linked, annuities and pension fund contracts (Hochreiter *et al.*, 2007); financial engineering applications (Hochreiter *et al.*, 2007; Consigli, 2007); personal financial advisory services and home accounts for individuals (Berger and Mulvey, 1998; Consiglio *et al.*, 2004; Consigli *et al.*, 2011).

Section 2 focuses on the individual asset-liability model with an underlying uncertainty modeled as a *scenario tree* (Dupačová *et al.*, 2001; Consigli *et al.*, 2010). In Section 3 a case study is presented to emphasise the information generated by the stochastic programming solution. Section 4 concludes.

2 The asset-liability management model

We consider a decision problem defined over the time stages $t_0, t_1, \ldots, t_n$, with conventionally $t_0 = 0$ as current time and $t_n = T$ as time horizon. In individual ALM problems, the decision horizon will in general coincide with the individuals' expected age of retirement and time stages may be nonhomogenous (Consigli, 2007; Medova *et al.*, 2008). We assume an individual seeking over a long-term horizon an optimal trade-off between a wealth objective and the risk not to achieve a set of targets distributed over time. In the case study developed in Section 3 we generalise previous work to allow for an horizon T different from the retirement age.

Individuals' optimal portfolios at the horizon, in presence of an explicit retirement goal, are likely to include pension products and annuities whose payoffs will cover their residual lifetime (Medova *et al.*, 2008; Consigli *et al.*, 2011). An adjustment is in this case introduced to take into account the difference, if any, between the horizon discounted annuities paid out over the retirement period and the current portfolio values. We call such horizon adjustment *end effect* on the individual wealth.

At the horizon the individual wealth will include the financial portfolio, the investment targets and the end effect. Through the planning horizon, for $j = 1, 2, \ldots, n$, in every subperiod $t_j - t_{j-1}$, the individual will assess his current financial wealth, preserve her consumption in real terms, face a set of predictable expenses and rebalance her investment portfolio. The rebalancing strategy will also be determined by the individual' goals and liabilities. Once a pension accumulation plan has been agreed, a constant nominal payment will be due at each stage.

We distinguish in the problem formulation *consumption* from *investment* goals: the former include those targets (university tuition, house renovation, new car, etc.) which once achieved are assumed will not increase the individual wealth; the latter, instead, include all targets (real estate, gold investments and so forth) that once achieved will become part of the individual wealth and will be exposed to market value fluctuations. Within a specific decision problem, investment goals may be further distinguished into levered or unlevered investments. Liabilities enter the decision problem in two ways: as possible short positions on the cash account and as borrowing decisions needed to fund an investment goal. The decision process is specified through holding, buying and selling decisions over an extended asset universe.

The described decision process may accommodate a wide range of financial plans and depending on the targets, the horizon stage partition and the stochastic models adopted for inflation, interest rates and equity instruments, the nature of the stochastic program may differ substantially.

2.1 *Individual wealth, consumption and investment targets*

As customary in MSP, the model specification requires the definition of an extended set of random coefficients whose evolution along a scenario tree represents the fundamental output of the scenario generator. All coefficients are specified exogenously per unit investment and passed onto the model algebraic generator (Consigli and Dempster, 1998) to derive an instance of the stochastic programming problem. We simplify the model description by introducing a time set $\mathcal{T} := \{t_0, t_1, \ldots, t_n\}$ with $t_n = T$ as planning horizon and H as retirement age, usually $H \geq T$. For $t \in \mathcal{T}$ we use in the model description $t-$ to denote the immediately preceding time stage and $t+$ to denote the immediately following time stage. The set of investment opportunities is denoted by $\mathcal{K}$ with $\mathcal{K} \equiv \{k|k = 1, \ldots, 18\}$. For modelling reasons the investment universe is also partitioned into four asset classes: $\mathcal{K} = \mathcal{K}_1 \cup \mathcal{K}_2 \cup \mathcal{K}_3 \cup \mathcal{K}_4$, where $\mathcal{K}_1$ includes all the mutual funds, $\mathcal{K}_2$ the unit-linked products, $\mathcal{K}_3$ the annuities and $\mathcal{K}_4$ all pension funds. We indicate with $j = 1, 2$ the liability classes (fixed and floating rate mortgages) and with $s = 1, \ldots, S$ the scenarios or sample paths from the root node to the horizon of the scenario tree (Bertocchi *et al.*, 2011).

We display in Figure 1 the set of investment opportunities and their management and fiscal costs, consistently with prevailing conditions in the European financial market.

The distinction into four asset classes is due either to different holding period performance conventions or to different contractual conditions,

	PRODUCTS	minimum age	minimum premium	Commitment fee	Managment fee% NAV	Accrual period tax	Income tax
1	Bank Account	18	500	0.00%	0.00%	25.00%	-
2	U-L European equity	18	500	0.60%	2.00%	12.00%	-
3	U-L International equity	18	500	0.50%	2.00%	11.50%	-
4	U-L Emerging equity	18	500	0.50%	2.00%	12.00%	-
5	U-L balanced	18	500	0.50%	2.00%	12.00%	-
6	U-L fixed income with guarantee	18	500	0.50%	1.50%	12.00%	-
7	U-L Corporate bonds	18	500	0.50%	1.50%	12.00%	-
8	U-L money market	18	500	0.50%	0.50%	12.00%	-
9	Insurance plan with revaluation	18	3000	1.50%	0.00%	12.00%	-
10	Annuity fixed income	35	50000	1.00%	2.00%	0.00%	12.00%
11	Annuity balanced	35	50000	1.00%	3.00%	0.00%	12.00%
12	Pension Fund equity	18	500	0.00%	2.50%	11.00%	14.00%
13	Pension Fund balanced	18	500	0.00%	2.00%	11.00%	14.00%
14	Pension Fund fixed income	18	500	0.00%	0.00%	11.00%	14.00%
15	Mutual Fund Equity	18	1000	1.00%	2.00%	12.00%	-
16	Mutual Fund balanced	18	1000	1.00%	2.00%	12.00%	-
17	Mutual Fund fixed income	18	1000	1.00%	1.00%	12.00%	-
18	Mutual Fund money market	18	1000	1.00%	1.00%	12.00%	-

Figure 1 The Investment Universe with Representative Management and Fiscal Costs.

management and operational costs. In general terms, the following relevant differences and financial structures may be recalled:

Mutual funds are typical collective investment products in which the investor can close the position at any point in time without facing any penalty nor fiscal cost.

Unit-linked products and revaluation plans are contracts with associated specific tax deduction and penalty schemes, which depend on the holding period. They have been introduced in the market to encourage mixed insurance-financial strategies.

Annuities are long term contracts with floating or fixed rate competing with pension funds to attract the demand for retirment protection. The range of annuity products is wide: we consider in the case study specifically deferred annuities benchmarked to the JPM EMU bond index and the MSCI World equity index, and a long maturity (e.g. 20–25 years). Annuities may in general start paying off prior to retirement.

Pension funds are canonical investment vehicles structured into an accumulation plan, with income deductions, and regular payments over the horizon and a payout that will start upon retirement.

The cash-flow pattern differ significantly from one class to the other as shown in Figure 2.

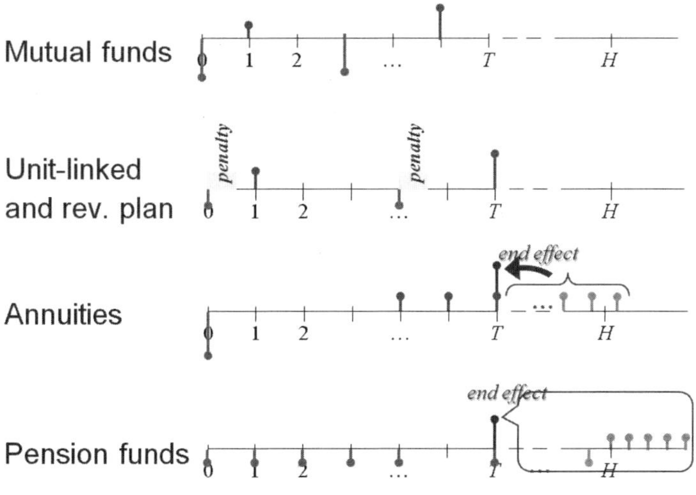

Figure 2 Asset Classes Cashflows.

Along each scenario $s = 1, 2, \ldots, S$ for $t \in \mathcal{T}$ the individual wealth will depend on the investment policy, exogenous consumption costs and revenues, and the household's goals. The following model variables are scenario dependent and allow the monitoring over time of the household financial performance:

$W_t(s)$	is the personal wealth at time t in scenario s;
$C_t(s)$	are the living costs at time t in scenario s;
$I_t(s)$	is the individual professional income at time t in scenario s;
$\widetilde{W}_t(s)$	is a target wealth at time t in scenario s;
$\Psi_t(s)$	includes the resources eligible at time t in scenario s to fund $\widetilde{W}_t(s)$.

For given inflation-adjusted consumption and targets and anticipated income over the remaining working life (the case of periods of unemployment as well as rapid carreer evolution can be accommodated here), the investment strategy will determine the individual wealth evolution. The following notation is adopted to describe the ALM model, for $k \in \mathcal{K}$ and $t \in \mathcal{T}$:

$x_{k,t}(s)$	amount held of asset k at time t in scenario s
$x_{k,h,t}(s)$	amount held of asset k at time t in scenario s, which was purchased in $h < t$
$\mathring{x}_{k,r,0}$	defines the initial (input) holdings at $t = 0$ held for r periods, prior to rebalancing. If the holding period is not relevant we just write $\mathring{x}_{k,0}$
$x^+_{k,t}(s)$	amount bought of asset k at time t in scenario s
$x^-_{k,h,t}(s)$	amount sold of asset k at time t in scenario s, which was purchased at time $h < t$
$y_{j,t}(s)$	liability j held at time t in scenario s
$y^+_{j,t}(s)$	increase of liability j at time t in scenario s
$y^-_{j,t}(s)$	liability j reduction at time t in scenario s.

We denote with $x^+_{1,t}(s)$ and $x^-_{1,t}(s)$ positive and negative cash positions respectively in scenario s at time t. At any point in time $x_{1,t}(s) = x^+_{1,t}(s) - x^-_{1,t}(s)$ represents the balance on the liquid cash position. As explained below, $r_{k,t}$ representes the price return of asset k at time t. Such return as shown in equation (1) may be reduced by the asset class management fee g_k determined by the fund manager.

The following personal wealth dynamics, for $t \in \mathcal{T}$ and $s = 1, \ldots, S$, can be considered:

$$\begin{aligned} W_t(s) = & \sum_{k \in \mathcal{K}_1 \cup 1} x_{k,t-}(s)(1 + r_{k,t})(1 - g_k) \\ & + \sum_{k \in K_2 \cup K_3} \sum_{h<t} x_{k,h,t-}(s)(1 + r_{k,t})(1 - g_k) \\ & + x^+_{1,t}(s) - x^-_{1,t}(s) - y_{1,t}(s) - y_{2,t}(s) \end{aligned} \tag{1}$$

At the horizon T, for $s = 1, 2, \ldots, S$

$$\begin{aligned} W_T(s) = & \sum_{k \in \mathcal{K}} x_{k,T-}(s)(1 + r_{k,T}(s))(1 - g_k) \\ & + x^+_{1,T}(s) - x^-_{1,T}(s) + E^1_T + E^2_T - y_{1,T}(s) - y_{2,T}(s) \end{aligned} \tag{2}$$

will denote the terminal financial wealth. In eq. (2), E^1_T and E^2_T denote the end effects for the pension fund portfolio and the annuities respectively. For pension funds we have for $t = T < H$ and $s = 1, \ldots, S$:

$$\begin{aligned} E^1_T(s) &= \sum_{k \in \mathcal{K}_4} x_{k,T}(s) e^{r_{T,H}(s)(H-T)} \left(\sum_{h \geq H} c_k \cdot e^{-(r_{T,h}(s)+\pi_k)(h-T)} - 1 \right) \\ &= \sum_{k \in \mathcal{K}_4} x_{k,T}(s) f_{k,T,H}(s) \end{aligned}$$

and for annuities, for $t = T$ and $s = 1, \ldots, S$:

$$\begin{aligned} E^2_T(s) &= \sum_{k \in \mathcal{K}_3} \sum_{h<T} \sum_{m \geq T} x_{k,h,T}(a_{k,m}(s) e^{-r_{T,m}(s)(m-T)} - 1) \\ &= \sum_{k \in \mathcal{K}_3} \sum_{h<T} x_{k,h,T} \cdot \alpha_{k,h,T}. \end{aligned}$$

At the horizon, we distinguish the individual financial wealth from the overall wealth, which include the targets $\widetilde{W}_T(s)$ revalued at the horizon:

$$\mathcal{W}_T(s) = W_T(s) + \widetilde{W}_T(s). \tag{3}$$

Assuming m investment goals over the horizon to be acquired at time τ_m, we have $\sum_m \widetilde{W}_{\tau_m,T}(s) = \widetilde{W}_T(s)$, where $\widetilde{W}_{\tau_m,T}(s)$ denotes the terminal value of the investment goal acquired at τ_m.

From equations (1) and (3) the individual wealth evolution will depend on the random returns $r_{k,t}(s)$ embodying the decision problem random structure. We have one such tree process for every investment opportunity in Table 1.

For $k \in \mathcal{K}$ and scenarios $s = 1, 2, \ldots, S$, the generic returns $r_{k,t}$ in equation (1) over a generic holding period $t = n\Delta t$ can be specified as

$$r_{k,t}(s) = \left[\prod_{m=1}^{n} (1 + r_{k,m\Delta t}(s)) \right]$$

$$r_{k,m\Delta t}(s) = \frac{v^k_{m\Delta t}(s)}{v^k_{(m-1)\Delta t}(s)} - 1 \; \forall m. \tag{4}$$

In (4) v^k_t denotes the unit *net-asset-value* (NAV) of contract k at time t. According to Figure 1 the asset payoffs are determined by a set of market indices or *benchmarks*. We clarify next how exogenous market returns will determine asset prices relevant to the solution of the MSP problem.

2.2 *Random coefficients and scenarios*

In addition to $r_{k,t}(s)$, the following extended set of random coefficients is required to specify the ALM model:

- $\zeta^{+/-}_t(s)$ are interest rates on the cash account surplus/deficit at time t in scenario s
- $b_{j,t}(s)$ is the mortgage rate at time t in scenario s for liability j
- $a_{k,h,t}(s)$ is the unit annuity at time t of asset k that was purchased in $h < t$
- $p_{k,h,t}$ is a penalty coefficient upon selling at time t asset k that was purchased in $h < t$
- $\alpha_{k,h,T}(s)$ is the end-effect multiplier per unit annuity bought in h and held in T in scenario s
- $f_{k,T,H}(s)$ is instead the end-effect multiplier per unit pension position available in T to mature in H
- $\pi_t(s)$ is the consumer price inflation at time t in scenario s.

The following class-dependent coefficients complete the model:

- u_k upfront fee for asset k
- g_k management cost per annum from holding asset k
- $\varphi_{k,h,t}(s)$ tax on capital gains from selling asset k at time t after a holding period $t - h$ in scenarios s

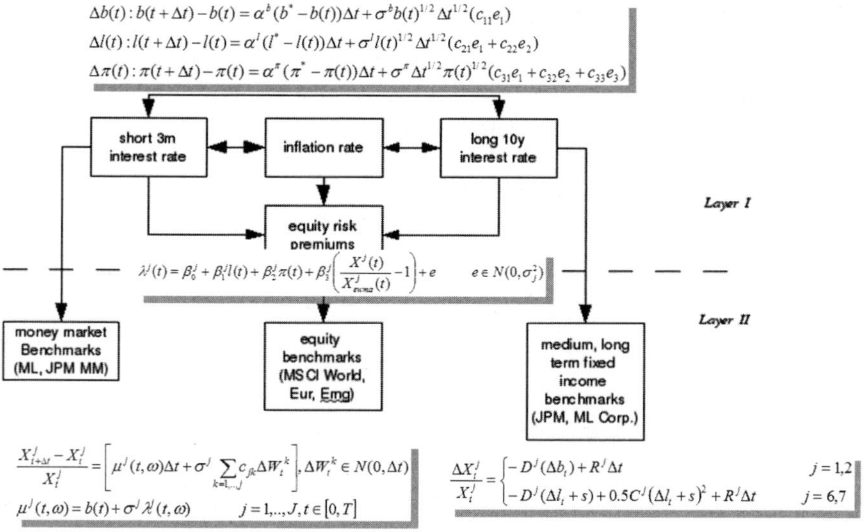

Figure 3 Stochastic Financial Model.

All the above scenario-dependent coefficients are derived from a core set of risk factors, cf. Figure 3 which include:

- the 3 month (short-term) interest rate
- the 10 year (long-term) interest rate
- the inflation rate
- the equity benchmarks

The three equity indices have a stochastic drift with a risk premium model plus a volatility component, whereas all processes are correlated and a constant correlation is assumed over the 10 years. Fixed-income and money-market-return-processes are determined according to a duration-convexity approximation. Figure 3 summarises the key elements of the financial model. For more detail we refer to (Consigli *et al.*, 2011).

The money market and fixed income benchmarks are functions of the 3-month and the 10-year interest rate. All three equity benchmarks are assumed to have constant volatility and a random drift (Campbell *et al.*, 1997; Mulvey, 1996; Dempster *et al.*, 2003; Consigli, 2007). The model structure in Figure 3 has been implemented to derive the benchmark returns. Each asset class will generate a return as a function of a given underlying dependence on the above benchmarks and the management fees and penalties associated with each specific asset class.

2.3 *The optimization problem*

The following multistage stochastic linear program is considered for an individual with an objective function based on a growth objective and an expected shortfall with respect to a target:

$$\max_{x \in X}[(1-\lambda)\mathbb{E}[W_T] - \lambda\mathbb{E}[\Psi_T | \Psi_T < \widetilde{W_T}]] \tag{5}$$

where:

$$\mathbb{E}[W_T] = \sum_{s \in S} p_s W_T^s, \quad \sum_s p_s = 1$$

and

$$\mathbb{E}[\Psi_T | \Psi_T < \widetilde{W_T}] = \sum_s p_s (\Psi_T^s | \Psi_T^s < \widetilde{W_T^s}).$$

The aim of the model is to maximize the final expected wealth under the penalty costs for target violations. The risk-reward trade-off function is determined by the coefficient λ, which for different values between 0 and 1 will span different individual relative risk aversions. Through $\Psi_t(s)$ we introduce in the problem a constraint on the set of asset classes and liabilities eligible to fund the target acquisition. In case of a pension wealth target, that can be funded through pension funds and/or annuities, $k \in \mathcal{K}_2 \cup \mathcal{K}_4$, we have:

$$\Psi_T(s) = \sum_{k \in \in \mathcal{K}_2 \cup \mathcal{K}_4} \sum_{h<t} x_{k,h,T^-}(s)(1 + r_{k,T}(s))(1 - g_k). \tag{6}$$

We show below results from a case study based on a sequence of intermediate goals over a 10 year horizon. The rebalancing strategy is governed by the set of inventory balance constraints spanning the decision horizon.

Inventory Balance Constraints The stage-by-stage portfolio evolution is recorded by the inventory balance constraints for $t \in \mathcal{T}$. At $t = 0$ the individual will change its current portfolio allocation to an optimal x_0: this is referred to as the root-node *implementable* solution to emphasise that in a multistage setting the only decision under full uncertainty is taken at time 0. For each asset $k \in \mathcal{K}$, we have:

$$\begin{aligned} x_{k,0} &= \mathring{x}_k + x_{k,0}^+(1 - u_k) - x_{k,r,0}^-(1 + p_{k,r,0} + \varphi_{k,r,0}) \\ y_0 &= 0. \end{aligned} \tag{7}$$

After rebalancing the individual portfolio value at time 0 will be already net of fees and penalties. For all subsequent rebalancing decisions the equations (11) apply with (9) and (10) for pension funds and annuities. A liability inventory balance is considered after the first stage $y_0 = 0$, to allow the funding of targets, for $j = 1, 2$:

$$y_{j,t}(s) = y_{j,t-}(s) + y^+_{j,t}(s) - y^-_{j,t}(s) \tag{8}$$

Inventory balance equations after the first stage depend on the asset class.

In (9) we first consider a decision taken at time t to start an accumulation plan. Then we assign $h = t$ and for all $t > h$ assume constant instalments $x^+_{i,t}(s)$ proportional to the elapsed period.

Pension Funds, $k \in \mathcal{K}_4$:

$$x_{k,t,t}(s) = x^+_{k,t}(s) \tag{9}$$

$h = t, \forall t > h$:

$$x^+_{k,t}(s) = x_{k,h}(s)$$

$$x_{k,t}(s) = x_{k,h,t-}(s)(1 + r_{k,t}(s))(1 - g_k) + x^+_{k,t}(s)(1 - u_k)$$

We have in (9) constant provisions in nominal terms and a pension value determined by the returns $r_{k,t}(s)$, consistently with a defined contribution scheme. At the horizon, the Pension fund value $x_{k,T}$ will generate a constant flow of pension payments depending on the adopted pension coefficient.

Variable life annuities, on the contrary, may be financed for $t \in \mathcal{T}$ via a lump sum investment and generate a sequence of variable, though stable, payments which over time will determine the annuity expiry.

Annuities, $k \in \mathcal{K}_3$:

$$x_{k,t,t}(s) = x^+_{k,t}(s)(1 - u_i) \tag{10}$$

$h = t, \forall t > h$:

$$A_{k,h,t}(s) = x_{k,h,h}(s)(1 + a_{i,h,t}(s))$$

$$x_{k,h,t}(s) = x_{k,h,t-}(s)(1 + r_{k,t}(s) - g_k) - A_{k,h,t}(s) - x^-_{k,h,t}(s)$$

Unlike pension funds, whose payout phase starts upon retirement, immediate annuities start paying off in the following period. Along each scenario, the annuity payments $A_{k,h,t}(s)$ are not reinvested but directly cashed. The combined effect of an increasing, though scenario dependent, annuity coefficient $a_{k,h,t}(s)$ and a descreasing residual annuity $x_{k,h,t}(s)$ preserves relatively stable cash flows to the annuity policy-holder.

For the other asset classes, $k \in \mathcal{K}_1 \cup \mathcal{K}_2$, a canonical set of inventory balance equations can be considered:

$$\begin{aligned} x_{k,h,t}(s) &= x_{k,h,t-}(s)(1 + r_{k,t}(s))(1 - g_k) \\ &\quad - x^-_{k,h,t}(s)(1 + p_{k,h,t} + \varphi_{k,h,t}(s)) \\ x_{k,t}(s) &= \sum_{h \le t-} x_{k,h,t}(s) + x^+_{k,t}(s)(1 - u_k). \end{aligned} \tag{11}$$

Finally for liabilities, $\forall j = 1, 2$ and $\forall s = 1, 2, \ldots, S$, we have:

$$y_{j,t}(s) = y_{j,t-}(s) + y^+_{j,t}(s) - y^-_{j,t}(s)$$

Cash Balance Contraints The cash balance constraints allow the tracking of all cash inflows and outflows at each stage. The first stage $(t = 0)$ is:

$$\mathring{x}_1 + \sum_{k,r} x^-_{k,r,0} - \sum_k x^+_{k,0} + x^-_{1,0} - x^+_{1,0} = 0$$

for $k \in \mathcal{K}$.

For the subsequent stages $(t > 0)$ and $s = 1, \ldots, S$:

$$\begin{aligned} & I_t(s) - C_t(s) + x^+_{1,t-}(s)(1 + \zeta^+_t(s)(1 - \vartheta_1)) \\ & \quad - x^-_{1,t-}(s)(1 + \zeta^-_t(s)) - x^+_{1,t}(s) + x^-_{1,t}(s) \\ & \quad + \sum_{k \in \mathcal{K}_3} \sum_h A_{k,h,t}(s) + \sum_{k \in \mathcal{K}_1} x^-_{k,t}(s) + \sum_{k \in \mathcal{K}_2 \cup \mathcal{K}_3} \sum_{h < t-} x^-_{k,h,t}(s) \\ & \quad - \sum_{k \in \mathcal{K}} x^+_{k,t}(s) + \sum_{j=1,2} y^+_{j,t}(s) - \sum_{j=1,2} y_{j,t-}(s) b_{j,t}(s) - \sum_{j=1,2} y^-_{j,t}(s) \\ & = \widetilde{W}_t(s) \end{aligned}$$

Bounds and Non-negativity Policy rules and specific portfolio constraints are included in the model using the following bounds.

$$\begin{aligned} l_k X_t(s) \le x_{k,t}(s) &\le u_k X_t(s) \\ X_t(s) &\ge (1 - \epsilon) X_{t-}(s) \\ 0 \le y_t(s) &\le \gamma \widetilde{W}_t(s) \end{aligned}$$

for $t = 1, \ldots, T$, $k = 1, \ldots, K$ and $s = 1, \ldots, S$.

$$\sum_{k=1,2} x_{k,t}(s) \geq M$$

$$x_{1,t}^{+} \geq 0$$

$$x_{1,t}^{-} \geq 0$$

where M is the minimal liquidity on the non-retirement portfolio share.

The presented ALM model can be specialised to formulate and solve households financial management problems as in (Consigli *et al.*, 2011), individual retirement planning problems (Consigli, 2007; Medova *et al.*, 2008), wealth management problems (Zenios and Ziemba, 2007a), and financial engineering applications. For each such problem class, long positions (individuals and households) will determine their optimal strategies and short positions (institutional investors) will determine their optimal replicating strategies (Consigli, 2007).

3 Numerical implementation and case study

We describe in this section the building blocks of a decision support tool development specifically for MSP problems and analyse its adoption within a practical case-study. All results have been collected on a PC with Intel Core Duo 1.2 GHz PC and 1GB of RAM. The described framework supports a real-world development with a I/O user interface agreed with the financial manager of a large corporation.

The tool combines the MATLAB mathematical and statistical environment as main computation engine and IDE, and GAMS as model generation tool, whereas a set of Excel spreadsheets are adopted as I/O user and internal interfaces. The end user is in this way free to customize the decision problem through the input interface and integrate the optimal solution and solution results within her operational database for additional analysis. Microsoft Excel has also been proven effective as data environment to download and update financial series from main data providers. MATLAB graphics tool is adopted to validate intermediate computational steps as well.

GAMS Rev 232 WEX-VIS 23.2.1 x86/MS Windows (www.gams.com) is adopted as core algebraic modeling language to generate files in MPS format and as vehicle to linear and nonlinear commercial solvers. The overall set of software interfaces is described in Figure 4.

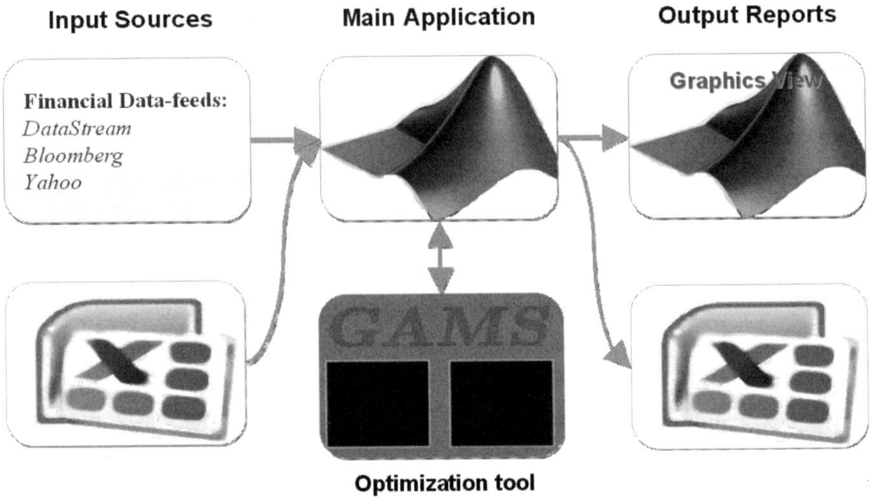

Figure 4 Main Framework and Interfaces.

Scenario generation is entirely performed in Matlab and the set of random coefficients passed on to GAMS to generate an instance of the multistage stochastic program. From the user perspective, the following steps define the procedure to follow on a Matlab interface from the problem instance definition to its solution:

1. specify the dynamic structure of the problem and the scenario tree branching sequence at each stage
2. introduce current and perspective income conditions and desirable target sequence over the decision horizon
3. identify the set of policy bounds on the portfolio strategy and the risk-reward trade-off coefficient
4. specify the investment universe and estimate the underlying statistical model
5. generate the scenario tree for the risk factors and the coefficients of the problem
6. generate the Mathematical Programming Standard (MPS) instance and solve the problem
7. analyse the solution output.

Each step will imply internal calls to Excel or GAMS or within Matlab to perform the required task. Figure 5 provides more insight into the specific tasks characterizing the system.

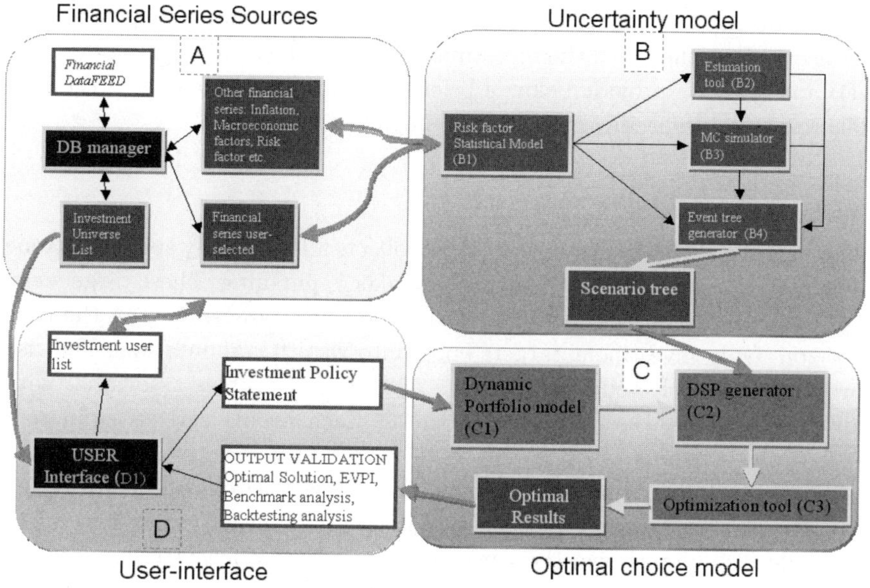

Figure 5 Decision Tool Modules.

3.1 *Decision tool modular structure*

The following four macro-modules describe in more detail the task sequence supported by the personal ALM tool:

(A) The *financial data feed module* collects the historical series from the financial providers and customize the user asset universe and related risk factors;

(B) the *uncertainty module* estimates the model parameters, provides some statistics and Monte Carlo trajectories, and generates the scenario tree;

(C) the *optimization module* collects the generated scenario tree and customizes the Dynamic Stochastic Problem (DSP) according to an investment policy statement (IPS) defined by the user;

(D) the Excel *user interface* is adopted for the individual-specific problem formulation and as solution output database.

Within each macro-module the sequence of steps from data collection to estimation, scenario generation and problem solution are displayed in Figure 5. (A), (B) and (C) define data structures and interface which do not depend on the specific problem instance and define a generic modeling environment. The policy statement in (D) is problem specific and will determine together with the investment universe the content of submodules (B) and (C). We provide below a brief summary of the content of the four macro-modules specifically employed in the ALM problem.

3.1.1 *Individual policy statement*

The Individual Policy Statement (IPS) collects all the family specific income information and the goals the decision maker is pursuing. These targets can be for consumption or investment and can occur at any stage over the time horizon. Only investment targets enter into wealth computation. The risk profile can be selected from a range of standard profiles associated with different λ values, as shown in Figure 6. Each profile carries a different risk/reward trade-off.

Targets are distributed over time and a final retirement wealth target can also be accommodated within this structure (Consigli *et al.*, 2011). The set of decision points over the horizon is determined *a priori* as the scenario tree expansion.

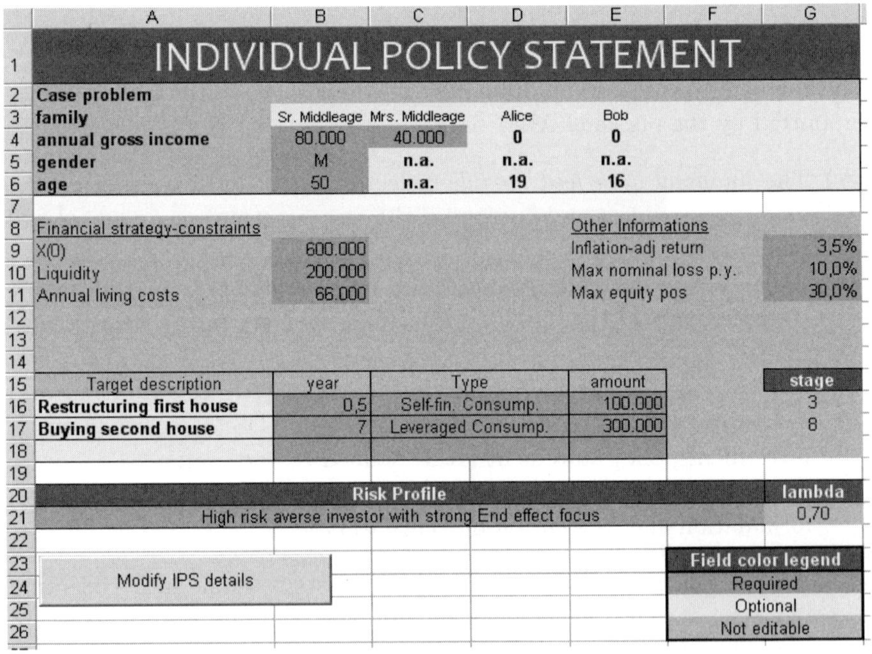

	A	B	C	D	E	F	G
1	INDIVIDUAL POLICY STATEMENT						
2	**Case problem**						
3	**family**	Sr. Middleage	Mrs. Middleage	Alice	Bob		
4	**annual gross income**	80.000	40.000	0	0		
5	**gender**	M	n.a.	n.a.	n.a.		
6	**age**	50	n.a.	19	16		
7							
8	Financial strategy-constraints				Other Informations		
9	X(0)	600.000			Inflation-adj return		3,5%
10	Liquidity	200.000			Max nominal loss p.y.		10,0%
11	Annual living costs	66.000			Max equity pos		30,0%
12							
13							
14							
15	Target description	year	Type		amount		stage
16	**Restructuring first house**	0,5	Self-fin. Consump.		100.000		3
17	**Buying second house**	7	Leveraged Consump.		300.000		8
18							
19							
20	Risk Profile						lambda
21	High risk averse investor with strong End effect focus						0,70
22							
23						Field color legend	
24	Modify IPS details					Required	
25						Optional	
26						Not editable	

Figure 6 Individual Policy Statement.

3.1.2 *Scenario manager*

In our results we consider a 256 scenario binary tree spanning a 10 year horizon according to the following stage structure:

Topology: 1 - 2 - 2 - 2 - 2 - 2 - 2 - 2 - 2

Horizon: 3m - 3m - 6m - 1y - 1y - 2y - 2y - 3y

Period: January 1, 2009 to 31-Dec-2018

The user interface in Figure 7 leads to the definition of the problem dynamic structure.

The scenario manager performs from data collection to the generation of the scenario coefficients to be incorporated in the optimization problem. From a set of data histories on each risk factor, statistical estimation is first performed to generate the set of coefficients needed by the Monte Carlo simulator. We present in Figure 8 the set of coefficients input to the scenario generator for the 10 year individual planning problem.

According to the model discretization and scenario tree structure, the scenario manager generates the risk-factors evolution and consequently the asset value dynamics.

The risk factors underlying the problem are correlated and generate the entire set of ALM coefficients, required by the algebraic formulation of the ALM problem. In Figure 9 from a given known current value we plot a set of future scenarios for the equity mutual fund.

Alternative modeling approaches will lead to different forward scenarios. The introduction of efficient sampling schemes as in (Consigli *et al.*, 2010; Pflug and Pichler, 2011; Heitsch and Römisch, 2011; Dempster

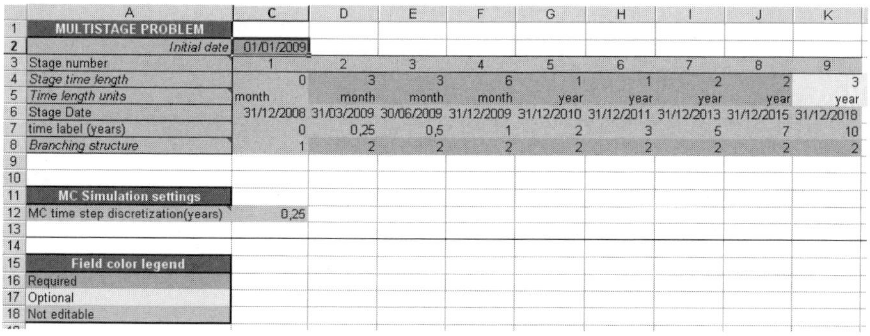

	A	C	D	E	F	G	H	I	J	K
1	MULTISTAGE PROBLEM									
2	*Initial date*	01/01/2009								
3	Stage number	1	2	3	4	5	6	7	8	9
4	*Stage time length*	0	3	3	6	1	1	2	2	3
5	*Time length units*	month	month	month	month	year	year	year	year	year
6	Stage Date	31/12/2008	31/03/2009	30/06/2009	31/12/2009	31/12/2010	31/12/2011	31/12/2013	31/12/2015	31/12/2018
7	time label (years)	0	0,25	0,5	1	2	3	5	7	10
8	*Branching structure*	1	2	2	2	2	2	2	2	2
9										
10										
11	MC Simulation settings									
12	MC time step discretization(years)	0,25								
13										
14										
15	Field color legend									
16	Required									
17	Optional									
18	Not editable									

Figure 7 Scenario Manager Interface.

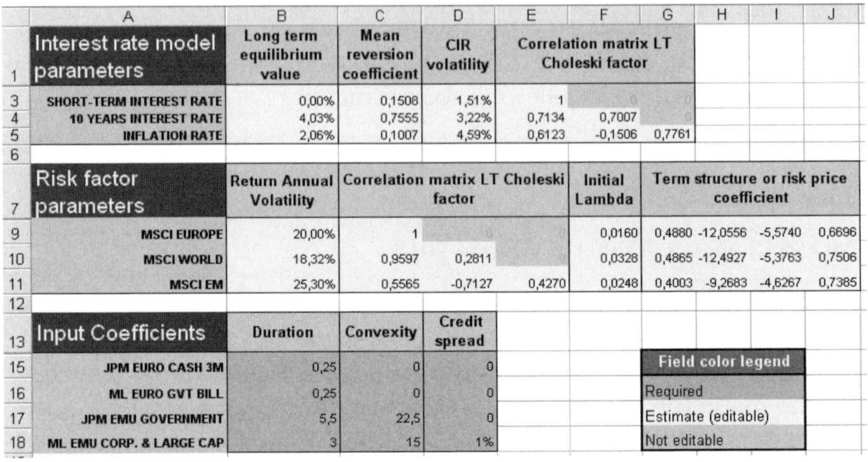

	A	B	C	D	E	F	G	H	I	J
1	Interest rate model parameters	Long term equilibrium value	Mean reversion coefficient	CIR volatility	Correlation matrix LT Choleski factor					
3	SHORT-TERM INTEREST RATE	0,00%	0,1508	1,51%	1	0	0			
4	10 YEARS INTEREST RATE	4,03%	0,7555	3,22%	0,7134	0,7007	0			
5	INFLATION RATE	2,06%	0,1007	4,59%	0,6123	-0,1506	0,7761			
6										
7	Risk factor parameters	Return Annual Volatility	Correlation matrix LT Choleski factor			Initial Lambda	Term structure or risk price coefficient			
9	MSCI EUROPE	20,00%	1	0	0	0,0160	0,4880	-12,0556	-5,5740	0,6696
10	MSCI WORLD	18,32%	0,9597	0,2811	0	0,0328	0,4865	-12,4927	-5,3763	0,7506
11	MSCI EM	25,30%	0,5565	-0,7127	0,4270	0,0248	0,4003	-9,2683	-4,6267	0,7385
12										
13	Input Coefficients	Duration	Convexity	Credit spread						
15	JPM EURO CASH 3M	0,25	0	0			Field color legend			
16	ML EURO GVT BILL	0,25	0	0			Required			
17	JPM EMU GOVERNMENT	5,5	22,5	0			Estimate (editable)			
18	ML EMU CORP. & LARGE CAP	3	15	1%			Not editable			

Figure 8 Model Parameters Interface.

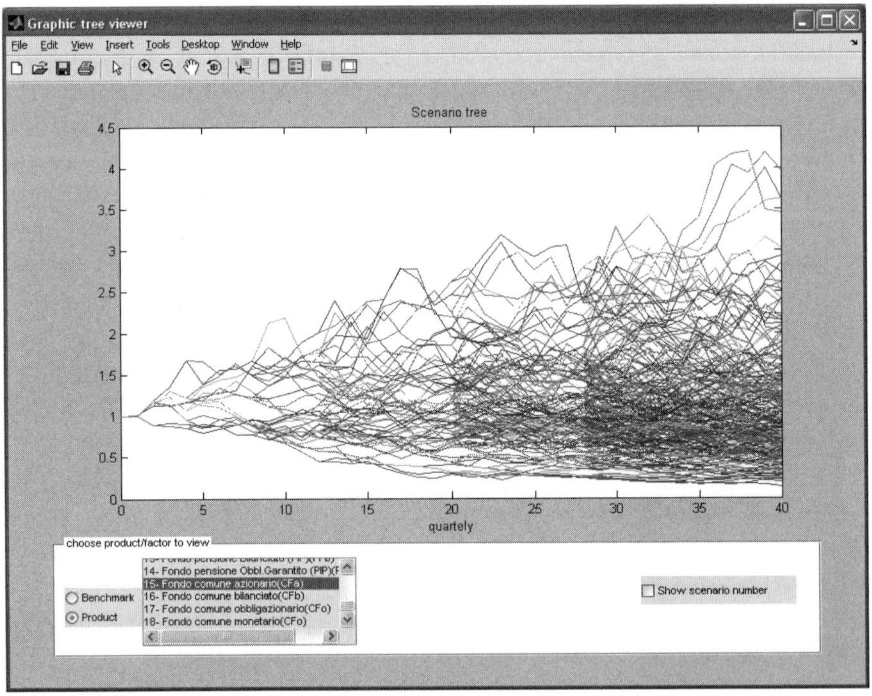

Figure 9 A Sample Scenario Tree.

et al., 2011) to populate the tree comes at this stage and would lead to a scenario representation under different methodological assumptions. Similarly by accessing directly the worksheet with the estimated coefficients the user will be able to perform stress-testing analysis on specific markets and analyse the sensitivity of the optimal strategy to alternative scenario trees.

3.1.3 *Output*

The output of a MSP is well known to be full of information, in fact, the optimal investment strategy is not limited to the here-and-now solution, but reports also the optimal decisions along each scenario, showing how the model reacts to different Economics conditions.

The generation of practical information to the end-user may be limited to the optimal here and now decision, but this would sacrify a rather valuable information set, or look for several outputs conveying the value of a dynamic solution without loosing its operational adoption. The figures that follow have been agreed with practitioners actually using such information according to a company-consistent template. This appears a requirement for the practical adoption of SP-based decision tools.

The focus in this application is on the evolution of the individual wealth and how from a given optimal initial decision, targets are achieved over the horizon and, say, which pension wealth will be accessible at the horizon. To this aim three scenario types have been introduced, conventionally refered to as the *best*, *mean* and *worst* scenarios, as representative of the scenario tree to analyse the optimal strategies and assess the financial and aggregate wealth available under different market conditions at the horizon. Scenario ranking is introduced from a pivot variable, here represented by the investment value, following the method in (Consigli *et al.*, 2010, 2011). Two relevant outputs can be considered to describe along a scenario, the optimal strategy on one side in terms of set of rebalancing decisions and the wealth evolution on the other with a distinction between the financial and aggregate wealth.

- the *logbook* will describe the portfolio composition and the trading strategies at each decision stage. Those strategies will differ along the three scenarios and agree at the root node;
- the optimal *wealth report* will highlight the wealth dynamics, consumption and investment targets. At the final stage the report shows the end effect amount associated with a retirement portfolio (if any).

We consider the above output info in a specific case problem agreed with the insurance company.

3.2 *Case study*

- 50-year-old man;
- € 120,000 annual family incomes;
- € 600,000 initial wealth;
- € 66,000 estimated annual living costs.

We show results under different targets' specification and for different degrees of risk aversion to convey the ability of the implemented tool to discriminate between different problem settings. Each test-problem is characterised by:

- the introduction of targets at either short-term (S), long-term (L) or both (B) horizons. The short period target is set after 6 months and includes a self-financing first-house renovation of € 100,000, while the long-term target comes after 7 years and concerns a real-estate investment of € 300,000. We write N for the case of no targets.
- the individual risk-aversion profile: we consider a growth investor (G) with high shortfall risk tolerance ($\lambda = 0.3$ in equation (5)), a medium risk averse investor (M) with target focus ($\lambda = 0.4$), a medium risk averse investor (L) with low shortfall risk tolerance ($\lambda = 0.6$) and a high risk averse investor (H) with strong target focus ($\lambda = 0.7$).

All results refer to a 10 year problem and the entire investment universe is considered along an average (A) scenario.

3.2.1 *Optimal solutions*

The solution of the Asset-Liability model leads to the definition of optimal decision sequences to reach the given targets. For different case problems we display the logbook table along the average scenario and the here-and-now decision defined by different investment universe partitions. Namely determined by the investment products, the asset classes, the underlying financial benchmarks and risk factors included in the root node solution.

Case (1): NGA investor. In absence of targets (N) with a growth investor (G) profile along the average scenario the investor prefers unit-linked investments, cf. Figure 10.

Mean Scenario — **Log book (per classes)**

Date	1	2	3	4	5	6	7	8	9
Products list	31-dic-08	31-mar-09	30-giu-09	31-dic-09	31-dic-10	31-dic-11	31-dic-13	31-dic-15	31-dic-18
1 Bank Account	long 1,880		short 1,553						long 27,848
2 Unit-Linked	buy 598,120 hold 595,129	buy 4,341 hold 659,197	buy 3,964 hold 818,571	buy 937,674 sell 934,522 hold 1,132,986	buy 978,880 sell 969,498 hold 1,173,986	buy 14,296 sell 4,948 hold 1,376,317	buy 1,732,129 sell 1,713,506 hold 1,923,469	buy 30,995 sell 12,401 hold 2,204,863	hold 2,562,434
3 Insurance Plans									
4 Annuities									
5 Pension Funds									
6 Mutual Funds									

(a) Logbook Representation

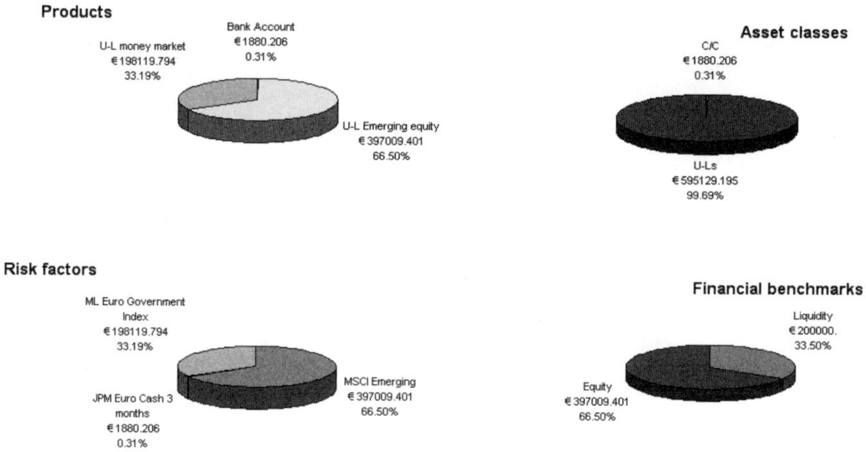

(b) Here-and-Now Solution Pie Representation

Figure 10 Case study 1 — NGA Investor.

Case (2): BGA. By introducing a short and a long term targets the here-and-now solution and optimal strategy change significantly towards liquid mutual-funds to cover the long term target, cf. Figure 11(b).

Case (3): BHA. We consider in this case study again the presence of a short and long-term target but reducing the tolerance on target violations (H). We observe along the average scenario an increasing share of the portfolio invested in pension-funds, cf. Figure 12. The here-and-now solution does not change in case (2).

For exemplificative purposes we present in Figure 13 the optimal wealth for case (3). In presence of pension funds or annuities in the terminal

Log book (per classes)

Mean Scenario

	Date	1	2	3	4	5	6	7	8	9
	Products list	31-dic-08	31-mar-09	30-giu-09	31-dic-09	31-dic-10	31-dic-11	31-dic-13	31-dic-15	31-dic-18
1	Bank Account			short 1,553						short 17,346
2	Unit-Linked	buy 378,782 hold 376,888	buy 205,132 sell 13,833 hold 594,684	sell 68,589 hold 614,199	buy 522,442 sell 519,291 hold 719,830	buy 14,330 sell 4,947 hold 746,611	buy 240,399 sell 231,048 hold 797,323	buy 274,885 sell 537,283 hold 690,803	buy 201,005 sell 17,820 hold 950,789	hold 1,130,016
3	Insurance Plans									
4	Annuities									
5	Pension Funds									
6	Mutual Funds	buy 221,218 hold 219,006	buy 40,985 sell 229,831 hold 40,575	sell 44,766				buy 281,024 hold 278,214	sell 292,727	

(a) Logbook Representation

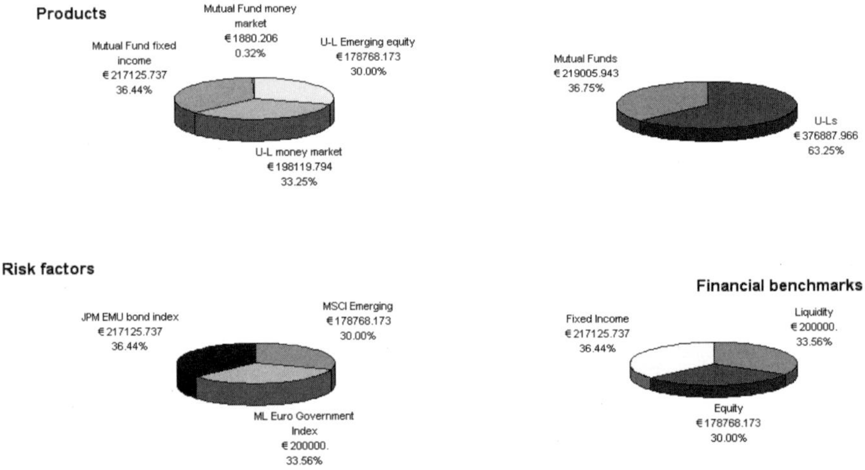

(b) Here-and-Now Solution Pie Representation

Figure 11 Case study 2 — BGA Investor.

portfolio, the wealth report displays a negative end-effect adjustment on the final wealth. Unlike the consumption goal, the investment goal adds up to the individual wealth to contribute to his overall welfare at the horizon.

We recall that a negative end-effect provides forward in time an assessment of the risk that indeed at the horizon the retirement portfolio may exceed the discounted value of all future pension payments. Over the 10 years the individual is neverthless expected to benefit substantially from the suggested strategy.

Mean Scenario — **Log book (per classes)**

Products list \ Date	1 31-dic-08	2 31-mar-09	3 30-giu-09	4 31-dic-09	5 31-dic-10	6 31-dic-11	7 31-dic-13	8 31-dic-15	9 31-dic-18
1 Bank Account			short 1,553						short 26,267
2 Unit-Linked	buy 378,782 hold 376,888	buy 205,132 sell 13,833 hold 594,684	sell 68,589 hold 614,199	buy 522,442 sell 519,291 hold 719,830	buy 14,330 sell 4,947 hold 746,611	buy 240,399 sell 231,048 hold 797,323	buy 274,885 sell 537,283 hold 690,803	buy 213,582 sell 780,784 hold 212,514	hold 228,551
3 Insurance Plans									
4 Annuities									
5 Pension Funds								buy 790,308 hold 790,308	hold 889,047
6 Mutual Funds	buy 221,218 hold 219,006	buy 40,985 sell 229,831 hold 40,575	sell 44,766				buy 281,024 hold 278,214	sell 296,995	

Figure 12 Case study 3 — BHA Investor, Logbook Representation.

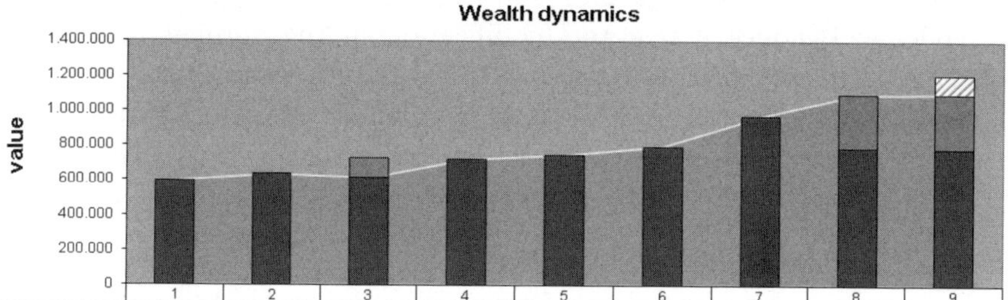

	1	2	3	4	5	6	7	8	9
Wealth +Cumul.Targets -End Effect	595.894	635.259	612.647	719.830	746.611	797.323	969.017	1.095.303	1.097.146
End Effects	0	0	0	0	0	0	0	0	108.527
Cumulative Target Investments	0	0	0	0	0	0	0	308.270	314.863
Targets (Consumption)	0	0	117.319	0	0	0	0	0	0
wealth	595.894	635.259	612.647	719.830	746.611	797.323	969.017	787.034	782.283

stage

Figure 13 Wealth Dynamics, Targets and End Effect Representation.

4 Conclusion

We have summarized the key elements of a recently developed decision tool for multistage stochastic programs, applied to a 10 year individual asset-liability management problem. The tool integrates within a computationally efficient framework a user interface, in which the specific problem instance is defined, a scenario generator for medium and long term financial planning problems, a model generator and solution platform. A case study focuses on a limited set of outputs explicitly selected to convey the importance of solution analysis to ease the practical adoption of SP-based ALM models.

From a modelling perspective the ALM model presents a set of interesting features, associated with the adopted investment universe:

- The inclusion of assets with financial and insurance features leads to the definition of asset class-specific inventory balance constraints, and the inclusion of a relevant set of management and fiscal fees, which will capture the specific market standard;
- Pension funds and annuities have associated payoffs which will generally fall outside the decision horizon: their inclusion in the optimal final portfolio is based on an end-effect adjustment;
- The introduction of a stochastic drift in the statistical models appears effective to span a long term horizon leading to reliable market dynamics,
- The presence of short, medium and long term goals within a multistage model will induce a relevant modification in the optimal investment policy, in favour of liquid assets before the target stage. An increased target focus will also lead to more conservative strategies in the short run. The selection of a long term pension accumulation plan may however be optimal also in presence of distributed targets.

The submodules dedicated to scenario generation and optimization are generic and may be adopted as reference modules for ALM problems other than for individuals.

Acknowledgements The authors acknowledge the support given by the research project PRIN2007 "Optimization of stochastic dynamic systems with applications to finance", no.20073BZ5A5 sci.resp. G. Consigli. The research has been partially supported by MIUR-ex60% 2008-2009-2010-2011 sci. resp. V. Moriggia. The chapter develops from an early development in cooperation with a leading insurance company, and we are grateful to Michele Gaffo and Massimo di Tria for their support and trust during that project.

References

Berger, A.J. and J.M. Mulvey. The home account advisor: Asset and liability management for individual investors. In W.T.Ziemba and J.M.Mulvey, eds., *Worldwide Asset and Liability Modeling*, pp. 634–665. Cambridge (UK): Cambridge University Press, 1998.

Bertocchi, M., G. Consigli and M.A.H. Dempster (eds.). *Handbook on Stochastic Optimization: Methods in Finance and Energy*. Fred Hillier on Operations Research and Management Science. Springer USA, 2011.

Birge, J.R. and F. Louveaux (eds.). *Introduction to Stochastic Programming. 2nd edition.* Springer Verlag, 2011.

Campbell, J.Y., W.A. Lo and A.C. MacKinlay. *The Econometrics of Financial Markets.* Princeton University Press, 1997.

Consigli, G. *Handbook of Asset and Liability Management*, volume 2, chapter Individual Asset Liability Management, pp. 751–827. North-Holland, 2007.

Consigli, G. and M.A.H. Dempster. Dynamic stochastic programming for asset-liability management. *Annals of Operations Research*, 81: pp. 131–161, 1998.

Consigli, G., G. Iaquinta and V. Moriggia. Path-dependent scenario trees for multistage stochastic programs in finance. *Quantitative Finance*: pp. 1–17, December 2010.

Consigli, G., G. Iaquinta, V. Moriggia and M. di Tria. Optimal management of life insurance household portfolios in the long run. *Pacific Journal of Optimization*, 7: pp. 197–219, May 2011.

Consiglio, A., F. Cocco and S.A. Zenios. www.personal_asset_allocation. *Interfaces*, 34(4): pp. 287–302, Jul–Aug 2004.

Dempster, M.A.H., M. Germano, E. A. Medova and M. Villaverde. Global asset liability management. *British Actuarial Journal*, 9(1): pp. 137–195, 2003.

Dempster, M.A.H., E.A. Medova and Y.S. Yong. Comparison of sampling methods for dynamic stochastic programming, *Handbook on Stochastic Optimization Methods in Finance and Energy: New Financial Products and Energy Market Strategies*, Operations Research and Management Science of Fred Hillier International Series, pp. 389–426. Springer Verlag, 2011.

Dupačová, J., G. Consigli and S.W. Wallace. Scenarios for multistage stochastic programmes. *Annals of Operations Research*, 100: pp. 25–53, 2001.

Heitsch, H. and W. Römisch. Scenario tree generation for multi-stage stochastic programs, *Handbook on Stochastic Optimization Methods in Finance and Energy: New Financial Products and Energy Market Strategies*, Operations Research and Management Science of Fred Hillier International Series, pp. 313–342. Springer, 2011.

Hochreiter, R., G. Pflug and V. Paulsen. *Handbook of Asset and Liability Management*, chapter Design and management of unit-linked life insurance contracts with guarantie. North-Holland, 2007.

Medova, E.A., J.K. Murphy, A.P. Owen and K. Rehman. Individual asset liability management. *Quantitative Finance*, 8(6): pp. 547–560, 2008.

Mulvey, J.M. Generating scenarios for the Towers Perrin Investment System. *Interfaces*, 26: pp. 1–13, 1996.

Mulvey, J.M. and W.T. Ziemba. Asset and liability management systems for long-term investors: discussion of the issues. In W.T. Ziemba and J.M. Mulvey (eds.), *Worldwide Asset and Liability Modeling*, pp. 3–38. Cambridge University Press, 1998.

Pflug, G. and A. Pichler. Approximations for probability distributions and stochastic optimization problems, *Handbook on Stochastic Optimization Methods in Finance and Energy: New Financial Products and Energy Market*

Strategies, Operations Research and Management Science of Fred Hillier International Series, pp. 343–388. Springer, 2011.

Zenios, S.A. and W.T. Ziemba (eds.). *Handbook of Asset and Liability Management*, Volume 1 of Handbooks Finance Series. North-Holland, 2007a.

Zenios, S.A. and W.T. Ziemba (eds.). *Handbook of Asset and Liability Management: Applications and Case Studies*, Volume 2 of Handbooks Finance Series. North-Holland, 2007b.

Chapter 4

Intertemporal Surplus Management with Jump Risks

Mareen Benk*

Summary

I have developed an intertemporal portfolio choice model with jump risks. It can be applied to pension and life insurance funds, and private investors. Following the model of Rudolf and Ziemba (2004), these long-term investors aspire to "maximize the intertemporal expected utility of the surplus", which is defined as "assets net of liabilities". Return on liabilities are modelled by a typical pure-diffusion process. Return on assets are assumed to follow a jump-diffusion process with two jump components. More specifically, the first jump component represents a systemic risk according to Das and Uppal (2004) and the second jump component represents an idiosyncratic risk according to Jarrow and Rosenfeld (1984). An investor's optimal portfolio consists of three funds: a market portfolio, a liability-hedging portfolio, and a riskless asset. In contrast to the results of Rudolf and Ziemba (2004), a market portfolio not only hedges diffusion risk, but it also hedges systemic risk and it takes into account idiosyncratic jump risk so that the investor is additionally protected against both a systemic risk and an idiosyncratic jump risk.

1 Introduction

Risk management for private investors is of current interest.[1] For example, the European central bank benefits of analyses of newly collected financial

---

*Department of Finance, WHU — Otto-Beisheim School of Management, D-56179 Vallendar, Germany, mbenk@whu.edu.

[1]Similar to the Federal Reserve Board, that has been publishing financial and consumptive information (i.e. data of debt, financial obligations, and income) of US citizens for many years, the European Central Bank has established a Household Finance and Consumption Network in 2006, that aims at collecting comparable data about European households regularly.

This paragraph draws upon the European Central Bank (2009), p. 5, p. 16, and p. 25, the Federal Reserve Board (2012), p. 1 ff., and Datamonitor (2006), p. 23 ff.

data of European households to measure the robustness of the financial system or the effects of monetary policy. According to the results of such analyses, individuals have been increasing their risk constantly by taking more debt for years. Especially the strong and positive correlation between the fortune of an individual and its indebtedness attracts one's attention. Comparably, well-off households are known to invest their assets more aggressively and participate in equity markets more often (Campbell, 2006). Some private debtors (and their creditors) underestimate the hazard of a high level of financial obligations, that can cause financial distress if the economic situation deteriorates. For instance, the countries of Western Europe and the USA reveal a not as yet known high level of consumer insolvencies in 2010 (Creditreform, 2011).[2]

In general, individuals do not match their assets and liabilities properly. More specifically, individuals are exposed to biometric, family, political, and economical risks and often fail to manage these risks correctly because this task demands too much of them (Börsch-Supan, 2005). Several developments of society and politics require individuals to provide to a greater extent for the future. For example, politicians have been reacting to demographic changes (Rudolf, 2004)[3] and have been realizing various reforms in industrialized countries for years. As a result many benefits of welfare states have disappeared (Soteriou and Zenios, 2003). Similarly, the traditional role of the family, that is supposed to care for its members by protecting them from any risk, diminishes (Börsch-Supan, 2005). The public and private pension provisions have to be regarded as insufficiently secured in several industrialized countries because of current uncertainties (Bertocchi *et al.*, 2010). For example, the high indebtedness of many industrialized nations has been causing a still persisting sovereign crisis since 2010, that leads rating agencies to both have a closer look at the creditworthiness of several states and downgrade some of them. Thus, the question arises whether the affected nations will be able to carry on with their public retirement system or whether difficulties to finance the usual pension provisions will emerge in the future. At the same time individuals fail to save a sufficient amount of money during the phase of employment although they are aware of the necessity to partially finance the phase

[2]384,895 citizens of European countries (please note that there are no data available for some European countries), and 1,573,000 citizens of the USA have filed for consumer insolvency in 2010.

[3]The key factors are a rising longevity and a diminishing gross reproduction rate.

of retirement (Creditreform, 2010; Ziemba, 2003). In order to explain the imperfect savings behaviour of human beings either psychological phenomena can be consulted (von Nitzsch and Hackethal, 2005) or potential misuses of established tools can be analyzed or shortcomings of employed methods can be evaluated (Elnekave, 2007). A strong demand for risk management services, that address individuals' needs, can be deduced.

In 2007 a global crisis began, that disclosed among other things significant weaknesses up to severe failures of risk management, that some providers of private banking services had offered (PricewaterhouseCoopers, 2009). The industry is being aware of the positive impact of high-quality risk management on reputation for some years (PricewaterhouseCoopers, 2005). According to a global survey, 28 percent of private banking providers regard reputation risk as essential and 13 percent even estimate it to be one of the greatest threats to their business (PricewaterhouseCoopers, 2004). In accordance with these judgements, more than a quarter of interviewed high net worth individuals (HNWIs) has withdrawn assets from their provider or entirely changed it in 2008 whereas the other part of the surveyed HNWIs remained loyal to their respective provider because of their risk management and due diligence services (Capgemini and Merrill Lynch, 2009).

As a result of the loss of assets under management, the private banking industry needs to respond to HNWIs' revealed preferences for high-quality risk management and currently discusses asset and liability management as an auspicious approach. In general, Asian and American providers have been using higher-quality risk management tools than European providers for years (PricewaterhouseCoopers, 2003; Reiche, 2005). Thus, the European financial service industry for HNWIs has begun to catch up with developing stochastic multi-period asset allocation models (Amenc *et al.*, 2009; Häger and Raffelsberger, 2005). However, these newly launched approaches still have the weakness of neglecting liabilities. 86 percent of interviewed private banking providers, whose majority practices its profession in Europe, perceived the advantages of asset and liability management while the quality of their prevailing risk management tools was regarded as low in 2008 (Amenc *et al.*, 2009). However, 55 percent of these surveyed managers have doubted whether the introduction of an asset and liability management tool could achieve an added value. They especially estimated the complexity of the approach as the main challenge by considering the qualification of their respective team.

Individuals (including HNWIs) strongly demand comprehensive asset and liability management models (Consiglio *et al.*, 2007). The majority of these tools is still developed for institutional investors (Consiglio *et al.*, 2007). Since the end of the nineteen-nineties, some attention has been shifting to the possibilities of modelling assets and liabilities for individuals because well-known experts had estimated the potential market to be huge (Mulvey and Ziemba, 1999). In general, individuals are confronted with the fact that only a few well-functioning, high-quality tools are designed for their needs, as for example "Financial Engines" for the US retirement market and an online system for the Italian banking market (Financial Engines, 2012; Consiglio *et al.*, 2004). Most of the asset and liability management models for individuals exclude illiquid assets (Consigli, 2007). The "Home Account Advisor" (Berger and Mulvey, 1999) and the model of Bertocchi *et al.* (2010) take into account real estate and can be regarded as an exception. Furthermore, research disregarded private banking until the beginning of the early 2000s (Schaubach, 2003).

The aim of this paper is the integration of two jump components into an intertemporal surplus management model. Following Rudolf and Ziemba (2004), the construction can be applied to pension and life insurance funds. Furthermore, the paper has the common methodical procedure to conceptually transfer the institutional asset and liability management approach to individuals' needs (Mulvey, 2005). As a result, the developed model can as well be applied to private investors (including HNWIs) (Consiglio *et al.*, 2007; Höcht *et al.*, 2008). The obtained liability-driven investment (LDI) technique especially satisfies the needs of wealthy individuals (Amenc *et al.*, 2009).[4] More specifically, the model solution reflects that rich households should pursue a risk averse behaviour because their possibilities to finance future goals strongly depend on the positive results of their investment strategy (Campbell and Viceira, 2002; Ziemba, 2003).

I protect an investor from the impact of systemic risk on an optimal portfolio choice by integrating the first jump component into an intertemporal surplus management model. There exist two methodologies in the literature to take account of market shocks and financial crises: Either researchers use the extreme value approach or they apply jump-diffusion models (Consigli, 2002). This paper uses the second one (i.e. I develop a jump-diffusion model). Furthermore, there are two approaches in the literature to model contagion or systemic risk: Either experts develop

[4] A liability-driven investment technique not only considers performance by selecting a respective portfolio but also takes into account risk management by choosing a liability hedging portfolio.

regime switching models or they resort to integrate joint poisson jumps (Branger *et al.*, 2009).[5] I take account of systemic risk by using the second approach. Research projects have found empirical evidence for the existence of jumps in returns of international equities.[6] According to Das and Uppal (2004), I define systemic risk as rare incidents, that are characterized by a high correlation across many assets.[7] For example, the correlation between assets has approached one during the systemic crisis in 2008 (Amenc and Martellini, 2011). Thus, this definition is plausible and of current interest.

Mr. Jean-Claude Trichet, who is the former President of the European Central Bank and the former Chairman of the European Systemic Risk Board, has warned on October 11, 2011, that the current [sovereign] crisis was of a systemic dimension and had leaped over from smaller to larger EU countries already (Handelsblatt, 2011). A historical perspective illustrates that the crisis, that began with a financial and economical crisis in the years 2007 until 2009, cannot be seen as an exception but as a regularly returning incident (Rudolf, 2010). Thus, such rare events should generally be considered by financial models. Various reasons for the rise of systemic risk in international financial markets exist. These causes comprise the impact of hedge funds, macroeconomic imbalances, that are unrecognized by political leaders, the increased complexity of various financial products, and spill-over effects from large companies to small ones (McKinsey Global Institute, 2007; International Monetary Fund, 2009; Capgemini and Merrill Lynch, 2009; Conrad *et al.*, 1991).

By integrating a second jump component into the intertemporal surplus management model, I take account of idiosyncratic jump risk. I define this type of risk as an unexpected shock of single assets' returns. According to the literature, there is no general agreement about the widely discussed matter whether a price has to be paid for idiosyncratic risk (Huang *et al.*, 2010). This paper assumes according to the CAPM-based literature that the respective shock is diversifiable in the market portfolio and that the investor does not receive any risk premium for idiosyncratic jump risk (Jarrow and

[5] In contrast to the definition of systemic risk, the definition of contagion risk does not imply an assumption about the correlation of assets.

[6] See for example the literature mentioned by Das and Uppal (2004) on page 2809 and their own important empirical outcome.

[7] I am aware that the International Monetary Fund, the Bank of International Settlement, and the Financial Stability Board use another definition of systemic risk (Rudolf, 2012). I regard my definition as correct in the given context because I comply with the common approach to integrate systemic risk into a portfolio selection model.

Rosenfeld, 1984). Although the investor does not obtain any risk premium for this type of risk there is an impact of it on the optimal portfolio selection strategy (Merton, 1976; Jarrow and Rosenfeld, 1984). Some portfolio selection models with idiosyncratic jumps were developed (Aase, 1984; Jeanblanc-Picqué and Pontier, 1990). However, these approaches neither take liabilities nor systemic risk into account.

According to the results of an empirical analysis, it is reasonable to assume that the shocks of returns of small companies' stock are diversifiable (Conrad *et al.*, 1991). Idiosyncratic shocks can be caused by information, that concern a single company or industry only, as for example the discovery of an oil field or the proceedings of a lawsuit (Merton, 1976).

Liu, Longstaff, and Pan (2003) find out that individuals with illiquid assets (as for instance real estate) invest similarly to those persons that use a jump-diffusion model for their investment strategy. The research results of Mizrach (2008) confirm the assumption that the Case-Shiller Repeat Housing Price index reveals significant jumps, that are mainly caused by information. The price and volatility process of real estate can be distorted due to the particular valuation procedure of this asset class (Mulvey, 2002). Thus, the second jump component of my intertemporal surplus management approach can be used to integrate real estate risk and potentially further illiquid asset classes into a portfolio selection model.

Due to the rise of oil investors there was a global increase of liquidity that has inflationary effects on some asset classes (i.e. especially illiquid ones are affected) besides some positive implications (McKinsey Global Institute, 2007). As a result there is a need for new techniques to get a deeper understanding of such a development. The model of this paper can incorporate real estate risk, which is among other things a significant risk for private banking customers (Amenc *et al.*, 2009).

In Section 2 I develop an intertemporal portfolio selection model with jump risks and obtain a three-fund theorem. Following Rudolf and Ziemba (2004), I introduce risk tolerance coefficients to the model and analyze their relationship to the funding ratio in Section 3. In Section 4 I conclude the paper. Derivations can be found in the appendixes.

2 An intertemporal surplus management model with jump risks — a three-fund theorem

Following Rudolf and Ziemba (2004) I develop an intertemporal surplus management model. This approach is suitable not only for pension funds and life insurance funds but also meets the requirements of private investors

(Consiglio *et al.*, 2007). Models designed for pension funds are especially feasible to be conceptually transferred to satisfy private investors' needs (Amenc *et al.*, 2009). The target groups aim at achieving the maximum of the expected lifetime utility. In particular, I develop a strategy that optimizes the surplus in such a way that it balances the current and the future surplus against one another (Campbell and Viceira, 2002). I make a contribution to the existing literature by integrating jump risks into an intertemporal surplus management model. I proceed similar to Das and Uppal (2004) by comparing a jump-diffusion model with a pure-diffusion model.

First of all, I assume that the liabilities, L(t), and the assets, A(t), follow conventional stochastic processes for t $\geq$ 0:

$$dL(t) = L(t)[\mu_L dt + \sigma_L dZ_L(t)], \tag{1}$$

$$dA(t) = A(t)[\hat{\mu}_A dt + \hat{\sigma}_A dZ_A(t)], \tag{2}$$

with

$$E_t\left[\frac{dA(t)}{A(t)}\right] = \hat{\mu}_A dt, \tag{3}$$

$$E_t\left[\frac{dA_i(t)}{A_i(t)} \times \frac{dA_j(t)}{A_j(t)}\right] = \hat{\sigma}_{ij} dt = \hat{\sigma}_i \hat{\sigma}_j \hat{\rho}_{ij} dt, \quad i = 1, \ldots, n, \quad i \neq j, \tag{4}$$

where $dA(t)/A(t)$ is the return process of the assets, $A_i(t)$ denotes the price of asset i, n is the entire number of risky assets, that an investor can choose for her portfolio, and the correlation between the Wiener processes $dZ_i(t)$ and $dZ_j(t)$ is defined as $\hat{\rho}_{ij} dt = E_t[dZ_i(t) \times dZ_j(t)]$ (Das and Uppal, 2004; Rudolf and Ziemba, 2004). Further, I assume a correlation between the Wiener processes $dZ_\mathrm{A}(t)$ and $dZ_\mathrm{L}(t)$, and denote this mathematical property by $\hat{\rho}_{AL} dt = E_t[dZ_A(t) \times dZ_L(t)]$. I specify parameters of the pure-diffusion process for the assets, and other quantities related to the pure-diffusion model, with a ˆ (carat) over the variable (Das and Uppal, 2004).

I enlarge the specified process for the assets in equation (2) by adding two jump components according to Jarrow and Rosenfeld (1984) and Merton (1976).

Following Das and Uppal (2004), the first jump component represents systemic risk and is restricted in two ways (Das and Uppal, 2004)[8]: First of

[8]Systemic risk implies both that the assets are strongly correlated and that correlated adjustments seldom occur whereas systematic risk only describes that the assets and a joint risk factor are correlated and does not restrict them with any further assumption (Das and Uppal, 2004).

all, I assume a simultaneous jump (i.e. of the total number of risky assets) for the processes of return. Secondly, I suppose a perfect correlation for the jump size if such a seldom jump occurs. In contrast to Das and Uppal's model, I thirdly assume a common jump size for all assets according to Jarrow and Rosenfeld and weight the jump size by an asset specific risk factor, $\mathrm{s_A}$, as shown in equation (5) below. I present the mathematical conditions for $\mathrm{s_A}$ in appendix I. I need the last restriction to obtain a portfolio strategy based on a three-fund theorem in contrast to the results of Das and Uppal.

The second jump component represents idiosyncratic risk, that is defined as unexpected shocks of returns of single assets. The investor does not receive any risk premium for this jump risk because I assume it to be diversifiable in the market portfolio according to Jarrow and Rosenfeld (1984). When compared with their approach, that focuses on the development of an instantaneous capital asset pricing model with a jump-diffusion process, I develop a portfolio selection model. Following Merton there is an impact on the optimal portfolio choice strategy although the investor does not receive any premium for this risk.

Enlarging the pure-diffusion process for the assets in equation (2) by two jump components, I derive:

$$\begin{aligned} dA(t) = A(t)\langle &\mu_A dt + \sigma_A dZ_A(t) + s_A \tilde{\pi}_A^S dX_A^S(t) \\ &+ \{-\lambda_A^I K_A^I dt + \tilde{\pi}_A^I dX_A^I(t)\}\rangle, \end{aligned} \tag{5}$$

where $dX_A^S(t)$ is the Poisson process of the systemic jump component with intensity λ_A^S, $\tilde{\pi}_A^S$ is the random jump amplitude of the systemic jump component, $dX_A^I(t)$ is the Poisson process of the idiosyncratic jump component with intensity λ_A^I, and $\tilde{\pi}_A^I$ is the random jump amplitude of the idiosyncratic jump component (Jarrow and Rosenfeld, 1984; Das and Uppal, 2004). $\tilde{\pi}_A^S$ and $\tilde{\pi}_A^I$ describe how much the asset price moves in percent if a systemic or an idiosyncratic jump occurs respectively. I assume that the logarithm of $\tilde{\pi}_A^S$ is distributed normal with mean K_A^S and variance $\langle v_A^S\rangle^2$. Comparably I assume that the logarithm of $\tilde{\pi}_A^I$ is distributed normal with mean K_A^I and variance $\langle v_A^I\rangle^2$. Moreover, I assume that the Wiener processes $\mathrm{dZ_A}(t)$ and $\mathrm{dZ_L}(t)$ are correlated, which is denoted by $\rho_{\mathrm{AL}}\mathrm{dt} = \mathrm{E_t}[\mathrm{dZ_A}(t) \times \mathrm{dZ_L}(\mathrm{t})]$. $\mathrm{dZ_A(t)}$, $dX_A^S(t)$, $dX_A^I(t)$, $\tilde{\pi}_A^S$, $\tilde{\pi}_A^I$ are assumed to be independent. $\mathrm{dZ_L}(t)$, $dX_A^S(t)$, $dX_A^I(t)$, $\tilde{\pi}_A^S$, $\tilde{\pi}_A^I$ are also assumed to be independent.

The total expected return of the assets, that follow the jump-diffusion process in equation (5), is described in equation (6). The investor receives a risk premium for both the diffusion risk, μ_{A}, and the systemic jump risk,

μ_A^J (Das and Uppal, 2004):

$$E_t\left[\frac{dA(t)}{A(t)}\right] = \mu_A dt + s_A \lambda_A^S K_A^S dt = \mu_A dt + \mu_A^J dt. \tag{6}$$

Further, the total covariance of the assets, that are assumed to follow the jump-diffusion process in equation (5), is described in equation (7). By calculating the co-moment I assume a perfect correlation of the systemic jump size, $\tilde{\pi}_A^S$, across the entire number of risky assets (Das and Uppal, 2004). Moreover, I assume the diversification of idiosyncratic jump risk in the market portfolio so that the idiosyncratic jump size, $\tilde{\pi}_A^I$, is uncorrelated with the market (Jarrow and Rosenfeld, 1984). More specifically, the total covariance equals the sum of the covariance between the diffusion elements, σ_{ij}, and the covariance between the systemic jump elements, σ_{ij}^J (Das and Uppal, 2004):

$$E_t\left[\frac{dA_i(t)}{A_i(t)} \times \frac{dA_j(t)}{A_j(t)}\right] = \sigma_i \sigma_j \rho_{ij} dt + s_i s_j \lambda_A^S (\langle K_A^S \rangle^2 + \langle v_A^S \rangle^2) dt = \sigma_{ij} dt + \sigma_{ij}^J dt. \tag{7}$$

The correlation between the Wiener processes $dZ_i(t)$ and $dZ_j(t)$ is defined as $\rho_{ij} dt = E_t[dZ_i(t) \times dZ_j(t)]$. I analyze the differences between a pure-diffusion model, whose processes for the liabilities and assets are specified in equations (1) and (2), and a jump-diffusion model, whose processes for the liabilities and assets are stated in equations (1) and (5). The latter model additionally protects an investor against systemic risk and considers idiosyncratic jump risk. In order to proceed with my analysis I adopt a proposition of Das and Uppal (2004), given in the equations (8) and (9) below, so that I select the first two moments of the assets' jump-diffusion process, that are specified in equations (6) and (7), to equal precisely the corresponding first two moments of the assets' pure-diffusion process, that are given in equations (3) and (4) (Jarrow and Rosenfeld, 1984; Merton, 1976).

$$\mu_A = \hat{\mu}_A - \mu_A^J, \tag{8}$$

$$\sigma_{ij} = \hat{\sigma}_{ij} - \sigma_{ij}^J. \tag{9}$$

The proposition implies that the first two moments of the pure-diffusion process of the assets and the jump-diffusion process of the assets are equalized. For example, I calculate the difference of the total expected return, $\hat{\mu}_A$, and μ_A^J because the systemic jump component, $s_A \tilde{\pi}_A^S dX_A^S(t)$, is going to add the latter term. Correspondingly, I can interpret the modification of the covariance. "In this way, she [the investor] reduces the expected return and covariance coming from the diffusion terms in order to offset exactly the contribution of the jump." (Das and Uppal, 2004). Despite of the equalizing proposition the models result in different portfolios because the jump-diffusion process of the assets is characterized by higher moments (see equations (22) and (23)).

Next, I state the surplus, $S(t)$, and the funding ratio, $F(t)$ (Rudolf and Ziemba, 2004):

$$\begin{aligned} S(t) &:= A(t) - L(t), \\ F(t) &:= A(t)/L(t). \end{aligned} \tag{10}$$

Subsequently, I denote the return process for the assets, $\mathrm{R_A(t)}$, and the one for the liabilities, $\mathrm{R_L(t)}$, by (Rudolf and Ziemba, 2004; Das and Uppal, 2004):

$$\begin{aligned} R_A(t) &:= dA(t)/A(t), \\ R_L(t) &:= dL(t)/L(t). \end{aligned}$$

Following Sharpe and Tint (1990), Rudolf and Ziemba (2004), and according to equation (10) I state the surplus process:

$$\begin{aligned} R_S(t) &= \frac{dS(t)}{A(t)} = R_A(t) - \frac{R_L(t)}{F(t)}, \\ R_S(t) &= \left[\mu_A - \frac{\mu_L}{F(t)}\right] dt + \sigma_A dZ_A(t) - \frac{\sigma_L}{F(t)} dZ_L(t) + s_A \tilde{\pi}_A^S dX_A^S(t) \\ &\quad + \left[-\lambda_A^I K_A^I dt + \tilde{\pi}_A^I dX_A^I(t)\right]. \end{aligned} \tag{11}$$

Based on $dS(t) = R_S(t)A(t)$ and applying equation (11) I derive

$$\begin{aligned} E_t[dS(t)] &= A(t)\left[\mu_A - \frac{\mu_L}{F(t)}\right] dt, \\ dS(t)dS(t) &= A^2(t)\left[\sigma_A^2 + \frac{\sigma_L^2}{F^2(t)} - 2\frac{\sigma_{AL}}{F(t)}\right] dt, \end{aligned} \tag{12}$$

where $\sigma_{\mathrm{AL}} := \sigma_{\mathrm{A}}\sigma_{\mathrm{L}}\rho_{\mathrm{AL}}$ denotes the covariance between the assets and the liabilities (Rudolf and Ziemba, 2004).

Following Gihman and Skorohod (1970, pp. 263–272), Malliaris and Brock (1990, pp. 121–124), and Piazzesi (2005) I apply $dS(t) = R_S(t)A(t)$ once again and use equation (11) so that I derive the systemic jump component of the surplus process, γ_S^S, and the idiosyncratic jump component of the surplus process, γ_S^I:

$$\begin{aligned}\gamma_S^S &= A(t)s_A\tilde{\pi}_A^S,\\ \gamma_S^I &= A(t)\tilde{\pi}_A^I.\end{aligned} \tag{13}$$

Following Rudolf and Ziemba (2004), the investor has the subsequent optimization problem:

$$\begin{aligned}J(S,t) &:= \max_{\omega} E_t\left(\int_t^T U(S,\tau)d\tau\right),\\ J(S,t) &= \max_{\omega} E_t\left(\int_t^{t+dt} U(S,\tau)d\tau + \int_{t+dt}^T U(S,\tau)d\tau\right),\end{aligned} \tag{14}$$

where "U is an additively separable, twice differentiable, concave utility function", T states the end of the planning horizon of the investor, "$\mathrm{E_t}$ denotes expectation with respect to the information set at time t", the J-function is the maximum of the intertemporal expected utility, that an investor derives from the surplus during her lifetime, and $\boldsymbol{\omega} \in \Re^n$ represents the vector of portfolio weights of the risky-assets.

According to Kushner (1967, Chapter 4), Jarrow and Rosenfeld (1984), and Rudolf and Ziemba (2004) the optimal control satisfying equation (14) exists and can be derived by using the Bellman principle and solving:

$$J(S,t) = U(S,t)dt + \max_{\omega} E_t[J(S+dS,t+dt)]. \tag{15}$$

Subsequently, I apply the generalized Itô's formula for jump-diffusion processes and combine it with equations (12) and (13) so that I derive (see appendix II for proof):

$$0 = U(t,S) + \max_{\omega}[\Delta J]$$

$$0 = U(t,S) + \max_{\omega}\begin{bmatrix} J_t + J_S A(t)\left(\mu_A - \dfrac{\mu_L}{F(t)}\right) \\ +\dfrac{1}{2}J_{SS}A^2(t)\left(\sigma_A^2 + \dfrac{\sigma_L^2}{F^2(t)} - 2\dfrac{\sigma_{AL}}{F(t)}\right) \\ +\lambda_A^S E_t\langle J(S + As_A\tilde{\pi}_A^S, t) - J(S,t)\rangle \\ +\lambda_A^I E_t\langle J(S + A\tilde{\pi}_A^I, t) - J(S,t)\rangle \end{bmatrix}, \tag{16}$$

where $\mathrm{J_t}$ is the first partial derivative of J with respect to t, $\mathrm{J_S}$ represents the first partial derivative of J with respect to S, and $\mathrm{J_{SS}}$ denotes the second partial derivative of J with respect to S (Rudolf and Ziemba, 2004; Jarrow and Rosenfeld, 1984).

I assume that $\boldsymbol{m_A} \in \Re^n$ is "the vector of expected asset returns of the risky assets", $\boldsymbol{e} \in \Re^n$ is the vector of ones, $\boldsymbol{v_{AL}} \in \Re^n$ is the vector of covariances between the n assets and the liabilities, $\boldsymbol{V} \in \Re^{n\times n}$ is the matrix of covariances between the n risky assets (Rudolf and Ziemba, 2004). $\boldsymbol{s} \in \Re^n$ is the vector of asset specific risk factors, that weight the systemic jump amplitude, $\tilde{\pi}_A^S$, and $\boldsymbol{j} \in \Re^n$ is the vector of random jump amplitudes of the idiosyncratic jump component. Subsequently, I transform equation (16) and derive (the proof and definitions of the matrix and the vectors can be found in the appendix III):

$$0 = U(t,S) + \max_{\omega} \begin{bmatrix} J_t + J_S A(t)\left(\boldsymbol{\omega}'(\boldsymbol{m_A} - r\boldsymbol{e}) + r - \dfrac{\mu_L}{F(t)}\right) \\ +\dfrac{1}{2} J_{SS} A^2(t)\left(\boldsymbol{\omega}'\boldsymbol{V}\boldsymbol{\omega} + \dfrac{\sigma_L^2}{F^2(t)} - 2\dfrac{\boldsymbol{\omega}'\boldsymbol{v_{AL}}}{F(t)}\right) \\ +\lambda_A^S E_t\left\langle J\left(S + A\boldsymbol{\omega}'\boldsymbol{s}\tilde{\pi}_A^S, t\right) - J(S,t)\right\rangle \\ +\lambda_A^I E_t\left\langle J\left(S + A\boldsymbol{\omega}'\boldsymbol{j}, t\right) - J(S,t)\right\rangle \end{bmatrix}, \tag{17}$$

where r denotes the return of the riskless asset (Rudolf and Ziemba, 2004). Then I differentiate equation (17) with respect to $\boldsymbol{\omega}$ and obtain:[9]

$$\begin{aligned} \boldsymbol{\omega} = & -\frac{J_S}{A(t)J_{SS}}\boldsymbol{V^{-1}}(\boldsymbol{m_A} - r\boldsymbol{e}) \\ & -\frac{\lambda_A^S}{A(t)J_{SS}}\boldsymbol{V^{-1}}E_t\langle J_S(S + A\boldsymbol{\omega}'\boldsymbol{s}\tilde{\pi}_A^S, t)\boldsymbol{s}\tilde{\pi}_A^S\rangle \\ & -\frac{\lambda_A^I}{A(t)J_{SS}}\boldsymbol{V^{-1}}E_t\langle J_S(S + A\boldsymbol{\omega}'\boldsymbol{j}, t)\boldsymbol{j}\rangle + \frac{\boldsymbol{V^{-1}}\boldsymbol{v_{AL}}}{F(t)}. \end{aligned} \tag{18}$$

In order to reduce complexity, I am going to use the abbreviations $J_S(S + A\boldsymbol{\omega}'\boldsymbol{s}\,\tilde{\pi}_A^S, t) = J_{S(\pi)}$ and $J_S(S + A\boldsymbol{\omega}'\boldsymbol{j}, t) = J_{S(j)}$. Furthermore, I define the vector of means of the jump amplitude of the idiosyncratic jump component as $\boldsymbol{K_I'} := (K_1^I, \ldots, K_n^I)'$. Next, I state the covariance between the term $J_{S(j)}$ and the random jump amplitude of the idiosyncratic jump

[9]
$$\begin{aligned} 0 = J_S(\boldsymbol{m_A} - r\boldsymbol{e}) + A(t)J_{SS}\boldsymbol{V}\boldsymbol{\omega} + \lambda_A^S E_t[J_S(S(t) + A(t)\boldsymbol{\omega}'\boldsymbol{s}\tilde{\pi}_A^S, t)\boldsymbol{s}\tilde{\pi}_A^S] \\ + \lambda_A^I E_t[J_S(S(t) + A(t)\boldsymbol{\omega}'\boldsymbol{j}, t)\boldsymbol{j}] - [A(t)/F(t)]J_{SS}\boldsymbol{v_{AL}} \end{aligned}$$

component of the ith risky asset, $\tilde{\pi}_i^I$, as $Cov[J_{S(j)}, \tilde{\pi}_i^I]$. I denote the vector of these covariances by $\boldsymbol{v}[J_{S(j)}, \boldsymbol{j}]' := (Cov[J_{S(j)}, \tilde{\pi}_i^I], \dots, Cov[J_{S(j)}, \tilde{\pi}_n^I])$. Transforming equation (18) while using the above defined terms I derive:

$$\begin{aligned}\boldsymbol{\omega} = &-\frac{J_S}{A(t)J_{SS}}\boldsymbol{V^{-1}}(\boldsymbol{m_A} - r\boldsymbol{e})\\ &-\frac{\lambda_A^S}{A(t)J_{SS}}\boldsymbol{V^{-1}}\boldsymbol{s}\{Cov[J_{S(\pi)}, \tilde{\pi}_A^S] + E_t[J_{S(\pi)}]K_A^S\}\\ &-\frac{\lambda_A^I}{A(t)J_{SS}}\boldsymbol{V^{-1}}\{\boldsymbol{v}[J_{S(j)}, \boldsymbol{j}] + E_t[J_{S(j)}]\boldsymbol{K_I}\} + \frac{\boldsymbol{V^{-1}}\boldsymbol{v_{AL}}}{F(t)}. \qquad (19)\end{aligned}$$

In order to derive a three-fund theorem it is necessary to apply the mutual fund theorem of Ross (1978, theorem 3) to equation (19). Jarrow and Rosenfeld (1984) illustrate and prove in detail that the mutual fund theorem of Ross is applicable to jump-diffusion processes if the jump components have the particular form given in equation (5) and if ΔJ, given in equation (16), is state independent (except surplus).

In the following I explain the investor's funds i.e. the components of the optimal portfolio. The terms on the right-hand side on the first line and the left term on the second line of equation (19) represent the first component of the optimal portfolio strategy, i.e. the market portfolio. In total, the market portfolio consists of three parts. First of all, the market portfolio has the capability to hedge diffusion risk. The respective component is equal to:

$$-\frac{J_S}{A(t)J_{SS}}\boldsymbol{V^{-1}}(\boldsymbol{m_A} - r\boldsymbol{e}).$$

The market portfolio not only hedges diffusion risk but also systemic risk. The relevant second component of the market portfolio is specified by:

$$-\frac{\lambda_A^S}{A(t)J_{SS}}\boldsymbol{V^{-1}}\boldsymbol{s}\{Cov[J_{S(\pi)}, \tilde{\pi}_A^S] + E_t[J_{S(\pi)}]K_A^S\},$$

where $J_{S(\pi)} = J_S(S + A\boldsymbol{\omega}'\boldsymbol{s}\tilde{\pi}_A^S, t)$ represents the above defined abbreviation. The last part of the market portfolio hedges idiosyncratic jump risk and is specified by

$$-\frac{\lambda_A^I}{A(t)J_{SS}}\boldsymbol{V^{-1}}\{\boldsymbol{v}[J_{S(j)}, \boldsymbol{j}] + E_t[J_{S(j)}]\boldsymbol{K_I}\},$$

where $J_{S(j)} = J_S(S + A\boldsymbol{\omega}'\boldsymbol{j}, t)$ represents the above defined abbreviation. The right term on the second line of equation (19) represents the hedge portfolio for the liabilities, which is independent of the investor's preferences, and is specified by:

$$\frac{\boldsymbol{V}^{-1}\boldsymbol{v}_{\mathbf{AL}}}{F(t)}.$$

In order to calculate the unconditional moments of the assets' return process, that is specified in equation (5), I proceed in accordance with the approach of Das and Uppal (2004). While doing so, I apply the knowledge of Curtiss (1942) about the interrelationship between the characteristic function and the moment generating function. In order to derive the moments of the continuously compounded returns of the assets i and j, I assume that $i = 1, \ldots, n$ and $i \neq j$ (Das and Uppal, 2004).

$$\text{Mean} = t\left(\mu_i - \frac{1}{2}\sigma_i^2 + s_i\lambda_A^S K_A^S\right), \tag{20}$$

$$\text{Covariance} = t\{\sigma_{ij} + s_i s_j \lambda_A^S(\langle K_A^S\rangle^2 + \langle v_A^S\rangle^2)\}, \tag{21}$$

$$\text{coskewness} = \frac{t\lambda_A^S s_i^2 s_j K_A^S\{3\langle v_A^S\rangle^2 + \langle K_A^S\rangle^2\}}{\text{variance}_i \text{variance}_j^{1/2}}, \tag{22}$$

$$\text{excess kurtosis} = \frac{t\lambda_A^S s_i^4\{3\langle v_A^S\rangle^4 + 6\langle K_A^S\rangle^2\langle v_A^S\rangle^2 + \langle K_A^S\rangle^4\}}{\text{variance}_i \text{variance}_j^{1/2}}. \tag{23}$$

In order to derive the optimal control equation (19) needs to be solved numerically:

$$\begin{aligned} 0 &= J_S(\boldsymbol{m_A} - r\boldsymbol{e}) + A(t)J_{SS}\boldsymbol{V\omega} + \lambda_A^S\boldsymbol{s}\{Cov[J_{S(\pi)}, \tilde{\pi}_A^S] \\ &\quad + E_t[J_{S(\pi)}]K_A^S\} + \lambda_A^I\{\boldsymbol{v}[J_{S(j)}, \boldsymbol{j}] \\ &\quad + E_t[J_{S(j)}]\boldsymbol{K_I}\} - [A(t)/F(t)]J_{SS}\boldsymbol{v}_{\mathbf{AL}} \end{aligned} \tag{24}$$

By assuming pure-diffusion processes for the liabilities and the assets, that are specified in equation (1) and equation (2), I derive the known closed-form solution of Rudolf and Ziemba (2004):

$$\hat{\boldsymbol{\omega}} = -\frac{J_S}{A(t)J_{SS}}\hat{\boldsymbol{V}}^{-1}(\hat{\boldsymbol{m}}_{\boldsymbol{A}} - r\boldsymbol{e}) + \frac{\hat{\boldsymbol{V}}^{-1}\hat{\boldsymbol{v}}_{\mathbf{AL}}}{F(t)}. \tag{25}$$

where I consider the particular case of a constant investment opportunity set (Jarrow and Rosenfeld, 1984). Similar to my proceeding the results of Das and Uppal (2004) can be compared to the closed-form solution of

Merton (1971) (i.e. Das and Uppal consider systemic risk and thus enlarge the result of Merton by a jump component).

3 Risk preference, and funding ratio

I assume a utility function from the HARA class (Rudolf and Ziemba, 2004). In accordance with the results of Merton (1971 and 1993, p. 140) my assumption is equivalent to the stating that the J-function (15) is part of the HARA class. I denote the risk tolerance coefficient by $\alpha < 1$. κ and θ are real constants and ρ is a utility deflator. Then

$$\begin{aligned} &U(S,t) \subset HARA \Leftrightarrow J(S,t) \subset HARA, \\ &J(S,t) = \frac{1-\alpha}{e^{\rho t}\alpha}\left(\frac{\kappa S}{1-\alpha} + \theta\right)^{\alpha}. \end{aligned} \tag{26}$$

Equation (26) contains the implicit assumption of a linear absolute risk tolerance, which can be seen by looking at the derivation $-J_S/J_{SS} = S/(1-\alpha) + \theta/\kappa$. If I assume that α approaches $-\infty$, $\theta = 1$, and $\rho = 0$ I obtain the special case of negative exponential utility functions $J = -e^{-aS}$. If I apply the particular conditions of $\theta = 0$, $\rho = 0$, and $\kappa = (1-\alpha)^{(\alpha-1)/\alpha}$ it results isoelastic power utility $J = S^{\alpha}/\alpha$ (Ingersoll, 1987). If I assume that α approaches 0 ($\theta = \rho = 0$) I obtain log utility $J = \ln(S)$, which is a special case of isoelastic power utility.

Given that I assume a utility function of the HARA class I derive the following terms:

$$\begin{aligned} J_S &= \frac{\kappa}{e^{\rho t}}\left(\frac{\kappa S}{1-\alpha} + \theta\right)^{\alpha-1}, \\ J_{SS} &= -\frac{\kappa^2}{e^{\rho t}}\left(\frac{\kappa S}{1-\alpha} + \theta\right)^{\alpha-2}, \\ J_S(S + A\boldsymbol{\omega}'\boldsymbol{s}\tilde{\pi}_A^S, t) &= \frac{\kappa}{e^{\rho t}}\left(\frac{\kappa[S + A\boldsymbol{\omega}'\boldsymbol{s}\tilde{\pi}_A^S]}{1-\alpha} + \theta\right)^{\alpha-1}, \\ J_S(S + A\boldsymbol{\omega}'\boldsymbol{j}, t) &= \frac{\kappa}{e^{\rho t}}\left(\frac{\kappa[S + A\boldsymbol{\omega}'\boldsymbol{j}]}{1-\alpha} + \theta\right)^{\alpha-1}. \end{aligned} \tag{27}$$

In order to enhance the simplicity I am going to use the abbreviations $\beta(\tilde{\pi}_A^S) = \left(\frac{\kappa[S+A\boldsymbol{\omega}'\boldsymbol{s}\tilde{\pi}_A^S]}{1-\alpha} + \theta\right)^{\alpha-1}$, and $\beta(\boldsymbol{j}) = \left(\frac{\kappa[S+A\boldsymbol{\omega}'\boldsymbol{j}]}{1-\alpha} + \theta\right)^{\alpha-1}$. I define the covariance between the term $\beta(\boldsymbol{j})$ and the idiosyncratic jump size of asset i, $\tilde{\pi}_i^I$, as $Cov[\beta(\boldsymbol{j}), \tilde{\pi}_i^I]$. I specify the corresponding vector by $\boldsymbol{v}[\beta(\boldsymbol{j}), \boldsymbol{j}]' := (Cov[\beta(\boldsymbol{j}), \tilde{\pi}_i^I], \ldots, Cov[\beta(\boldsymbol{j}), \tilde{\pi}_n^I])$. Subsequently, I insert the terms of equation (27) in equation (19) and obtain:

$$\omega = \left\{\frac{1}{1-\alpha}\left[1 - \frac{1}{F}\right] + \frac{\theta}{A\kappa}\right\} \boldsymbol{V^{-1}}(\boldsymbol{m_A} - r\boldsymbol{e})$$
$$+\frac{1}{A\kappa}\left(\frac{\kappa S}{1-\alpha} + \theta\right)^{2-\alpha}$$
$$\times \boldsymbol{V^{-1}} \left\langle \begin{array}{l} \lambda_A^S \boldsymbol{s}\{Cov[\beta(\tilde{\pi}_A^S), \tilde{\pi}_A^S] + E_t[\beta(\tilde{\pi}_A^S)]K_A^S\} \\ +\lambda_A^I\{\boldsymbol{v}[\beta(\boldsymbol{j}), \boldsymbol{j}] + E_t[\beta(\boldsymbol{j})]\boldsymbol{K_I}\} \end{array} \right\rangle + \frac{\boldsymbol{V^{-1}}\boldsymbol{v_{AL}}}{F}. \tag{28}$$

The first two terms on the right-hand side of equation (28) belong to the market portfolio. The third term on the right-hand side of equation (28) is the hedge portfolio for the liabilities, which is independent of preferences.

Next, I assume the special case of log utility:

$$J(S,t) = \ln(S). \tag{29}$$

Under the assumption of log utility I obtain the following derivations:

$$\begin{aligned} J_S = S^{-1}, \quad J_{SS} &= -S^{-2}, \\ J_S(S + A\boldsymbol{\omega}'\boldsymbol{s}\tilde{\pi}_A^S, t) &= (S + A\boldsymbol{\omega}'\boldsymbol{s}\tilde{\pi}_A^S)^{-1}, \\ J_S(S + A\boldsymbol{\omega}'\boldsymbol{j}, t) &= (S + A\boldsymbol{\omega}'\boldsymbol{j})^{-1}. \end{aligned} \tag{30}$$

In order to enhance simplicity I am going to use the abbreviations $\delta(\tilde{\pi}_A^S) = (S + A\boldsymbol{\omega}'\boldsymbol{s}\tilde{\pi}_A^S)^{-1}$, and $\delta(\boldsymbol{j}) = (S + A\boldsymbol{\omega}'\boldsymbol{j})^{-1}$. I define the covariance between the term $\delta(\boldsymbol{j})$ and the idiosyncratic jump size of asset i, $\tilde{\pi}_i^I$, as $Cov[\delta(\boldsymbol{j}), \tilde{\pi}_i^I]$. I specify the corresponding vector by $\boldsymbol{v}[\delta(\boldsymbol{j}), \boldsymbol{j}]' := (Cov[\delta(\boldsymbol{j}), \tilde{\pi}_1^I], \ldots, Cov[\delta(\boldsymbol{j}), \tilde{\pi}_n^I])$. Subsequently, I insert the terms of equation (30) in equation (19) and obtain:

$$\omega = \left[1 - \frac{1}{F}\right] \boldsymbol{V^{-1}}$$
$$\times \left\langle \begin{array}{l} (\boldsymbol{m_A} - r\boldsymbol{e}) + S\lambda_A^S \boldsymbol{s}\{Cov_t[\delta(\tilde{\pi}_A^S), \tilde{\pi}_A^S] + E_t[\delta(\tilde{\pi}_A^S)]K_A^S\} \\ +\lambda_A^I S\{\boldsymbol{v}[\delta(\boldsymbol{j}), \boldsymbol{j}] + E_t[\delta(\boldsymbol{j})]\boldsymbol{K_I}\} \end{array} \right\rangle$$
$$+\frac{\boldsymbol{V^{-1}}\boldsymbol{v_{AL}}}{F} \tag{31}$$

Under the assumption of log utility it is possible to make an investment in the market portfolio, that is as well independent of preferences.

4 Conclusions

Individuals increasingly demand high-quality risk management services, that are tailored to their personal needs. On the one hand, households have more serious been getting into debt for years and they are at higher risk of getting into financial distress. On the other hand, individuals face the facts that they need to provide to a greater extend for the future. For example, it has become an absolute necessity to take out private retirement provisions. Asset and liability management gives private investors the benefit of allocating their assets optimally while planning the secure financing of future liabilities and goals at the same time. Thus, the development of such models for individuals is of strong current interest and further innovation in this area of research can be expected.

In 2007 an economic and financial crisis began, that had among other things an impact on the private banking industry. Providers of private banking services became aware of the added value of high-quality risk management. They witnessed that a failure of such tools damaged the reputation of the respective bank institutions as well as it induced affected HNWIs to withdraw some or all of their assets. Due to the crisis providers of private banking services learnt that high-quality risk management constitutes a key success factor. Private banking advisors perceive the benefits of asset and liability management ever since. Consequently, providers of private banking services are interested in offering advice, that is based on asset and liability management models. They want to benefit from the positive impact of these risk management tools on customer loyalty during times of crisis. However, the bank institutions recognize the complexity of these tools and are inhibited from designing and implementing them straight away.

I have responded to the strong interest, that individuals (including HNWIs) and providers of private banking services have in high-quality risk management. I have used a LDI technique, so that the model can be applied to pension and life insurance funds, and individuals (including HNWIs).

I have derived a three-fund theorem. The optimal portfolio of an investor consists of three funds: a market portfolio, a liability-hedging portfolio, and a riskless asset. In contrast to the results of Rudolf and Ziemba (2004), a market portfolio not only hedges diffusion risk, but it also hedges systemic risk and it takes into account idiosyncratic jump risk so that the investor is additionally protected against both a systemic risk and an idiosyncratic jump risk.

The 2007 to 2009 financial and economical crisis, that has turned into a sovereign crisis in 2010, illustrates how important systemic risk is to players in international financial markets. Therefore, it is of current interest to intertemporal surplus optimizers that a model takes account of systemic risk. In addition, long-term investors are attracted to alternative investments. Thus, intertemporal surplus optimizers also benefit from a model that considers idiosyncratic jump risk because it allows to take account of real estate risk among other things.

Appendix I: Derivation of the asset specific risk factor of the first jump component

The following mathematical derivation goes back to Jarrow and Rosenfeld (1984). I apply the original procedure to an intertemporal surplus management model. In particular, the jump-diffusion process of the assets, that is stated in equation (5), is split up into two parts. The first one is assumed not to be diversifiable in the market portfolio whereas the second one is supposed to be diversifiable in the market portfolio.

In the following I denote the jump-diffusion process of asset i and let n be the entire number of risky assets:

$$dA_i(t) = A_i(t)\langle \mu_i dt + \sigma_i dZ_i(t) + s_i \tilde{\pi}_A^S dX_A^S(t) + \{-\lambda_A^I K_i^I dt + \tilde{\pi}_i^I dX_i^I(t)\}\rangle.$$

The diffusion part of asset i (dD_i) of the equation above consists of two components:

$$dD_i = \mu_i dt + \sigma_i dZ_i(t).$$

In agreement with the arguments of Ross (1978) I deduce from the preceding equation the following mathematical relationship (Jarrow and Rosenfeld, 1984).

$$dD_i(t) = \mu_i dt + s_i d\psi(t) + g_i d\eta_i(t),$$

where I assume the existence of both the constants (φ_i, s_i, and g_i) and the Wiener processes ($d\psi(t)$ and $d\eta_i(t)$). Moreover, the following assumptions are valid:

$$s_i^2 + g_i^2 = \sigma_i^2,$$

$$E_t[d\psi(t) d\eta_i(t)] = 0,$$

$$\sum_{i=1}^{n} \phi_i = 1,$$

$$\sum_{i=1}^{n} \phi_i (g_i d\eta_i(t)) = 0,$$

$$\sum_{i=1}^{n} \varphi_i \mu_i > r,$$

where r denotes the riskless rate of interest.

It is possible to split a final number of normally distributed random variables into a common factor, $d\psi(t)$, and residuals, $d\eta_i(t)$. Both the common factor and the residuals are normally distributed. Since I use normally distributed returns I benefit from the elementary property that returns are statistically independent if their covariance is equal to zero. I assume that the Wiener processes ($d\psi(t)$ and $d\eta_i(t)$), the Poisson processes of both jump components ($dX_A^S(t)$ and $dX_i^I(t)$), and both stochastic jump amplitudes ($\tilde{\pi}_A^S$, and $\tilde{\pi}_i^I$) are independent of each other. As a result I can interpret $d\psi(t)$ and $d\eta_i(t)$ as risk factors. The first one is not diversifiable whereas the second one is diversifiable in the market portfolio. Next, I assume that the Wiener processes, $d\psi(t)$, and $dZ_L(t)$, are correlated and denote this relationship by $d\psi(t)dZ_L(t) = \rho_{\psi L}dt$. Finally, I substitute the modifications in the jump-diffusion process of asset i and I derive the following equivalent expression:

$$\begin{aligned} dA_i(t) = A_i(t)\langle \mu_i dt + s_i d\psi(t) + g_i d\eta_i(t) + s_i \tilde{\pi}_A^S dX_A^S(t) \\ + \{-\lambda_A^I K_i^I dt + \tilde{\pi}_i^I dX_i^I(t)\}\rangle, \end{aligned}$$

where K_i^I denotes the expected value of the logarithm of the idiosyncratic jump amplitude, $\tilde{\pi}_i^I$. Consequently, the asset specific risk factor, s_i, belongs to the part of the jump-diffusion process in equation (5), that is not diversifiable in the market portfolio.

Appendix II: Derivation of equation (16)

In order to reduce complexity, I leave out the arguments of the J-function (Rudolf and Ziemba, 2004). I define o(dt) as a variable that adds up all components of dt, whose order is higher than one. Accordingly, I have o(dt):$\lim_{\mathrm{dt}\to 0}$ o(dt)/dt = 0. First of all, I apply the generalized Itô's formula for jump-diffusion processes (Gihman and Skorohod, 1970; Kushner, 1967; Malliaris and Brook, 1990):

$$\begin{aligned} J(S,t) &= U(S,t)dt + \max_{\omega} E_t[J(S+dS, t+dt)] \\ &= U dt + \max_{\omega} E_t \end{aligned}$$

$$\times \begin{bmatrix} J + J_t dt + J_S dS + \frac{1}{2} J_{SS} dS^2 + \lambda_A^S \{J(S + \gamma_S^S, t) - J(S,t)\} dt \\ + \lambda_A^I \{J\,(S + \gamma_S^I, t) - J(S,t)\} dt + o(dt) \end{bmatrix},$$

$$0 = U dt + \max_{\omega} E_t$$

$$\times \begin{bmatrix} J_t dt + J_S dS + \frac{1}{2} J_{SS} dS^2 + \lambda_A^S \{J(S + \gamma_S^S, t) - J(S,t)\} dt \\ + \lambda_A^I \{J(S + \gamma_S^I, t) - J(S,t)\} dt + o(dt) \end{bmatrix}.$$

Next, I modify the equation above and derive

$$0 = U dt + \max_{\omega}$$

$$\times \begin{bmatrix} J_t dt + J_S E_t(dS) + \frac{1}{2} J_{SS} dS dS + \lambda_A^S E_t \{J(S + \gamma_S^S, t) - J(S,t)\} dt \\ + \lambda_A^I E_t \left\{J\ \left(S + \gamma_S^I, t\right) - J(S,t)\right\} dt \end{bmatrix}.$$

Finally, I substitute in the equation above equations (12) and (13) and derive equation (16).

Appendix III: Derivation of equation (17)

I describe the process of the price of asset $\mathrm{A}_i(t)$, $i = 1, \ldots, n$, as Rudolf and Ziemba (2004):

$$dA_i(t) = A_i(t) \langle \mu_i dt + \sigma_i dZ_i(t) + s_i \tilde{\pi}_A^S dX_A^S(t) \\ + \{-\lambda_A^I K_i^I dt + \tilde{\pi}_i^I dX_i^I(t)\} \rangle.$$

I define the vector of the expected asset returns of the n risky assets as $\mathbf{m'}_A := (\mu_1, \ldots, \mu_n)$. I weight the systemic jump amplitude, $\tilde{\pi}_A^S$, that is common to all assets, by a vector of asset specific risk factors, that I define as $\mathbf{s'} := (s_1, \ldots, s_n)$. Moreover, I define the vector of the random jump amplitudes of the idiosyncratic jump component as $\boldsymbol{j'} := (\tilde{\pi}_1^I, \ldots, \tilde{\pi}_n^I)$. The first part of the total covariance given by equation (7) is the covariance between the diffusion components of assets' returns, σ_{ij}. I define the corresponding matrix, that contains the covariance terms between the diffusion components, by:

$$\mathbf{V} = \begin{bmatrix} \sigma_{11} & \dots & \sigma_{1n} \\ \vdots & \vdots & \vdots \\ \sigma_{n1} & \dots & \sigma_{nn} \end{bmatrix}.$$

The second part of the total covariance given by equation (7) is the covariance between the systemic jump components of assets' returns, σ_{ij}^J. I denote the resultant matrix, that contains the covariance terms between the systemic jump components, by Das and Uppal (2004):

$$\boldsymbol{V}^J = \begin{bmatrix} \sigma_{11}^J & \dots & \sigma_{1n}^J \\ \vdots & \vdots & \vdots \\ \sigma_{n1}^J & \dots & \sigma_{nn}^J \end{bmatrix}.$$

According to equation (8), I adjust the total expected return of the assets' jump-diffusion process as follows:

$$\mu_i = \hat{\mu}_i - \mu_i^J.$$

Similarly, the second moment of the assets' jump-diffusion process is modified as stated in equation (9), i.e.:

$$\sigma_{ij} = \hat{\sigma}_{ij} - \sigma_{ij}^J.$$

I assume that $\mathrm{x_i(t)}$ represents the share of asset i, that an investor owns at the moment t. I define B(t) to be equal to the amount of cash, that is invested in the asset portfolio at the instant t. Next, I derive the worth of the asset portfolio A(t):

$$A(t) = \sum_{i=1}^{n} x_i(t)A_i(t) + B(t).$$

In the next infinitesimal period of time, dt, the value of the asset portfolio develops as follows:

$$dA(t) = \sum_{i=1}^{n} x_i(t)dA_i(t) + dB(t) = \sum_{i=1}^{n} x_i(t)dA_i(t) + rB(t)dt,$$

where r represents the riskless rate of interest. I denote the vector of risky assets portfolio weights, $\boldsymbol{\omega}' = (\omega_1, \dots, \omega_n)$, by:

$$\omega_i(t) := \frac{x_i(t)A_i(t)}{A(t)}, \quad i = 1, \dots, n.$$

If I add up both the risky and the riskless investment the result is assumed to be one. Furthermore, I suppose that $\mathbf{e}' = (1, \ldots, 1)$ represents the vector of ones, $\boldsymbol{e} \in \Re^n$. Thus, the investor holds a share of $B(t)/A(t) = 1 - \boldsymbol{\omega}'\boldsymbol{e}$ of the riskless asset. Consequently, I modify the equation above for dA and derive:

$$dA(t) = A(t)\begin{bmatrix} \sum_{i=1}^{n} [\omega_i(\mu_i - r) + r]dt + \sum_{i=1}^{n} \omega_i \sigma_i dZ_i(t) \\ + \sum_{i=1}^{n} \omega_i s_i \tilde{\pi}_A^S dX_A^S(t) \\ + \left\{ - \sum_{i=1}^{n} \omega_i \lambda_A^I K_i^I dt + \sum_{i=1}^{n} \omega_i \tilde{\pi}_i^I dX_i^I(t) \right\} \end{bmatrix}.$$

I compare the equation above with equation (5), i.e. $dA(t) = A(t) \langle \mu_A dt + \sigma_A dZ_A(t) + s_A \tilde{\pi}_A^S dX_A^S(t) + \{-\lambda_A^I K_A^I dt + \tilde{\pi}_A^I dX_A^I(t)\}\rangle$, and obtain similar to Rudolf and Ziemba (2004):

$$\mu_A = \boldsymbol{\omega}'(\boldsymbol{m_A} - r\boldsymbol{e}) + r,$$

$$\sigma_A dZ_A(t) = \sum_{i=1}^{n} \omega_i \sigma_i dZ_i(t).$$

Subsequently, I transform the preceding equation and it results according to Rudolf and Ziemba (2004):

$$\sigma_A^2 = \boldsymbol{\omega}' \boldsymbol{V} \boldsymbol{\omega},$$

where $\boldsymbol{V}$ represents the previously defined covariance matrix. The covariance between the risky asset i and the liabilities is denoted by σ_{iL}. I define the corresponding vector by $\mathbf{v'_{AL}} := (\sigma_{\mathrm{1L}}, \ldots, \sigma_{\mathrm{nL}})$. As a result, I obtain the relationship $\sigma_{AL} = \boldsymbol{\omega}'\boldsymbol{v}_{\mathbf{AL}}$.

I compare the expression above for the process of assets, dA(t), with equation (5), i.e. $dA(t) = A(t)\langle \mu_A dt + \sigma_A dZ_A(t) + s_A \tilde{\pi}_A^S dX_A^S(t) + \{-\lambda_A^I K_A^I dt + \tilde{\pi}_A^I dX_A^I(t)\}\rangle$, because I need to take into account both jump components as well. I obtain the following relationships for both the compounded Poisson processes of the systemic jump component and the compounded Poisson process of the idiosyncratic jump component, respectively:

$$A(t)s_A\tilde{\pi}_A^S dX_A^S(t) = A(t)\sum_{i=1}^{n}\omega_i s_i \tilde{\pi}_A^S dX_A^S(t),$$

$$A(t)\tilde{\pi}_A^I dX_A^I(t) = A(t)\sum_{i=1}^{n}\omega_i \tilde{\pi}_i^I dX_i^I(t).$$

Subsequently, I derive the following expressions for the systemic jump component of the surplus process, γ_S^S, and the idiosyncratic jump component of the surplus process, γ_S^I:

$$\gamma_S^S = A(t)s_A\tilde{\pi}_A^S = A(t)\boldsymbol{\omega}'\boldsymbol{s}\tilde{\pi}_A^S,$$

$$\gamma_S^I = A(t)\tilde{\pi}_A^I = A(t)\boldsymbol{\omega}'\boldsymbol{j}.$$

I assume a simultaneous occurrence of systemic jumps across all assets so that $dX_i^S(t) = dX_j^S(t) = dX_A^S(t)$ for all $i = 1,\ldots,n$, $j = 1,\ldots,n$ and $i \neq j$ (Das and Uppal, 2004). Consequently, the intensity of the systemic jump component, λ_A^S, is identical for all assets, i.e. $\lambda_i^S = \lambda_j^S = \lambda_A^S$. Similarly, I assume that the intensity of the idiosyncratic jump component, λ_A^I, is identical for all assets, i.e. $\lambda_i^I = \lambda_j^I = \lambda_A^I$.

I use the final expressions of expected returns, variances, covariances, jump components and intensities in order to derive equation (17).

References

Aase, K. K. Optimum Portfolio Diversification in a General Continuous-Time Model. *Stochastic Processes and their Applications*, 18: pp. 81–98, 1984.

Amenc, N., F. Goltz and D. Schröder, Private Bankers on Private Banking: Financial Risks and Asset/Liability Management. *The Journal of Wealth Management*, 12: pp. 39–50, 2007.

Amenc, N. and L. Martellini. In Diversification We Trust? *The Journal of Portfolio Management*, 37: pp. 1–2, 2011.

Amenc, N., L. Martellini, V. Milhau and V. Ziemann. Asset-Liability Management in Private Wealth Management. *The Journal of Portfolio Management*, 36: pp. 100–120, 2009.

Berger, A. J. and J. M. Mulvey. The Home Account Advisor: Asset and Liability Management for Individual Investors. In W. T. Ziemba and J. M. Mulvey, eds.: *Worldwide Asset and Liability Modeling*, Cambridge: Cambridge University Press, 1999.

Bertocchi, M., S. L. Schwartz and W. T. Ziemba. *Optimizing the Aging, Retirement, and Pensions Dilemma.* Hoboken: John Wiley and Sons, 2010.

Börsch-Supan, A. Risiken im Lebenszyklus: Theorie und Evidenz. *Perspektiven der Wirtschaftspolitik*, 6: pp. 449–469, 2005.

Branger, N., H. Kraft and C. Meinerding. What is the Impact of Stock Market Contagion on an Investor's Portfolio Choice? *Insurance: Mathematics and Economics*, 45: pp. 94–112, 2009.

Campbell, J. Y. Household Finance. *The Journal of Finance*, 61: pp. 1553–1604, 2006.

Campbell, J. Y. and L. Viceira. *Strategic Asset Allocation: Portfolio Choice for Long-Term Investors.* Oxford: Oxford University Press, 2002.

Capgemini Merrill Lynch. World Wealth Report 2009. Address: http://www.ml.com/media/113831.pdf. (printout of January 5, 2011), 2009.

Conrad, J., M. N. Gultekin and G. Kaul. Asymmetric Predictability of Conditional Variances. *The Review of Financial Studies*, 4: pp. 597–622, 1991.

Consigli, G. Tail Estimation and Mean-VAR Portfolio Selection in Markets Subject to Financial Instability. *Journal of Banking and Finance*, 26: pp. 1355–1382, 2002.

Consigli, G. Asset-Liability Management for Individual Investors. In S. A. Zenios and W. T. Ziemba, (eds.): *Handbook of Asset and Liability Management: Volume 2: Applications and Case Studies*, Amsterdam: Elsevier, 2007.

Consiglio, A., F. Cocco and S. A. Zenios. "www.Personal_Asset_Allocation." *Interfaces*, 34: pp. 287–302, 2004.

Consiglio, A., F. Cocco and S. A. Zenios. Scenario Optimization Asset and Liability Modelling for Individual Investors. *Annals of Operations Research*, 152: pp. 167–191, 2007.

Creditreform. "Schuldneratlas Deutschland 2010." Address: http://www.creditreform.de/Deutsch/Creditreform/Presse/Archiv/SchuldnerAtlas_Deutschland/2010/Analyse_SchuldnerAtlas_Deutschland_2010.pdf. (printout of April 29, 2010), 2010.

Creditreform. "Insolvenzen in Europa: Jahr 2010/2011." Address: http://www.creditreform.de/Deutsch/Creditreform/Presse/Archiv/Insolvenzen_Europa/2010-11/Insolvenzen_in_Europa_2010-11_de.pdf. (printout of April 29, 2011), 2011.

Curtiss, J. H. A Note on the Theory of Moment Generating Functions. *The Annals of Mathematical Statistics*, 13: pp. 430–433, 1942.

Das, S. R. and R. Uppal. Systemic Risk and International Portfolio Choice. *The Journal of Finance*, 59: pp. 2809–2834, 2004.

Datamonitor. Wealth Management Business Models in 20 Years: The Long Term View, 2006.

Elnekave, R. The Mathematics of Savings and Retirement Planning. *The Journal of Wealth Management*, 10: pp. 87–92, 2007.

European Central Bank. "Survey Data on Household Finance and Consumption." (European Household Finance and Consumption Network) Address: http://www.ecb.int/pub/pdf/scpops/ecbocp100.pdf. (printout of January 11, 2011), 2009.

Federal Reserve Board. "Household Debt Service and Financial Obligation Ratios." (Statistics) Address: http://www.federalreserve.gov/releases/housedebt/default.htm. (printout of January 16, 2012), 2012.

Financial Engines. "Retirement Help for Life." (an independent investment advisor) Address: www.financialengines.com (printout of January 19, 2012), 2012.

Gihman, I. I. and A. V. Skorohod. *Stochastic Differential Equations*. Würzburg: Springer Verlag, 1970.

Häger, J. and A. Raffelsberger. Risikomanagement für vermögende Privatpersonen. In M. Spreiter (ed.): *Private Banking: Kundenbindung und Ertragssteigerung in der Praxis*, Wiesbaden: Gabler Verlag, 2005.

Handelsblatt. Trichet bangt um Euro-Zone (News am Abend), 2011.

Höcht, S., N. Kah Hwa, C. G. Rösch and R. Zagst. "Asset Liability Management in Financial Planning." *The Journal of Wealth Management*, 11: pp. 29–46, 2008.

Huang, W., Q. Liu, S. G. Rhee and L. Zhang. "Return Reversals, Idiosyncratic Risk, and Expected Returns." *The Review of Financial Studies*, 23: pp. 147–168, 2010.

Ingersoll J. E. *Theory of Financial Decision Making*. Savage: Rowman and Littlefield, 1987.

International Monetary Fund. "IMF Urges Rethink of How to Manage Global Systemic Risk." (IMF Survey Magazine: Policy) Address: www.imf.org/external/pubs/ft/survey/so/2009/pol030609a.htm (printout of April 12, 2011), 2009.

Jarrow, R. A. and E. R. Rosenfeld. Jump Risks and the Intertemporal Capital Asset Pricing Model. *Journal of Business*, 57: pp. 337–351, 1984.

Jeanblanc-Picqué, M. and M. Pontier. Optimal Portfolio for a Small Investor in a Market Model with Discontinuous Prices. *Applied Mathematics and Optimization*, 22: pp. 287–310, 1990.

Kushner, H. J. *Stochastic Stability and Control*. New York: Academic Press, 1967.

Liu, J., F. A. Longstaff and J. Pan. Dynamic Asset Allocation with Event Risk. *The Journal of Finance*, 58: pp. 231–259, 2003.

Malliaris, A. G. and W. A. Brock. *Stochastic Methods in Economics and Finance*. Fifth edition, Amsterdam: Elsevier, 1990.

McKinsey Global Institute. *The New Power Brokers: How Oil, Asia, Hedge Funds, and Private Equity are Shaping Global Capital Markets*. Address: http://www.econ.pucrio.br/mgarcia/Seminario/textos_preliminares/MGI_New_Power_Brokers.pdf (printout of May 27, 2012), 2007.

Merton, R. C. Optimum Consumption and Portfolio Rules in a Continuous-Time Model. *Journal of Economic Theory*, 3: pp. 373–413, 1971.

Merton, R. C. Option Pricing when Underlying Stock Returns are Discontinuous. *Journal of Financial Economics*, 3: pp. 125–144, 1976.

Merton, R. C. *Continuous-Time Finance*. Third edition, Oxford: Blackwell, 1993.

Mizrach, B. "Jump and Cojump Risk in Subprime Home Equity Derivatives." Working Paper, Rutgers University, 2008.

Mulvey, J. M. Research on Alternative Investments at Princeton. *Quantitative Finance*, 2: pp. 174–176, 2002.

Mulvey, J. M. "Essential Portfolio Theory." (A Rydex Investments White Paper) Address: http://www.rydexfundsfp.com/pdf/whitepaper_EPT. pdf. (printout of February 23, 2007), 2005.

Mulvey, J. M. and W. T. Ziemba. Asset and Liability Management Systems for Long-Term Investors: Discussion of the Issues. In W. T. Ziemba and J. M. Mulvey, (eds.): *Worldwide Asset and Liability Modeling*, Cambridge: Cambridge University Press, 1999.

Piazzesi, M. Bond Yields and the Federal Reserve. *Journal of Political Economy*, 113: pp. 311–344, 2005.

PricewaterhouseCoopers. "Global Private Banking/Wealth Management Survey 2003." Address: http://www.pwc.com/images/gx/eng/fs/pbwm03execsummary. pdf. (printout of August 17, 2006), 2003.

PricewaterhouseCoopers. "Uncertainty Tamed? The Revolution of Risk Management in the Financial Services Industry." Address: http://www.pwc.com/fr/ pwc_ pdf/ pwc_ eiurisk_ 2004-07-27.pdf. (printout of August 17, 2006), 2004.

PricewaterhouseCoopers. "Global Private Banking/Wealth Management Survey 2005." Address: http://www.pwc.com/at/pdf/publikationen/Global_Private_ Banking_Survey_2005.pdf. (printout of September 8, 2009), 2005.

PricewaterhouseCoopers. "Global Private Banking/Wealth Management Survey 2009." Address: http://www.pwc.com/gx/en/private-banking-wealth-mgmt-survey/pbwm09.html. (printout of January 18, 2011), 2009.

Reiche, L. "Die "Mystery-Shopper" und die Superreichen." Address: http://www. manager-magazin.de/geld/geldanlage/0,2828,387714-7,00.html. (printout of April 12, 2006), 2005.

Ross, S. A. Mutual Fund Separation in Financial Theory: The Separating Distributions. *Journal of Economic Theory*, 17: pp. 254–286, 1978.

Rudolf, M. Market Risk and Credit Risk of Old-age Security Systems. In B. Scherer (ed.): *Asset and Liability Management Tools*, Pamplona: Risk Waters Group, 2004.

Rudolf, M. Lehren aus der Finanzmarkt- und Wirtschaftskrise für das Finanzsystem. *Zeitschrift für Controlling und Management*, pp. 2–3, 2010.

Rudolf, M. "Sovereigns as Counterparties: Systemic Risk." Course Materials, WHU-Otto Beisheim School of Management, 2012.

Rudolf, M. W. T. Ziemba. Intertemporal Surplus Management. *Journal of Economic Dynamics and Control*, 28: pp. 975–990, 2004.

Schaubach, P. *Family Office im Private Wealth Management: Konzeption und empirische Untersuchung aus Sicht der Vermögensinhaber.* Bad Soden: Uhlenbruch Verlag, 2003.

Scherer, B. (ed.). *Asset and Liability Management Tools.* Pamplona: Risk Waters Group, 2004.

Sharpe, W. F. and L. G. Tint. Liabilities: A New Approach. *Journal of Portfolio Management*, 16: pp. 5–10, 1990.

Soteriou, A. C. and S. A. Zenios. "Delivering e-Banking Services: An Emerging Internet Business Model and a Case Study." Working Paper, Universität Zypern, 2003.

Spreiter, M. (ed.). *Private Banking: Kundenbindung und Ertragssteigerung in der Praxis.* Wiesbaden: Gabler Verlag, 2005.
von Nitzsch, R. and A. Hackethal. "Kundennutzenorientierte Gestaltung der Anlageberatung im Privatkundengeschäft." Working Paper, Forschungsinstitut für Asset Management, 2005.
Zenios, S. A. and W. T. Ziemba (eds.). *Handbook of Asset and Liability Management: Volume 2: Applications and Case Studies.* Amsterdam: Elsevier, 2007.
Ziemba, W. T. and J. M. Mulvey (eds.). *Worldwide Asset and Liability Modeling.* Second edition, Cambridge: Cambridge University Press, 1999.
Ziemba, W. T. *The Stochastic Programming Approach to Asset, Liability, and Wealth Management.* Charlottesville: Research Foundation of CFA Institute, 2003.

Chapter 5

Jump-Diffusion Risk-Sensitive Benchmarked Asset Management

Mark Davis* and Sébastien Lleo†

Summary

In earlier works (Davis and Lleo, 2011b; Davis and Lleo, 2012), we showed that jump-diffusion risk-sensitive asset management problem without benchmark admit a unique classical ($C^{1,2}$) solution. In this article we extend these solution techniques to a benchmarked asset management problem with jumps. Benchmarked asset management problems are highly relevant to the financial industry: most investment funds have a benchmark, such as a financial index or a customized portfolio, against which their performance is assessed. We show here under two different sets of assumptions that the stochastic control problem associated with the benchmarked aset management problem admits a unique $C^{1,2}$ solution and that the optimal investment strategy exists and is unique.

Keywords: Asset management, benchmarks, risk-sensitive stochastic control, jump diffusion processes, Poisson point processes, Lévy processes, Hamilton Jacobi Bellman partial integro-differential equation, policy improvement, parabolic PDE, classical solutions, viscosity solutions.

1 Introduction

Historically, the portfolio selection literature has taken the perspective of an investor who aims at maximizing either the risk-adjusted return or the utility generated by a portfolio of assets. This specification may work well for individual investors and a number of fund managers, but it is inadequate for institutional investors whose performance is assessed

*Department of Mathematics, Imperial College London, London SW7 2AZ, England, mark.davis@imperial.ac.uk.

†Finance Department and Value and Persuasion Research Centre, Reims Management School, 59 rue Pierre Taittinger, 51100 Reims, France, sebastien.lleo@reims-ms.fr.

with respect to an investment benchmark. In these benchmarked investment problems, the investor aims at selecting an asset allocation to outperform a given investment benchmark such as a financial index (the S&P 500, MSCI World or BarCap Aggregate Bond Index) or a custom-designed investment benchmark (i.e., a mix of 30% S&P 500, 30% MSCI World and 40% BarCap Aggregate Bond Index).

Several authors have addressed the benchmarked investment problem in a continuous time setting using diffusion processes. In a complete market setting, Browne (1999) showed that solving the benchmarked investment problem was equivalent to solving a nonlinear Dirichlet problem and addressed a number of related questions. Pham (2003) developed a risk-sensitive approach similar to ours but limited his analysis to one risky asset and one factor and solved the resulting problem using a large deviation approach. Davis and Lleo (2008, 2011a) used a stochastic control-orientated approach due to Kuroda and Nagai (2002) to solve explicitly the risk-sensitive benchmarked investment management problem. Benchmarked problems are also related to asset and liability management (ALM) problems (see for example Rudolph and Ziemba, 2004; Benk, 2012), with a subtle difference: while benchmarked investors attempt to outperform their benchmark, ALM investors try to ensure that a given liability is met. The most a benchmarked investor has to lose if they underperform are asset management mandates, clients, and reputation. On the other hand, failure by an ALM investor to at least replicate the liability could result in bankruptcy.

We consider a benchmark investment management problem in a jump-diffusion setting under two sets of assumptions on the coefficients of the processes. In both cases, the asset prices are allowed to depend on a random factor process X_t, the components of which can be interpreted either as macroeconomic factors or simply as a statistical representation of the uncertainty of asset returns. Under the first set of assumptions, the growth rates of assets are an affine function of the factors, while the asset diffusions are constant and the asset jumps are independent from the factors. The factors are modelled as a Gaussian diffusion process. Under the second set of assumptions, both the asset growth rates and volatility are allowed to depend on the factors, making it possible to incorporate stochastic volatility in our analysis. Moreover, the factors are now modelled as jump-diffusion processes.

We formulate the benchmarked investment management problem as a risk-sensitive control problem. In risk-sensitive control, the decision maker's objective is to select a control policy $h(t)$ to maximize the criterion

$$J(\theta, h) := -\frac{1}{\theta} \ln \mathbf{E}[e^{-\theta R_T(h)}] \tag{1}$$

where $R_T(h)$ is a reward function at a fixed final time T corresponding to a control process h, and the exogenous parameter $\theta > 0$ represents the decision maker's degree of risk aversion. A Taylor expansion of J around $\theta = 0$ gives

$$J(\theta, h) = \mathbf{E}[R_T(h)] - \frac{\theta}{2}\mathbf{Var}[R_T(h)] + O(\theta^2). \tag{2}$$

In optimal investment problems (as in Bielecki and Pliska, 1999 or Kuroda and Nagai, 2002) it is customary to take $R_T = \log V_T$, where V_T is the value of the investment portfolio corresponding to an asset allocation strategy h; then (2) shows that risk-sensitive control amounts to 'dynamic Markowitz': we maximize the expected return subject to a constraint on the variance of returns. In benchmarked investment management, $R_T = \ln V_T/L_T$, measures the outerpformance of the investor's portfolio with respect to the benchmark at a level L_T. With this choice of reward function, risk-sensitive control amounts to maximizing the expected outperformance subject to a constraint on the variance of outperformance.

The solution technique we present here extends the approaches proposed by Davis and Lleo (2011b, 2012) in an investment management context without benchmark. The first step is to use a change-of-measure idea introduced by Kuroda and Nagai (2002) that reduces the risk-sensitive optimization problem to a stochastic control problem in the factor process. Under the first set of assumptions, affine asset growth rates with constant volatility, the factor process X_t has no jumps and the associated Bellman equation is a partial differential equation (PDE) which can be shown to admit a unique classical ($C^{1,2}$) solution.

Under the second set of assumptions, we now have a fully nonlinear controlled jump-diffusion, and the Bellman equation is a partial integro-differential (PIDE) for which no analytical solution exists. In such situation, viscosity solutions are generally used to show that the value function is the unique continuous solution of the Hamilton Jacobi Bellman partial integro-differential equation (HJB PIDE); see in particular Crandall Ishii and Lions (1992), Barles and Imbert (2008) for an overview of viscosity solutions and Fleming and Soner (2006), Øksendal and Sulem (2005) or Touzi (2002) for a discussion of their application to stochastic control, as well as Davis and Lleo (2010) for a viscosity approach to risk-sensitive asset management. In the context of optimal control, a limitation of viscosity solutions is that they are weak solutions: they do not generally satisfy the smoothness assumption required to obtain a sufficient condition via a classical verification theorem. Proving that the HJB PIDE admits a unique classical ($C^{1,2}$) solution requires the development of a more sophisticated argument combining viscosity solutions and classical solutions. The key references are due to Pham (1998), Davis, Guo and Wu (2009) and Davis and Lleo (2012). This line of reasoning differs from more abstract results on classical solutions to PIDEs (see for example Mikulyavichyus and Pragarauskas, 1994) in that it is directly tailored to the problem at hand. It is therefore of particular interest for the type of applied problems related to optimal control.

The article is organised as follows. We introduce the analytical setting in Section 2 before formulating the control problem in Section 3. The main result, Theorem 3.1, is stated in Section 3.3 and proved the next three sections: in Sections 4, 5 and 6, we respectively address the questions of the existence of classical solution to the HJB PIDE under both sets of assumptions, and present a verification theorem establishing optimality of our candidate control.

2 Analytical setting

The asset market comprises m risky securities S_i, $i = 1, \dots m$ and a money market account process S_0. The investor selects an asset allocation with the objective to outperform a given investment benchmark L. The dynamics of the assets and benchmark depend on n factors $X_1(t), \dots, X_n(t)$ which follow the dynamics given in the jump diffusion equation (5) below.

Let $(\Omega, \mathcal{F}, \mathbb{P})$ be the underlying probability space. On this space is defined an $\mathbb{R}^M$-valued $(\mathcal{F}_t)$-Brownian motion $W(t)$ with components $W_k(t)$, $k = 1, \dots, M$, and $M := m + n$. Moreover, let N be a $(\mathcal{F}_t)$-Poisson point process on $(0, \infty) \times \mathbf{Z}$, independent of $W(t)$, where $(\mathbf{Z}, \mathcal{B}_{\mathbf{Z}})$ is a given Borel space. Define

$$\mathfrak{Z} := \{U \in \mathcal{B}_{\mathbf{Z}}, \mathbf{E}[N(t, U)] < \infty \; \forall t\} \tag{3}$$

For notational convenience, we fix throughout the paper a set $\mathbf{Z}_0 \in \mathcal{B}_{\mathbf{Z}}$ such that $\nu(\mathbf{Z} \backslash \mathbf{Z}_0) < \infty$ and define, as in Øksendal and Sulem (2005)

$$\bar{N}(dt, dz) = \begin{cases} N(dt, dz) - \hat{N}(dt, dz) = N(dt, dz) - \nu(dz)dt =: \tilde{N}(dt, dz) & \text{if } z \in \mathbf{Z}_0 \\ N(dt, dz) & \text{if } z \in \mathbf{Z} \backslash \mathbf{Z}_0 \end{cases} \tag{4}$$

For $t \in [0,T]$ let $\mathcal{F}_t$ be the σ-field generated by the Brownian motions $W_k(s)$ and Poisson processes $N(A,s)$ for $k = 1,\ldots,M$, $A \in \mathcal{B}_{\mathbf{Z}}$ and $0 \leq s \leq t$, completed with all null sets of $\mathcal{F}_T$. It is well known that the filtration $\{\mathcal{F}_t\}_{t\in[0,T]}$ satisfies the 'usual conditions'.

2.1 *Factor dynamics*

The factor process $X(t) \in \mathbb{R}^n$ is allowed to have a full jump-diffusion dynamics, satisfying the stochastic differential equation:

$$\begin{aligned} dX(t) &= b(t, X(t^-))dt + \Lambda(t, X(t))dW(t) \\ &\quad + \int_{\mathbf{Z}} \xi(t, X(t^-), z)\bar{N}(dt, dz), \quad X(0) = x_0 \in \mathbb{R}^n. \end{aligned} \tag{5}$$

We will consider two different sets of standing assumptions:

- affine drift and constant diffusion with no jumps in the factor;
- standard stochastic control assumptions of bounded drift and diffusion, with jumps in both assets and factors.

Assumption 2.1 (Affine Drift and Constant Diffusion).

(i) $b(t,x) = b + Bx$ where $b \in \mathbb{R}^n$, $B \in \mathbb{R}^{n\times n}$
(ii) $\Lambda \in \mathbb{R}^{n\times n}$,
(iii) $\Lambda\Lambda' > 0$
(iv) $\xi(t,x,z) = 0$

Assumption 2.2 (Standard Control Assumptions).

(i) The function $b : [0,T] \times \mathbb{R}^n \to \mathbb{R}^n$ is bounded and Lipshitz continuous

$$|b(t,y) - b(s,x)| \leq K_b(|t-s| + |y-x|) \tag{6}$$

for some constant $K_b > 0$.
(ii) The function $\Lambda : [0,T] \times \mathbb{R}^n \to \mathbb{R}^{n\times M}$ is bounded and Lipschitz continuous,

$$|\Lambda(t,y) - \Lambda(s,x)| \leq K_\Lambda(|t-s| + |y-x|) \tag{7}$$

for some constant $K_\Lambda > 0$.

(iii) There exists $\eta_\Lambda > 0$ such that

$$\zeta'\Lambda\Lambda'(t,x)\zeta \geq \eta_\Lambda|\zeta|^2 \tag{8}$$

for all $\zeta \in \mathbb{R}^n$

(iv) There exists $K'_b > 0$ and $K'_\Lambda > 0$ such that

$$|b_t| + |b_x| \leq K'_b \tag{9}$$

$$|\Lambda_t| + |\Lambda_x| \leq K'_\Lambda \tag{10}$$

(v) The function $\xi : [0,T] \times \mathbb{R}^n \times \mathbf{Z} \to \mathbb{R}$ is bounded and Lipshitz continuous, i.e.

$$|\xi(t,y,z) - \xi(s,x,z)| \leq K_\xi(|t-s| + |y-x|) \tag{11}$$

for some constant $K_\xi > 0$.

(vi) The vector valued function $\xi(t,x,z)$ satisfies:

$$\int_{\mathbf{Z}} |\xi(t,x,z)|\nu(dz) < \infty, \quad \forall (t,x) \in [0,T] \times \mathbb{R}^n \tag{12}$$

The minimal condition on ξ under which the factor equation (5) is well posed is

$$\int_{\mathbf{Z}_0} |\xi(t,x,z)|^2\nu(dz) < \infty,$$

see Definition II.4.1 in Ikeda and Watanabe (1981). However, for this paper it is essential to impose the stronger condition (12) in order to connect the viscosity solution of the HJB PIDE to the viscosity solution of a related parabolic PDE. The same condition is imposed in Davis, Guo and Wu (2009).

***Remark* 2.1.** Note that (9) and (10) follow respectively from (6) and (7) when b and λ are differentiable.

2.2 *Asset market dynamics*

Let S_0 denote the wealth invested in the money market account with dynamics given by the equation:

$$\frac{dS_0(t)}{S_0(t)} = a_0(t, X(t))dt, \quad S_0(0) = s_0 \tag{13}$$

and let $S_i(t)$ denote the price at time t of the ith security, with $i = 1, \ldots, m$. The dynamics of risky security i can be expressed as:

$$\frac{dS_i(t)}{S_i(t^-)} = [a(t, X(t^-))]_i dt + \sum_{k=1}^{N} \Sigma_{ik}(t, X(t))dW_k(t) + \int_{\mathbf{Z}} \gamma_i(t, z)\bar{N}(dt, dz), \quad S_i(0) = s_i, \quad i = 1, \ldots, m \tag{14}$$

The standing assumptions are as follows.

Assumption 2.3 (Affine Drift and Constant Diffusion).

(i) $a_0(t, x) = a_0 + A_0 x$ where $a_0 \in \mathbb{R}^m$, $A_0 \in \mathbb{R}^{m \times n}$
(ii) $a(t, x) = a + Ax$ where $a \in \mathbb{R}^m$, $A \in \mathbb{R}^{m \times n}$
(iii) $\Sigma(t, x) = \Sigma$ with $\Sigma \in \mathbb{R}^{m \times M}$
(iv) $\Sigma\Sigma' > 0$
(v) $\gamma(t, z) = \gamma(z) \in \mathbb{R}^m$ satisfies:

$$\int_{\mathbf{Z}_0} |\gamma(z)|^2 \nu(dz) < \infty \tag{15}$$

Assumption 2.4 (Standard Control Assumptions).

(i) the function a_0 defined as $a_0 : [0, T] \times \mathbb{R}^n \to \mathbb{R}$ is bounded, of class $C^{1,1}([0, T] \times \mathbb{R}^n)$ and is Lipshitz continuous

$$|a_0(t, y) - a_0(s, x)| \leq K_0(|t - s| + |y - x|) \tag{16}$$

for some constant $K_0 > 0$.

(ii) There exists $K_0' > 0$ such that

$$\left|\frac{\partial a_0}{\partial t}\right| + |Da_0| \leq K_0' \tag{17}$$

(iii) the function $a : [0, T] \times \mathbb{R}^n \to \mathbb{R}^m$ is bounded and Lipshitz continuous, i.e.

$$|a(t, y) - a(s, x)| \leq K_a(|t - s| + |y - x|) \tag{18}$$

for some constant $K_a > 0$.

(iv) the function $\Sigma : [0, T] \times \mathbb{R}^n \to \mathbb{R}^{m \times M}$ is bounded and Lipshitz continuous, i.e.

$$|\Sigma(t, y) - \Sigma(s, x)| \leq K_\Sigma(|t - s| + |y - x|) \tag{19}$$

for some constant $K_\Sigma > 0$.

(v) There exists $\psi_\Sigma > 0$ such that

$$\zeta'\Sigma\Sigma'(t,x)\zeta \geq \psi_\Sigma|\zeta|^2 \tag{20}$$

for all $\zeta \in \mathbb{R}^m$

(vi) There exists $K'_a > 0$ and $K'_\Sigma > 0$ such that

$$|a_t| + |a_x| \leq K'_a \tag{21}$$

$$|\Sigma_t| + |\Sigma_x| \leq K'_\Sigma \tag{22}$$

(vii) The function $\gamma : [0,T] \times \mathbf{Z} \to \mathbb{R}^m$ is bounded, continuous and satisfies the growth condition

$$|\gamma(t,z) - \gamma(s,z)| \leq K_\gamma(|t-s|) \tag{23}$$

for some constant $K_\gamma > 0$.

(viii) The vector valued function $\gamma(t,z)$ satisfy:

$$\int_{\mathbf{Z}_0} |\gamma(t,z)|^2 \nu(dz) < \infty, \quad \forall (t,x) \in [0,T] \times \mathbb{R}^n \tag{24}$$

(ix) We also require

$$|\Lambda\Sigma'(t,y) - \Lambda\Sigma'(s,x)| \leq K_{\Lambda\Sigma}(|t-s| + |y-x|) \tag{25}$$

for some constant $K_{\Lambda\Sigma} > 0$

***Remark* 2.2.** Note that (21) and (22) follow respectively from (18) and (19) when b and λ are differentiable.

2.3 *Benchmark modelling*

We assume that the dynamics of the benchmark follows the jump-diffusion process the asset prices. Specifically,

$$\frac{dL(t)}{L(t)} = c(t, X(t^-))dt + \varsigma'(t, X(t))dW(t) + \int_{\mathbf{Z}} \eta(t,z)\bar{N}(dt,dz), \quad L(0) = l$$

where α is a scalar constant, β is a n-element column vector, and γ is a N-element column vector.

Assumption 2.5 (Affine Drift and Constant Diffusion).

(i) $c(t,x) = c + Cx$ where $c \in \mathbb{R}$, $C \in \mathbb{R}^n$

(ii) $\varsigma(t,x) = \varsigma$ with $\varsigma \in \mathbb{R}^M$

(iii) $\eta(t,z) = \eta(z) \in \mathbb{R}$ satisfies:

$$\int_{\mathbf{Z}_0} |\eta(z)|^2 \nu(dz) < \infty \tag{26}$$

Assumption 2.6 (Standard Control Assumptions).

(i) the function $c : [0,T] \times \mathbb{R}^n \to \mathbb{R}^m$ is bounded and Lipshitz continuous, i.e.

$$|c(t,y) - c(s,x)| \leq K_c(|t-s| + |y-x|) \tag{27}$$

for some constant $K_c > 0$.

(ii) the function $\varsigma : [0,T] \times \mathbb{R} \to \mathbb{R}^M$ is bounded and Lipshitz continuous, i.e.

$$|\varsigma(t,y) - \varsigma(s,x)| \leq K_\varsigma(|t-s| + |y-x|) \tag{28}$$

for some constant $K_\varsigma > 0$.

(iii) $\varsigma(t,x) > 0 \forall (t,x) \in [0,T] \times \mathbb{R}^n$

(iv) There exists $K_c' > 0$ and $K_\varsigma' > 0$ such that

$$|c_t| + |c_x| \leq K_c' \tag{29}$$

$$|\varsigma_t| + |\varsigma_x| \leq K_\varsigma' \tag{30}$$

(v) the function $\eta : [0,T] \times \mathbf{Z} \to \mathbb{R}$ is bounded, continuous and satisfies the growth condition

$$|\eta(t,z) - \eta(s,z)| \leq K_\eta\left(|t-s|\right) \tag{31}$$

for some constant $K_\eta > 0$.

(vi) The vector valued function $\eta(t,z)$ satisfy:

$$\int_{\mathbf{Z}_0} |\eta(t,z)|^2 \nu(dz) < \infty, \quad \forall (t,x) \in [0,T] \times \mathbb{R}^n \tag{32}$$

(vii) We also require

$$|\varsigma'\Sigma'(t,y) - \varsigma'\Sigma'(s,x)| \leq K_{\varsigma\Sigma}\left(|t-s| + |y-x|\right) \tag{33}$$

for some constant $K_{\varsigma\Sigma} > 0$

We need a further assumption relating the jumps in the asset prices, in the benchmark level and in the factors.

Assumption 2.7. $\gamma(t,z)\xi'(t,x,z) = \eta\xi'(t,x,z) = 0$, $\forall(t,x,z) \in [0,T] \times \mathbb{R}^n \times \mathbf{S}$.

This assumption implies that there are no simultaneous jumps in the factor process and the asset price process. This imposes some restriction, but appears essential in the argument below.

2.4 *Portfolio dynamics*

The function γ appearing in (14) is assumed to satisfy the following conditions.

Assumption 2.8. Define

$$\mathbf{S} := \text{supp}(\nu) \in \mathcal{B}_{\mathbf{Z}}$$

and

$$\tilde{\mathbf{S}} := \text{supp}(\nu \circ \gamma^{-1}) \in \mathcal{B}(\mathbb{R}^m)$$

where supp$(\cdot)$ denotes the support of the measure, and let $\prod_{i=1}^{m}[\gamma_i^{min}, \gamma_i^{max}]$ be the smallest closed hypercube containing $\tilde{\mathbf{S}}$. We assume that $\gamma(t,z) \in \mathbb{R}^m$ satisfies

$$-1 \leq \gamma_i^{min} \leq \gamma_i(t,z) \leq \gamma_i^{max} < +\infty, \quad i = 1, \ldots, m$$

and

$$\gamma_i^{min} < 0 < \gamma_i^{max}, \quad i = 1, \ldots, m$$

for $i = 1, \ldots, m$.

We also assume that $\eta(t,z) \in \mathbb{R}$ satisfies

$$-1 < \eta^{min} \leq \eta_i(t,z) \leq \eta^{max} < +\infty$$

and

$$\eta^{min} < 0 < \eta^{max}$$

Define the set $\mathcal{J}_0$ as

$$\mathcal{J}_0 := \{h \in \mathbb{R}^m : 1 + h'\psi > 0 \quad \forall \psi \in \tilde{\mathbf{S}}\} \tag{34}$$

For a given $z \in \mathbf{S}$, the equation $h'\gamma(t,z) = -1$ describes a hyperplane in $\mathbb{R}^m$. Under Assumption 2.8, $\mathcal{J}_0$ is a convex subset of $\mathbb{R}^m$ for all $(t,x) \in [0,T] \times \mathbb{R}^n$.

Let $\mathcal{G}_t := \sigma((S(s), X(s)), 0 \leq s \leq t)$ be the sigma-field generated by the security and factor processes up to time t.

Definition 2.1. *An $\mathbb{R}^m$-valued control process $h(t)$ is in class $\mathcal{H}_0$ if the following conditions are satisfied:*

(i) *$h(t)$ is progressively measurable with respect to $\{\mathcal{B}([0,t]) \otimes \mathcal{G}_t\}_{t\geq 0}$ and is càdlàg;*
(ii) *$h(t) \in \mathcal{J}_0 \quad \forall t$ a.s.*

We note that under Assumption 2.8, a control process $h(t)$ satisfying (ii) is bounded.

By the budget equation, the proportion invested in the money market account is equal to $h_0(t) = 1 - \sum_{i=1}^{m} h_i(t)$. This implies that the wealth $V(t)$ of the investor in response to an investment strategy $h(t) \in \mathcal{H}_0$, follows the dynamics

$$\frac{dV(t)}{V(t^-)} = (a_0(t, X(t)))dt + h'(t)\hat{a}(t, X(t))dt + h'(t)\Sigma(t, X(t))dW_t + \int_{\mathbf{Z}} h'(t)\gamma(t,z)\bar{N}(dt,dz) \tag{35}$$

where $V(0) = v_0$ is the initial endowment and $\hat{a} := a - a_0\mathbf{1}$, $\mathbf{1} \in \mathbf{R}^m$ denotes the m-element unit column vector.

We define the asset portfolio's log excess return its benchmark, $F(t)$, as:

$$F(t) = \ln\frac{V(t)}{L(t)} = \ln V(t) - \ln L(t), \quad F(0) = f_0$$

By Itô's lemma,

$$\begin{aligned} F(t) = \int_0^t \Bigg[&(a_0(s,x) - c(s,x)) + h'\hat{a}(s,x) - \frac{1}{2}(h'\Sigma\Sigma'(s,x)h - \varsigma'\varsigma(s,x)) \\ &+ \int_{\mathbf{Z}} \left\{ \ln\left(\frac{1 + h'(s)\gamma(s,z)}{1+\eta(s,z)}\right) - (h'(s)\gamma(s,z) - \eta(s,z)) \right\} \nu(dz) \Bigg] ds \\ &+ \int_0^t (h'(s)\Sigma - \varsigma')(s,x)dW(s) \\ &+ \int_0^t \int_{\mathbf{Z}} \ln\left(\frac{1+h'(s)\gamma(s,z)}{1+\eta(s,z)}\right) \bar{N}(dt,dz) \end{aligned} \tag{36}$$

2.5 *Investment constraints*

We consider $r \in \mathbb{N}$ fixed investment constraints expressed in the form

$$\Upsilon' h(t) \leq \upsilon \tag{37}$$

where $\Upsilon \in \mathbb{R}^m \times \mathbb{R}^r$ is a matrix and $\upsilon \in \mathbb{R}^r$ is a column vector. For the constrained control problem to be sensible, we need Υ and υ to satisfy the following condition:

Assumption 2.9. The system

$$\Upsilon' y \leq \upsilon$$

for the variable $y \in \mathbb{R}^m$ admits at least two solutions.

We define the feasible region $\mathcal{J}$ as

$$\mathcal{J} := \{h \in \mathcal{J}_0 : \Upsilon' h \leq \upsilon\} \tag{38}$$

The feasible region $\mathcal{J}$ is a a convex subset of $\mathbb{R}^m$ and as a result of Assumption 2.9, $\mathcal{J}$ has at least one interior point.

2.6 *Problem formulation*

The class $\mathcal{A}$ of admissible investment strategies is defined as follows.

Definition 2.2. *A control process $h(t)$ is in class $\mathcal{A}$ if the following conditions are satisfied:*

(i) $h \in \mathcal{H}$, *where*

$$\mathcal{H} := \{h(t) \in \mathcal{H}_0 : h(t) \in \mathcal{J} \ \forall t \in [0, T], \text{a.s.}\} \tag{39}$$

(ii) $\mathbf{E}\chi^h(T) = 1$ *where $\chi^h(t)$ is the Doléans exponential defined for $t \in [0, T]$ by*

$$\begin{aligned}\chi^h(t) := \exp\Bigg\{ &-\theta \int_0^t [h(s)'\Sigma(s, X(s)) - \varsigma'(s, X(s)] dW_s \\ &- \frac{1}{2}\theta^2 \int_0^t [h(s)'\Sigma(s, X(s)) - \varsigma'(s, X(s)] \\ &\times [h(s)'\Sigma(s, X(s)) - \varsigma'(s, X(s)]' ds\end{aligned}$$

$$+ \int_0^t \int_{\mathbf{Z}} \ln\left(1 - G(s,z,h(s))\right) \tilde{N}(ds,dz)$$
$$+ \int_0^t \int_{\mathbf{Z}} \left\{\ln\left(1 - G(s,z,h(s))\right) + G(s,z,h(s))\right\} \nu(dz)ds \Big\}, \tag{40}$$

with

$$G(t,z,h) = 1 - \left(\frac{1 + h'\gamma(t,z)}{1 + \eta(t,z)}\right)^{-\theta} \tag{41}$$

The investor's objective is to maximise the risk-sensitive criterion $J(h,v)$ of (1) with $R_T = \ln F(T)$. From (35) and the general Itô formula we find that the term $e^{-\theta \ln F(T)}$ can be expressed as

$$e^{-\theta \ln F(T)} = f_0^{-\theta} \exp\left\{\theta \int_0^T g(t, X_t, h(t))dt\right\} \chi^h(T) \tag{42}$$

where

$$\begin{aligned} g(t,x,h) = & \frac{1}{2}(\theta+1)h'\Sigma\Sigma'(t,x)h - \theta h'\Sigma\varsigma(t,x) + \frac{1}{2}(\theta-1)\varsigma'\varsigma'(t,x) \\ & - (a_0(t,x) - c(t,x)) - h'\hat{a}(t,x) \\ & + \int_{\mathbf{Z}} \left\{ \frac{1}{\theta}\left[\left(\frac{1 + h'\gamma(t,z)}{1+\eta(t,z)}\right)^{-\theta} - 1\right] \right. \\ & \left. + \ (h'\gamma(t,z) - \eta(t,z)) 1_{\mathbf{Z}_0}(z) \right\} \nu(dz) \end{aligned} \tag{43}$$

and the Doléans exponential $\chi^h(T)$ is given by (40).

***Remark* 2.3.** For a given, fixed h, the functional g is bounded and Lipschitz continuous in the state variable x. This follows easily by boundedness and Lipschitz continuity of the coefficients a_0, a, Σ and γ.

For $h \in \mathcal{A}$ and $\theta > 0$ let $\mathbb{P}^h$ be the measure on $(\Omega, \mathcal{F}_T)$ defined via the Radon-Nikodým derivative

$$\frac{d\mathbb{P}^h}{d\mathbb{P}} = \chi^h(T), \tag{44}$$

and let $\mathbf{E}^h$ denote the corresponding expectation. Then from (42) we see that the criterion J is given by

$$J(h, v_0) = \ln v_0 - \frac{1}{\theta} \ln \mathbf{E}^h \left[\exp \left(\theta \int_0^{\prime} g(t, X_t, h(t)) dt \right) \right]. \tag{45}$$

Evidently, the value v_0 plays no role in the optimization process. Throughout the rest of the paper we normalize to $v_0 = 1$. Moreover, under $\mathbb{P}^h$,

$$W_t^h = W_t + \theta \int_0^t (\Sigma(s, X(s))' h(s) - \varsigma) ds$$

is a standard Brownian motion and the $\mathbb{P}^h$-compensated Poisson random measure is given by

$$\begin{aligned}
&\int_0^t \int_{\mathbf{Z}_0} \tilde{N}^h(ds, dz) \\
&\quad = \int_0^t \int_{\mathbf{Z}_0} N(ds, dz) - \int_0^t \int_{\mathbf{Z}_0} \{1 - G(s, z, h(s))\} \nu(dz) ds \\
&\quad = \int_0^t \int_{\mathbf{Z}_0} N(ds, dz) - \int_0^t \int_{\mathbf{Z}_0} \left\{ \left(\frac{1 + h'\gamma(s, z)}{1 + \eta(s, z)} \right)^{-\theta} \right\} \nu(dz) ds
\end{aligned}$$

As a result, under $\mathbb{P}^h$ the factor process $X(s)$, $0 \leq s \leq t$ satisfies the SDE:

$$\begin{aligned}
dX(s) &= f(s, X(s), h(s)) ds + \Lambda(s, X(s)) dW_s^\theta \\
&\quad + \int_{\mathbf{Z}} \xi(s, X(s^-), z) \tilde{N}^h(ds, dz), \quad X(0) = x_0
\end{aligned} \tag{46}$$

where

$$\begin{aligned}
f(t, x, h) &:= b(t, x) - \theta \Lambda (\Sigma(t, x)' h - \varsigma(t, x)) \\
&\quad + \int_{\mathbf{Z}} \xi(t, x, z) \left[\left(\frac{1 + h'\gamma(t, z)}{1 + \eta(t, z)} \right)^{-\theta} - 1_{\mathbf{Z}_0}(z) \right] \nu(dz)
\end{aligned} \tag{47}$$

and b is the $\mathbb{P}$-measure drift of the factor process (see (5)).

***Remark* 2.4.** The drift function f is Lipschitz continuous with coefficient $K_f = K_b + \theta K_{\Lambda\Sigma} + K_\xi K_0$ where $K_0 > 0$ is a constant. For a constant control h the state process $X(t)$ is a Markov process with generator

$$\mathcal{L}u(t,x) := f(t,x,h)'Du + \frac{1}{2}\mathrm{tr}(\Lambda\Lambda'(t,X)D^2u) + \int_{\mathbf{Z}} \{u(x+\xi(t,x,z)) - u(x) - \xi(t,x,z)'Du\}\nu(dz)ds \quad (48)$$

In summary, we have shown that the risk-sensitive asset allocation problem is equivalent to the stochastic control problem of minimizing the cost criterion

$$\tilde{J}(h) = \mathbf{E}^h\left[\exp\left(\theta\int_0^{t'} g(t,X_t,h(t))dt\right)\right] \quad (49)$$

over the control set $\mathcal{A}$ for a controlled process X_t satisfying (in 'weak solution' form) the jump-diffusion SDE (46). The remainder of the paper is devoted to solving the stochastic control problem (46),(49).

3 Dynamic programming and the value function

We will solve the control problem by studying the Hamilton-Jacobi-Bellman (HJB) equation of dynamic programming, which involves embedding the original problem in a family of problems indexed by time-space points (s,x), the starting time and position of the controlled process X_t. The description here is in the same spirit as Bouchard and Touzi (2011).

For fixed $s \in [0,T]$ we define the filtration $\{\mathcal{F}_t^s, t \in [s,T]\}$ by

$$\mathcal{F}_t^s = \sigma\{W_k(r) - W_k(t), N(A,r) - N(A,t), k = 1,\dots,M, A \in \mathcal{B}_{\mathbf{Z}}, s \le r \le t\}$$

and note that $\mathcal{F}_t^s$ is independent of $\mathcal{F}_t$. $X(t)$ will denote the solution of (5) on $[s,t]$ with initial condition $X(s) = x$ and $\mathbb{P}_{s,x}$ the measure on $\mathcal{F}_T^s$ such that $\mathbb{P}_{s,x}[X_s = x] = 1$. The class of admissible controls $\mathcal{A}^s$ is defined analogously to $\mathcal{A}$ above with h adapted to $\mathcal{F}_t^s$, leading to a change of measure on $\mathcal{F}_T^s$ defined by the Radon-Nikodým derivative

$$\frac{d\mathbb{P}_{s,x}^h}{d\mathbb{P}_{s,x}} = \chi_s^h(T).$$

We will now introduce the following two auxiliary criterion functions under the measure $\mathbb{P}^h_{s,x}$:

$$\tilde{I}(s,x,h) = \mathbf{E}^h_{s,x}\left[\exp\left\{\theta\int_s^T g(t,X_t,h(t))dt\right\}\right] \tag{50}$$

$$I(s,x,h) = -\frac{1}{\theta}\ln\tilde{I}(s,x,h). \tag{51}$$

***Remark* 3.1.** The criterion $\tilde{I}$ defined in (50), which is the cost function for our stochastic control problem, can be interpreted as a payoff of 1 at the terminal time T 'discounted' at a stochastic controlled rate of $-\theta g(\cdot)$ (which is however not necessarily ≥ 0).

The corresponding value functions are

$$\tilde{\Phi}(s,x) = \inf_{h\in\mathcal{A}^s}\tilde{I}(s,x,h); \quad \Phi(s,x) = \sup_{h\in\mathcal{A}^s} I(s,x,h). \tag{52}$$

Lemma 3.1. $\tilde{\Phi}(s,x) = \inf_{h\in\mathcal{A}}\tilde{I}(s,x,h)$. *That is, the infimum is unchanged if the class* $\mathcal{A}^s$ *is replaced by the larger class* $\mathcal{A}$.

Proof. This uses exactly the argument of Remark 2, page 958 of Bouchard and Touzi (2011). We condition on the initial filtration and use the independence of $\mathcal{F}_s$ and $\mathcal{F}^s_t$. □

3.1 *The risk-sensitive control problems under* $\mathbb{P}_h$

We will show that the value function Φ defined in (52) satisfies the HJB PIDE

$$\frac{\partial\Phi}{\partial t} + \sup_{h\in\mathcal{J}} L^h(t,x,\Phi,D\Phi,D^2\Phi) = 0 \tag{53}$$

where $\mathcal{J}$ is defined in (38),

$$\begin{aligned} L^h(t,x,u,p,M) = {} & f(t,x,h)'p + \frac{1}{2}\mathrm{tr}(\Lambda\Lambda'(t,x)M) - \frac{\theta}{2}p'\Lambda\Lambda'(t,x)p \\ & - g(t,x,h) + \mathcal{I}_{NL}[t,x,u,p] \end{aligned} \tag{54}$$

with

$$\mathcal{I}_{NL}[t,x,u,p] = \int_{\mathbf{Z}}\left\{-\frac{1}{\theta}(e^{-\theta[u(t,x+\xi(t,x,z))-u(t,x)]} - 1) - \xi(t,x,z)'p\right\}\nu(dz) \tag{55}$$

and subject to the terminal condition (recall our normalization $v_0 = 1$)

$$\Phi(T, x) = 0, \quad x \in \mathbb{R}^n. \tag{56}$$

Condition (12) ensures that $\mathcal{I}_{NL}$ is well defined, at least for bounded u.

For $\tilde{\Phi}$, the corresponding HJB PIDE is

$$\begin{aligned}
&\frac{\partial \tilde{\Phi}}{\partial t}(t,x) + \frac{1}{2}\mathrm{tr}(\Lambda\Lambda'(t,x)D^2\tilde{\Phi}(t,x)) + H(t,x,\tilde{\Phi},D\tilde{\Phi}) \\
&\quad + \int_{\mathbf{Z}} \{\tilde{\Phi}(t,x+\xi(t,x,z)) - \tilde{\Phi}(t,x) - \xi(t,x,z)'D\tilde{\Phi}(t,x)\}\nu(dz) = 0
\end{aligned} \tag{57}$$

subject to terminal condition

$$\tilde{\Phi}(T, x) = 1 \tag{58}$$

where for $r \in \mathbb{R}$, $p \in \mathbb{R}^n$

$$H(s,x,r,p) = \inf_{h \in \mathcal{J}} \{f(s,x,h)'p + \theta g(s,x,h)r\} \tag{59}$$

***Remark* 3.2.** The function H satisfies a Lipschitz condition as well as the linear growth condition

$$|H(s,x,r,p)| \leq C\left(1 + |p|\right), \quad \forall (s,x) \in Q_0$$

The value functions Φ and $\tilde{\Phi}$ are related through the strictly monotone continuous transformation $\tilde{\Phi}(t,x) = \exp\{-\theta\Phi(t,x)\}$. Thus an admissible (optimal) strategy for the exponentially transformed problem is also admissible (optimal) for the risk-sensitive problem. In the remainder of the article, we will refer to the control problem and HJB PIDE related to the value function Φ as the *risk sensitive* control problem and the *risk sensitive* HJB PIDE, and to the control problem and HJB PIDE related to the value function $\tilde{\Phi}$ as the *exponentially transformed* control problem and the *exponentially transformed* HJB PIDE.

3.2 *Properties of the value function* $\tilde{\Phi}$

We start by establishing two *a priori* properties of the value function.

Proposition 3.1. *The exponentially transformed value function $\tilde{\Phi}$ is positive and bounded, i.e. there exists $M > 0$ such that*

$$0 \leq \tilde{\Phi}(t,x) \leq M \quad \forall (t,x) \in [0,T] \times \mathbb{R}^n$$

Proof. By definition,

$$\tilde{\Phi}(t,x) = \inf_{h\in\mathcal{A}} \mathbf{E}^h_{t,x}\left[\exp\left\{\theta\int_t^T g(s,X_s,h(s))ds - \theta\ln v\right\}\right] \geq 0$$

Moreover, the strategy of investing only in the money-market account, i.e. taking $h \equiv 0$ is sub-optimal, and hence

$$\tilde{\Phi}(t,x) \leq \mathbf{E}^0_{t,x} e^{\theta\int_t^T g(X(s),0)ds} = \mathbf{E}^0_{t,x} e^{\theta\int_t^T a_0(s,X(s)ds} \leq e^{\theta\breve{a}_0(T-t)},$$

where $\breve{a}_0$ is a bound for $|a_0(t,x)|$ (see Assumption 2.4(i)). This concludes the proof. □

Proposition 3.2. *The value function $\tilde{\Phi}$ is Lipschitz continuous in the state variable x.*

Proof. See Davis and Lleo (2012) for the proof. □

Proposition 3.3. *Under either Assumption* 2.3(v) *or both Assumption* 2.4(v) *and Assumption* 2.7, *the supremum in* (53), (54) *admits a unique Borel measurable maximizer $\hat{h}(t,x,p)$ for $(t,x,p) \in [0,T] \times \mathbb{R}^n \times \mathbb{R}^n$.*

Proof. We present the proof under Assumption 2.4(v) and Assumption 2.7. The proof under Assumption 2.3(v) follows as a special case.

The supremum in (53) can be expressed as

$$\begin{aligned}
&\sup_{h\in\mathcal{J}} L^h(t,x,u,p,M) \\
&= \sup_{h\in\mathcal{J}}\Bigg\{\left(b(t,x) + \int_{\mathbf{Z}}\xi(t,x,z)\left[\left(\frac{1+h'\gamma(t,z)}{1+\eta(t,z)}\right)^{-\theta} - 1_{\mathbf{Z}_0}(z)\right]\nu(dz)\right)' p \\
&\quad - \theta(h'\Sigma(t,x) - \varsigma')\Lambda'(t,x)p \\
&\quad + \frac{1}{2}\mathrm{tr}(\Lambda\Lambda'(t,x)'M) - \frac{\theta}{2}p'\Lambda\Lambda'(t,x)'p + \mathcal{I}_{NL}[t,x,u,p] \\
&\quad - \frac{1}{2}(\theta+1)h'\Sigma\Sigma'(t,x)h + \theta h'\Sigma\varsigma - \frac{1}{2}(\theta-1)\varsigma'\varsigma(t,x) \\
&\quad + (a_0(t,x) - c(t,x)) + h'\hat{a}(t,x) \\
&\quad - \int_{\mathbf{Z}}\left\{\frac{1}{\theta}\left[\left(\frac{1+h'\gamma(t,z)}{1+\eta(t,z)}\right)^{-\theta} - 1\right]\right. \\
&\quad \left.+ (h'\gamma(t,z) - \eta(t,z))1_{\mathbf{Z}_0}(z)\right\}\nu(dz)\Bigg\}
\end{aligned}$$

$$
\begin{aligned}
&= b'(t,x)p + \frac{1}{2}\mathrm{tr}(\Lambda\Lambda'(t,x)M) - \frac{\theta}{2}p'\Lambda\Lambda'(t,x)'p + (a_0(t,x) - c(t,x)) \\
&\quad - \frac{1}{2}(\theta - 1)\varsigma'\varsigma(t,x) + \mathcal{I}_{NL}[t,x,u,p] \\
&\quad + \sup_{h\in\mathcal{J}}\left\{ -\frac{1}{2}(\theta+1)h'\Sigma\Sigma'(t,x)h + \theta h'\Sigma\varsigma \right. \\
&\quad - \theta(h'\Sigma(t,x) - \varsigma')\Lambda'(t,x)p + h'\hat{a}(t,x) \\
&\quad -\frac{1}{\theta}\int_{\mathbf{Z}}\left\{ (1 - \theta\xi(t,x,z)'p)\left[\left(\frac{1+h'\gamma(t,z)}{1+\eta(t,z)}\right)^{-\theta} - 1\right]\right. \\
&\quad \left.\left. + \theta(h'\gamma(t,z) - \eta(t,z))\mathit{1}_{\mathbf{Z}_0}(z) \right\}\nu(dz)\right\}
\end{aligned} \tag{60}
$$

Define the auxiliary functional

$$
\begin{aligned}
\ell(h;x,p) &= \frac{1}{2}(\theta+1)h'\Sigma\Sigma'(t,x)h - \theta h'\Sigma\varsigma \\
&\quad + \theta(h'\Sigma(t,x) - \varsigma')\Lambda'(t,x)p - h'\hat{a}(t,x) \\
&\quad + \frac{1}{\theta}\int_{\mathbf{Z}}\left\{ (1 - \theta\xi(t,x,z)'p)\left[\left(\frac{1+h'\gamma(t,z)}{1+\eta(t,z)}\right)^{-\theta} - 1\right]\right. \\
&\quad \left. + \theta(h'\gamma(t,z) - \eta(t,z))\mathit{1}_{\mathbf{Z}_0}(z) \right\}\nu(dz)
\end{aligned}
$$

for $h \in \mathbb{R}^m$, $x \in \mathbb{R}^n$, $p \in \mathbb{R}^n$ and $\theta \in (0,\infty)$. Under Assumption (20), for any $p \in \mathbb{R}^n$ the term

$$
\begin{aligned}
&\frac{1}{2}(\theta+1)h'\Sigma\Sigma'(t,x)h - \theta h'\Sigma\varsigma + \theta(h'\Sigma(t,x) - \varsigma')\Lambda'(t,x)p - h'\hat{a}(t,x) \\
&\quad + \int_{\mathbf{Z}}(h'\gamma(t,z) - \eta(t,z))\mathit{1}_{\mathbf{Z}_0}(z)\nu(dz)
\end{aligned}
$$

is strictly convex in h $\forall(t,x,z) \in [0,T] \times \mathbb{R}^n \times \mathbf{Z}$ a.s. $d\nu$. Under Assumption 2.7, the nonlinear jump-related term

$$
\frac{1}{\theta}\int_{\mathbf{Z}}\left\{ (1 - \theta\xi'(t,x,z)p)\left[\left(\frac{1+h'\gamma(t,z)}{1+\eta(t,z)}\right)^{-\theta} - 1\right]\right\}\nu(dz)
$$

simplifies to

$$\frac{1}{\theta}\int_{\mathbf{Z}}\left\{\left[\left(\frac{1+h'\gamma(t,z)}{1+\eta(t,z)}\right)^{-\theta}-1\right]\right\}\nu(dz)$$

which is also convex in h $\forall(t,x,z)\in[0,T]\times\mathbb{R}^n\times\mathbf{Z}$ a.s. $d\nu$.

As a function of the variable h, $\ell(h;x,p)$ can be defined more precisely as a mapping from the vector space $\mathbb{R}^m$ into $\mathbb{R}$. Moreover, ℓ is continuous in h $\forall h\in\mathbb{R}^m$, twice differentiable and with continuous derivatives. Finally, f attains its infimum, and the infimum is finite.

Looking at the constraints, the matrix Υ defines a mapping from the vector space $\mathbb{R}^m$ into the normed space generated by associating to the constraint vector space $\mathcal{U}$ the Euclidian norm. Under Assumption 2.9, there exists an h_1 such that $\Upsilon'h<\upsilon$.

As a result, we can conclude that the auxiliary constrained optimization problem

$$\min_{h\in\mathcal{U}}\ell(h;x,p)$$

is a convex programming problem satisfying the assumptions of Lagrange Duality (see for example Theorem 1 in Section 8.6 in Luenberger, 1969). We therefore conclude that the supremum is reached for a unique maximizer $\hat{h}(t,x,p)$, which is an interior point of the set $\mathcal{J}$ defined in equation (34), and the supremum, evaluated at $\hat{h}(t,x,p)\in\mathbb{R}^n$, is finite. By measurable selection, $\hat{h}$ can be taken as a Borel measurable function on $[0,T]\times\mathbb{R}^n\times\mathbb{R}^n$. □

3.3 *Main result*

We now come to the main result of this paper.

Theorem 3.1. *Under either of*

1. *Affine drift assumptions* 2.1, 2.3, 2.7 *and* 2.9; *or*
2. *Standard control assumptions* 2.2, 2.4, 2.7, 2.8 *and* 2.9;

the following hold:

1. *The exponentially transformed value function* $\tilde{\Phi}$ *defined at* (52) *is the unique* $C^{1,2}([0,T]\times\mathbb{R}^n)$ *solution of the RS HJB PIDE* (57)–(58).
2. *The value function* Φ, *also defined at* (52), *is the unique* $C^{1,2}([0,T]\times\mathbb{R}^n)$ *solution of the RS HJB PIDE* (53)–(65).

3. *The asset allocation* $h^*(t) = \hat{h}(t, X_t, D\Phi(t, X_t))$, *where* $\hat{h}$ *is the function introduced in Proposition* 3.3, *is optimal in the class* $\mathcal{A}$ *of admissible controls.*

***Proof of Theorem* 3.1.** *Existence of a classical* ($C^{1,2}$) *solution* — proving that $\tilde{\Phi}$ is a $C^{1,2}([0,T] \times \mathbb{R}^n)$ solution of the RS HJB PDE (57)–(58) requires a different approach for each set of assumptions:

- Section 4 outlines the approach under the affine drift and constant diffusion assumptions 2.1, 2.3, 2.7 and 2.9.
- Section 5 presents an overview of the approach under standard control assumptions 2.2, 2.4, 2.7, 2.8 and 2.9

Existence of an optimal control — by Proposition 3.3, the supremum in (53) admits a unique Borel measurable maximizer. Moreover, by Proposition 6.1, this maximizer is admissible and by Proposition 6.2 it is also a maximizer with respect to the $\mathbb{P}$-measure criterion J defined in (45). Thus, we can take this maximizer as our optimal asset allocation.

Verification and uniqueness of the classical solution — $\tilde{\Phi}$ is bounded by Proposition 3.1. Part (i). of Verification Theorem 6.2 therefore applies. Choosing as optimal control the unique maximizer of the supremum (60), part (ii). of Theorem 6.1 also applies: $\tilde{\Phi}$ is the unique solution to the HJB PIDE.

It then follows that Φ is the unique classical solution to the HJB PIDE (53) with terminal condition (65). □

4 Existence of a classical ($C^{1,2}$) solution under affine drift assumptions

Under Assumptions 2.1, 2.3, 2.7 and 2.9, the problem reduces to solving a certain stochastic control problem in the factor process, which has no jumps. As a result, the HJB equation is a PDE, not a PIDE:

$$\frac{\partial \Phi}{\partial t} + \sup_{h \in \mathcal{J}} L^h(t, x, \Phi, D\Phi, D^2\Phi) = 0 \tag{61}$$

where $\mathcal{J}$ is defined in (38),

$$L^h(t, x, u, p, M) = f(t, x, h)'p + \frac{1}{2}\text{tr}\,(\Lambda\Lambda'(t, x)M) - \frac{\theta}{2}p'\Lambda\Lambda'(t, x)p \\ -g(t, x, h) \tag{62}$$

where

$$\begin{aligned} g(t,x,h) &= \frac{1}{2}(\theta+1)h'\Sigma\Sigma'h - \theta h'\Sigma\varsigma + \frac{1}{2}(\theta-1)\varsigma'\varsigma' \\ &\quad -[(a_0 - c) + (A_0 - C_0)x] - h'(a+Ax) \\ &\quad + \int_{\mathbf{Z}} \left\{ \frac{1}{\theta}\left[\left(\frac{1+h'\gamma(z)}{1+\eta(z)}\right)^{-\theta} - 1\right] \right. \\ &\quad \left. + (h'\gamma(z) - \eta(z)) 1_{\mathbf{Z}_0}(z) \right\} \nu(dz) \end{aligned} \tag{63}$$

$$f(t,x,h) := b + Bx - \theta\Lambda\left(\Sigma'h - \varsigma\right) \tag{64}$$

and subject to the terminal condition

$$\Phi(T,x) = 0, \quad x \in \mathbb{R}^n. \tag{65}$$

For $\tilde{\Phi}$, the corresponding HJB PDE is

$$\frac{\partial\tilde{\Phi}}{\partial t}(t,x) + \frac{1}{2}\mathrm{tr}\left(\Lambda\Lambda'(t,x)D^2\tilde{\Phi}\right) + H(t,x,\tilde{\Phi},D\tilde{\Phi}) = 0 \tag{66}$$

subject to terminal condition

$$\tilde{\Phi}(T,x) = 1 \tag{67}$$

where for $r \in \mathbb{R}$, $p \in \mathbb{R}^n$

$$H(s,x,r,p) = \inf_{h\in\mathcal{J}} \{f(s,x,h)'p + \theta g(s,x,h)r\} \tag{68}$$

The proof of existence of a classical solution follows a similar arguments to those developed by Fleming and Rishel (1975) (Theorem 6.2 and Appendix E) based on PDE results from Ladyzenskaja *et al.* (1968). Namely, we can use an approximation in policy space alongside results on linear parabolic partial differential equations to prove that the exponentially transformed value functions $\tilde{\Phi}$ is of class $C^{1,2}((0,T)\times\mathbb{R}^n)$. Then it follows that the value functions Φ is also of class $C^{1,2}((0,T)\times\mathbb{R}^n)$. The approximation in policy space algorithm was originally proposed by Bellman in the 1950s (see Bellman (1957) for details) as a numerical method to compute the value function.

The approach proposed in Davis and Lleo (2011b) in an asset management context without a benchmark can be used directly with very minor modifications to solve the benchmarked asset management problem. This approach has two steps. First, use the approximation in policy space algorithm to show existence of a classical solution in a bounded region. Then, extend this argument to unbounded state space. To derive this second result Davis and Lleo follow a different argument than Fleming and Rishel (1975) which makes more use of the actual structure of the control problem. This argument requires the definition of 'zero beta' policies.

Definition 4.1 (0β-policy). *By reference to the definition of the function g in equation* (43), *a* 'zero beta' (0β) control policy $\check{h}(t)$ *is an admissible control policy for which the function g is independent from the state variable x.*

The term 'zero beta' is borrowed from financial economics (see for instance Black, 1972). To avoid assuming the existence of a globally risk-free rate in factor models such as the CAPM, the APT or in ad-hoc valuation models, it is customary to build portfolios without any exposure to the factor(s) as a substitute for the risk-free rate. These special portfolios are referred to as 'zero beta' portfolios by reference to the slope coefficient β used to measure the sensitivity of asset returns to the valuation factor(s).

In the risk sensitive benchmarked asset management model, if $A_0 = 0$, then the policy $h^0 = 0$, i.e. invest all the wealth in the risk-free asset, is a 0β-policy. In the general case when $A_0 \neq 0$, we see from 35 that the set $\mathcal{Z}$ of 0β-policies is the set of admissible policies $\check{h}$ which satisfy the equation

$$\check{h}'\hat{A} = -A_0$$

Note that since $m > n$, there is potentially an infinite number of 0β-policies as long as the following assumption is satisfied

Assumption 4.1. The matrix $\hat{A}$ has rank n.

Assumption 4.2.

$$\mathcal{Z} \cap \mathcal{J} \neq \{\emptyset\}$$

Assumption 4.2 ensures that at least one zero beta policy $\check{h}$ is in $\mathcal{J}$. This assumption forces some consistency between the drift coefficients $A_0, \hat{A}$ and the jump coefficient γ, but is consistent with the 'spirit' of zero beta policies: zero beta policies are proxies for the risk-free asset and should not result in a highly risky portfolio allocation.

Without loss of generality, we can fix a 0β control $\check{h}$ as a constant function of time so that $g(x, \check{h}) = \check{g}$, where $\check{g}$ is a constant.

5 Existence of a classical ($C^{1,2}$) solution under standard control assumptions

The objective of is to prove that the value functions Φ and $\tilde{\Phi}$ are smooth. The process involves six steps [1] which are detailed in Davis and Lleo (2012) in the context of an investment management problem without benchmark. The addition of a benchmark only requires minor changes, but neither affects the line of reasoning nor the key results:

Step 1: **$\tilde{\Phi}$ is a Lipschitz Continuous Viscosity Solution (VS-PIDE) of** (57)

First, change notation and rewrite the HJB PIDE as

$$-\frac{\partial \tilde{\Phi}}{\partial t}(t, x) + F(t, x, \tilde{\Phi}, D\tilde{\Phi}, D^2\tilde{\Phi}, \mathcal{I}[t, x, \tilde{\Phi}]) = 0 \tag{69}$$

subject to terminal condition

$$\tilde{\Phi}(t, x) = 1 \tag{70}$$

where

$$\begin{aligned} &F(t, x, \tilde{\Phi}, D\tilde{\Phi}, D^2\tilde{\Phi}, \mathcal{I}[t, x, \tilde{\Phi}]) \\ &\quad = H_v(t, x, \tilde{\Phi}, D\tilde{\Phi}) - \frac{1}{2}\mathrm{tr}(\Lambda\Lambda'(t, x) D^2\tilde{\Phi}(t, x)) - \mathcal{I}[t, x, \tilde{\Phi}], \end{aligned} \tag{71}$$

$$\begin{aligned} \mathcal{I}[t, x, \tilde{\Phi}] := \int_{\mathbf{Z}} \Big\{ &\tilde{\Phi}(t, x + \xi(t, x, z)) - \tilde{\Phi}(t, x) \\ &- \xi(t, x, z)' D\tilde{\Phi}(t, x) 1_{\mathbf{Z}_0} \Big\} \nu(dz) \end{aligned} \tag{72}$$

[1] Davis and Lleo (2012) include a seventh step to show that the unique maximizer of the supremum (60) is the optimal control. We address this question in Section 6.

and

$$H_v(s, x, r, p) = -H(s, x, r, p)$$
$$= \sup_{h \in \mathcal{A}} \{-f_v(s, x, h)'p - \theta g(s, x, h)r\} \qquad (73)$$

for $r \in \mathbb{R}$, $p \in \mathbb{R}^n$ and where

$$f_v(t, x, h) := f(t, x, h) - \int_{\mathbf{Z} \setminus \mathbf{Z}_0} \xi(t, x, z)\nu(dz)$$
$$= b(t, x) - \theta\Lambda(\Sigma(t, X(t))'h(t) - \varsigma)$$
$$+ \int_{\mathbf{Z}} \xi(t, x, z) \left[\left(\frac{1 + h'\gamma(t, z)}{1 + \eta(t, z)} \right)^{-\theta} - 1 \right] \nu(dz) \quad (74)$$

We can show that $\tilde{\Phi}$ is *a* Lipschitz continuous solution of (69).

The application of viscosity solution techniques to stochastic control problem require two lines of argument. The first is to prove that the value function is a viscosity solution of the associated HJB equation. Theorem 3.5 of Bouchard and Touzi (2011) provides the weak Dynamic Programming Principle (DPP) we needed to prove this result. The second line of argument relates to the the comparison result required to show uniqueness. It is widely appreciated that the heart of viscosity solution theory lies in the uniqueness theorems and, as seen for example in Barles and Imbert (2008), to prove the necessary comparison theorems stronger conditions are generally required than those needed for existence. Our strategy is to by-pass this question entirely by taking a route that only requires uniqueness of viscosity solutions for PDEs — where a large literature exists (starting with Crandall *et al.*, 1992) — rather than PIDEs where results are sparser.

Step 2: **From PIDE to PDE** — Change notation and rewrite the HJB PIDE as a parabolic PDE *à la* Pham (1998):

$$\frac{\partial u}{\partial t}(t, x) + \frac{1}{2}\mathrm{tr}(\Lambda\Lambda'(t, x)D^2 u) + H_a(t, x, u, Du) + d_a^{\tilde{\Phi}}(t, x) = 0 \qquad (75)$$

subject to terminal condition $u(T, x) = 1$ and with

$$H_a(s, x, r, p) = \inf_{h \in \mathcal{U}} \{f_a(s, x, h)'p + \theta g(s, x, h)r\} \qquad (76)$$

for $r \in \mathbb{R}$, $p \in \mathbb{R}^n$ and where

$$\begin{aligned} f_a(s,x,h) &:= f(s,x,h) - \int_{\mathbf{Z}} \xi(s,x,z)\nu(dz) \\ &= b(t,x) - \theta\Lambda(\Sigma(t,X(t))'h(t) - \varsigma) \\ &\quad + \int_{\mathbf{Z}} \xi(s,x,z)\left[\left(\frac{1+h'\gamma(t,z)}{1+\eta(t,z)}\right)^{-\theta} - 1_{\mathbf{Z}_0}(z) - 1\right]\nu(dz) \end{aligned} \tag{77}$$

and

$$d_a^{\tilde{\Phi}}(t,x) = \int_{\mathbf{Z}} \{\tilde{\Phi}(t,x+\xi(t,x,z)) - \tilde{\Phi}(t,x)\}\nu(dz) \tag{78}$$

In particular, f_a is Lipschitz continuous and bounded and $\check{\mathcal{I}}$ is continuous. The second property stem from Assumption (12).

Step 3: **Viscosity Solution to PDE** (75) — consider a viscosity solution u of the semi-linear PDE (76) (always interpreted as an equation for 'unknown' u with the last term prespecified, with $\tilde{\Phi}$ defined as in Step 1.)

$$\frac{\partial u}{\partial t}(t,x) + \frac{1}{2}\mathrm{tr}(\Lambda\Lambda'(t,x)D^2u) + H_a(t,x,u,Du) + d_a^{\tilde{\Phi}}(t,x) = 0 \tag{79}$$

$\tilde{\Phi}$ is a viscosity solution of the PDE (75) — this is due to the fact that, in definition (78), PIDE (69) and PDE (75) are in essence the same equation. Hence, if $\tilde{\Phi}$ solves one of them, then it solves both.

The definition of viscosity solution we need to use is due to Pham (1998) and formalized by Davis, Guo and Wu (2009). This new definition is not standard, although it is equivalent to standard definitions based on semi-jets and test functions.

Step 4: **Uniqueness of the Viscosity Solution to the PDE** (75) — If a function u solves the PDE (75) it does not mean that u also solves the PIDE (69) because the term d_a in the PDE (75) depends on $\tilde{\Phi}$ regardless of the choice of u. Thus, if we were to show the existence of a classical solution u to PDE (75), we would not be sure that this solution is the value function $\tilde{\Phi}$ unless we can show that PDE (75) admits a unique solution. Because we proved earlier on that f_a is Lipschitz continuous and boundedand that $d_a^{\tilde{\Phi}}$ is continuous, this step only requires applying a "classical" comparison result for viscosity solutions (see Theorem 8.2 in Crandall, Ishii and Lions (1992)).

Step 5: **Existence of a Classical Solution to the HJB PDE** (75) — Now that we have been able to rewrite the HJB PIDE as a parabolic PDE, we have access to the literature addressing the existence of a strong solution to the HJB PDE, such as Fleming and Rishel (1975) or Krylov (1980) and Krylov (1987). The crucial point in this argument is that this new PDE has a unique viscosity solution which also solves the initial PIDE. From there, we only need to prove existence of a classical solution to the PDE in order to show that the value functions Φ is also of class $C^{1,2}((0,T) \times \mathbb{R}^n)$.

However, the control-based argument used in the Section 4 cannot be used here. The reason for this is that reinterpreting the PIDE as a PDE removes the natural connection between the PDE and the dynamics of the factor process. It therefore becomes more effective to consider the PDE in abstraction from the control problem and prove existence directly through standard PDE arguments.

Step 6: **Existence of a Classical Solution to the HJB PIDE** (57) — Combining Steps 4 and 5, we conclude that $\tilde{\Phi}$ and Φ are respectively a classical ($C^{1,2}$) solution of (3.8) and a classical ($C^{1,2}$) solution of (57).

6 Verification

In this section, we prove a verification theorem to the effect that if (53) has a $C^{1,2}$ solution then that solution is equal to Φ defined by (52) and the control $h^*(t) = \hat{h}(t, x, D\Phi)$ is optimal.

6.1 *The unique maximizer of the supremum* (60) *is the optimal control, i.e.* $h^*(t, X_t) = \hat{h}(t, X_t, D\Phi(t, X_t))$

By measurable selection, $\hat{h}$ can be taken as a Borel measurable function on $[0, T] \times \mathbb{R}^n \times \mathbb{R}^n$ (see Proposition 3.3). Moreover,

Proposition 6.1. *The maximizing control is admissible:* $\hat{h}(t, X_t, D\Phi(t, X_t)) \in \mathcal{A}$.

Proof. The proof follows closely Proposition 4.3 in Davis and Lleo (2011b). □

***Remark* 6.1.** The argument used in Davis and Lleo (2011b) is based on a result by Mémin (1979). One could also derive a similar argument using the elegant result by Cheridito *et al.* (2005).

Applying Proposition 6.1 we deduce that the control $\hat{h}$ is optimal for the auxiliary problems (50) and (51) resulting from the change of measure. However, this proposition is not sufficient to conclude that $\hat{h}(t)$ is optimal for the original problem (45) set under the $\mathbb{P}$-measure. The next result shows that this is indeed the case.

Proposition 6.2. *The optimal control* $\hat{h}(t, X_t, D\Phi(t, X_t))$ *for the auxiliary problem* $\sup_{h\in\mathcal{A}} I(t, x, h; \theta; T; v)$ *where* I *is defined in* (51) *is optimal for the initial problem* $\sup_{h\in\mathcal{H}} J(t, x, h; \theta; T; v)$ *where* J *is defined in* (45), *i.e.* $h^*(t, X_t) = \hat{h}(t, X_t, D\Phi(t, X_t))$.

Proof. The proof follows closely Proposition 4.4 in Appendix A of Davis and Lleo (2011b). □

6.2 *Verification*

We will first prove a verification theorem for the exponentially transformed problem with HJB PDE (57) and value function $\tilde{\Phi}(t, x)$. As a corollary, we will obtain a verification theorem for the risk sensitive control problem with HJB PDE (53) and value function $\Phi(t, x)$. We define the first order operator

$$\tilde{L}_t^h \varphi(t, x) := f(t, x, h)' D\varphi(t, x) + \theta g(x, h)\varphi(t, x) \tag{80}$$

Theorem 6.1 (Verification Theorem for the Exponentially Transformed Control Problem).

(i) *Assume that there exists a* $C^{1,2}([0, T] \times \mathbb{R}^n)$ *solution* $\tilde{\Phi}$ *to the HJB PIDE* (57):

$$\begin{aligned}&\frac{\partial\tilde{\Phi}}{\partial t}(t, x) + \frac{1}{2} tr(\Lambda\Lambda'(t, x)D^2\tilde{\Phi}(t, x)) + H(t, x, \tilde{\Phi}, D\tilde{\Phi})\\ &\qquad + \int_{\mathbf{Z}} \left\{\tilde{\Phi}(t, x + \xi(t, x, z)) - \tilde{\Phi}(t, x) - \xi(t, x, z)' D\tilde{\Phi}(t, x)\right\} \nu(dz)\\ &= 0\end{aligned}$$

then $\tilde{\Phi}(t, x) \leq \tilde{I}(v, x, \tilde{h}; t; T; \theta)$ *for any* $\tilde{h} \in \mathcal{A}(T)$.

(ii) *Further assume that there exists a Borel-measurable minimizer* $h^*(t,x)$ *of* $\tilde{h} \mapsto \tilde{L}^{\tilde{h}}\tilde{\phi}$ *defined in* (80) *then* $\tilde{\Phi}(t,x) = \tilde{I}(v,x,h^*;t;T;\theta)$ *and* $h^*(t,x)$ *is optimal.*

Proof. The proof is standard. It can be found for example in Davis and Lleo (2012). □

Corollary 6.1 (Verification Theorem for the Risk-Sensitive Control Problem)

(i) *Assume that there exists a* $C^{1,2}([0,T] \times \mathbb{R}^n)$ *solution* Φ *to the HJB PIDE* (53):

$$\frac{\partial \Phi}{\partial t} + \sup_{h \in \mathcal{J}} L_t^h \Phi(t, X(t)) = 0$$

where L *is defined in* (54) *then* $\Phi(t,x) \geq I(t,x,h;\theta;T;v)$ *for any* $\tilde{h} \in \mathcal{A}(T)$.

(ii) *Further assume that there exists a Borel-measurable minimizer* $h^*(t,x)$ *of* $\tilde{h} \mapsto L^h\Phi$ *defined in* (54) *then* $\Phi(t,x) = I(t,x,h^*;\theta;T;v)$ *and* $h^*(t,x)$ *is optimal.*

Proof. This corollary follows from the relation between Φ and $\tilde{\Phi}$ and from the fact that an admissible (optimal) strategy for the exponentially transformed problem is also admissible (optimal) for the risk-sensitive problem. □

7 Conclusion

In this article, we have extended the approach proposed in Davis and Lleo (2011b) and Davis and Lleo (2012) to solve a risk-sensitive jump-diffusion benchmarked investment management problem under two sets of assumptions. Under the first set of assumptions, affine asset growth rates with constant volatility, the factor process X_t has no jumps and the associated Bellman equation is a partial differential equation (PDE) which can be shown to admit a unique classical ($C^{1,2}$) solution.

Under the second set of assumptions, we now have a fully nonlinear controlled jump-diffusion, and the Bellman equation is a partial integro-differential equation for which no analytical solution exists. Proving that the Hamilton Jacobi Bellman partial integro-differential equation (HJB PIDE) admits a unique classical solution requires the development of a more sophisticated argument combining viscosity solutions and classical solutions. The argument used in the derivation hinges on only three key points. First, the Lipshitz continuity of the value function provides us with the ability to rewrite the HJB PIDE as a PDE. Second, viscosity solutions give us existence and uniqueness of a weak solution to both of these equations. A proof of existence by Fleming and Rishel based on a policy improvement originally due to Bellman completes the analysis by providing a smooth solution. The robustness of this approach is a clear advantage for control problems: solving benchmarked or ALM investment management problems is not more difficult than addressing an asset only investment problem.

References

Barles, G. and C. Imbert. Second-order elliptic integro-differential equations: Viscosity solutions' theory revisited. *Annales de l'Institut Henri Poincaré*, 25(3): pp. 567–585, 2008.

Bellman, R. *Dynamic Programming.* Princeton University Press, 1957.

Benk, M. Intertemporal surplus management with jump risks. In H. Gassmann, S. Wallace, and W. Ziemba, (eds.), *Stochastic Programming: Applications in Finance, Energy and Logistics*, 2012.

Bielecki, T.R. and S.R. Pliska. Risk-sensitive dynamic asset management. *Applied Mathematics and Optimization*, 39: pp. 337–360, 1999.

Black, F. Capital market equilibrium with restricted borrowing. *Journal of Business*, 45(1): pp. 445–454, 1972.

Bouchard, B. and N. Touzi. Weak dynamic programming principle for viscosity solutions. *SIAM Journal on Control and Optimization*, 49: pp. 948–962, 2011.

Browne, S. Beating a moving target: Optimal portfolio strategies for outperforming a stochastic benchmark. *Finance and Stochastics*, 3: pp. 275–294, 1999.

Cheridito, P., D. Filipovic and M. Yor. Equivalent and absolutely continuous measure changes for jump-diffusion processes. *The Annals of Applied Probability*, 15(3): pp. 1713–1732, 2005.

Crandall, M., H. Ishii and P.-L. Lions. User's guide to viscosity solutions of second order partial differential equations. *Bulletin of the American Mathematical Society*, 27(1): pp. 1–67, July 1992.

Davis, M.H.A., X. Guo and G. Wu. Impulse control of multi-dimensional jump diffusions. *SIAM Journal on Optimization and Control* (to appear), 2009.

Davis, M.H.A. and S. Lleo. Risk-sensitive benchmarked asset management. *Quantitative Finance*, 8(4): pp. 415–426, June 2008.

Davis, M.H.A. and S. Lleo. Risk-sensitive asset management and affine processes. In Masaaki Kijima, Chiaki Hara, Keiichi Tanaka, and Yukio Muromachi (eds.), *KIER-TMU International Workshop on Financial Engineering 2009*, pp. 1–43. World Scientific, 2010.

Davis, M.H.A. and S. Lleo. Fractional kelly strategies for benchmarked asset management, *The Kelly Capital Growth Investment Criterion: Theory and Practice*, pp. 385–407. World Scientific, 2011a.

Davis, M.H.A. and S. Lleo. Jump-diffusion risk-sensitive asset management I: Diffusion factor model. *SIAM Journal on Financial Mathematics*, 2: pp. 22–54, 2011b.

Davis, M.H.A. and S. Lleo. Jump-diffusion risk-sensitive asset management II: Jump-diffusion factor model. *working paper*, 2012.

Fleming, W.H. and R.W. Rishel. *Deterministic and Stochastic Optimal Control.* Springer-Verlag, Berlin, 1975.

Fleming, W.H. and H.M. Soner. *Controlled Markov Processes and Viscosity Solutions*, volume 25 of *Stochastic Modeling and Applied Probability.* Springer-Verlag, 2nd edition, 2006.

Ikeda, N. and S. Watanabe. *Stochastic Differential Equations and Diffusion Processes.* North-Holland Publishing Company, 1981.

Krylov, N.V. *Controlled Diffusion Processes.* Springer Verlag, New York, 1980.

Krylov, N.V. *Nonlinear Elliptic and Parabolic Equations of the Second Order.* Kluwer Academic Publishers, 1987.

Kuroda, K. and H. Nagai. Risk-sensitive portfolio optimization on infinite time horizon. *Stochastics and Stochastics Reports*, 73: pp. 309–331, 2002.

Ladyzenskaja, O.A., V.A. Solonnikov and O.O. Uralceva. *Linear and Quasilinear Equations of Parabolic Type.* American Mathematical Society, Providence RI, 1968.

Luenberger, D. *Optimization by Vector Space Methods.* John Wiley & Sons, 1969.

Memin, J. *Séminaires de Probabilités XII. Lecture Notes in Mathematics*, Chapter 4, Décomposition multiplicative de semimartingales exponentielles et applications, Vol. 649, pp. 35–46. Springer-Verlag, Berlin, 1979.

Mikulyavichyus, R. and H. Pragarauskas. Classical solutions of boundary value problems for some nonlinear integro-differential equations. (Russian). *Lietuvos Matematikos Rinkinys*, 34(3): pp. 347–361, 1994. translation in Lithuanian Math. J. 34 (1994), no. 3, pp. 275–287 (1995).

Øksendal, B. and A. Sulem. *Applied Stochastic Control of Jump Diffusions.* Springer, 2005.

Pham, H. Optimal stopping of controlled jump diffusion processes: A viscosity solution approach. *Journal of Mathematical Systems, Estimation and Control*, 8(1): pp. 1–27, 1998.

Pham, H. A large deviations approach to optimal long term investment. *Finance and Stochastics*, 7: pp. 169–195, 2003.

Rudolf, M. and W.T. Ziemba. Intertemporal surplus management. *Journal of Economic Dynamics & Control*, 28: pp. 975–990, 2004.

Touzi, N. Stochastic control and application to finance. http://www.cmap.polytechnique.fr/~touzi/pise02.pdf, 2002. Special Research Semester on Financial Mathematics, Scuola Normale Superiore, Pisa, April 29–July 15, 2002.

Chapter 6

Dynamic Portfolio Optimization under Regime-Based Firm Strength

Chanaka Edirisinghe* and Xin Zhang†

Summary

This chapter presents a dynamic optimization model under regime-switching market uncertainty to manage stock portfolios. The novelty of the approach is that stocks for long and short investments are selected based on the notion of fundamental business strength and consistent with specific regime scenarios. Firm-strength is determined via the quarterly financial statements of a firm, relative to other firms in the same market sector/industry. Referred to as Relative Firm Strength (RFS), it is determined using Data Envelopment Analysis (DEA) in which various financial metrics from accounting statements are specified as inputs and outputs. It is shown that such an RFS measure has predictive power of stock returns in the historical data period 1971–2010. The regime-switching market model is combined with the RFS metric of stock selections to test the performance of a two-quarter stochastic portfolio optimization model during Jan–Jun, 2011. The out-of-sample results demonstrate that the proposed methodology is superior to using sector-based ETF portfolios or the market index itself.

1 Introduction

The problem of selecting an optimal portfolio of financial assets, such as equities, is critically important in the management of mutual funds, retirement and pension funds, bank and insurance portfolio management. Such problems are characterized by the difficulties in choosing an investment horizon, how often the fund needs to be re-balanced to account for the

*College of Business, University of Tennessee, Knoxville, TN 37996, chanaka@utk.edu, corresponding author.
†College of Business, Austin Peay State University, Clarksville, TN 37044, zhangx@apsu.edu.

dynamic evolutions in the economy, inclusion of transactions costs, presence of legal and policy constraints, etc. Considering investments in the equity markets, the fund manager is also required to choose market sectors or industry groups that are expected to outperform in a given time period, and then, individual firms that are likely to display strong performance in a competitive market. As economic changes occur, they are likely to affect firm-specific business conditions, thus, requiring rebalancing in the underlying portfolios to manage the investment risks. This chapter presents a methodology to deal with these issues in a two-period investment framework under trading costs and regime-switching market uncertainty that affects firm fundamentals.

The fact that a firm's stock return is quite closely related to the firm's business strength, implied by the strength in the financial statements, is an area of study generally referred to as Fundamental Analysis, which dates back to Dodd and Graham (1934). A large body of evidence demonstrates that there is a link between fundamental (accounting) information and stock prices. For example, Ball and Brown (1968) highlight this connection by showing that significant returns can be earned with perfect foresight of the future earnings. Ohlson (1995) developed and analyzed a model of a firm's market value as it relates to future earnings, book values, and dividends. Frankel and Lee (1998) examined the usefulness of an analyst-based valuation model in predicting cross-sectional stock returns using the accounting information. Hirschey (2003) concluded that in the long run, trends in stock prices mirror real changes in business prospects as measured by revenues, earnings, dividends, etc. Samaras *et al.* (2008) developed a multi-criteria decision support system to evaluate the Athens Stock Exchange stocks within a long term horizon based on accounting analysis.

The above strong-link between fundamental financial data and stock returns is often used to guide investment selections by either using the so-called discounted cash flow model such as Residual Income Valuation (RIV) model (Ali *et al.*, 2003), or using a limited view of the available data under the so-called Value-Growth (VG) paradigm. The RIV model attempts to attach an intrinsic valuation for a firm based on estimated book value of owner's equity and earnings, as well as forecasted return on equity and growth rates. Such valuation methodologies attempt to determine undervalued or over-valued stocks based on current market prices. On the other hand, the VG paradigm utilizes standard financial ratios such as Book-to-Market-equity (B/M), Earnings-to-Price (E/P), or Dividend-to-Price (D/P) directly to avoid firms that are relatively risky or select those with long-term value. Typically, Value shares are characterized by high B/M, E/P and D/P ratios while Growth shares have low ratios. In support of this view, several empirical studies have shown that Value shares have outperformed Growth shares (Fama and French, 1993; Haugen, 1997; Lakonishok *et al.*, 1994).

We contend that long-term forecasts of growth rates and earnings as required by RIV are quite complicated and lack accuracy, and the limited view of firm fundamentals advocated by the VG paradigm is inadequate in capturing dynamically evolving market competitive forces. In this chapter, we utilize a Data Envelopment Analysis (DEA) based fundamental strength indicator that uses several aspects of financial performance of a firm to guide the process of individual stock selections for further portfolio analysis, herein termed Relative Firm Strength (RFS) analysis.

Often investors seek fund rebalancing because the economic state of the market changes from time to time, or because certain assets have fallen out of the favorite list due to their performance shortfalls, or perhaps because risk attitudes and policy parameters have undergone changes. Incorporating market regime changes in financial markets has evidenced a resurgence of attention in the recent years. The idea of modeling financial markets using regime switching models can be traced back to Hamilton (1989). Naik (1993) used Markov jump models to value European options under continuous-time models. Also, see Wahab *et al.* (2010) for a non-financial application of a 3-regime based option pricing model under geometric Brownian motions coupled with a finite-state Markov chain for the electricity markets.

Graflund and Nilsson (2003) investigate the questions of dynamic portfolio selection and intertemporal hedging within a Markovian regime-switching framework. They model the data generating processes via regime-switching models as mixtures of Gaussian distributions and find that taking the specific regimes into consideration has a strong influence on optimal portfolio selections. Ang and Bekaert (2002) show that regime-switching models with CRRA utility can successfully capture the asymmetric correlations found in international equity returns during volatile and stable markets, i.e., two market regimes, and confirm the benefits of international diversification. Guidolin and Timmermann (2008) allow both the exposure of local markets to global risk factors and the world price of covariance, co-skewness, and co-kurtosis risk to vary across regimes. They find evidence of two regimes in the distribution of international stock returns, namely, a bear regime with low mean returns and high volatility and a bull regime associated with less volatile returns and more attractive investment opportunities.

Yin and Zhou (2004) study a discrete-time version of Markowitz's mean-variance portfolio optimization model where the market parameters depend on the market regime that switches according to a finite-state Markov chain. Sarno and Valente (2000) use regime-switching model to examine the time-varying relation between spot and future prices. They utilize autoregressive technique to estimate parameters and solve the model by EM algorithm. Brooks and Katsaris (2005) use a three-regime switching model to explain the dynamics of S&P500. They use bubble deviation and abnormal volume to define the regime. Powell *et al.* (2007) use 125 years monthly market index returns to identify the mechanism of shifts in regimes and support a three-regime hypothesis.

The remainder of the chapter is organized as follows. Section 2 begins with an introduction to DEA and then we develop the notion of RFS metric for a given firm using accounting data. Section 3 presents the market regime model based on quarterly data. It develops that a three-regime model is optimal for the data period. The RFS metric and the regime model are combined in Section 4 to identify long/short pools of stocks under each regime scenario. The two-quarter stochastic programming portfolio model is developed using these stock selections. Model application in the U.S. markets is reported in Section 5. Concluding remarks are in Section 6.

2 DEA-based relative firm strength

A firm's business strength is related to both its internal productivity as well as its competitive status within the industry it operates in. Internal productivity gains via, say, lower production costs, as well as the firm's effectiveness in dealing with supply competition and in marketing its products and services are reflected by the data in the firm's balance sheets and cash flow statements. We use this data within the so-called Data Envelopment Analysis (DEA) technique to evaluate and rank a firm's business strength, relative to other firms within a market sector, in an attempt to identify firms that are potentially less-risky (or more-risky) as long-term holdings.

Based on the economic notion of Pareto optimality, Data Envelopment Analysis (DEA) employs a non-parametric approach for estimating the production frontier based on the specification of the production possibility set rather than a production functional form itself. It originated from the work by Farrell (1957), which was later popularized by Charnes *et al.* (1978), called the CCR model. The latter model constructs the efficient frontier by optimizing the ratio of a linear combination of multi-outputs to a linear combination of multi-inputs from available production technologies pertaining to a group of similar firms (say, in a given industry or market sector). The CCR-DEA model necessarily implies a constant returns to scale (CRS) relationship between inputs and outputs. Thus, the resulting firm efficiency captures not only the productivity inefficiency of a firm at its actual scale size, but also any inefficiency due to its actual scale size being different from the optimal scale size, see Banker (1984). Those firms with production input/output mixes that lie on the efficient frontier are operationally-efficient, and therefore, DEA can be used as a tool to identify "best-practice" firms that can be used as benchmarks of strong performance.

2.1 *Financial DEA model*

CCR-DEA efficiency scores are bounded within 0 and 1, and thus, firms with efficiency score of 1 cannot differentiate among themselves. Similarly, firms with efficiency of 0 cannot be differentiated for their relative weakness either. However, when DEA scores are computed for the purpose of quantifying business performance, such an artificial 0–1 truncation of its value may adversely affect its correlation with the external stock return processes. We eliminate the truncation from above by essentially allowing for further differentiation of firms with DEA scores of 1 by checking if such a firm can further increase its inputs proportionately without sacrificing the firms's efficiency, an idea originally attributed to Andersen and Petersen (1993). Applying this idea in the context of CCR-DEA, the so-called *super-efficiency* CCR-DEA model is proposed by Edirisinghe and Zhang (2008).

Let $\mathcal{I}$ and $\mathcal{O}$ denote disjoint sets (of indices) of input and output financial parameters, respectively, for computing a firm's efficiency. For a given time period (say, a quarter), the value of financial parameter P_i for firm j is denoted by ξ_{ij}, where $i \in \mathcal{I} \cup \mathcal{O}$, $j = 1, \ldots, J$, and J is the number of firms under investigation. Then, the DEA-based relative performance of firm k, η_k, relative to the remaining $J-1$ firms, is determined by the linear programming (LP) model,

$$\begin{aligned}\eta_k := \max_u & \sum_{i\in\mathcal{O}} \xi_{ik}u_i \\ \text{s.t.} & \sum_{i\in\mathcal{I}} \xi_{ik}u_i \le 1 \\ & -\sum_{i\in\mathcal{I}} \xi_{ij}u_i + \sum_{i\in\mathcal{O}} \xi_{ij}u_i \le 0, \quad j\neq k, j=1,\dots,J \\ & -U\sum_{i\in\mathcal{I}} \xi_{ik}u_i + \sum_{i\in\mathcal{O}} \xi_{ik}u_i \le 0 \\ & u_i \ge 0,\ i\in\mathcal{I}\cup\mathcal{O},\end{aligned} \tag{1}$$

where U is a sufficiently large positive constant no less than 1 (we set $U =$ 100 in our experiments), and the LP model is solved for each $k = 1,\dots,J$. The DEA score above falls between 0 and U for all firms. The fact that the specific value of U only affects those firms that are *efficient* is evident from the following result:

Proposition 2.1. *For a given value of $U \ge 1$, define $\eta_k(U) := \eta_k$.*

(i) *If $\eta_k(1) < 1$, then $\eta_k(U) = \eta_k(1) < 1$ holds for all $U \ge 1$.*
(ii) *$\eta_k(U) \ge 1$ holds if only if $\eta_k(1) = 1$.*

For the DEA score to be representative of the firm's fundamental strength, η_k must be strongly and positively correlated with market returns of the underlying stock. To facilitate statistical analysis of the latter correlation, the underlying random processes are preferred to be (approximately) *normally* dispersed. Here, we apply the logarithmic transformation of DEA-scores (in general, the Box-Cox transformation $\frac{s^\theta-1}{\theta}$ for $\theta\neq 0$ may be applied) under a minimum DEA score of $\eta_k \leftarrow \max\{\eta_k, 0.01\}$. The metric so-computed, i.e., $s_k := \ell n(\eta_k)$, is referred to as a "historical strength" because the financial data ξ must be known prior to evaluating the firm-performance. Note that, thus, $s_k \in [-4.61, +4.61]$ where we have set $U = 100$. The conditional expectation of s_k, given the historical evolution $\mathcal{H}$ of ξ, is termed the Relative Firm Strength (RFS) of firm k, and denoted by $\bar{s}_k(\mathcal{H}) := E[s_k|\xi \in \mathcal{H}]$. As market regimes change, so does the historical evolution $\mathcal{H}$, and thus, the RFS of a firm changes.

2.2 *Parameters of RFS*

The RFS determining model in (1) requires specification of relevant input and output parameters for measuring the firm strength. Moreover, such scores need to be validated using correlation analysis with stock returns. The type of parameters needed in (1) are the fundamental financial metrics of firm operation. In general, such financial parameters span various operational perspectives, such as profitability, asset utilization, liquidity, leverage, valuation, and growth, see Edirisinghe and Zhang (2008). Such data on firms are obtained from the publicly-available financial statements. In this chapter, we focus on 4 input parameters and 4 output parameters, as given in Table 1.

Accounts Receivables (AR) represent the money yet to be collected for the sales made on credit to purchasers of the firm's products and services, and thus, it is preferable for these short-term debt collections to be small. Long-term Debt (LD) is the loans and financial obligations lasting over a year, and a firm wishes to generate revenues with the least possible LD. Capital Expenditure (CAPEX) is used to acquire or upgrade physical assets such as equipment, property, or industrial buildings to generate future benefits, and thus, a firm prefers to use smaller amounts of CAPEX to generate greater benefits. Cost of Goods Sold (COGS) is the cost of the merchandise that was sold to customers which includes labor, materials, overhead, depreciation, etc. Obviously, a smaller COGS is an indicator of managerial excellence. On the other hand, Revenue (RV) and Earnings Per Share (EPS) represent the metrics of profitability of a firm which are necessarily objectives to be maximized. Net income growth rate (NIG) is the sequential rate of growth of the income from period to period, the increase of which is a firm's operational strategy. Price to Book (P/B), which is

Table 1 Input/Output Parameters for Fundamental Financial Strength.

i	**Financial Parameter**		**Status**
1	Accounts Receivables	(AR)	Input
2	Long-term Debt	(LD)	Input
3	Capital Expenditure	(CAPEX)	Input
4	Cost of Goods Sold	(COGS)	Input
5	Revenue	(RV)	Output
6	Earnings per Share	(EPS)	Output
7	Price to Book ratio	(P/B)	Output
8	Net Income Growth Rate	(NIG)	Output

a stock's market value to its book value, is generally larger for a growth company. A lower P/B ratio could mean that the stock is undervalued or possibly there is something fundamentally 'wrong' in the company. Hence, we seek firms in which P/B ratio is large enough by considering it as an output parameter. Accordingly, referring to Table 1, the parameter index sets for the DEA model are $\mathcal{I} = \{1, 2, 3, 4\}$ and $\mathcal{O} = \{5, 6, 7, 8\}$.

2.3 *Correlation analysis*

To validate if the historical strength scores s_j are strongly correlated with stock returns r_j, the (firm j) correlation between s_j and r_j is computed using the historical periods in $\mathcal{H}$, say H periods, and it is denoted by γ_j. Then, the average of the firm-correlations over all J firms is used as the statistic, i.e., $\bar{\gamma}(\mathcal{H}) := \frac{1}{J}\sum_j \gamma_j$, to ascertain the predictive power of the DEA-based strength metric. It must be noted that the group of J firms are supposed to be similar in the sense that they belong to the same industry, market segment, or national economy. Accordingly, the model (1) needs to be applied within each such group separately. How such groups are chosen is discussed later in the chapter.

To check if a computed $\bar{\gamma}$ is statistically significant, let $\bar{\Gamma}$ be the associated population correlation. For establishing predictive strength, given a required minimum positive correlation $\rho \in (0, 1)$, consider the hypothesis test:

$$H_0 : \bar{\Gamma} \le \rho_0; \quad H_1 : \bar{\Gamma} > \rho_0, \tag{2}$$

and the test statistic:

$$\Phi := \frac{1}{2J}\sum_{j=1}^{J} \ell n \left[\frac{1+\gamma_j}{1-\gamma_j}\right]. \tag{3}$$

Proposition 2.2. *H_0 is not rejected at α-significance level if*

$$\Phi \le \frac{1}{2}\ell n \left[\frac{1+\rho_0}{1-\rho_0}\right] + \frac{\mathcal{Z}^{-1}(1-\alpha)}{\sqrt{J(H-3)}}. \tag{4}$$

Moreover, H_0 is rejected if

$$\Phi > \frac{1}{2}\left(1 - \frac{\hat{\nu}}{J}\right) \ell n \left[\frac{1 + \frac{\rho_0}{1-\hat{\nu}/J}}{1 - \frac{\rho_0}{1-\hat{\nu}/J}}\right] + \frac{\mathcal{Z}^{-1}(1-\alpha)}{\sqrt{J(H-3)}}, \tag{5}$$

where $\hat{\nu}$ is given by

$$\hat{\nu} := \max_{\nu=0,1,2,\ldots,J-1} \{\nu : \nu < J(1-\rho_0)\}. \tag{6}$$

Proof. The result in (4) is shown in Edirisinghe and Zhang (2007). The result in (5) can be shown following the derivation in Edirisinghe and Zhang (2010).

It must be noted that firm performance can also be impacted by the specific market regime that the firms are operating in. For example, in weak economies the potential for revenue growth may become substantially diminished for those firms with inferior managerial strategy. Consequently, the significance of $\bar{\gamma}$ needs to be associated with history $\mathcal{H}$ belonging to a specific market regime. For this purpose, the process of determining market regimes is discussed next.

3 Modeling market regimes

We subscribe to the notion that time-variation in stock returns embeds changes in expected returns arising from cyclical changes of economic states, see Campbell (2000) and Campbell and Cochrane (1999). In the absence of regime changes, one has stock returns with constant mean and volatility (a random walk with constant drift). Under market regime switching, the average level of returns can be modeled as random switching from one mean level to another as the stock market switches states in response to underlying changes such as a shift in the required risk premium or economic states. Each return regime can be associated with a different mean and volatility. There have been several multi-regime models of stock returns as stated earlier, but in this section our analysis follows that in Powell *et al.* (2007). In our case, given the RFS of a firm is evaluated on a quarterly (financial reporting) period, capital market returns on a quarterly basis are used to identify an optimal number of regimes to be employed. Then the quarterly expected RFS can be conditioned on each market regime so-identified. We use the historical quarterly periods from 1971 to 2010, resulting in $n = 160$ quarters of data.

Suppose there are M market regimes (states), denoted by $m = 1,\ldots,M$. These regimes are generally hidden and the value of M is unknown. We consider a discrete-time stochastic process of regime evolution which is assumed to be Markovian. Accordingly, the transition probability $\pi_{m_1 m_2}$ of the economy transitioning from state m_1 in quarter t to state m_2

in quarter $t+1$ is invariant of the states of the economy in quarters prior to time t.

Define R_t as the random variable representing the quarterly returns of a broader market index. The assumed functional form for the global distribution of quarterly returns corresponds to

$$f(R_t;\boldsymbol{\theta}) := \sum_{m=1}^{M} \delta_m f_m(R_t;\boldsymbol{\theta}_m) = \sum_{m=1}^{M} \frac{\delta_m}{\sigma_m\sqrt{2\pi}} e^{\frac{1}{2\sigma_m^2}(r_t-\mu_m)^2}, \qquad (7)$$

where r_t is the realized quarterly return, M is the number of Gaussian distributions (regimes), and $\boldsymbol{\theta}_m = (\mu_m, \sigma_m, \delta_m)$ is a vector of unknown parameters representing the mean, variance and proportion of observations that correspond to each regime m. This equation describes the aggregated, collective behavior of returns, useful for examining how many regimes underlie the global frequency distribution and the parameters that govern the regimes. Alternative component specifications of the model can be referred to as alternative hypotheses, where each value of $M(=1,2,\ldots)$ corresponds to one hypothesis, see Hamilton (1994). The n (=160) quarterly market returns are classified into $L(=50)$ discrete categories of equal width (=0.01), indexed $l = 1,\ldots,L$.

If the observed proportion of returns in category l is $\hat{p}_l$ and the expected proportion as a function of parameter estimates is $p_l(\boldsymbol{\theta})$, the likelihood function is $\Pi_{l=1}^{L}(p_l(\boldsymbol{\theta}))^{n\hat{p}_l}$, and thus, the log-likelihood function is

$$\log \mathcal{L}(\boldsymbol{\theta}|\hat{p}) = n\sum_{l=1}^{L} \hat{p}_l \log p_l(\boldsymbol{\theta}).$$

Subtracting a constant term from $\log \mathcal{L}$, the equivalent function

$$\begin{cases} 2n\displaystyle\sum_{l=1}^{L} \hat{p}_l \log\left[\frac{p_l(\boldsymbol{\theta})}{\hat{p}_l}\right] & \text{if } \hat{p}_l > 0 \\ 0 & \text{if } \hat{p}_l = 0 \end{cases} \qquad (8)$$

can be maximized as a function of $\boldsymbol{\theta}$ and asymptotically it has the χ^2 distribution with $L - 3M - 1$ degrees of freedom (assuming all parameter estimates in (7) are random variables). The hypothesis that an M regime-model fits the data will be accepted if the *p-value* of χ^2 is greater than 5% (of significance level). The hypothesis test is performed for $M = 1, 2, 3, 4$, and 5 to choose the number of regimes with the largest *p-value*.

The regime present in a given quarter is determined using the conditional probability based on the estimated parameter vector ($\boldsymbol{\theta}$).

If Q_t denotes the random variable representing the regime of the quarter t, this conditional probability is

$$Prob\,(Q_t = m|R_t; \boldsymbol{\theta}) = f_m(R_t; \boldsymbol{\theta}_m)/f(R_t; \boldsymbol{\theta}). \tag{9}$$

Then, the regime present in quarter t, denoted m_t^*, is defined as the regime that maximizes the conditional probability, see Powell *et al.* (2007):

$$m_t^* = \arg\max_m \{Prob\,(Q_t = m|R_t; \boldsymbol{\theta})\}. \tag{10}$$

3.1 *Regime analysis (1971–2010)*

For the time series of 160 historical quarters from 1971 to 2010, see Figure 1(a), the resulting frequency distributions are shown in Figure 1(b). The log-likelihood estimation results with varying number of regimes are reported in Table 2, which shows that the highest *p-value* in the goodness of fit test is achieved with the mixture of three Gaussian regimes (when up to five regimes are tested). For the 3-regime case, it is evident that there is a dominant "Normal" (N) regime (with a weight of 75.2%) with a positive quarterly mean return of 2.9%. A less frequent "Weak" (W) regime (with a weight of 7.7%) is characterized by the negative average quarterly returns of −10.6%, while the "Strong" (S) regime pertains to about 17% of quarters with an average return of 8.9%.

Applying the conditional probability in (9), m_t^* is determined to be either W, N, or S for each quarter t, $t = 1, \ldots, 160$. This results in 13, 119, and 28 quarters of regimes W, N, and S, respectively, thus roughly agreeing with the weights δ of the mixer distribution for the 3-regime model. Quarterly regime transition probability is defined by

$$\pi_{m_1 m_2} = Prob(Q_{t+1} = m_2|Q_t = m_1), \tag{11}$$

which is assumed to be stationary. They are empirically estimated and given in Table 3. These probabilities indicate, as expected, that quarters with returns that are largely from either Strong or Weak regimes are not highly persistent. Unconditional regime probabilities, $\pi_m^* := Prob(Q_t = m)$, are in the right-most column of Table 3.

3.2 *Regime-based firm-RFS*

In Section 2.1, a given firm j's relative financial strength is defined as the expectation of the (logarithm of) DEA score from model (1) given a history $\mathcal{H}$ of the accounting data ξ of the parameters in Table 1, i.e., RFS value

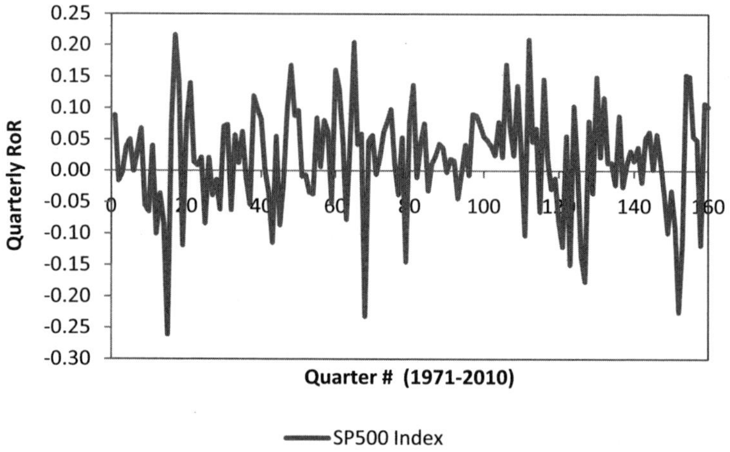

(a) Index quarterly return series

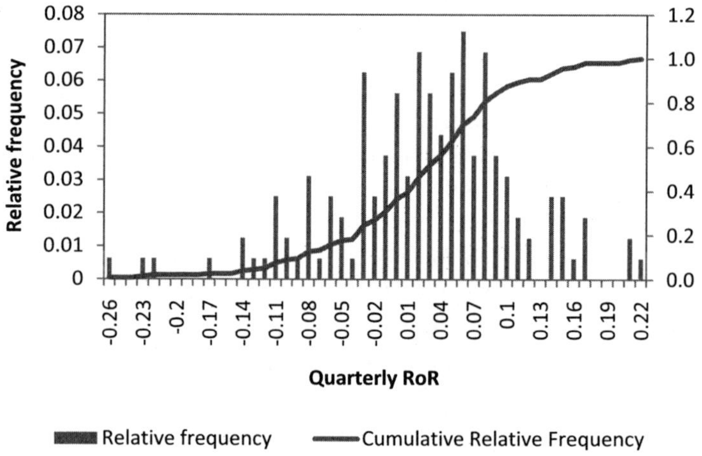

(b) Index return distribution

Figure 1 SP500 Market Index Quarterly Returns (1971–2010).

$\bar{s}_j(\mathcal{H}) := E[s_j|\xi \in \mathcal{H}]$. Indeed, as the regime of the history changes, so does the RFS. Accordingly, we define a "regime-based RFS" to represent firm fundamental strength given that the firm operates in an economy endowed with a specified regime. That is,

$$\bar{s}_j(m) := E[s_j|\xi \in \mathcal{H}(m)] \tag{12}$$

Table 2 Goodness of Fit of Market Regimes.

Regimes (M)	Parameter Vector ($\boldsymbol{\theta}$)	χ^2 value	p-value
1	$\boldsymbol{\theta}_1 = [0.0198, 0.0831, 1.0]$	68.4927	0.0174
2	$\boldsymbol{\theta}_1 = [0.0080, 0.0967, 0.648]$ $\boldsymbol{\theta}_2 = [0.0410, 0.0383, 0.352]$	59.0527	0.0642
3	$\boldsymbol{\theta}_1 = [-0.1061, 0.0554, 0.0769]$ $\boldsymbol{\theta}_2 = [0.0289, 0.0815, 0.7515]$ $\boldsymbol{\theta}_3 = [0.0891, 0.0260, 0.1716]$	53.9972	0.084
4	$\boldsymbol{\theta}_1 = [-0.2024, 0.0168, 0.0226]$ $\boldsymbol{\theta}_2 = [0.0539, 0.0297, 0.2826]$ $\boldsymbol{\theta}_3 = [0.1562, 0.0325, 0.0831]$ $\boldsymbol{\theta}_4 = [-0.0032, 0.0672, 0.6117]$	55.3638	0.0341
5	$\boldsymbol{\theta}_1 = [-0.2417, 0.0163, 0.0187]$ $\boldsymbol{\theta}_2 = [0.0642, 0.0195, 0.1876]$ $\boldsymbol{\theta}_3 = [0.1246, 0.0468, 0.1736]$ $\boldsymbol{\theta}_4 = [-0.1035, 0.0326, 0.1188]$ $\boldsymbol{\theta}_5 = [0.0065, 0.0400, 0.5013]$	49.7015	0.0510

Table 3 Quarterly Transition Probabilities for 3-Regime Model.

Regime in quarter t	Regime in quarter $t+1$			State probability (π_m^*)
	Weak	Normal	Strong	
Weak	0.1667	0.5833	0.2500	0.0813
Normal	0.0750	0.8000	0.1250	0.7437
Strong	0.0371	0.6296	0.3333	0.1750

is a regime-based RFS given that state of the economy is in regime m, where $m = W, N, S$. The historical sequence of data $\mathcal{H}(m)$ is identified by

$$\mathcal{H}(m) = \{\xi^t, t = 1, \ldots | m_t^* = m\}, \tag{13}$$

where ξ^t is the vector of data (for parameters in Table 1) in historical quarter t.

4 Portfolio optimization under regime-based RFS

We develop a dynamic portfolio optimization model in two quarters that accommodates stock selections under firm strength. The notable feature of the model is that stock (universe) selections for the two quarters are guided by the regime-based RFS, see (12), and the presence of each regime is specified in the portfolio optimization model using a regime-scenario tree.

Under the usual mean-variance risk/return trade-off, we consider a two-stage stochastic optimization model with transactions costs of trading. It may be considered an extension of the standard Markowitz model to two periods under firm selections via regime-based RFS.

4.1 *RFS-based stock selections*

Given the universe of J firms, suppose the regime-based RFS of firm j for a future quarter in regime m is computed as $\bar{s}_j(m)$, where $j = 1, \ldots, J$ and $m = W, N, S$, having chosen the history $\mathcal{H}(m)$ appropriately. The unconditional RFS for firm j, denoted $\bar{s}_j^*$, is regime-independent in the sense that it utilizes historical data irrespective of specific regimes, i.e., $\bar{s}_j^* := E[s_j | \xi \in \cup_m \mathcal{H}(m)]$. The use of either $\bar{s}_j(m)$ or $\bar{s}_j^*$ is contingent upon whether the regime of the future quarter is known (to be m) or unknown.

A large (small) RFS value is indicative of the firm being a candidate for "long" ("short") investment. Accordingly, we specify long (short) thresholds for identifying sets of stocks for long/short investments under both unconditional or regime-based RFS. Let the long-threshold be denoted by κ_L, where $\kappa_L \in (-4.61, +4.61)$, and a short-threshold be denoted by κ_S, where $\kappa_S \in (-4.61, \kappa_L)$. Then, the firm j is declared a potential long-investment for a future quarter if either $\bar{s}_j^* \geq \kappa_L$ when the regime is unknown for the quarter, or $\bar{s}_j(m) \geq \kappa_L$ when the future is known to be in regime m. Short investments are defined in a similar manner when RFS falls below κ_S. This process identifies unconditional long/short stock lists for further evaluation as:

$$\begin{aligned} &\text{Unconditional Long Set (ULS)} : \mathcal{N}_L^* := \{j : \bar{s}_j^* \geq \kappa_L,\ j = 1, \ldots, J\} \\ &\text{Unconditional Short Set (USS)} : \mathcal{N}_S^* := \{j : \bar{s}_j^* \leq \kappa_S,\ j = 1, \ldots, J\}, \end{aligned} \tag{14}$$

and the regime-based long/short stock lists, under regime m:

$$\begin{aligned} &\text{Regime-based Long Set (RLS)} : \mathcal{N}_L(m) := \{j : \bar{s}_j(m) \geq \kappa_L,\ j = 1, \ldots, J\} \\ &\text{Regime-based Short Set (RSS)} : \mathcal{N}_S(m) := \{j : \bar{s}_j(m) \leq \kappa_S,\ j = 1, \ldots, J\}. \end{aligned} \tag{15}$$

Moreover, the sets of "neutral" (or non-invested) stocks are denoted by:

$$\begin{aligned} &\bar{\mathcal{N}}^* := \{1, \ldots, J\} \setminus \{\mathcal{N}_L^* \cup \mathcal{N}_S^*\} \\ &\text{and}\quad \bar{\mathcal{N}}(m) := \{1, \ldots, J\} \setminus \{\mathcal{N}_L(m) \cup \mathcal{N}_S(m)\}, \quad \text{for } m = W, N, S. \end{aligned} \tag{16}$$

4.2 *Decisions under regime-scenarios*

Suppose an investor has an initial position x_j^0 (in stock j) at time '0' and revises her portfolio at the beginning of quarter 1. At the beginning of quarter 2, she rebalances her portfolio again after observing the market conditions in quarter 1, with the objective of maximizing the total net return. At time '0', the actual market regime in quarter 1 is unknown; however, the current quarter's regime at time '0' (before quarter 1) is known, denoted by $\circ$. Then, the market regime in quarter 1 is only known in probability, i.e., $m = W, N, S$ with regime Markovian probabilities $\pi_{\circ m}$, see (11) and the values in Table 3. Investment position at the beginning of quarter 1 in stock j, $\$x_j^1$, must be made *nonanticipatively*, that is, independently of a specific regime in quarter 1. However, stock investments at the beginning of quarter 2 are made contingent upon market regime m in quarter 1, and it is denoted by $\$x_j^2(m)$.

The underlying scenario tree consists of three outcomes for stock returns in quarter 1 representing the three market regimes. At time '0', the best guesses of these three outcomes are set to be the expected stock returns under each regime. Moreover, we assume that the mean return under each regime is stationary, and thus, it is independent of the quarter. That is, if the random return on stock j in quarter 1 is r_j^1, the three outcomes of returns are given by $\mu_j(m) = E[r_j^1 | Q_1 = m]$, for $m = W, N, S$. The unconditional mean return at the 'root' node (at time '0') is $\mu_j^0 = \sum_m \pi_{\circ m}\mu_j(m)$.

The regime-RFS based *investment-direction (ID) rule* is expressed, for quarter 1, as

$$x_j^1 \geq 0, \forall j \in \mathcal{N}_L^*, \quad x_j^1 \leq 0, \forall j \in \mathcal{N}_S^*, \quad \text{and} \quad x_j^1 = 0, \forall j \in \bar{\mathcal{N}}^*. \tag{17}$$

The ID-rule in quarter 2, for $m = W, N, S$, is given by

$$\begin{gathered} x_j^2(m) \geq 0, \forall j \in \mathcal{N}_L(m), \quad x_j^2(m) \leq 0, \forall j \in \mathcal{N}_S(m), \quad \text{and} \\ x_j^2(m) = 0, \forall j \in \bar{\mathcal{N}}(m). \end{gathered} \tag{18}$$

The so-called stock trade sizes in quarter 1 are $y_j^1 = |x_j^1 - x_j^0|$, and it is denoted by the vector $y^1 \in \Re^J$. Similarly, for quarter 2, the trade vector is $y^2(m) \in \Re^J$ where $y_j^2(m) = |x_j^2(m) - x_j^1|$, $j = 1, \ldots, J$, for each regime m.

4.3 *Transactions cost model*

Portfolio rebalancing is not costless. Placing a trade with a broker for execution entails a direct cost per share traded, as well as a fixed cost independent of the trade size. In addition, there is also a significant cost due to the size of the trading volume y, as well as the broker's ability to place the trading volume on the market. If a significant volume of shares is traded (relative to the market daily traded volume in the stock), then the trade execution price may be adversely affected. Such profit erosion due to market impact generally depends on the price at which the trade is desired, trade size relative to the market daily volume in the stock, and other company specifics such as market capitalization, and the beta of the security.

Our trading cost model has two parts: proportional transactions costs and market impact costs. The former cost per unit of trade in stock j is α_{0j}. The latter cost is expressed per unit of trade and it depends directly on the fraction of market daily volume of the stock that is traded in the portfolio. Ignoring the fixed costs of trading, the total trade loss function in quarter 1 is then

$$F_1(y^1) = \sum_{j=1}^{J} y_j^1 \left(\alpha_{0j} + \alpha_{1j} \frac{y_j^1}{V_j^1} \right), \tag{19}$$

where V_j^1 is the expected daily (market dollar) volume in stock j at the beginning of quarter 1. The constants α_{0j} and α_{1j} are calibrated to the market data. Similarly, the trade loss function in quarter 2 given regime m in quarter 1 is

$$F_2(y^2(m)) = \sum_{j=1}^{J} y_j^2(m) \left(\alpha_{0j} + \alpha_{1j} \frac{y_j^2(m)}{V_j^2(m)} \right). \tag{20}$$

4.4 *Budget constraints*

Suppose the initial cash position is $\$C^0$, and thus the initial wealth is $C^0 + \sum_j x_j^0$. Denoting the cash position after rebalancing in quarter 1 by C^1, we impose the self-financing budget constraint:

$$\sum_{j=1}^{J} (x_j^1 - x_j^0) + F_1(y^1) + C^1 = C^0. \tag{21}$$

Quarter 2 budget constraints are

$$\sum_{j=1}^{J} [x_j^2(m) - (1 + \mu_j(m))x_j^1] + F_2(y^2(m)) + C^2(m) = (1 + r_0)C^1,$$

$$m = W, N, S, \tag{22}$$

where $C^2(m)$ is the accumulated cash after rebalancing at the beginning of quarter 2 if the market regime m is observed in quarter 1 and r_0 is the quarterly risk-free rate of return.

4.5 *Risk-return framework*

The expected portfolio gain in quarter 1 is

$$G_1(x^1) := \sum_{j=1}^{J} \mu_j^0 x_j^1 + r_0 C^1 - F_1(y^1). \tag{23}$$

The conditional expectation of portfolio gain in quarter 2, under regime m in quarter 1, is $\sum_{j=1}^{J} \mu_j(m) x_j^2(m) + r_0 C^2(m) - F_2(y^2(m))$. Thus, the expected portfolio gain in quarter 2 is

$$\bar{G}_2(x^2) := \sum_m \pi_{om} \sum_{j=1}^{J} \mu_j(m) x_j^2(m) + r_0 C^2(m) - F_2(y^2(m)). \tag{24}$$

Risks in the decisions x^1 and $x^2(m)$ are measured by portfolio quarterly variances. The conditional covariance of stock returns, given regime m, is denoted by $\sigma_{jk}(m) := E[(r_j^1 - \mu_j(m))(r_k^1 - \mu_k(m))|Q_1 = m]$. The unconditional covariance of returns between stocks j and k is denoted by $\sigma_{jk}^0 = \sum_m \pi_{om} \sigma_{jk}(m)$. Then, the risk metric for quarter 1 is

$$\mathcal{R}_1(x^1) = \sum_{j,k} \sigma_{jk}^0 x_j^1 x_k^1, \tag{25}$$

and the expected value of risk in quarter 2 is

$$\mathcal{R}_1(x^2) = \sum_m \pi_{om} \sum_{j,k} \sigma_{jk}(m) x_j^2(m) x_k^2(m). \tag{26}$$

The investor wishes to maximize the total expected two-quarter portfolio gain consisting of G_1 and $\bar{G}_2$, such that the total risk consisting of $\mathcal{R}_1(x^1)$ and $\mathcal{R}_1(x^2)$ is controlled.

4.6 *Two-period optimization model*

The preceding model components are collected here to present the dynamic investment model in two-stage format. Let $d(\leq 1)$ represent the quarterly *discount* factor. The decision model in quarter 1 is:

$$\max \sum_{j=1}^{J} \mu_j^0 x_j^1 + r_0 C^1 - F_1(y^1) - \lambda \sum_{j,k} \sigma_{jk}^0 x_j^1 x_k^1 + d \sum_{m=W,N,S} \pi_{om} \psi_m(x^1, C^1)$$

$$\text{s.t.} \sum_{j=1}^{J} x_j^1 + F_1(y^1) + C^1 = C^0 + \sum_{j=1}^{J} x_j^0$$

$$y^1 = |x^1 - x^0|, \quad F_1(y^1) = \sum_{j=1}^{J} y_j^1 \left(\alpha_{0j} + \alpha_{1j} \frac{y_j^1}{V_j^1} \right)$$

$$x_j^1 \geq 0, \forall j \in \mathcal{N}_L^*; \quad x_j^1 \leq 0, \forall j \in \mathcal{N}_S^*; \quad x_j^1 = 0, \forall j \in \bar{\mathcal{N}}^*, \tag{27}$$

where the quarter 2 value function, under regime m $(= W, N, S)$, is given by

$$\psi_m(x^1, C^1)$$

$$= \max \sum_{j=1}^{J} \mu_j(m) x_j^2(m) + r_0 C^2(m) - F_2(y^2(m)) - \frac{\lambda}{d} \sum_{j,k} \sigma_{jk}(m) x_j^2(m) x_k^2(m)$$

$$\text{s.t.} \sum_{j=1}^{J} x_j^2(m) + F_2(y^2(m)) + C^2(m) = (1 + r_0) C^1 + \sum_{j=1}^{J} (1 + \mu_j(m)) x_j^1$$

$$y^2(m) = |x^2(m) - x^1|, \quad F_2(y^2(m)) = \sum_{j=1}^{J} y_j^2(m) \left(\alpha_{0j} + \alpha_{1j} \frac{y_j^2(m)}{V_j^2(m)} \right)$$

$$x_j^2(m) \geq 0, \forall j \in \mathcal{N}_L(m); x_j^2(m) \leq 0, \forall j \in \mathcal{N}_S(m); x_j^2(m) = 0, \forall j \in \bar{\mathcal{N}}(m). \tag{28}$$

5 Model application

The preceding two-stage optimization model using stock selections under regime-based RFS is applied in the US markets. We consider the S&P 500 index stocks individually, as well as the 9 separate sector Exchange-Traded Funds (ETF) spanning the index, with ticker symbols: XLK (Technology), XLV (Health Care), XLB (Basic Materials), XLI (Industrial Goods), XLE (Energy), XLY (Consumer Discretionary), XLP (Consumer Staples), XLU (Utilities), and SLF (Financials). The number of firms in each sector for performance analysis is 86, 57, 31, 53, 29, 90, 38, 32, and 84, respectively. For these 500 firms, quarterly accounting data for parameters in Table 1 were obtained for the period 1971–2010. Not all firms have complete data for the entire history period. Also, financial sector firms are excluded from our analysis following the common practice in many empirical studies in finance. The basic argument is that financial stocks are not only sensitive to the standard business risks of most industrial firms but also to changes in interest rates. In this regard, the famous Fama and French (1992) study also noted: "we exclude financial firms because the high leverage that is normal with these firms probably does not have the same meaning as for non-financial firms, where high leverage more likely indicates distress."

The above sectors have distinct characteristics that make them unique representatives of the overall market. Therefore, evaluation of the fundamental strength of a firm j must be made relative to a representative set of firms of the market sector to which the firm j belongs. Accordingly, the model (1) is applied within each market sector separately. The quarterly historical strength scores s_j so-computed are then subject to the correlation analysis in Section 2.3 where the history is all quarters, $\mathcal{H}_0 = \{t | t \in [71Q1, 10Q4]\}$. Using quarterly stock returns in $\mathcal{H}_0$, the average firm correlation metric $\bar{\gamma}(\mathcal{H}_0)$ is computed for each sector, see Table 4. Also, we determine the correlations $\bar{\gamma}(\mathcal{H}(m))$ for each regime $m = W, N, S$, identified by the histories:

$$\mathcal{H}(m) = \{t | m_t^* = m, t \in [71Q1, 10Q4]\}, \quad \text{for } m = W, N, S, \tag{29}$$

where regime status m_t^* of a quarter is established in (10) under market regime analysis. The test statistic Φ in (3), using $\rho_0 = 10\%$ and significance level $\alpha = 5\%$, confirms that all sectors in all regimes yield significant correlations, implying that model (1) is a viable tool for determining firm-fundamental strength.

Table 4 Sector Firm-Correlations under Market Regimes.

		Regimes: Weak, Normal, Strong		
Sector name	**Overall $\bar{\gamma}(\mathcal{H}_0)$**	$\bar{\gamma}(\mathcal{H}(W))$	$\bar{\gamma}(\mathcal{H}(N))$	$\bar{\gamma}(\mathcal{H}(S))$
Technology	0.329	0.310	0.341	0.291
Health Care	0.212	0.199	0.271	0.220
Basic Mat.	0.187	0.176	0.197	0.181
Industrial Goods	0.267	0.246	0.269	0.234
Energy	0.169	0.160	0.186	0.165
Consumer Discre.	0.248	0.226	0.250	0.256
Consumer Stap.	0.298	0.268	0.311	0.276
Utilities	0.204	0.189	0.221	0.205

5.1 *RFS estimation and firm selections*

The out-of-sample investment period is set to quarter 1 and quarter 2 of 2011, assuming that historical data is available only up to the end of 2010. Thus, the relative firm strength (RFS) in (12) is needed to specify stock selections in 11Q1 and 11Q2. More precisely, for the nonanticipative 11Q1 investment, we need unconditional RFS for each firm, while for 11Q2 we need the conditional RFS contingent upon the specific regime scenario being followed in 11Q1. To compute the expectation in (12) for a future period t, a variety of time series techniques may be employed; however, for the purposes of this chapter, a weighted moving average method is applied, using a historical t_0 quarters, as follows:

$$\bar{s}_j(m) := \sum_{\tau \in \mathcal{H}(m;t_0)} v_\tau s_j^\tau(m). \tag{30}$$

$\mathcal{H}(m;t_0)$ represents the most-recent t_0 quarters in history $\mathcal{H}(m)$, $s_j^\tau(m)$ is the historical firm strength in quarter τ, and the convex multipliers v_τ satisfy $v_\tau = 2v_{\tau-1}$, i.e., historical strength in period τ is considered to be twice as influential as that in period $\tau - 1$, and thus, v forms a geometric progression. Setting $t_0 = 4$, the multipliers are $\frac{1}{15}$, $\frac{2}{15}$, $\frac{4}{15}$, and $\frac{8}{15}$. Moreover, the resulting historical periods for conditional RFS are: $\mathcal{H}(W;4) = \{02Q3, 08Q4, 09Q1, 10Q2\}$, $\mathcal{H}(N;4) = \{08Q2, 08Q3, 09Q4, 10Q1\}$, and $\mathcal{H}(S;4) = \{09Q2, 09Q3, 10Q3, 10Q4\}$. The unconditional RFS, $\bar{s}_j^*$, is calculated using the four-quarter history $\mathcal{H} = \{10Q1, 10Q2, 10Q3, 10Q4\}$. The unconditional RFS values are plotted in Figure 2 for each sector.

Referring to (14) and (15), candidate long and short stocks are chosen setting the thresholds $\kappa_L = -0.36$ and $\kappa_S = -1.61$ (corresponding to DEA-efficiencies of 70% and 20%, respectively). For 11Q1 investment using the unconditional RFS, this yields a total of $|\mathcal{N}_L^*| = 97$ long stocks and $|\mathcal{N}_S^*| = 77$ short stocks out of the 416 stocks in all 8 sectors (excluding Financials sector). The sectoral breakdown is given in Table 5. In the same table, the regime-based stock selections are also reported for all sectors. It is evident that a transition from Weak to Strong regime results in an increase in long selections along with a decrease in short selections. Also note that roughly 60% of the stocks are 'neutral' in each market regime.

5.2 *Portfolio analysis*

The portfolio model in (27)–(28) is implemented for the two quarters 2011Q1 and 2011Q2, where the current time '0' is the end of 2010Q4. The

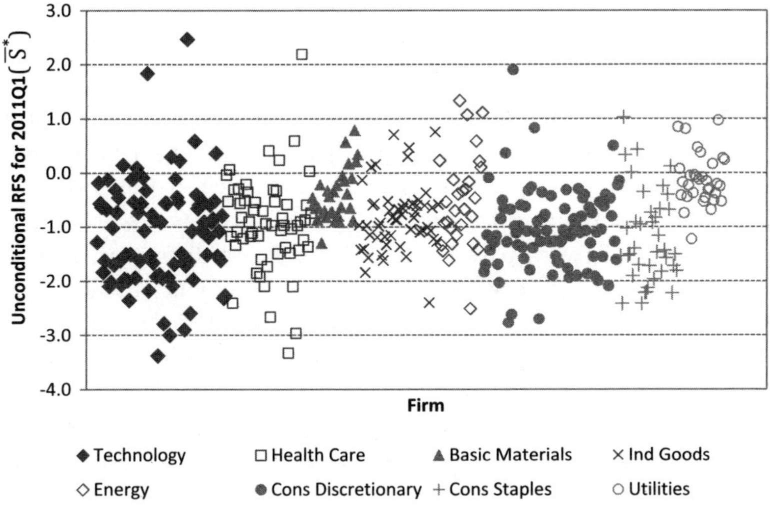

Figure 2 Unconditional RFS for Firms in Each Sector for 2011Q1.

Table 5 Stock Selections under Market Regimes.

	Unconditional		Weak		Normal		Strong	
Sector (# stocks)	Long	Short	Long	Short	Long	Short	Long	Short
Technology (86)	15	32	10	36	14	22	23	16
Health Care (57)	13	9	10	13	11	8	15	8
Basic Mat (31)	12	0	9	4	12	2	15	0
Ind Goods (53)	7	3	3	8	5	3	9	2
Energy (29)	11	2	5	4	9	2	14	3
Cons Discre (90)	11	18	7	20	8	16	11	11
Cons Staples (38)	7	13	6	16	8	12	12	9
Utilities (32)	21	0	14	5	18	2	24	0
Total (416)	97	77	64	106	85	67	123	49

unconditional stock selections are applied in 11Q1 and the regime-based stock selections are considered for 11Q2 under the three regime scenarios, see Table 5.

Regime-based mean return $\mu_j(m)$ is estimated from the historical returns in $\mathcal{H}(m)$ and the unconditional mean μ_j^0 is set to $\sum_m \pi_{\circ m}\mu_j(m)$ where the current regime (in 10Q4) is 'Strong', i.e., $\circ = S$. The history $\mathcal{H}(m)$ is also used in computing the covariances σ_{jk}^0 and $\sigma_{jk}(m)$. The transaction cost parameters $\alpha_{0j} = 2\%$ and $\alpha_{1j} = 1$ for all stocks, while the discount factor $d = 1$ and the risk-free rate $r_0 = 0$ is set. The trading volumes V_j^1 and $V_j^2(m)$ are computed using the average daily volume in a typical quarter 1 and 2, respectively, using the relevant data history. If $x_j^1 x_j^2(m) < 0$ holds for some stock j, then it is assumed that the total traded volume in quarter 2 is simply the difference in the positions, for the purpose of computing trading costs. In reality, such a transaction is often the result of two separate trades in opposite directions. In essence, thus, the model slightly over-estimates the transactioncs costs. Portfolios resulting from (27)–(28) are referred to as 'regimeRFS' portfolios.

On the other hand, under no firm-specific fundamental data, sector ETFs are used to form ETF portfolios under the regime-based two-quarter optimization. In this case, all sectors (except the Financials) are used in both quarters for long/short investments. All required parameters are estimated in a similar manner to the RFS-based portfolio model using the relevant regime histories. The resulting portfolios are referred to as 'regimeETF' portfolios. Furthermore, non-regime oriented static portfolios are also constructed where the model component for 11Q2 is dropped from (27)–(28) and the positions created in quarter 1, i.e., x_j^1, are held unchanged through the end of quarter 2. These are referred to as 'staticRFS' and 'staticETF' portfolios.

All initial stock positions are set to zero at the beginning of Jan 2011 for the model execution. Then, the model is rolled-over to the end of 11Q1 to determine optimal positions at the beginning of 11Q2. In the case of regime-based two-quarter models, for the second rollover of the model, it is assumed that long/short stock selections remain the same as in the first model run. Only the moments of the stock returns are updated for the second model run (at the beginning of 11Q2) using the actual data of 11Q1. Optimal positions so-obtained via all four models are out-of-sample simulated using the actual prices in Jan–Jun, 2011. These out-of-sample mean-standard deviation efficient frontiers for the four portfolios, regimeRFS, regimeETF, staticRFS, and staticETF, are in Figure 3. It is evident that regime-based portfolios are more efficient than their static counterparts. Moreover, RFS portfolios are superior to ETF portfolios both in regime-based and static cases. The performance of RFS-based portfolios becomes more pronounced at higher levels of variance risk.

Since the market index has an annualized volatility of 13.4% during Jan–Jun, 2011, the risk tolerance parameter λ is adjusted such that the resulting ETF-based and RFS-based portfolios also will have an

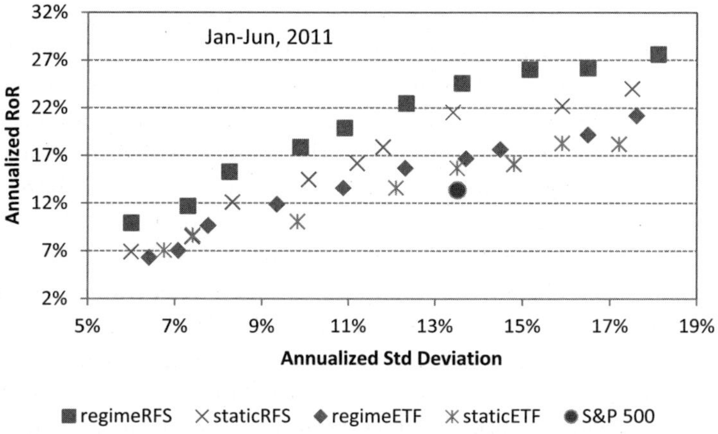

Figure 3 Portfolio (out-of-Sample) Efficient Frontiers.

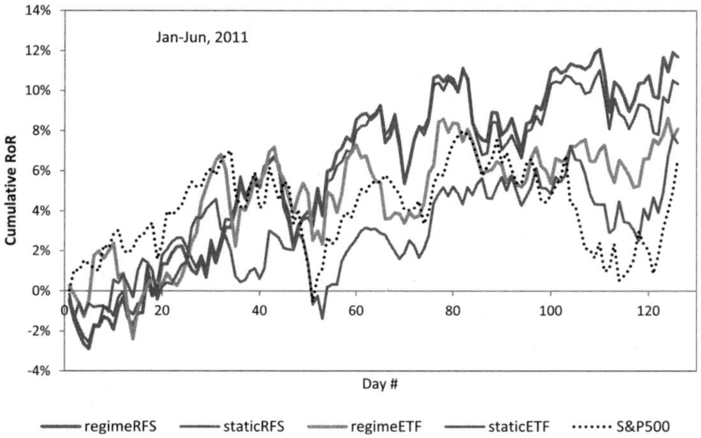

Figure 4 Out-of-sample Performance of ETF and RFS Portfolios.

out-of-sample annualized volatility of roughly 13.4%. Performance comparison of these four portfolios is presented in Figure 4.

6 Conclusions

This chapter presents an integrated methodology that combines stock selection based on firm fundamentals, regime-switching nature of the markets, and dynamic optimization modeling of risk-return trade-off. Using quarterly data from 1971–2010, it is concluded that there are three main market regimes, and their parameters and transition probabilities are estimated. Given a certain regime, we present a DEA-based methodology to determine a firm's relative firm strength (RFS), by comparing its fundamental financials with those of other firms operating in a similar industry or the same market sector. This process allows the analyst to identify which stocks are likely strong long or short investments and under what market regime should they be employed. Therefore, as economic conditions change, so do the long/short stock selections. Such a scheme is able to reduce the long-term (in this case, quarterly) risk exposure of a stock portfolio. These stock selections are then subject to portfolio analysis within a two-period mean-variance stochastic optimization model in which regime-based scenarios are incorporated. The out-of-sample portfolio results demonstrate that this methodology yields superior performance in comparison to using sector-based ETF portfolios or the market index itself. Also, it is shown that the proposed dynamic model outperforms a static model of investments in our out-of-sample testing.

The portfolio analysis in this chapter may be improved in several ways. For example, although the risk-return optimization model utilized the standard mean-variance framework, various other risk measures may be utilized. Also, for portfolio performance measurement, this chapter relied on the standard deviation risk; however, there are several other important dimensions of performance, such as maximum drawdown, reward-to-drawdown, etc., see Edirisinghe (2007). Finally, in the two-quarter rollover of the regime-based stochastic optimization model, we assumed RFS-based stock selections to be invariant in the second quarter. However, in practice, it is desirable to re-compute RFS in the second rollover using the actual accounting information of the firms in the already observed first quarter. It is conjectured that in such a setting, the resulting portfolio performance may be further improved.

Acknowledgement

The authors wish to thank Wei Wu, graduate student at the University of Tennessee, for computational assistance provided for the work in this chapter.

References

Andersen, P. and N.C. Petersen. A procedure for ranking efficient units in Data Envelopment Analysis. *Management Science*, 39: pp. 1261–1264, 1993.

Ang, A. and G. Bekaert. International Asset Allocation with Regime Shifts. *Review of Financial Studies*, 15: pp. 1137–1187, 2002.

Ball, R. and P. Brown. An Empirical Evaluation of Accounting Income Numbers. *Journal of Accounting Research*, 6: pp. 159–178, 1968.

Banker, R.D. Estimating most productive scale size using data envelopment analysis. *European Journal of Operational Research*, 17: pp. 35–44, 1984.

Brooks, C. and A. Katsaris. A Three-Regime Model of Speculative Behaviour: Modelling the Evolution of the S&P 500 Composite Index. *The Economic Journal*, 115: pp. 767–797, 2005.

Campbell, J.Y. Asset pricing at the millennium. *Journal of Finance*, 55: pp. 1515–1567, 2000.

Campbell, J.Y. and J.H. Cochrane. By force of habit: A consumption-based explanation of aggregate stock market behavior. *The Journal of Political Economy*, 107: pp. 205–251, 1999.

Charnes, A., W.W. Cooper, and E. Rhodes. Measuring the efficiency of decision-making units. *European Journal of Operational Research*, 2: pp. 429–444, 1978.

Edirisinghe, N.C.P. Integrated risk control using stochastic programming ALM models for money management. In S.A. Zenios and W.T. Ziemba, eds., *Handbook of Asset and Liability Management*, volume 2, chapter 16, pp. 707–750. Elsevier Science BV, 2007.

Edirisinghe, N.C.P. and X. Zhang. Generalized DEA model of fundamental analysis and its application to portfolio optimization. *Journal of Banking and Finance*, 31: pp. 3311–3335, 2007.

Edirisinghe, N.C.P. and X. Zhang. Portfolio selection under DEA-based relative financial strength indicators: case of US industries. *Journal of the Operational Research Society*, 59: pp. 842–856, 2008.

Edirisinghe, N.C.P. and X. Zhang. Input/output selection in DEA under expert information with application to financial markets. *European Journal of Operational Research*, 207: pp. 1669–1678, 2010.

Fama, E.F. and K.R. French. The Cross-Section of Expected Stock Returns. *Journal of Finance*, 47: pp. 427–465, 1992.

Farrell, M.J. The Measurement of Efficiency of Production. *Journal of the Royal Statistical Society (Series A)*, 120: pp. 251–281, 1957.

Frankel, R. and C.M.C. Lee. Accounting valuation, market expectations, and cross-sectional stock returns. *Journal of Accounting and Economics*, 25: pp. 283–319, 1998.

Graflund, A. and B. Nilsson. Dynamic Portfolio Selection: the Relevance of Switching Regimes and Investment Horizon. *European Financial Management*, 9: pp. 179–200, 2003.

Guidolin, M. and A. Timmermann. International asset allocation under regime switching, skew, and kurtosis preferences. *Review of Financial Studies*, 21: pp. 889–935, 2008.

Hamilton, J.D. A new approach to the economic analysis of nonstationary time series and the business cycle. *Econometrica*, 57: pp. 357–384, 1989.

Hamilton, J.D. *Time Series Analysis*. Princeton, NJ: Princeton University Press, 1994.

Hirschey, M. Extreme Return Reversal in the Stock Market- Strong support for insightful fundamental analysis. *The Journal of Portfolio Management*, 29: pp. 78–90, 2003.

Naik, V. Option valuation and hedging strategies with jumps in the volatility of asset returns. *Journal of Economic Theory*, 48: pp. 1969–1984, 1993.

Ohlson, J. Earnings, book values, and dividends in security valuation. *Contemporary Accounting Research*, 11: pp. 661–687, 1995.

Powell, J., R. Roa, J. Shi, and V. Xayavong. A Test for Long-Term Cyclical Clustering of Stock Market Regimes. *Australian Journal of Management*, 32: pp. 205–221, 2007.

Samaras, G.D., N.F. Matsatsinis, and C. Zopounidis. A multi-criteria DSS for stock evaluation using fundamental analysis. *European Journal of Operational Research*, 187: pp. 1380–1401, 2008.

Sarno, L. and G. Valente. The cost of carry model and regime shifts in stock index futures markets: An empirical investigation. *Journal of Futures Markets*, 20: pp. 603–624, 2000.

Wahab, M.I.M., Z. Yin and N. C. P. Edirisinghe. Pricing swing options in the electricity markets under regime-switching uncertainty. *Quantitative Finance*, 10: pp. 975–994, 2010.

Yin, G. and X. Y. Zhou. Markowitz's Mean-Variance Portfolio Selection With Regime Switching: From Discrete-Time Models to Their Continuous-Time Limits. *IEEE Transactions on Automatic Control*, 49: pp. 349–360, 2004.

Chapter 7

Option Portfolio Management as a Chance Constrained Problem

Dmitry Golembiovsky* and Anatoliy Abramov†

Summary

In a risk-neutral world the price of an option is equal to its expected payoff discounted at the risk-free interest rate. So, the expected return from investment in any static option portfolio corresponds with the risk-free rate. However, it is possible to manage the portfolio dynamically in such a way that it provides higher return with a probability close to unity and lower return (possibly large negative return) with a given low probability. For this purpose stochastic programming with chance constraints can be used.

In the paper the problem of option portfolio management is investigated. The hypothetical market considered includes European options expiring in the nearest month. It is a risk-neutral market where the underlying asset follows a geometric Brownian motion and prices of all options are calculated on the basis of the Black-Scholes formula. A stochastic chance constrained program for option portfolio management is developed along with the corresponding multinomial scenario tree.

The results of a Monte-Carlo simulation of the portfolio management are presented. They confirm that it is possible to make more than the risk-free return in a risk-neutral options market.

1 Introduction

Usually exchange-traded options on an underlying asset expire on a day of the month, often the third Friday or the last business day preceding the third Friday. There are three four-month cycles that govern which expiration dates are traded on a given day in a given month. The January

*Department of Risks, Bank ZENIT, Moscow Financial-Industrial University "Sinergy" 9, Banny Lane, Moscow 129110, Russia, d.golembiovsky@zenit.ru.

†Department of Risks, Bank ZENIT, Moscow Financial-Industrial University "Sinergy" 9, Banny Lane, Moscow 129110, Russia, a.abramov@zenit.ru.

four month cycle includes January, April, July and October. The February cycle consists of February, May, August and November. The March cycle's months are March, June, September and December (Hull, 2005).

The options traded usually expire during some consecutive months following the trading date, as well as the next several months in the month's trading cycle. For instance, options trading on 8 March 2012, might have expiration dates in March 2012, April 2012, May 2012 (three consecutive months) as well as June 2012, September 2012, December 2012 and March 2013 (the next four months in the March cycle). Options of several strikes for each expiration month are traded at the same time.

Every day exchanges calculate the settlement price of each option after completion of trading. The fundamental works on stock option pricing were published by Black and Scholes (1973) and Merton (1973). The formula for futures options was suggested by Black (1976). Although an enormous number of alternative option pricing models have been developed since that time, exchanges nevertheless usually use these models for calculating the settlement prices of European options. To price American options the scheme of Cox, Ross and Rubinstein (1979) is commonly applied. Some exchanges use this scheme for European options also.

One of the main assumptions of the Black, Scholes, Merton's and Cox, Ross, Rubinstein's models is the constancy of underlying asset volatility. To take into account a variable volatility of real underlying assets, the concept of implied volatility is used. So, each option has its own implied volatility under the settlement price calculation. Determining the implied volatility, exchanges try to ensure fair settlement prices for all options and make the market arbitrage-free.

Settlement prices are used for evaluation of option positions and for calculation of collateral. To hold derivative positions some collateral is required by exchanges. Derivative portfolio collateral is the sum of two components: the amount of money which is necessary for immediate closing option positions (*option liquidation value*) and estimation of maximal possible one day losses. The second component is usually named *maintenance margin*. If the option liquidation value is negative (money will be received from the closing of the positions), it decreases the needed collateral. Some exchanges also calculate the so called *initial margin* which is required to open positions. The initial margin is simply the maintenance margin multiplied by some coefficient which is more than unity.

In 1988 the Chicago Mercantile Exchange (CME) developed the SPAN methodology (Standard Portfolio ANalysis of Risk) for maintenance margin calculation. The first step of SPAN is *Scanning Risk*. 16 scenarios of underlying price and implied volatility are considered. Portfolio losses corresponding to each scenario are calculated. The maximum of these 16 figures is the Scanning Risk of the portfolio. The next steps of SPAN make the estimation of possible one-day portfolio losses more precise. They take into account divergence of the same derivatives for different maturities, spread credits from derivatives on different underlying assets, risk increasing close to delivery and short option minimal charge (Chicago Mercantile Exchange, 1999). There are some modifications of SPAN; one of them is from London Clearing House (LCH Clearnet, 2008).

In this paper we consider the hypothetical exchange market where the cash settled European options of one expiration month are traded simultaneously. The underlying asset does not pay a dividend. We can think about it as a non-dividend share. The price of the underlying asset follows a geometric Brownian motion with constant volatility and drift equals the risk-free rate. The options are quoted on the basis of the Black-Scholes formula. Some commission is paid for buying and selling each option. Each trading day is considered as one moment in time. For simplicity, the working days are considered only. There are 21 working days in each month. Options expire on the 21st day of the relevant month. We don't consider trading underlying asset but only options trading. Option portfolio maintenance margin and initial margin coincide. They are calculated according to the first step of SPAN methodology, so, it is the Scanning Risk of a portfolio.

Excluding the commission, it is a risk-neutral market, because an option price is its expected payoff discounted at the risk-free rate of interest. An expected return of any investment in a portfolio in the considered market accords with the risk-free interest rate if the portfolio is held until the options expire. To show this, let us consider the simplified case of a zero risk-free rate. Then the expected return from any bought option is equal to zero, as is the expected return from option selling. Consequently, the expected return from any static option portfolio is zero again. If the commission is taken into account, the expected return of a portfolio in the considered market is less than zero.

However, it is possible to restructure the portfolio dynamically in such a way that it provides higher return with probability close to unity and lower return (possibly a large negative return) with a given low probability. For developing such a strategy the appropriate technique is stochastic programming with chance constraints, i.e., the constraints which have to be satisfied with a high probability (Prékopa, 2003). In general, the chance constrained problems are non-convex and difficult to solve. However, there is a convex approximation for a relatively significant class of chance constrained problems with continuous random variables (Nemirovski and Shapiro, 2006). The precise algorithm for the case of discrete distribution was given by Ruszczyński (2002). Recently some papers appeared which provide effective heuristics for solving large chance constrained problems with discrete random variables (Watson *et al.*, 2010; Tanner and Beier, 2008).

An important factor for the success of a stochastic program is a scenario model. Scenario trees are commonly used for option pricing. Since the underlying asset price follows a Wiener process in the considered market, it would be natural to use the binomial recombining tree of Cox, Ross and Rubinstein as a scenario model. However, research shows that two successors for each of the nodes are not enough for option portfolio management. Using a binomial tree, the real portfolio value on the next stage can differ substantially from the planned values (Golembiovsky and Abramov, 2011). This leads to poor results in the portfolio management.

Laurent developed a method of tree construction for a real option market based on the Wasserstein distance (Laurent, 2006). It takes into account variable volatility and real option prices on the market. As long as we consider a hypothetical option market with constant underlying price volatility, some kind of simple "bracket mean" approximation of the Gaussian distribution is used for the tree construction.

The outline of the paper is as follows. In the next section we present the model of the options market for portfolio optimization and simulation. One of the essential features of the market is collateral, which is calculated using the Scanning Risk of a portfolio. Section 3 presents a scenario model for portfolio optimization as a multinomial tree. Section 4 develops the problem of portfolio optimization as a stochastic chance constrained problem. In section 5 some portfolio simulation results are presented. The next section concludes and provides some directions of future research.

2 Model of options market

Let us consider a simplified model of a typical option market. It includes European cash-settled options on some non-dividend underlying asset. The options expiring in the closest month are traded simultaneously.

For simplicity the working days are considered only. There are 21 working days in each month. So, in this respect a year consists of 252 days. Each trading day is considered as one moment; we do not consider online trading. The options expire on the 21st day of the relevant month.

Settlement on the expired options is fulfilled immediately after trading. The new options appear on the 1st day of each month.

It is also assumed that underlying asset spot price follows a geometric Brownian motion. The underlying price volatility σ is a constant. The drift equals the risk-free rate. Options are priced based on the Black-Scholes formula. Some commission is collected for each trade. For simplicity, underlying asset trading is not considered; an investor can buy and sell options only.

The options are traded from the 1st to 11th day only. There is no trading during the last 10 days. On any day only such options are traded for which the logarithms of the ratios of their strikes over the current underlying price belong to the interval

$$(r - \sigma^2/2)\,t \ \pm 2.326\sigma\sqrt{t} \tag{1}$$

where r is a risk-free rate, t is time to expiry. The fractile 2.236 corresponds to a probability of 0.99. These conditions simulate low options liquidity near expiry and for strikes which are far from the money.

To trade options on the market, collateral is required. In our market the maintenance margin is determined according to the first step of Standard

Table 1 Scanning Risk scenarios. An exchange determines a standard underlying price changing interval and implied volatility changing interval for each underlying asset. Under scenarios 1-14 the underlying price and volatility are changing inside these intervals. Now CME considers underlying price increasing for 3 standard intervals and decreasing for 3 standard intervals for the 15th and 16th scenarios accordingly. Further, only 33% of losses of 15th and 16th scenarios are taken into account.

No. of Scenario	Scenario
1	Underlying unchanged, volatility up
2	Underlying unchanged, volatility down
3	Underlying up 1/3, volatility up
4	Underlying up 1/3, volatility down
5	Underlying down 1/3, volatility up
6	Underlying down 1/3, volatility down
7	Underlying up 2/3, volatility up
8	Underlying up 2/3, volatility down
9	Underlying down 2/3, volatility up
10	Underlying down 2/3, volatility down
11	Underlying up 3/3, volatility up
12	Underlying up 3/3, volatility down
13	Underlying down 3/3, volatility up
14	Underlying down 3/3, volatility down
15	Underlying up extremely (cover part of loss)
16	Underlying down extremely (cover part of loss)

Portfolio ANalysis of Risk (SPAN) – the so called "Scanning Risk" (Chicago Mercantile Exchange, 1999; LCH Clearnet, 2008). To calculate the Scanning Risk SPAN uses 16 scenarios which are presented in Table 1. (Do not confuse these SPAN scenarios with a scenario tree of a stochastic programming problem!). Day losses of all derivatives in a portfolio are calculated for each of the 16 scenarios and summarized separately. The highest of the obtained 16 figures is the Scanning Risk of the portfolio.

An example is given in Table 2. The so called "End of Month" (EOM) European options on futures of the S&P 500 index are traded at the CME. They expire on the last day of each month like options of our hypothetical market. Day losses for each traded derivative are provided by a clearing house at the end of each trading session in the so called Risk Parameter File. It may be downloaded from the CME web site. SPAN presents losses as positive figures and gains as negative figures.

The example considers the general case of options with different expiration dates. The underlying future settlement price on 11.01.2011

Table 2 Scanning Risk example on 11.01.2011. The portfolio includes 2 long Call options expired on 31.01.2011 and one short Put option expired on the 28.02.2011. The greatest loss of the portfolio is provided by the 14th scenario. The portfolio Scanning Risk is $42524.

No. of Scenario	S&P 500 EOM long Call option. Expiring on 31.01.2011. Strike 1230. Losses, $	S&P 500 EOM short Put option. Expiring on 28.02.2011. Strike 1310. Losses, $	S&P 500 EOM option portfolio. Losses, $
1	−1423	−2767	2(−1423)−1(−2767)= −79
2	1048	2166	2(1048)−1(2166)= −70
3	−7487	1647	2(−7487)−1(1647)= −16621
4	−6330	8078	2(−6330)−1(8078)= −20738
5	3543	−8077	2(3543)−1(−8077)=15163
6	7342	−5069	2(7342)−1(−5069)=19753
7	−14285	5122	2(−14285)−1(5122)= −33692
8	−13827	11275	2(−13827)−1(11275)= −38929
9	7153	−14156	2(7153)−1(−14156)= 28462
10	10582	−12553	2(10582)−1(−12553)= 33717
11	−21485	7702	2(−21485)−1(7702)= −50672
12	−21327	12207	2(−21327)−1(12207)= −54861
13	9405	−20819	2(9405)−1(−20819)= 39629
14	11236	−20052	2(11236)−1(−20052)= 42524
15	−21888	4074	2(−21888)−1(4074)= −47850
16	3720	−21467	2(3720)−1(−21467)= 28907

is 1270.4. So, both options are in the money. The settlement price of a Call is \$11350; the settlement price of a Put is \$12375. So the portfolio liquidation value is $\$12375 - \$11350 \times 2 = -\$10325$. The money will be received from portfolio if the positions are closed. The total collateral comes to $\$42524 - \$10325 = \$32199$. The calculation was checked by the program PC-SPAN 4.01 which uses the Risk Parameter File as input information.

3 A scenario model for stochastic programming

At the expiration moment the option value has a piece-wise linear dependence on the underlying price. Hence the dependence of any option portfolio value on the underlying price is also piece-wise linear. A portfolio of options with the same expiry has local extrema of its pay-off function in the points of the strikes at the expiration moment. An example is given by Figure 1. So the underlying prices equal to all strikes of options that can be traded have to be represented on the tree.

Before the expiration the option portfolio value shows a non-linear dependence on the underlying asset price. Consider again Figure 1. For the

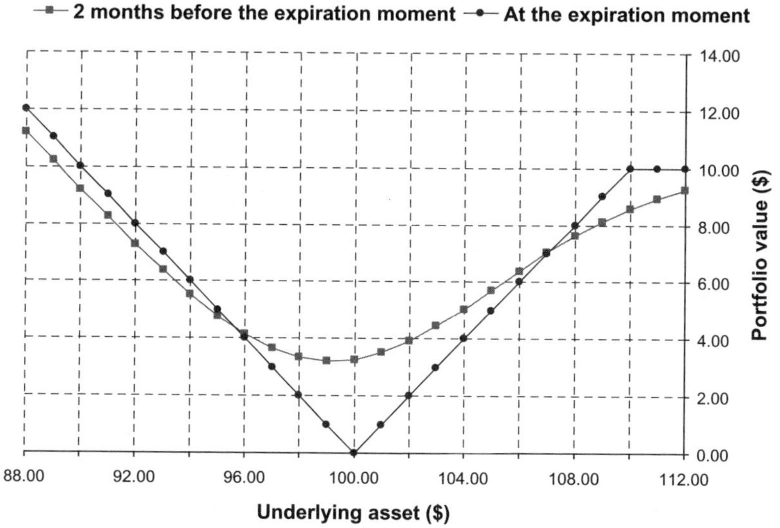

Figure 1 The portfolio consists of one bought Call option with the strike of \$100, one bought Put option with the same strike and one sold Call option with the strike of \$110. There are two months to expiry. The risk-free rate is equal to 0.05 and the price volatility is equal to 0.1.

day two months before expiry the smallest portfolio value is realized for the underlying price \$99. So it is not enough to represent only all traded strikes by the nodes of the tree. The intermediate underlying price values must be represented on the tree also with a rather small step. Denote this step by Δg_τ, where τ is the number of the tree stage. In the general case the step size can be varied for different stages.

Since the underlying price follows a geometric Brownian motion, the probability of the j-th successor of a node number n on the stage τ (later on denoted by (τ, n)) is given by

$$p^j_{(\tau,n)} = N\left(x_{+\frac{\Delta g_\tau}{2}}\right) - N\left(x_{-\frac{\Delta g_\tau}{2}}\right), \text{ as } j \in 2, \ldots, F_{(\tau,n)} - 1 \tag{2}$$

$$p^j_{(\tau,n)} = N\left(x_{+\frac{\Delta g_\tau}{2}}\right), \quad \text{as } j = 1 \tag{3}$$

$$p^j_{(\tau,n)} = 1 - N\left(x_{-\frac{\Delta g_\tau}{2}}\right), \quad \text{as } j = F_{(\tau,n)} \tag{4}$$

where $N(x)$ is the cumulative normal distribution function with parameters $(r - \sigma^2/2)\, t_\tau$ and $\sigma\sqrt{t_\tau}$; r, σ and t_τ are the risk-free rate, underlying price volatility and time to the next stage, respectively. The parameters $x_{\pm\Delta g_\tau/2}$ are the logarithms of the ratios of the underlying prices according to the successor nodes divided by the underlying price in the current node:

$$x_{\pm\frac{\Delta g_\tau}{2}} = \ln\left(\frac{\left(S^j_{(\tau,n)} \pm \frac{\Delta g_\tau}{2}\right)}{S_{(\tau,n)}}\right). \tag{5}$$

Here $S_{(\tau,n)}$ is the price of the underlying asset in the current node (τ, n), $S^j_{(\tau,n)}$ is the price of the underlying asset in the j-th successor of the node (τ, n). Each node gets $F_{(\tau,n)}$ successors such that the logarithms of ratios of the corresponding prices over the price in the node (τ, n) cover the interval

$$(r - \sigma^2/2)\, t_\tau \pm 2.326\sigma\sqrt{t_\tau}. \tag{6}$$

As mentioned, options are traded from the 1st to the 11th day of month at the considered market. We simulated everyday trading during this period of each month. From this reason we used a 4-stage scenario tree for solving the multistage stochastic programming problem for the 1st to the 9th day. Its stages correspond to the current day, the next day, the 11th day of the current month and the last (21st) day of the month. If the present day is the 10th day of a current month, a three-stage tree was used. The stages of

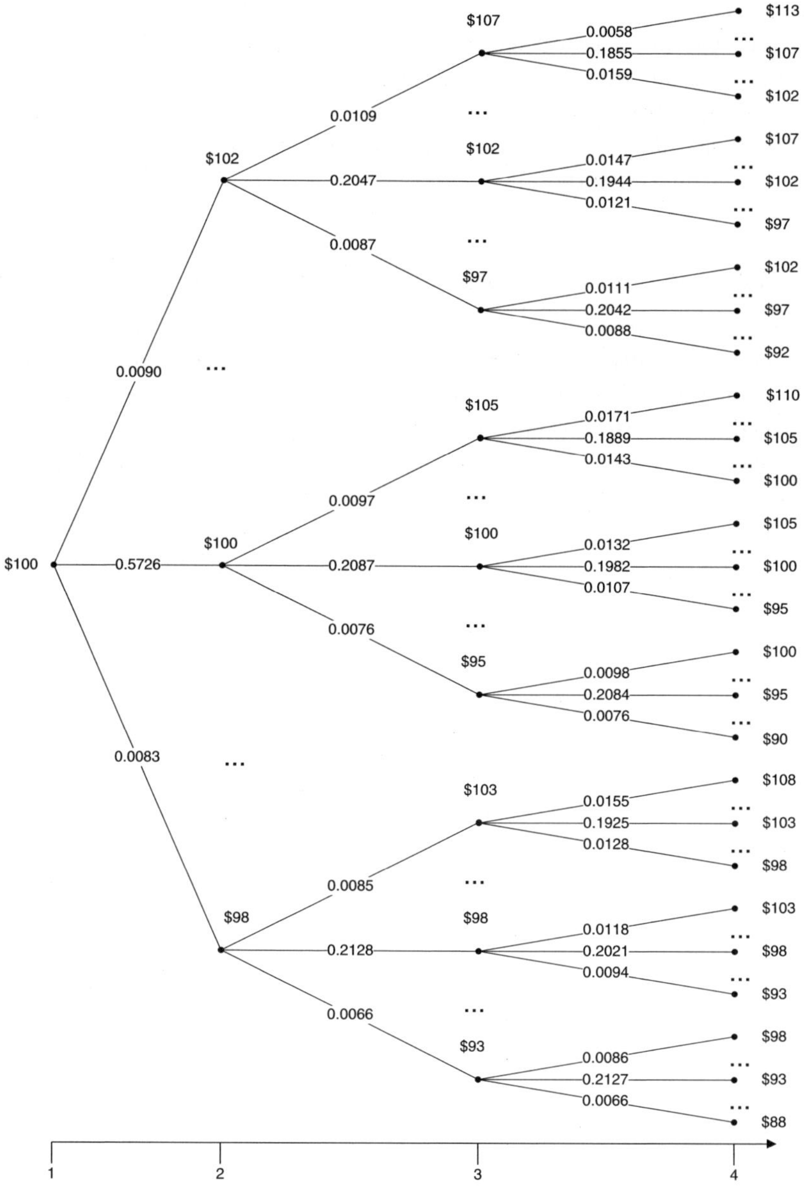

Figure 2 An underlying asset price scenario tree for the 1st day of month. The risk-free rate $r = 0.0$, the price volatility $\sigma = 0.1$. The distance between prices corresponded with neighboring successor nodes is $\Delta g_\tau = 1.0$ for all τ. The strikes from \$88 to \$112 with the step $\Delta h = \$2$ are considered. So, each node except terminal stage has successors which correspond with all strikes that can be reached from it and with intermediate price values. The whole amount of the scenarios is 608.

such a tree correspond to the days numbered 10, 11 and 21. Finally, on the 11th day we built a two-stage tree.

For the simulation described in section 5 the initial underlying price was \$100, the risk free rate $r = 0.0$, the price volatility $\sigma = 0.1$. We used the constant step $\Delta g_\tau = 1.0$ for all stages of all trees. A schematic illustration of a scenario tree for the first day of a month is shown in Figure 2.

4 The problem of option portfolio optimization

Let c be a commission collected for each option trade. As mentioned, options are traded at Black-Scholes prices in the considered market. Using the Black-Scholes prices under optimization must lead to arbitrage opportunities in the scenario tree. However, the research has shown that possible arbitrage is eliminated by a definite value of the commission c and a rather small price step Δg_τ.

Denote by $\nu = 1, \dots, \mathrm{N}$ the number of a scenario in the scenario tree. Let $w_{\tau\nu}^{ih}$ be the price of some option calculated on the basis of the Black-Scholes formula. Index τ, $\tau = 1, \dots, \mathrm{T}$ is the number of a stage of the scenario tree. In accordance with section 3, the number of stages T can be equal to 4, 3 or 2. The option type is defined by the index $i \in I = \{C, P\}$, where "C" denotes Call option and "P" denotes Put option. h is the strike of the option; $h \in H$ where H is the set of strikes of traded options.

By $x_{\tau\nu}^{ih}$ denote the number of options of type i and strike h which are recommended for buying in stage τ, $\tau = 1, \dots, \mathrm{T} - 1$ under scenario ν, $\nu = 1, \dots, \mathrm{N}$. The analogous variables for option selling are denoted by $y_{\tau\nu}^{ih}$.

Let $z_{\tau\nu}^{ih}$ be the number of options in the portfolio before its adjustment under scenario ν at stage τ, $\tau = 1, \dots, \mathrm{T}$, such that $i \in I = \{C, P\}$, $h \in H$. The variable $z_{\tau\nu}^{ih}$ is positive for a long position and negative for a short position; $z_{\tau\nu}^{ih}$ is a known constant for $\tau = 1$.

Denote by $A_{\tau\nu}$, $\tau = 1, \dots, \mathrm{T}$, $\nu = 1, \dots, \mathrm{N}$ the cash balance under scenario ν at stage τ before the portfolio adjustment; for the root node of the tree $A_{\tau\nu} = A_{1\nu}$ is the current cash balance of the portfolio. $A_{\tau\nu}$ takes into account paid commissions and option premiums which are calculated on the basis of the Black-Scholes formula.

Let $L_{\tau\nu}$, $\tau = 1, \dots, \mathrm{T}$, $\nu = 1, \dots, \mathrm{N}$ be the liquidation value of options in the portfolio. It is the sum of money needed to close all the positions.

To introduce non-anticipativity constraints the following notation is needed also. At each stage τ, $\tau = 1, \dots, \mathrm{T} - 1$ some subsets of the scenarios coincide. We let Π_τ be the number of such subsets and denote the subsets by $U_{\tau\pi}$, where $\pi = 1,\ldots,\Pi_\tau$. Note that $\Pi_1 = 1$ and U_{11} includes all the scenarios of the tree.

The option portfolio optimization program includes the following constraints.

Nonnegativity constraints:

$$x^{ih}_{\tau\nu} \geq 0, y^{ih}_{\tau\nu} \geq 0$$
$$\forall i \in I, \quad ,h \in H, \quad \tau = 1,\ldots,\mathrm{T}-1, \quad \nu = 1,\ldots,\mathrm{N}. \tag{7}$$

Non-anticipativity constraints: If scenarios ν_1 and ν_2 $(\nu_1 \neq \nu_2)$ coincide at stage τ then

$$x^{ih}_{\tau\nu_1} = x^{ih}_{\tau\nu_2}, y^{ih}_{\tau\nu_1} = y^{ih}_{\tau\nu_2}, \forall \nu_1, \nu_2 \in U_{\tau\pi},$$
$$i \in I, \quad h \in H, \tau = 1,\ldots,\mathrm{T}-1, \pi = 1,\ldots,\Pi_\tau. \tag{8}$$

The number of options in stage τ under scenario ν is given by the following:

$$z^{ih}_{\tau\nu} = z^{ih}_{(\tau-1)\nu} + x^{ih}_{(\tau-1)\nu} - y^{ih}_{(\tau-1)\nu},$$
$$\forall i \in I, \quad h \in H, \quad \tau = 2,\ldots,\mathrm{T}, \quad \nu = 1,\ldots,\mathrm{N}. \tag{9}$$

Portfolio adjustment cost:

$$Z_{\tau\nu} = \sum_{i\in I}\sum_{h\in H} w^{ih}_{\tau\nu}\left(x^{ih}_{\tau\nu} - y^{ih}_{\tau\nu}\right) + c\sum_{i\in I}\sum_{h\in H} x^{ih}_{\tau\nu} + c\sum_{i\in I}\sum_{h\in H} y^{ih}_{\tau\nu} \tag{10}$$

$\tau = 1,\ldots,\mathrm{T}-1, \nu = 1,\ldots,\mathrm{N}.$

Portfolio liquidation value:

$$L_{\tau\nu} = -\sum_{i\in I}\sum_{h\in H} w^{ih}_{\tau\nu} z^{ih}_{\tau\nu}, \quad \tau = 1,\ldots,\mathrm{T}, \quad \nu = 1,\ldots,\mathrm{N}. \tag{11}$$

Account balance in stage τ:

$$A_{\tau\nu} = A_{(\tau-1)\nu} - Z_{(\tau-1)\nu}, \quad \tau = 2,\ldots,\mathrm{T}, \quad \nu = 1,\ldots,\mathrm{N}. \tag{12}$$

Let $\lambda^{ih}_{\tau\nu k}$ be the loss of the option which refers to SPAN scenario number k, $k = 1,\ldots,16$. The SPAN initial margin for scenario ν at stage τ is given by the variable $\mu_{\tau\nu}$:

$$\mu_{\tau\nu} \geq \sum_{i\in I}\sum_{h\in H} \lambda^{ih}_{\tau\nu k} z^{ih}_{\tau\nu} + \sum_{i\in I}\sum_{h\in H} \lambda^{ih}_{\tau\nu k} x^{ih}_{\tau\nu} - \sum_{i\in I}\sum_{h\in H} \lambda^{ih}_{\tau\nu k} y^{ih}_{\tau\nu},$$
$$k = 1,\ldots,16, \quad \tau = 1,\ldots,\mathrm{T}-1, \quad \nu = 1,\ldots,\mathrm{N}. \tag{13}$$

For each stage τ and scenario ν the right sides of constraints (13) are losses referring to the 16 SPAN scenarios. The active constraint determines the appropriate value of the variable $\mu_{\tau\nu}$ which is equal to the portfolio Scanning Risk (Golembiovskii and Dolmatov, 2001).

To hold option positions, the account balance after the portfolio adjustment has to be not less than the collateral. Accordingly there are the budget constraints:

$$A_{\tau\nu} - Z_{\tau\nu} \geq L_{\tau\nu} - \sum_{i\in I}\sum_{h\in H} w_{\tau\nu}^{ih} x_{\tau\nu}^{ih} + \sum_{i\in I}\sum_{h\in H} w_{\tau\nu}^{ih} y_{\tau\nu}^{ih} + \beta\, \mu_{\tau\nu}, \quad \tau = 1,\ldots,\mathrm{T}-1, \quad \nu = 1,\ldots \mathrm{N} \tag{14}$$

where β is a stock coefficient. Since trading is not considered at the expiration moment, for the corresponded tree stage constraints (14) take the form

$$A_{\mathrm{T}\nu} \geq L_{\mathrm{T}\nu}, \quad \nu = 1,\ldots \mathrm{N}. \tag{15}$$

The cash balance must not be negative:

$$A_{\tau\nu} \geq 0,\ \tau = 2,\ldots,\mathrm{T}, \quad \nu = 1,\ldots \mathrm{N}. \tag{16}$$

Estimation of the portfolio value at the terminal stage T:

$$W_{\mathrm{T}\nu} = A_{\mathrm{T}\nu} - L_{\mathrm{T}\nu}, \nu = 1,\ldots,\mathrm{N}. \tag{17}$$

Chance constraints:

$$W_{\mathrm{T}\nu} \geq u\ \forall\nu \text{ such that } d_\nu = 1, d_\nu \in \{0,1\},\ \nu = 1,\ldots,\mathrm{N}. \tag{18}$$

Optimization criterion:

$$\max \sum_{\nu=1}^{\mathrm{N}} d_\nu \rho_\nu. \tag{19}$$

Here u is a minimal demanded portfolio value at the expiration moment; d_ν is an indicator of active scenarios; ρ_ν is the probability of the scenario ν.

An alternative optimization criterion could be max u under constraints on the minimal value of the sum in (19). However, as experiments have shown, the problem with criterion (19) is solved much faster.

5 Solving the problem and some simulation results

Constraints (18) were defined as indicator constraints for CPLEX 12.2 which we used to solve problem (7)–(19). The computer used was an Intel Pentium Dual-Core 2.20 GHz. The parameters of the problem are demonstrated in Table 3. The time of the portfolio optimization on the first day of the month was approximately 5 minutes including database operations. Solving the mixed-integer problem by branch and bound the node limit was set equal to 1000 for the solver CPLEX.

The results of solving the problem for the root node of the scenario tree are presented in Table 4. The corresponding value of the objective is 0.9755. The recommended portfolio looks like a complex asymmetrical long strangle. Profiles of the portfolio value for different time moments are illustrated by Figure 3. For the current moment the portfolio profile shows no loss for almost all underlying price values.

We ran 100 simulated experiments on one month option portfolio management in our hypothetical market. Every time we started from the portfolio of Table 4. Then a new value of underlying price was simulated using Monte-Carlo method on the basis of a geometric Brownian motion equation with parameters given in Table 3, and the problem of option

Table 3 Parameters of the problem of option portfolio optimization on the 1st day of the month. The scenario tree is described by Figure 2. Options of 13 strikes are considered.

Parameter	Value
Volatility of underlying asset price, σ	0.10
Risk-free rate, r	0.00
Lower and upper bounds of option strikes	\$88; \$112
Distance between strikes, Δh	\$2.00
Underlying asset price on the day 1	\$100.00
Initial account balance	\$100.00
Commission for one trade, c	\$0.001
Margin stock coefficient, β	5
Minimal demanded portfolio value at expiration, u	\$100.50

Table 4 Option trading recommendations for the 1st day of month. The parameters of the optimization problem in Table 3.

Strike, \$	Option	Market price, \$	Position	Buy	Sell
102	Put	2.40	0.00	13.42	0.00
100	Call	1.12	0.00	2.72	0.00
94	Call	6.01	0.00	8.91	0.00

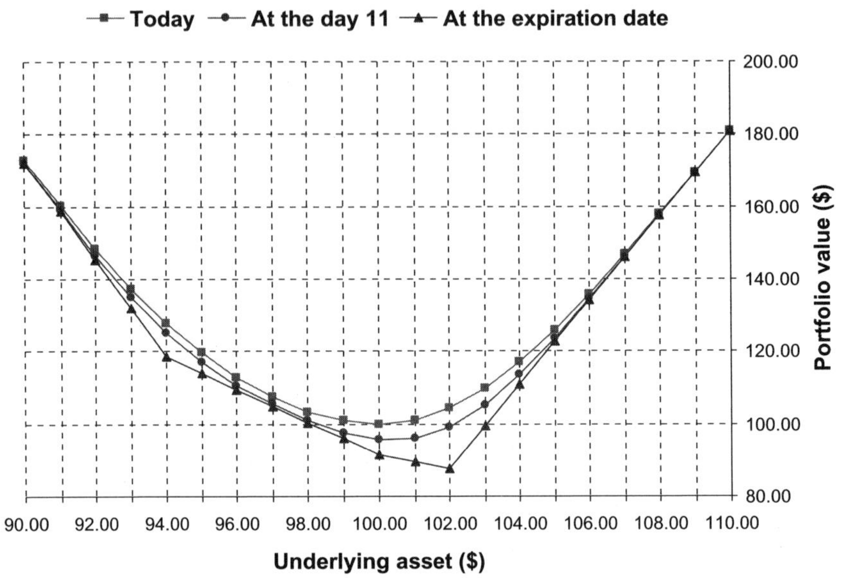

Figure 3 Dependence of the value of the Table 4 portfolio on the underlying price for different time moments. Today the current portfolio value is \$99.9749.

portfolio optimization was solved again. The execution of the received trade recommendations was simulated. Trading of any fraction of a contract was permitted. The last day when the portfolio was adjusted in such a way was the 11th day according to the previous description. The entire simulation took approximately 98 hours and 30 minutes on an Intel Pentium computer with Dual-Core 2.20 GHz processors.

Figure 4 shows 100 realized tracks of the portfolio value. 94 of them gave a positive return that is more than the risk-free rate, which is equal to zero. However, only 89 tracks gave a final portfolio value not less than 100.5 as was demanded according to Table 3. So the reliability of the portfolio management is 0.89 which is less than the level 0.9755 promised under the problem solved on the 1st day of the month. This decreasing reliability is connected with the price step between neighboring nodes. Subsequent experiments have shown that if the step Δg_T is less than 1.0. the real reliability is closer to the theoretical one.

The average final portfolio value for these 100 experiments is \$100.95. As was mentioned, the theoretical expected average of the final portfolio value is less than \$100 on the average sum of paid commission. This discrepancy evidently can be explained by an effect of randomness.

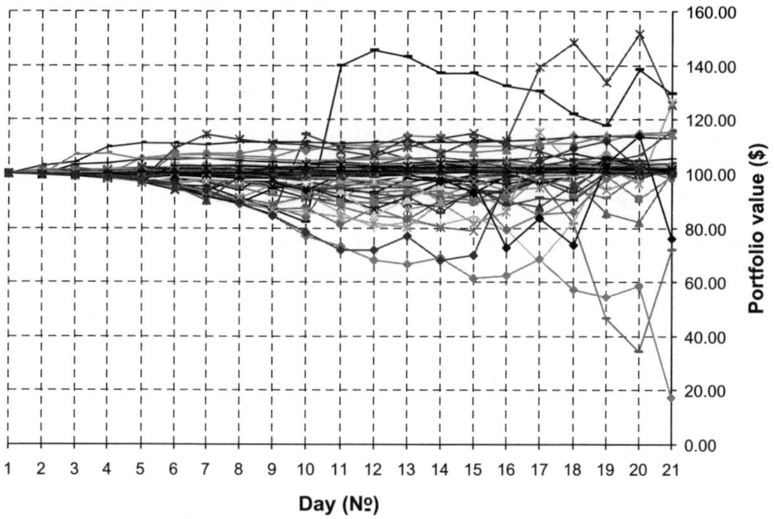

Figure 4 100 tracks of the option portfolio value. The initial account balance and initial underlying price are both equal to \$100 according to Table 3. The portfolio structure on the 1st day of any month is given by Table 4.

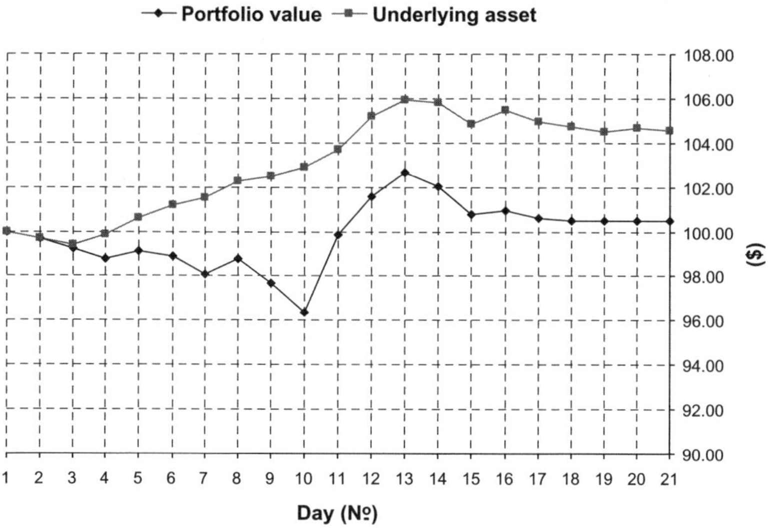

Figure 5 Results of one experiment on option portfolio management. The portfolio structure and the adjustment recommendations on the 11th day correspond with Table 5. The final portfolio value is \$100.5.

One successful track is illustrated by Figure 5. As can be seen, the portfolio value had been decreasing to the 10th day. After this it started to increase. The final portfolio value is \$100.5. The structure of the portfolio and the recommended trades on day 11 are presented in Table 5. The portfolio profiles after the execution of the trades are shown in Figure 6.

Table 5 The portfolio before the adjustment and trades for the 11th day of the month.

Strike,\$	Option	Market price,\$	Position	Buy	Sell
108	call	0.02	0.00	15.83	0.00
106	call	0.14	−1.05	0.00	0.00
106	put	2.47	1.05	0.00	0.00
104	call	0.67	5.15	0.00	0.00
104	put	1.00	−5.15	0.00	0.00
102	call	1.91	−31.59	20.77	0.00
102	put	0.24	10.82	0.00	0.00
100	call	3.70	9.01	0.00	18.22
100	put	0.03	−26.92	0.00	0.00
98	call	5.68	7.84	0.00	0.00
98	put	0.00	65.17	0.00	0.00
96	call	7.67	−0.74	0.00	0.00
94	call	9.67	8.84	0.00	0.00
94	put	0.00	−8.91	0.00	0.00

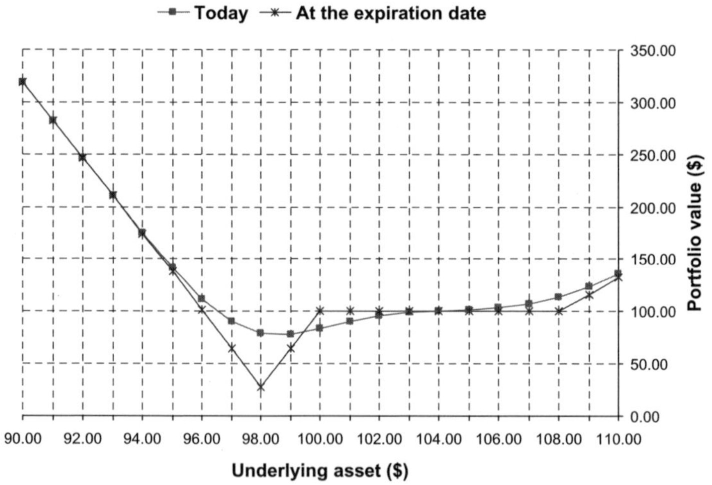

Figure 6 Dependence of the value of the Table 5 adjusted portfolio on the underlying price on the 11th day and on the 21st day. Today's portfolio value is \$99.987, underlying price is \$103.674.

6 Conclusions and future research

Model of a risk-neutral option market suitable for portfolio management research was suggested. A multinomial scenario tree of the underlying price representing all option strike prices was constructed. We developed the model of option portfolio management as a chance constrained stochastic programming problem. The portfolio simulation showed that it is possible to make more than the risk-free return with some high probability.

The first subject for future research is using a faster method for solving the portfolio optimization problem. The progressive hedging algorithm of Watson *et al.* (2010) and the heuristic method of Tanner and Beier (2008) are among the possibilities. To solve a higher dimensional problem that includes options of different expirations, the use of parallel computers is necessary.

To move toward real option markets, it is necessary to use a multinomial scenario tree representing risk-neutral underlying price distribution taking into account real option prices.

References

Black, F. and M. Scholes. The Pricing of Options and Corporate Liabilities. *Journal of Political Economy*, 3: pp. 637–654, 1973.

Black, F. The pricing of commodity contracts. *Journal of Financial Economics*, 3: pp. 167–179, 1976.

Chicago Mercantile Exchange. The Standard Portfolio Analysis of Risk (SPAN) performance bond system at the Chicago Mercantile Exchange. Technical specification, 1999.

Cox, J. C., R. A. Ross and M. Rubinstein. Option Pricing: a Simplified Approach. *Journal of Financial Economics*, 7: pp. 229–263, 1979.

Golembiovsky D. Yu. and A.M. Abramov. Model of Option Portfolio Management. *Risk Management*, 4: pp. 43–56, 2011. (In Russian).

Golembiovskii D. Yu. and A. S. Dolmatov. An Optimization Model for a Portfolio of Financial Derivative Instruments with Pledge Limitations. *Journal of Computer and System Sciences International*, 40(3): pp. 425–435, 2001.

Hull, J. C. *Options, Futures and Other Derivatives.* 8th edition (Hardcover), 2011.

Laurent, A. A Scenario Generation Algorithm for Multistage Stochastic Programming: Application for Asset Allocation Models with Derivatives. Ph.D Thesis in economics. Faculty of Economics, Department of Finance. University of Lugano, Lugano, Switzerland, 2006.

LCH.Clearnet. London SPAN Version 4. Technical Information Package, www.lchclearnet.com, 2008.

Merton, R. C. Theory of rational option pricing. *Bell Journal of Economics and Management Science*, 4: pp. 141–183, 1973.

Nemirovski, A. and A. Shapiro. Convex approximations of chance constrained programs. *SIAM Journal on Optimization*, 17: pp. 969–996, 2006.

Prékopa, A. *Stochastic Programming. Handbooks in Operations Research and Management Science*, Chapter 5, Probabilistic Programming, Vol. 10, Elsevier, 2003.

Ruszczynski, A. Probabilistic programming with discrete distributions and precedence constrained knapsack polyhedra. *Mathematical Programming*, 93: pp. 195–215, 2002.

Tanner, M. W. and E. B. Beier. A General Heuristics Method for Joint Chance-Constrained Stochastic Programs with Discretely Distributed Parameters, http://www.optimization-online.org/DB_HTML/2007/08/1755.html, 2008.

Watson, J.-P., R. J.-B. Wets and D. L. Woodruff. Scalable Heuristics for a Class of Chance-Constrained Stochastic Programs. *Informs Journal on Computing*, 4: pp. 543–554, 2010.

Chapter 8

Stochastic Models for Optimizing Immunization Strategies in Fixed-Income Security Portfolios under Some Sources of Uncertainty[1]

Larraitz Aranburu*, Laureano F. Escudero†,
M. Araceli Garín** and Gloria Pérez‡

Summary

In this paper we present a set of approaches for stochastic optimizing of immunization strategies based on risk averse measures as alternatives to the optimization of the objective function expected value, i.e., in the so-called risk neutral environment. The risk averse measures to consider whose validity is analyzed in this work are as follows: two-stage mean-risk immunization, two-stage and multistage Value-at-Risk strategy, two-stage and multistage stochastic dominance strategy, and a new measure as a mixture of the multistage VaR & stochastic dominance strategies. The common characteristic of these measures is that they require from the modeler a threshold for the objective function value related to each scenario (the recent ones even allow a set of thresholds) and a failure probability when not reaching it. Uncertainty is represented by a scenario tree and is dealt with by both two-stage and multi-stage stochastic (linear and) mixed integer models with complete recourse. We will test the different risk averse strategies presented in the paper by using, as a pilot case, the

*Dpto. de Economía Aplicada III, Universidad del País Vasco, Bilbao (Vizcaya), Spain, larraitz.aranburu@ehu.es.

†Dpto. Estadística e Investigación Operativa, Universidad Rey Juan Carlos, Móstoles (Madrid), Spain, laureano.escudero@urjc.es.

**Dpto. de Economía Aplicada III, Universidad del País Vasco, Bilbao (Vizcaya), Spain, mariaaraceli.garin@ehu.es.

‡Dpto. de Matemática Aplicada, Estadística e Investigación Operativa, Universidad del País Vasco, Leioa (Vizcaya), Spain, gloria.perez@ehu.es.

[1]This research has been partially supported by the projects ECO2008-00777/ECON from the Ministry of Education and Science, PLANIN MTM2009-14087-C04-01 from the Ministry of Science and Innovation, Grupo de Investigación IT-347-10 from the Basque Government, and RIESGOS CM from Comunidad de Madrid, Spain.

optimization of the immunization strategies in fixed-income security portfolios under some sources of uncertainty. One of the main differences in the bond portfolio strategies that are proposed in the work and the models that have been encountered in the literature is that we consider an investor who wishes to invest in a market with coupon bonds that have a different level of risk of default.

Keywords: Stochastic optimization, risk measures and management, fixed income-assets, immunized portfolios, interest rate, default probability.

1 Introduction

Stochastic optimization models have been proposed and studied extensively since the 1950s by Beale (1955), Dantzig (1955), Charnes and Cooper (1959), Wets (1966), Kall and Wallace (1994), Birge and Louveaux (1997), Wets and Ziemba (1999), and Shapiro, Dentcheva and Ruszczynski (2011), among many others. A stochastic vision is proposed for the financial models dealt with in this work, rather than the traditional deterministic vision, such that the uncertain parameters that are not controlled by the modeler are considered as random variables whose known or estimated probability distributions are independent of the decision variables.

The majority of the financial models proposed until the last decade are static and single-period. However, in cases where uncertainty prevails at all the stages of the planning horizon, then stochastic optimization models become more appropriate. Such models are not very common at present in practical financial applications due to their complexity and the complex requirement for input data. Nevertheless, some very interesting models have appeared in the literature in recent years. There are many ALM (Asset and Liability Management) stochastic optimization models such as those in Ziemba and Mulvey (1998), Wallace and Ziemba (2005) and Zenios and Ziemba, (2006), that are generally preferred by pensions, insurance companies, wealthy individuals and hedge funds; see also Ziemba (2003) and Escudero *et al.* (2009), among many others. An advantage of these scenario based models is that the parameters are not assumed to be known but are scenario dependent, hence they are uncertain. Bradley and Crane, (1972), present a multistage decision tree model for bond portfolio management. Kusy and Ziemba (1986), compare their stochastic optimization model for the Vancouver Savings Credit Union with that of Bradley and Crane, and argue in favor of the stochastic optimization model on computational and performance grounds. Both of these models are now easily solved with current optimization technology, see (2005), for example. The Bradley and Crane model ushered in bond portfolio management and the management of fixed income securities in the literature; for example, see the early book by Dempster (1980) and also Golub *et al.* (1995) and Zenios *et al.* (1998). See also Bertocchi, Moriggia and Dupačová (2006) and Consigli, Bertocchi and Dempster (2011) for the current state of the literature on this subject. Since the early papers, computation methods have advanced spectacularly, so that large-scale LP optimization problems, at least, can now be resolved efficiently. These advances have enabled up to apply stochastic optimization increasingly more to real-life financial problems. Some of these financial applications are collected in (1998, 2005, 2006, 2011). See also recent results in (2008, 2010), among others.

A significant contribution to the above line of research has been made possible thanks to the flexibility of stochastic optimization models to integrate through scenarios diverse multi-dimensional risk factors for risk management. However, the optimization models generated still become intractable when a large number of variables must be combined, particularly if they are 0–1 variables, with exponential increases in scenarios. In this case procedures are needed to break down the problem and reduce the number of scenarios, see (2000, 2001, 2001, 2003, 2009), among others.

In this paper we present an approach for stochastic modelization of different immunization strategies in fixed-income security portfolios under some sources of uncertainty, such as two-stage mean-risk immunization, two-stage and multistage Value-at-Risk strategy, two-stage and multistage stochastic dominance strategy and, in addition, the new measure as a mixture of the multistage VaR & stochastic dominance strategies. The common characteristic of these measures is that they require from the modeler a threshold, at least, for the objective function value and a probability failure. We introduce different models that allow us to consider transaction costs. Uncertainty is represented by a scenario tree and is dealt with by both two-stage and multistage stochastic linear and mixed integer models with complete recourse. The main difference in the bond portfolio strategies that are proposed and the models that have been encountered in the literature is that we consider an investor who wishes to invest in a market with coupon bonds that have a different level of risk of default. Then, there are two sources of uncertainty, or two risks, associated with the model, namely, interest rate risk and credit risk or risk of default. The latter is concerned with the solvency of their issuers and, therefore, the bonds themselves.

The remainder of the paper is organized as follows. Section 2 introduces the classical deterministic LP model to fix notation and to set up the prototype to be improved by introducing uncertainty in the main parameters, namely, interest rate path and credit rating ranking (see Section 3). The traditional two-stage stochastic optimization approach is presented in Section 4. Section 5 is devoted to presenting the two-stage risk averse strategies whose validity is tested in this work, namely, mean-risk immunization (Section 5.1), Value-at-Risk (VaR) strategy (Section 5.2) and stochastic dominance strategy (Section 5.3). Section 6 presents our two-stage approach for fixed-income security portfolio immunization based on the VaR strategy. Section 7 presents our multistage stochastic scheme for portfolio optimization in a risk adverse environment by using the strategies based on the concepts maxmin (Section 7.1) and stochastic dominance (Section 7.2), and the new multistage strategy as the mixture of the VaR & stochastic dominance strategies (Section 7.3). Section 8 introduces a pilot case to show the performance of the different immunization strategies that have been presented in this work. Finally, Section 9 concludes.

2 The deterministic LP optimization model

In this section we introduce the basic deterministic LP optimization model used in fixed-income security portfolio restructuring. In the next sections we introduce potential or actual extensions of this basic model. We now present the basic components and common notation of the mathematical prototype.

Let us consider a partition of the planning horizon (PH) into k subintervals of equal length $[t_0, t_1], [t_1, t_2], \ldots, [t_{k-1}, t_k]$, being t_0 the starting and t_k the end of the PH. We also assume that portfolio rebalancing is only allowed at the beginning of each subinterval. $|\mathcal{T}| = k + 1$ is the number of time periods, and t_k is the final period. So, set $\mathcal{T}$ is defined as the discretization or splitting of the time horizon, i.e., $\mathcal{T} = \{t_0, t_1, \ldots, t_k\}$. Let $\mathcal{I}$ denotes the set of securities to be included in the portfolio. For the sake of simplicity, and without loss of generality (wlog), we assume that there are $|\mathcal{I}|$ different coupon bonds available at t_0, each of them maturing at t^i, such that $t^i \in \{t_1, \ldots, t_k\}$, but not necessarily for all the bonds, since there might be some bond i^* with period of maturity, t^{i*}, later than the final period of PH, i.e., $t^{i*} > t_k$. Then, coupon payments are also due at rebalancing points, where t^i is the maturity period of bond $i \in \mathcal{I}$.

Let I_0 denote the initial budget to be invested in the portfolio. As decision variables, x^+_{it} denotes the volume of security $i \in \mathcal{I}$ purchased in period t, and x^-_{it} denotes the volume of security i sold in period t. Variable z_{it} denotes the volume of security i to be held in the portfolio after the transactions conducted in period t. And variable V_t is defined as the final value of the portfolio in period t, for $t_0 \leq t \leq t_k$. Let β be a parameter that denotes a percentage of the volume negotiated, which represents the transaction costs affecting each readjustment. Let us also assume that the nominal figure and the coupon payments do not generate transaction costs. P_{i0} denotes the unit price on the market of security i at the starting of the PH, i.e., initial period t_0, for $i \in \mathcal{I}$. c_i is the annual coupon for security i. And C_{it} denotes the payment stream generated by one unit of security $i \in \mathcal{I}$ in period t, $t \in T - \{t_0\}$. It is expressed as follows,

$$C_{it} = \begin{cases} c_i \cdot h, & t_0 < t < t^i, \\ F_i + c_i \cdot h, & t = t^i, \\ 0, & t > t^i, \end{cases} \tag{1}$$

where h is the constant length of each sub-interval (fraction of a year) and F_i is the nominal value of security i, in this case, $F_i = P_{i,0}$, $\forall i \in \mathcal{I}$.

Finally, P_{it} denotes the unit price of security i at time period t. If we consider the transaction costs as a percentage β then P^-_{it}, which denotes the unit selling price of security i at period t, is computed as $P^-_{it} = (1-\beta)P_{it}$; and P^+_{it} is the unit purchase price of security i at period t, so that $P^+_{it} = (1+\beta)P_{it}$.

The optimization model is a LP problem to maximize the final value of the portfolio,

$$\text{(LP)} \qquad \max V_{t_k} \tag{2}$$

$$\text{s.t.} \quad x^+_{i0} = z_{i0} \quad \forall i \in \mathcal{I} \tag{3}$$

$$\sum_{i \in \mathcal{I}} P^+_{i0} x^+_{i0} = I_0 \tag{4}$$

$$z_{i,t-1} + x^+_{it} - x^-_{it} = z_{it} \quad \forall i \in \mathcal{I}, \quad t = t_1, \ldots, t_{k-1} \tag{5}$$

$$x^-_{i:t^i=t,t} = z_{i:t^i=t,t-1} \quad \forall i \in \mathcal{I}, \quad t = t_1, \ldots, t_k \tag{6}$$

$$x^-_{it_k} = z_{i,t_{k-1}} \quad \forall i \in \mathcal{I} : t_k < t^i \tag{7}$$

$$\sum_{i \in \mathcal{I}:t<t^i} P^+_{it} x^+_{it} - \sum_{i \in \mathcal{I}:t<t^i} P^+_{it} x^-_{it} = \sum_{i \in \mathcal{I}:t \leq t^i} C_{it} z_{i,t-1} \quad \forall t = t_1, \ldots, t_{k-1} \tag{8}$$

$$\sum_{i\in\mathcal{I}:t_k<t^i} P^-_{it_k}x^-_{it_k} + \sum_{i\in\mathcal{I}:t_k\leq t^i} C_{it}z_{it_{k-1}} = V_{t_k} \tag{9}$$

$$0 \leq z_{i0}, x^+_{i0} \quad \forall i \in \mathcal{I} \tag{10}$$

$$0 \leq x^+_{it} \quad \forall i \in \mathcal{I}, \quad t = t_1, \ldots, t_{k-1} \tag{11}$$

$$0 \leq z_{it}, x^-_{it} \quad \forall i \in \mathcal{I}, \quad t = t_1, \ldots, t_k \tag{12}$$

$$0 \leq V_{t_k}. \tag{13}$$

Equations (3)–(4) are the first period constraints, representing the structural ones. In general these constraints establish that the portfolio must be formed at the initial period, t_0. The initial budget constraint assigns values to the initial investment in each security. In particular, equation (3) forces the portfolio to be built at the initial period. Equation (4) establishes the initial investment of the portfolio holder. Constraints (5)–(7) are balance equations that link the volume of securities purchased and sold in each period with the volume of securities in the portfolio. Equations (8) are the constraints that ensure that the portfolio is self-financing, i.e., the funds to purchase a new security at each period must be obtained by selling securities and from the yield from coupons on the bonds in the portfolio. Moreover, equation (9) forces the portfolio to be dismantled at the end of the PH, so the final value is the sum of the cash obtained from selling securities and the yield obtained from the coupons paid on bonds.

Solutions considerations: Given the state of the art in LP optimization solvers, model (2)–(13) should not present major difficulties for problem solving, even for large scale instances. However, the optimal solution may not be suitable, given the uncertainty of the main parameters, namely the interest rate path and credit rating ranking.

3 Two sources of uncertainty: interest rate and credit risk

Uncertainty will be represented in terms of random experiments with outcomes (i.e., scenarios) denoted by ω. The relevant set of outcomes is clearly problem-dependent. Also, it is usually not very important to be able to define those outcomes accurately because the focus is mainly on their impact on some (random) variables. The set of all outcomes is represented by Ω, that can be visualized as a scenario tree and it can be dealt with by using stochastic optimization approaches. Each node in the scenario tree depicts a juncture for rendering decisions. A scenario is defined as a complete path from the root node to a leaf, and defines a single realization of all the uncertain parameters.

Throughout the paper we will define a set of strategies from two independent sources of uncertainty, namely, interest rates and ratings of the issuers of each bond. For the sake of simplicity and wlog we assume that there are n different coupon bonds available at t_0, each of them maturing at $t^1, t^2, \ldots, t^n$ respectively, that is, coupon payments are also due at rebalancing points. For each coupon bond in a given class j, $i \in \mathcal{I}_j$ with maturity period t^i, we will consider a replica $i' \in \mathcal{I}_{j'}$ in a different class j', with the same maturity period $t^i = t^{i'}$, but with different level of risk of default. Then, $\mathcal{J}$ is the set of classes of securities under consideration, where the securities in each class have the same credit risk but different maturity periods. The risk in this case concerns the solvency of their issuers and, therefore, the bonds themselves.

The price of a bond is directly related to the interest rate level all over the PH. The interest rate is the percentage of the par value or nominal (the money you invest) that you will receive as payment when holding a bond. The interest rate depends on the maturity of the bond (when it will pay the par value) and it does not change in a linear way. In fact, the Term Structure of Interest Rates (for short, TSIR) changes much more in the short term and is more stable in the long term. In this way, interest rates may change independently for different bond maturities. As this is really difficult to modelize and following one of the assumptions of Khang's immunization theory (1983), we will consider flat changes in interest rate structure. This means that every interest rate will change in a common way, whatever its maturity is. So, if the annual interest rate is r, the corresponding interest rate for the maturity Δt ($r_{\Delta t}$) can be expressed,

$$r_{\Delta t} = \Delta t \cdot r,$$

where Δt is the time in years. We will assume that, at the time when we invest, i.e., initial period t_0, the interest rate is at a certain level r_{t_0}. In the next period it might change into any of the scenarios under consideration, see for example Figure 1. For simplicity, the tree assumes that change can only happen after the portfolio has been compiled.

Moreover, the following definitions are provided to facilitate notation. All the illustrative examples that will be used to illustrate the definitions are from Standard&Poor's "2009 Annual Global Corporate

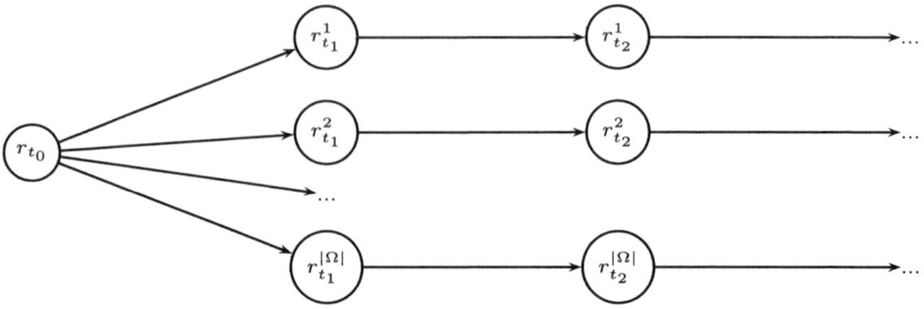

Figure 1 Two-stage Scenario Tree for the Interest Rate Uncertainty.

Table 1 S&P Ranking Definition.

Rating	Bond Quality
AAA	Prime rating
AA	High grade rating
A	Upper medium grade rating
BBB	Lower medium grade rating
BB	Speculative
B	Highly speculative
CCC	Very highly speculative
CC	Extremely speculative
C	Very extremely speculative
D	Likely to default on capital, interest

Default Study And Rating Transitions", 17th March 2010, see the urn http://www.standardandpoors.com.

Definition 1. *Credit rating means the rating given to an individual or a company to indicate their solvency as debtors in the issuing of short- or long-term securities. Rating agencies examine companies that issue bonds, as well as the situation of the bonds issued at regular intervals, and may upgrade or downgrade their ratings whenever they see fit.*

Taking the S&P ranking as a reference point, bonds can be rated as shown in Table 1. The rating agency would examine the quality of the companies every year, so, their rating might change. As an example, Table 2 shows the global corporate average transition rates (1981–2009) in percentage.

Definition 2. *Default means failure to pay back a loan on maturity, or when the terms of an agreement are fulfilled.*

Table 2 Examples of Rate Transition Probabilities.

	AAA	AA	A	BBB	BB	B	CCC	Default
AAA	88.21	7.73	0.52	0.06	0.08	0.03	0.06	0
AA	0.56	86.6	8.10	0.55	0.06	0.09	0.02	0.02
A	0.04	1.95	87.05	5.47	0.40	0.16	0.02	0.08
BBB	0.01	0.14	3.76	84.16	4.13	0.70	0.16	0.26
BB	0.02	0.05	0.18	5.17	75.52	7.48	0.79	0.97
B	0.00	0.04	0.15	0.24	5.43	72.73	4.61	4.93
CCC	0.00	0.00	0.21	0.31	0.88	11.28	44.98	27.98

Table 3 Examples of Risks of Default.

	Years since emission						
Credit Rating	1	2	3	4	5	7	10
AAA	0.00	0.03	0.14	0.26	0.39	0.58	0.82
AA	0.02	0.07	0.14	0.24	0.33	0.52	0.74
A	0.08	0.21	0.35	0.53	0.72	1.22	1.97
BBB	0.26	0.72	1.23	1.86	2.53	3.80	5.60
BB	0.97	2.94	5.27	7.49	9.51	13.19	17.45
B	4.93	10.76	15.65	19.46	22.30	26.47	30.82
CCC	27.98	36.95	42.40	45.57	48.05	50.26	53.41

The likelihood *risk of default* is closely linked to the credit rating of an organization at a given instant in time. By way of example, consider Table 3, which represents the risks of default, q_i, calculated as cumulative mean probabilities in percentage according to the S&P report.

Definition 3. *Recovery rate, z_i, is the proportion of the money owed that the issuer undertakes to pay to the purchaser in case of default.*

For example, S&P gives the figures shown in Table 4 for the period 1987–2009 and for the discounted recovery rates by instrument type.

4 Traditional two-stage stochastic optimization approach

The stochastic version of the problem provides a general purpose-modeling framework, which embraces many real-world features, such as turnover constraints, transaction costs, limits on groups of assets, risk aversion, immunization constraints and other considerations. In particular, two-stage stochastic linear problems provide a suitable framework for modeling decision problems under uncertainty arising in several financial applications.

Table 4 Examples of Recovery Rates.

Type of instrument	Recovery rate (%)
Term Loans	**69.4**
Revolving credit	**78**
All loans	**73.8**
Senior secured bonds	**57.2**
Senior unsecured bonds	**43**
Senior subordinated bond	**28.3**
All other subordinated bond	**19.4**
All bonds	**37.4**

The flexibility of these models is related to their dynamic nature, i.e., besides the first stage variables representing decisions made in face of uncertainty, the model includes second-stage decisions, i.e., recourse actions, which may be taken once a specific realization of the random parameters is observed.

The model for portfolio selection described below seeks to obtain the optimal readjustment, independently of changes in interest rates. This model enables transaction costs to be factored in, and takes into account different credit ratings for bonds. Thus, decision-makers can obtain the optimal portfolio taking into account the risk associated with the likelihood of bankruptcy or default on the part of the issuer of each bond and the weight that they themselves attribute to that risk.

Let the following notation for sets, parameters and variables which define the Deterministic Equivalent Model (DEM) for the stochastic scheme in the case of bonds portfolio optimization.

Sets

- $\mathcal{I}$, set of securities i to be included in the portfolio. $|\mathcal{I}|$ is the number of securities in set $\mathcal{I}$.
- $\mathcal{J}$, set of classes of fixed-income security considered j, i.e., the set of different credit ratings considered.
- $\mathcal{I}_j$, set of securities i that belong to class j, $\mathcal{I}_j \subset \mathcal{I}$, such that $\mathcal{I} = \cup_{j\in\mathcal{J}}\mathcal{I}_j$, and $\mathcal{I}_j \cap \mathcal{I}_g = \emptyset$, $j, g \in \mathcal{J} : j \neq g$. $|\mathcal{I}_j|$ is the number of securities in set $\mathcal{I}_j$, $j \in \mathcal{J}$.
- $\mathcal{T}$, set of periods in PH plus the initial period t_0, such that it is the discretization or splitting of the time horizon, i.e., $\mathcal{T} = \{t_0, t_1, \ldots, t_k\}$.
- Ω, set of scenarios ω under consideration. In this case, they show the different joint situations for interest rates and default. $|\Omega|$ denotes the number of scenarios.

<u>*Parameters*</u>

I_0, initial investment budget.

F_i, nominal value of security i, for $i \in \mathcal{I}$.

t^i, maturity period of security i, such that $t^i \in \{t_1, \ldots, t_k, \ldots, t_n\}$, for $i \in \mathcal{I}$.

β, percentage of the volume negotiated, which represents the transaction costs affecting each readjustment. It is also considered that the nominal figure and the coupon payments do not generate transaction costs.

P_{i0}, unit price on the market of security i at the starting of PH, period t_0, for $i \in \mathcal{I}$.

w^ω, likelihood of scenario ω, given by the modeler.

r_t^ω, risk-free interest rate in period t under scenario ω. At period $t = t_0$, $r = r_{t_0}^\omega = r_{t_0}^{\omega'}$, for $\omega \neq \omega'$, $\omega, \omega' \in \Omega$.

q_{jt}^ω, measure of risk calculated at period t as the probability of default of the securities of class j under scenario ω. At period $t = t_0$, $q_j = q_{jt_0}^\omega = q_{jt_0}^{\omega'}$, for $\omega \neq \omega'$, $\omega, \omega' \in \Omega$, $j \in \mathcal{J}$.

z_i, recovery rate (as a fraction of one) for security i, for $i \in \mathcal{I}_j$, $j \in \mathcal{J}$.

R_{it}^ω, real interest rate (as a fraction of one) for security i of class j at time period t under scenario ω, for $i \in \mathcal{I}_j$, $j \in \mathcal{J}$, $t \in \mathcal{T}$, $\omega \in \Omega$. Following Hull (1989), we calculate it as follows,

$$R_{it}^\omega = r_t^\omega + \frac{(1 + r_t^\omega)(1 - z_i)q_{jt}^\omega}{1 - (1 - z_i)q_{jt}^\omega}.$$

At period $t = t_0$, $R_i = R_{i0}^\omega = R_{i0}^{\omega'}$, for $\omega \neq \omega'$, $\omega, \omega' \in \Omega$, $i \in \mathcal{I}$. Moreover, $R_{it}^\omega = r_t^\omega$, for $i \in \mathcal{I}_j$, such that $q_{jt}^\omega = 0$.

c_i, annual coupon for security i. This coupon is calculated as $R_i \cdot F_i$, where R_i is the real interest rate at initial period t_0, and F_i is the nominal value of security i, for $i \in \mathcal{I}$.

C_{it}^ω, payment generated by one unit of security i in period t under scenario ω, for $i \in \mathcal{I}_j$,$j \in \mathcal{J}$, $t \in \mathcal{T} - \{t_0\}$, $\omega \in \Omega$. It is calculated as follows,

$$C_{it}^\omega = \begin{cases} c_i \cdot h, & t_0 < t < t^i : q_{jt}^\omega < 1 \\ F_i + c_i \cdot h, & t = t^i : q_{jt}^\omega < 1 \\ z_i \cdot F_i, & t : q_{j,t-1}^\omega < 1 \quad \text{and} \quad q_{jt}^\omega = 1 \\ 0, & \text{otherwise,} \end{cases} \tag{14}$$

where h is the constant length of each sub-interval (fraction of a year) and q_{jt}^{ω} is the default probability for class j at period t under scenario ω. It is worth to point out that the third expression in (14) will only occur, if any, for period t such that $t_0 < t \leq t^i$. Notice that if there is a period t and a scenario ω such that $q_{jt}^{\omega} = 1$, it means that the class j of securities is in default and, then, $C_{it}^{\omega} = z_i \cdot F_i$, and $C_{i\tau}^{\omega} = 0$, $\forall \tau = t+1, \cdots, t_k$, for $i \in \mathcal{I}_j$. In another case, what happens is that for each period t and each scenario ω, $q_{jt}^{\omega} < 1$, i.e., the class of securities j is not in default, and the payment stream for all the securities in that class coincides with that defined in expression (1).

P_{it}^{ω}, unit price of security i at time period t under scenario ω, if R_{it}^{ω} is the real interest for security i at period t under scenario ω, for $i \in \mathcal{I}_j, j \in \mathcal{J}$, $t \in \mathcal{T} - \{t_0\}$, $\omega \in \Omega$. It is obtained as follows,

$$P_{it}^{\omega} = \sum_{t<\tau\leq t^i} C_{i\tau}^{\omega} \Re_i^{\omega}(t,\tau),$$

where $\Re_i^{\omega}(t,\tau) = (1 + R_{it}^{\omega} \cdot h)^{-(\tau-t+1)}$.

$P_{it}^{+\omega}$, unit purchase price of security i at period t under scenario ω when transaction costs (β) are considered.

$$P_{it}^{+\omega} = (1+\beta)P_{it}^{\omega}. \tag{15}$$

$P_{it}^{-\omega}$, unit selling price of security i at period t, under scenario ω.

$$P_{it}^{-\omega} = (1-\beta)P_{it}^{\omega}. \tag{16}$$

<u>*Variables*</u>

$x_{it}^{+\omega}$, volume of security i purchased in period t under scenario ω, for $i \in \mathcal{I}$, $t \in \mathcal{T}$, $\omega \in \Omega$.

$x_{it}^{-\omega}$, volume of security i sold in period t under scenario ω, for $i \in \mathcal{I}$, $t \in \mathcal{T} - \{t_0\}$, $\omega \in \Omega$.

z_{it}^{ω}, volume of security i to be held in the portfolio following the transactions conducted in period t under scenario ω, for $i \in \mathcal{I}$, $t \in \mathcal{T}$, $\omega \in \Omega$.

V_t^{ω}, final value of the portfolio in period t under scenario ω, for $t \in \mathcal{T}$, $\omega \in \Omega$.

The two-stage stochastic approach for maximizing the expected final value of the portfolio over the set of scenarios (i.e., in a risk neutral environment) is as follows,

$$\text{(DEM1)} \quad \max \sum_{\omega\in\Omega} w^{\omega} V_{t_k}^{\omega} \tag{17}$$

$$\text{s.t.} \quad x_{i0}^{+} = z_{i0} \quad \forall i \in \mathcal{I} \tag{18}$$

$$\sum_{i\in\mathcal{I}} P_{i0}^{+} x_{i0}^{+} = I_0 \tag{19}$$

$$z_{i,t-1}^{\omega} + x_{it}^{+\omega} - x_{it}^{-\omega} = z_{it}^{\omega} \quad \forall i \in \mathcal{I}, \quad t = t_1, \ldots, t_{k-1}, \quad \omega \in \Omega \tag{20}$$

$$x_{i:t^i=t,t}^{-\omega} = z_{i:t^i=t,t-1}^{\omega} \quad \forall i \in \mathcal{I}, \quad t = t_1, \ldots, t_k, \quad \omega \in \Omega \tag{21}$$

$$x_{i,t_k}^{-\omega} = z_{i,t_{k-1}}^{\omega} \quad \forall i \in \mathcal{I} : \quad t_k < t^i, \quad \omega \in \Omega \tag{22}$$

$$\sum_{i\in\mathcal{I}:t<t^i} P_{it}^{+\omega} x_{it}^{+\omega} - \sum_{i\in\mathcal{I}:t<t^i} P_{it}^{+\omega} x_{it}^{-\omega} = \sum_{i\in\mathcal{I}:t\le t^i} C_{it}^{\omega} z_{i,t-1}^{\omega} \quad \forall t = t_1, \ldots, t_{k-1}, \quad \omega \in \Omega \tag{23}$$

$$\sum_{i\in\mathcal{I}:t_k<t^i} P_{it_k}^{-\omega} x_{it_k}^{-\omega} + \sum_{i\in\mathcal{I}:t_k\le t^i} C_{it}^{\omega} z_{i,t_{k-1}}^{\omega} = V_{t_k}^{\omega} \quad \forall \omega \in \Omega \tag{24}$$

$$0 \le z_{i0}, x_{i0}^{+} \quad \forall i \in \mathcal{I} \tag{25}$$

$$0 \le x_{it}^{+\omega} \quad \forall i \in \mathcal{I}, \quad t = t_1, \ldots, t_{k-1}, \quad \omega \in \Omega \tag{26}$$

$$0 \le z_{it}^{\omega}, x_{it}^{-\omega} \quad \forall i \in \mathcal{I}, \quad t = t_1, \ldots, t_k, \quad \omega \in \Omega \tag{27}$$

$$0 \le V_{t_k}^{\omega} \quad \forall \omega \in \Omega. \tag{28}$$

Solution considerations: It is well known that for a large set of scenarios, this kind of problems becomes quite big. Methods that ignore the special structure of stochastic linear programs become quite inefficient. Taking benefit from the model's structure is specially beneficial in stochastic programs. For large dimensions, perhaps the method most frequently used is based on building an outer linearization of the recourse cost function and a solution of the first stage problem (the primal problem) plus its linearization.

The Benders decomposition (1962) has been widely extended in stochastic optimization to take care of feasibility questions, see the well known L-shaped method introduced Van Slyke and Wets (1969); see also Birge and Louveaux (1997), Higle and Sen (1996) and Kall and Wallace (1994), among many others. However, generating feasibility cuts by the mere application of the scenario related feasibility problem solutions may not be very efficient in some cases. Instead, we have proposed a scenario cluster decomposition approach for dealing with the feasibility problem, which generates tighter feasibility cuts to add to the master problem, see (2011).

5 Risk adverse models

5.1 *Mean-risk immunization strategy*

Let us consider the general two-stage stochastic linear problem for the maximization of the objective function expected value,

$$\begin{aligned} \mathcal{Q}_E = \max\ & cx + E_\psi[\max w^\omega (q^{\omega T} y^\omega)] \\ \text{s.t.}\quad & b_1 \leq Ax \leq b_2 \\ & h_1^\omega \leq T^\omega x + W^\omega y^\omega \leq h_2^\omega \quad \forall \omega \in \Omega \\ & x, y^\omega \geq 0, \quad \forall \omega \in \Omega, \end{aligned} \tag{29}$$

where c is a known vector of the objective function coefficients for the x variables in the first stage, b_1 and b_2 are the left and right hand side vectors for the first stage constraints, respectively, and A is the known matrix for the first stage constraints; w^ω is the likelihood attributed to scenario ω, h_1^ω and h_2^ω are the left and right hand side vectors for the second stage constraints, respectively, and q^ω is the vector of the objective function coefficients for the y variables, while T^ω and W^ω are the technology matrices under scenario ω, for $\omega \in \Omega$.

Piecing together the stochastic components of the problem, we have the vector $\xi^\omega = (q^\omega, h_1^\omega, h_2^\omega, T^\omega, W^\omega)$. Finally, E_ξ represents the mathematical expectation with respect to ξ over the set of scenarios Ω.

The main criticism that can be made against this very popular mean strategy is that it ignores the variance on the objective function value over the scenarios and, in particular, the "left" queue of the non-wanted scenarios. However, there are some other approaches that in addition deal with new risk measures (in a risk adverse environment) by also considering, e.g., semi-deviations (1999), excess probabilities (2004), value-at-risk and conditional value-at-risk and variants (1999, 2005, 2002, 2006), and first- and second-order stochastic dominance constraints (1903, 1903, 2008, 2011). Those approaches are more amenable than the classical mean-variance due to Markowitz (1952, 1959), mainly in the presence of 0–1 variables. In this work, we can use the excess probability approach, such that, $\mathcal{Q}_P = P(\omega \in \Omega : cx + q^\omega y^\omega > \phi)$ where ϕ is a prescribed threshold for the excess probability $\mathcal{Q}_P$. So, alternatively to maximize $\mathcal{Q}_E$, where $\mathcal{Q}_E = cx + \sum_{\omega \in \Omega} w^\omega q^\omega y^\omega$, the mean-risk function to maximize is

$$\mathcal{Q}_E + \psi \mathcal{Q}_P, \tag{30}$$

where ψ is a positive weighting parameter. Following this formulation, and using a more amenable expression of (30) for computational purposes at least, see (2004), let us consider a mean-risk model by changing the objective function, and appending a constraint and the 0–1 variable per each scenario ω in model (DEM1) (17)–(28). So, the variable ν^ω will take the value 1 if the final wealth under scenario ω is below threshold ϕ and 0, otherwise, then,

$$\nu^\omega = \begin{cases} 1 & \text{for } V_{t_k}^\omega < \phi \\ 0 & \text{otherwise.} \end{cases}$$

So, the new mean-risk model becomes a parametric model defined as follows,

$$\text{(DEM-MR)} \quad \max \sum_{\omega \in \Omega} w^\omega V_{t_k}^\omega + \psi \sum_{\omega \in \Omega} w^\omega (1 - \nu^\omega) \tag{31}$$

$$\text{s.t. constraints} \quad (18)\text{–}(28) \quad \text{and}$$

$$V_{t_k}^\omega + M\nu^\omega \geq \phi \quad \forall \omega \in \Omega \tag{32}$$

$$\nu^\omega \in \{0, 1\} \quad \forall \omega \in \Omega. \tag{33}$$

Notice that the parameters ψ, ϕ and M must be adequately chosen to calibrate the model the first two ones, and M for reducing the computational effort.

5.2 *Value-at-Risk (VaR) strategy*

Recent theoretical research on risk management suggests that the measures based on quantiles are good functions to measure the risk. Between them, the value at risk (VaR) has become a reference for many financial applications, even if it is not a coherent measure. (Note: It is said that a measure is a coherent one if it satisfies the 4 axioms introduced in

(1999)). However, the VaR approach is very attractive since it is easy to interpret, it measures up in monetary units and can be used to carry out an estimation of the necessary volume of own funds to cover the risk of market in activities of business developed by the financial institutions. A very useful approximation in the context of stochastic optimization is the maximization of an objective function defined as the value at risk, i.e.,

$$\begin{aligned}
(\text{VaR})\ \ & \max VaR \\
\text{s.t.}\ \ & cx + q^\omega y^\omega + M\nu^\omega \geq VaR \quad \forall \omega \in \Omega \\
& b_1 \leq Ax \leq b_2 \\
& h_1^\omega \leq T^\omega x + W^\omega y^\omega \leq h_2^\omega \quad \forall \omega \in \Omega \\
& \sum_{\omega \in \Omega} w^\omega \nu^\omega \leq \alpha \\
& x, y^\omega \geq 0 \quad \forall \omega \in \Omega \\
& VaR \geq 0 \\
& \nu^\omega \in \{0, 1\} \quad \forall \omega \in \Omega,
\end{aligned} \tag{34}$$

where ν^ω is a 0–1 variable such that its value is 1 if the objective function value $cx + q^\omega y^\omega$ for scenario ω is below VaR and, otherwise, is 0, and M is a given parameter. Notice that the optimization is realized over $(1-\alpha)\%$ of the scenarios, at least α being the accepted probability of the scenarios to occur whose function value $cx + q^\omega y^\omega$ is smaller than VaR. The advantage of the VaR strategy over the traditional maxmin strategy, see model (36), is that this requires that the objective function value should not be smaller than VaR for all the scenarios, no matter how representative and how many they are. So, its value may not be too high given this very restrictive constraint. The advantage of the VaR approach over the maxmin strategy and the expected value strategy (29) is obvious, in that it takes into account the probability of non-desired scenarios. However, it does not consider how bad accepted scenarios can be. The so-called CVaR strategy takes into account the conditional expectation of the objective function value $cx+q^\omega y^\omega$ above VaR, see Rockeafellar and Uryasev (2000, 2002) and Schultz and Tiedemann (2006). Although the improvement on risk management of the CVaR strategy over the VaR one has been substantial, we believe that the recent stochastic dominance constraints approach (see below) is an advance on the aim of increasing the objective function immunization. See some computational results in Section 8, and more extensive computational testing in Aranburu (2011).

5.3 *Stochastic dominance strategy*

The concept of stochastic dominance constraints aims to identify acceptable solutions for the problem under uncertainty and optimizing over them. "The random variable X is said to be stochastically smaller in first order (resp. in second order) than a random variable Y, i.e., $X \preceq_1 Y$ (resp. $X \preceq_2 Y$) iff $Eh(X) \leq Eh(Y)$ for all nondecreasing (resp. nondecreasing convex) functions h for which both expectations exist", see Gollmer, Neiser and Schultz (2008) (resp. Gollmer, Gotzes and Schultz (2011)).

Let us consider the following model whose aim is to maximize an objective function, whose vector of coefficients could be given by a general function $g(x, y)$ that, in our case, has been defined as the objective function expected value over the set of scenarios. Additionally, the function value $cx + q^\omega y^\omega$ must not be below the threshold, say ϕ^p for a finite number of thresholds, say $|\mathcal{P}|$, with a probability of failure, say α^p given for threshold ϕ^p, for $p \in \mathcal{P}$, where $\mathcal{P}$ is the so-called set of profiles.

$$
\begin{aligned}
\text{(S-D)} \max\ & cx + \textstyle\sum_{\omega\in\Omega} w^\omega q^\omega y^\omega \\
\text{s.t.}\quad & cx + q^\omega y^\omega + M\nu^{\omega p} \geq \phi^p \quad \forall \omega \in \Omega, \quad p \in \mathcal{P} \\
& b_1 \leq Ax \leq b_2 \\
& h_1^\omega \leq T^\omega x + W^\omega y^\omega \leq h_2^\omega \quad \forall \omega \in \Omega \\
& \sum_{\omega\in\Omega} w^\omega \nu^{\omega p} \leq \alpha^p \quad \forall p \in \mathcal{P} \\
& x, y^\omega \geq 0 \quad \forall \omega \in \Omega \\
& \nu^{\omega p} \in \{0,1\} \quad \forall \omega \in \Omega, \quad p \in \mathcal{P},
\end{aligned}
\tag{35}
$$

where the 0–1 variable $\nu^{\omega p}$ is defined as the above 0–1 variables, i.e., it takes the value 1 provided that the function value $cx + q^\omega y^\omega$ does not reach the threshold ϕ^p for the pth profile under scenario ω and 0 otherwise, for $\omega \in \Omega$, $p \in \mathcal{P}$.

Notice that the stochastic dominance strategy, $(S - D)$ model (35), has the advantage over the other strategies presented above, in that it forces the function $cx + q^\omega y^\omega$ not to be below than given thresholds of the objective function over the scenarios for given probabilities of failure.

Solution considerations: We must point out that the models (34) and (35) have a computational disadvantage when comparing them with model (31), since they have constraints linking 0–1 variables from different scenarios. Notice that this disadvantage is stronger for model (35) with $|\mathcal{P}| > 1$ than for model (34). In any case, a decomposition approach must be used for problem solving in large-scale instances. So, we propose to use a Branch-and-Fix Coordination approach that has proved to provide good results, see (2007, 2009, 2010a).

6 Two-stage fixed-income security portfolio immunization based on the VaR strategy

Traditional portfolio immunization is a strategy based on the maxmin approach to be used to match interest-rate risk of an asset portfolio against future stream of liabilities, in order to achieve net zero market exposure. There is a large amount of literature on portfolio immunization, see for instance Bierwag (1987) and Bierwag and Khang (1979), among many others. A bond portfolio is said to be immunized under parallel interest-rate shifts when the market value of the portfolio is, at least, as great as the present value (V) of the liabilities to be satisfied and the portfolio duration is matched with the liability duration. Bierwag and Khang prove that immunization can be seen as a maximization strategy throughout all states of the art in which the objective of investors is to achieve a guaranteed minimum return over the PH. According to Dantzig (1971), this maximum solution can be found by solving an equivalent LP problem that depends on the TSIR hypothesis. With this terminology, *immunization* is the maximin strategy obtained as the optimal solution of the following deterministic LP model,

$$\begin{aligned} & \max V \\ \text{s.t.} \quad & \sum_{i=1}^{n} x_i = I_0 \\ & \sum_{i=1}^{n} v_i^{\omega} x_i \geq V \quad \forall \omega \in \Omega \\ & V, x_i \geq 0 \quad \forall i = 1, \ldots, n, \end{aligned} \tag{36}$$

where the variable x_i indicates the percentage of investment I_0 that has to be assigned to buy asset i, v_i^{ω} denotes the portfolio value at the end of the PH if an amount of $\mathcal{I}$ budget is invested at the starting of the PH in asset i and, shortly afterwards, the interest rate becomes r^{ω} remaining unchanged until the end of the PH; and, finally, V is the minimum portfolio value guaranteed at the end of the PH. In any case, it must be pointed out that the solution given by the LP DEM model (36) is not exactly an immunized portfolio but an approximation to it. This is due to the fact that only a finite number of scenarios (i.e., interest rate shifts) is considered. One way to obtain a more accurate result is to include not only the biggest interest rate fluctuations but also, sufficiently small shifts in interest rates. One of the most important factors for the development of portfolio strategies to combat interest rate risk is the so-called Dynamic Global Portfolio Immunization Theorem put forward by Khang (1983), under which the optimum strategy for assuring the final value of a portfolio at the end of a time period, regardless of changes in interest rates, is that which matches Macaulay's term planning horizon[1] for that portfolio at all times. Given the nature of the term of a portfolio, this strategy would imply continuous readjustment. In any event, the optimality of the strategy is based on the following assumptions, among others:

a. The interest rate structure has parallel changes, i.e. if the structure changes from $g(t)$ to $g^*(t, \beta)$, then

$$g^*(t, \beta) = g(t) + \beta.$$

b. There are no transaction costs. This assumption is extemely important, since if there are transaction costs and those costs are high, then the continuous readjustment of the portfolio would not be viable.

c. The yield on bonds depends only on the current interest rate, with no need to take into account any risk of default.

The first of these assumptions avoids the risk of underestimating the performance of the structure, known as the *Risk of Immunization*, see Fong and Vasicek (1983). The assumption of no transaction costs is crucial in a dynamic context, since the strategy of continuous readjustment may not be optimum if transaction costs are considered, given the high costs that it would entail. Furthermore, given the third assumption, there is no analysis in the context of bonds with different credit ratings and, therefore, with a positive associated risk of default.

A two-stage stochastic model extension of the immunization model requires de following notation,

v_{i0}^{ω}, final value of an investment of P_{i0} monetary units in security i made at the initial period, t_0, if the instantaneous spot interest rate changes, shortly afterwards from R_{it_0} to $R_{it_1}^{\omega}$ and no additional unexpected interest

---

[1]The Macaulay duration is the weighted-average term to maturity of the cash flows from a bond, where the weights are the present value of the cash flow divided by the price:

$$d_i = \frac{\sum_{s=1}^{i}(t_s - t_0)\,Coupon \cdot Discount}{\sum_{s=1}^{i} Coupon \cdot Discount}.$$

In a zero-coupon bond the Macaulay duration is equal to the bonds maturity.

rate change takes place until the end of the PH under scenario ω. It is calculated as follows,

$$v_{i0}^{\omega} = \frac{\sum_{\tau=t_1}^{t_k} C_{i\tau}^{\omega} \Re_i^{\omega}(t_0,\tau)}{\Re_i^{\omega}(t_0,t_k)}, \tag{37}$$

where $i \in \mathcal{I}$, $\omega \in \Omega$, and $\Re_i^{\omega}(t_0,\tau) = \prod_{s=t_0}^{\tau}(1 + R_{is}^{\omega} \cdot h)^{-1}$.

Then, the immunization at period t_0 is formulated by the constraints

$$\sum_{i\in\mathcal{I}} v_{i0}^{\omega} z_{i0} \geq V_0 \quad \forall \omega \in \Omega,$$

such that it is the big substantive difference with the two-stage stochastic model introduced in Section 4. So, the two-stage maxmin DEM is as follows,

$$\text{(DEM2)} \qquad \max V_0 + \sum_{\omega\in\Omega} w^{\omega} V_{t_k}^{\omega} \tag{38}$$

$$\text{s.t. constraints (18)–(28) and}$$

$$\sum_{i\in\mathcal{I}} v_{i0}^{\omega} z_{i0} \geq V_0 \quad \forall \omega \in \Omega \tag{39}$$

$$V_0 \geq 0, \tag{40}$$

where the objective function (38) to maximize gives the investment value at the end of the initial period t_0 and the expected final value of the portfolio over the set of scenarios. Additionally, constraints (39) are the key to the immunization strategy at the initial period, since they ensure a minimum wealth under all the future uncertainties.

As a new two-stage approach in the risk averse environment, we will introduce a hybrid mean-risk and Value-at-Risk model for optimizing immunization strategies over the scenarios, such that the probability of failure does not exceed a given α%. For this purpose, we change the objective function (38) and constraints (39). Firstly, we consider a 0-1 variable per scenario, ν^{ω}, such that, as before, it will take value 1 if the value of the portfolio at the end of the initial period under scenario ω, $\sum_{i\in\mathcal{I}} v_{i0}^{\omega} z_{i0}$, is smaller than V_0, and 0, otherwise. We point out that we introduce these 0-1 variables in order to control by some means the strict behavior of constraints (39). The new set of constraints that replace (39) and since now they are second stage constraints, are given in (42). Level α will depend on the risk averseness of the investor. To ensure this, the constraint (43) is introduced in the model. Then, the new VaR DEM is as follows,

$$\text{(DEM3)} \qquad \max V_0 + \sum_{\omega\in\Omega} w^\omega V_{t_k}^\omega \tag{41}$$

$$\text{s.t. constraints (18)–(28) and}$$

$$\sum_{i\in\mathcal{I}} v_{i0}^\omega z_{i0} + M\nu^\omega \geq V_0 \quad \forall\omega\in\Omega \tag{42}$$

$$\sum_{\omega\in\Omega} w^\omega \nu^\omega \leq \alpha \tag{43}$$

$$V_0 \geq 0 \tag{44}$$

$$\nu^\omega \in \{0,1\} \quad \forall\omega\in\Omega. \tag{45}$$

Note: The parameters M and α must be adequately chosen to calibrate the model. We have experimented with the choice of M as the maximum value between V_0 and $V_{t_k}^\omega$, $\forall\omega\in\Omega$, and the results are satisfactory.

Solution considerations: Model (DEM3) is a two-stage stochastic mixed 0–1 optimization one and, then, a further computational effort must be made, since plain use of even state-of-the-art optimization engines cannot be affordable given the large-scale nature of the problem. Alternatively, we propose some types of decomposition approaches based on Benders decomposition, Branch-and-Fix Coordination and others that have been proved to provide good results in the field, see (2007, 2009, 2009, 2010a).

7 A more accurate solution: a multistage stochastic scheme

7.1 *Multistage maxmin immunization strategy*

In the previous sections we have presented two-stage models where the immunization condition is imposed on itself precisely at the initial period, t_0. However, the immunization strategy should prevail at all the periods of the PH, given the uncertainty of the main parameters all over the horizon. So, furthermore, some corrective action should be implemented between periods t and $t+1$, then multistage stochastic models become more appropriate.

In the general formulation of a multistage model, decisions on each stage have to be made stage-wise. For example, first stage variables are selected before observing the realization of uncertain parameters, through

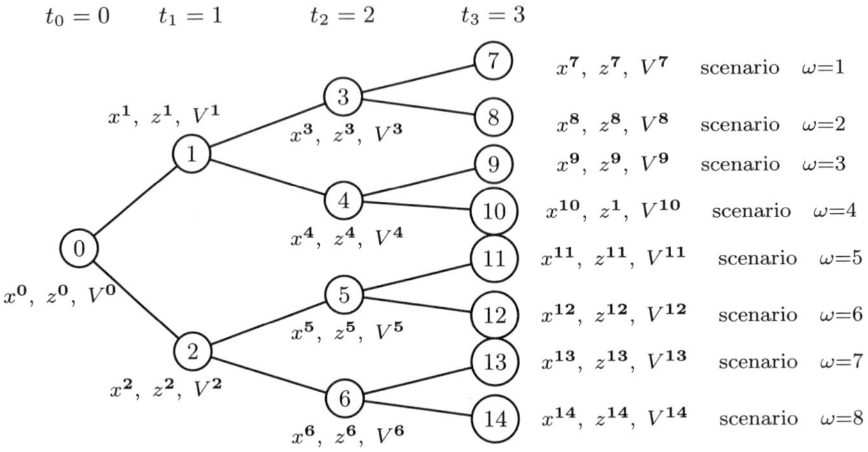

Figure 2 Multistage scenario tree

the occurrence of the scenarios. After having decided on first stage and having observed each realization of uncertain parameters, the second stage (or recourse) decision has to be made. After having decided on a second stage and having observed each realization of uncertain parameters, the next (third) stage decision has to be made, and so on. At each stage, there are variables which correspond to decisions that have to be made without anticipation of some of future problem data, i.e., they take the same value under each scenario that belong to the same group for each stage (i.e., the so-called nonanticipativity constraints must be satisfied).

Each node, say g, in Figure 2 represents a point in time where a decision can be made. Once a decision is made, some contingencies may occur, and information related to these contingencies is available at the beginning of each stage. In this context, a stage is a point in time where a decision is made and, in some cases, can be included by a subset of consecutive time periods. In this example, there are $|\mathcal{T}| = 4$ stages, and $\mathcal{T} = \{t_0 = 0, \cdots, t_k = 3\}$. At each stage, there are some types of vectors of decision variables, namely, x, z and V. Let also $\mathcal{G}$ denote the set of scenario groups, and $\mathcal{G}_t$ the subset of scenario groups that belong to stage t, such that $\mathcal{G} = \cup_{t \in \mathcal{T}} \mathcal{G}_t$. The structure of this information is visualized as a tree where each root-to-leaf path represents one specific scenario, and corresponds to one realization of the whole set of the uncertain parameters. Here, we have $|\Omega| = 8$ scenarios and $|\mathcal{G}| = 15$ scenario groups.

Notice that the scenario group concept corresponds to the node concept in the underlying scenario tree. However, it can sometimes be useful to work with scenario groups instead of nodes if, for example, we want to split the set of scenarios into different subsets. Two scenarios belong to the same group in a given stage provided that they have the same realizations of the uncertain parameters up to the stage. Ω_g denotes the set of scenarios that belong to group g, for $g \in \mathcal{G}$. Notice also that there is an identity between the scenario groups of the last stage and the scenarios, then scenario group $g = 7$ corresponds to the first scenario, $\Omega_7 = \{1\}$, and scenario group $g = 14$ is exactly the last scenario, $\Omega_{14} = \{8\}$. We will use the boldface letter to denote the corresponding group and distinguish it from the scenarios. $t(g)$ denotes the stage of scenario group g, such that, $g \in \mathcal{G}_{t(g)}$. And, finally, $\pi(g)$ indicates the immediate ancestor group of group g, such that $\pi(g) \in \mathcal{G}_{t(g)-1}$, for $g \in \mathcal{G} - \{\mathcal{G}_0\}$. The following additional notation is used in the model.

Parameters

$w^{\mathbf{g}}$, likelihood of scenario group g, for $g \in \mathcal{G}_t$, $t \in \mathcal{T}$, computed as $\sum_{\omega \in \Omega_g} w^{\omega}$.

$r^{\mathbf{g}}$, risk-free interest rate under scenario group g, for $g \in \mathcal{G}_t$, $t \in \mathcal{T}$.

$q_j^{\mathbf{g}}$, measure of risk calculated as the probability of default of the securities of class j under scenario group g, for $g \in \mathcal{G}_t$, $t \in \mathcal{T}$, $j \in \mathcal{J}$.

$R_i^{\mathbf{g}}$, real interest rate (as a fraction of one) for security i of class j under scenario group g, for $g \in \mathcal{G}_t$, $t \in \mathcal{T}$, $i \in \mathcal{I}_j$, $j \in \mathcal{J}$. It is calculated as follows,

$$R_i^{\mathbf{g}} = r^{\mathbf{g}} + \frac{(1 + r^{\mathbf{g}})(1 - z_i)q_j^{\mathbf{g}}}{1 - (1 - z_i)q_j^{\mathbf{g}}}.$$

$C_i^{\mathbf{g}}$, payment stream vector generated by one unit of security i under scenario group g, for $g \in \mathcal{G}_t$, $t \in \mathcal{T} - \{t_0\}$, $i \in \mathcal{I}_j$, $j \in \mathcal{J}$. Its element for period τ is calculated as follows,

$$C_{i\tau}^{\mathbf{g}} = \begin{cases} c_i \cdot h, & t(g) < \tau < t^i : q_j^{\mathbf{g}} < 1 \\ F_i + c_i \cdot h, & \tau = t^i : q_j^{\mathbf{g}} < 1 \\ z_i \cdot F_i, & \tau = t(g) : q_j^{\pi(\mathbf{g})} < 1 \quad \text{and} \quad q_j^{\mathbf{g}} = 1 \\ 0, & \text{otherwise,} \end{cases} \tag{46}$$

where h is the constant length of each sub-interval (fraction of a year) and $q_j^{\mathbf{g}}$ is the default probability for class j under scenario group g. Notice that $q_j^{\mathbf{g}} = 1$ means that in stage $t(g)$, the class of securities j is in default and, then, $C_{i\tau}^{\mathbf{g}} = z_i F_i$ at period $\tau = t(g)$, and $C_{i\tau}^{\mathbf{g}} = 0$ for $\tau = t(g) + 1, \ldots, t_k$, for all the securities $i \in \mathcal{I}_j$.

$P_i^{\mathbf{g}}$, unit price of security i under scenario group g, for $g \in \mathcal{G}_t$, $t \in \mathcal{T} - \{t_0\}$, $i \in \mathcal{I}_j$, $j \in \mathcal{J}$. It is calculated as follows,

$$P_i^{\mathbf{g}} = \sum_{t(g) < \tau \leq t^i} C_{i\tau}^{\mathbf{g}} \Re_i^{\mathbf{g}}(t(g), \tau),$$

where $t(g)$ is the period to which scenario group g belongs to, i.e., $g \in \mathcal{G}_{t(g)}$, and τ denotes the corresponding period in the payment $C_{i\tau}^{\mathbf{g}}$. Finally, $\Re_i^{\mathbf{g}}(t(g), \tau) = (1 + R_i^{\mathbf{g}})^{-(\tau - t(g) + 1)}$.

$P_i^{+\mathbf{g}}$, unit purchase price of security i under scenario group g, for $i \in \mathcal{I}_j$, $j \in \mathcal{J}$, $g \in \mathcal{G}_t$, $t \in \mathcal{T} - \{t_k\}$, where $P_i^{+\mathbf{g}} = (1 + \beta) P_i^{\mathbf{g}}$.

$P_i^{+\mathbf{g}}$, unit selling price of security i under scenario group g, for $i \in \mathcal{I}_j$, $j \in \mathcal{J}$, $g \in \mathcal{G}_t$, $t \in \mathcal{T} - \{t_0\}$, where $P_i^{-\mathbf{g}} = (1 - \beta) P_i^{\mathbf{g}}$.

$v_{i,t(g)}^{\omega}$, final value under scenario group g of an investment of $P_i^{\mathbf{g}}$ monetary units in security i made at period $t(g)$, if the instantaneous real interest rate changes, just afterwards, from $R_i^{\mathbf{g}}$ until R_i^{ω} **at the end** of the PH, for $i \in \mathcal{I}$, $g \in \mathcal{G}_{t(g)}$, $\omega \in \Omega_g$. It is calculated as follows,

$$v_{i,t(g)}^{\omega} = \frac{\sum_{\tau=t(g)+1}^{t_k} C_{i,\tau}^{\mathbf{g}} \Re_i^{\omega}(t(g), \tau)}{\Re_i^{\omega}(t(g), t_k)}, \tag{47}$$

where $\Re_i^{\omega}(t(g), \tau) = \prod_{s=t(g)}^{\tau} (1 + R_{i,s}^{\omega} \cdot h)^{-1}$, $\omega \in \Omega_g$.

<u>*Variables*</u>

$x_i^{+\mathbf{g}}$, volume of security i purchased under scenario group g, for $g \in \mathcal{G}_t$, $t \in \mathcal{T}$, $i \in I$.

$x_i^{-\mathbf{g}}$, volume of security i sold under scenario group g, for $g \in \mathcal{G}_t$, $t \in \mathcal{T} - \{t_0\}$, $i \in I$.

$z_i^{\mathbf{g}}$, volume of security i to be held in the portfolio following the transactions conducted under scenario group g, for $g \in \mathcal{G}_t$, $t \in \mathcal{T}$, $i \in I$.

$V^{\mathbf{g}}$, final value of the portfolio under scenario group g, for $g \in \mathcal{G}_t$, $t \in \mathcal{T}$.

The multistage DEM that includes the immunization constraints under each scenario group to maximize the investment value at the end of the initial period t_0 and the expected final value portfolio over the set of scenarios in a multistage approach and risk averse environment with the maxmin strategy, is given as follows,

$$\text{(DEM4)} \quad \max V^{\mathbf{0}} + \sum_{t=t_1}^{t_k} \sum_{g \in \mathcal{G}_t} w^{\mathbf{g}} V^{\mathbf{g}} \tag{48}$$

$$\text{s.t.} \quad x_i^{\mathbf{0}} = z_i^{\mathbf{0}} \quad \forall i \in \mathcal{I} \tag{49}$$

$$\sum_{i \in \mathcal{I}} P_i^{+\mathbf{0}} x_i^{+\mathbf{0}} = I_0 \tag{50}$$

$$\sum_{i \in \mathcal{I}} v_{i,0}^{\omega} z_i^{\mathbf{0}} \geq V^{\mathbf{0}} \quad \forall \omega \in \Omega_{\mathbf{0}} = \Omega \tag{51}$$

$$z_i^{\pi(\mathbf{g})} + x_i^{+\mathbf{g}} - x_i^{-\mathbf{g}} = z_i^{\mathbf{g}} \quad \forall i \in \mathcal{I} \quad g \in \mathcal{G}_t, \quad t = t_1, \ldots, t_{k-1} \tag{52}$$

$$x_{i:t^i=t}^{-\mathbf{g}} = z_{i:t^i=t}^{\pi(\mathbf{g})} \quad \forall i \in \mathcal{I}, \quad g \in \mathcal{G}_t, \quad t = t_1 \ldots, t_k \tag{53}$$

$$x_i^{-\mathbf{g}} = z_i^{\pi(\mathbf{g})} \quad \forall i \in \mathcal{I} : t_k < t^i, \quad g \in \mathcal{G}_{t_k} \tag{54}$$

$$\sum_{i \in \mathcal{I}: t < t^i} v_{i,t(g)}^{\omega} z_i^{\mathbf{g}} \geq V^{\mathbf{g}} \quad \forall \omega \in \Omega_{\mathbf{g}}, \quad g \in \mathcal{G}_t, \quad t = t_1, \ldots, t_{k-1} \tag{55}$$

$$\sum_{i \in \mathcal{I}: t < t^i} P_i^{+\mathbf{g}} x_i^{+\mathbf{g}} - \sum_{i \in \mathcal{I}: t < t^i} P_i^{+\mathbf{g}} x_i^{-\mathbf{g}} = \sum_{i \in \mathcal{I}: t \leq t^i} C_{it}^{\mathbf{g}} z_i^{\pi(\mathbf{g})}$$
$$\forall g \in \mathcal{G}_t, \quad t = t_1, \ldots, t_{k-1} \tag{56}$$

$$\sum_{i \in \mathcal{I}: t_k < t^i} P_i^{-\mathbf{g}} x_i^{-\mathbf{g}} + \sum_{i \in \mathcal{I}: t_k \leq t^i} C_{it_k}^{\mathbf{g}} z_i^{\pi(\mathbf{g})} = V^{\mathbf{g}} \quad \forall g \in \mathcal{G}_{t_k} \tag{57}$$

$$0 \leq x_i^{+\mathbf{g}} \quad \forall i \in \mathcal{I}, \quad g \in \mathcal{G}_t, \quad t = t_0, \ldots, t_{k-1} \tag{58}$$

$$0 \leq z_i^{\mathbf{g}}, x_i^{-\mathbf{g}} \quad \forall i \in \mathcal{I}, \quad g \in \mathcal{G}_t, \quad t = t_0, \ldots, t_k \tag{59}$$

$$0 \leq V^{\mathbf{g}} \quad \forall g \in \mathcal{G}_t, \quad t = t_0, \ldots, t_k. \tag{60}$$

Equations (49)–(49) are the same as (3)–(4) in model (DEM1), but for the multistage environment. Constraints (51) are the key to the immunization strategy at initial period, i.e., they ensure a minimum wealth under all the future uncertainties. Constraints (52)–(54) are balance equations for each scenario group that link the volume of securities purchased and sold in each period with the volume of securities in the portfolio. Constraints (55) are the key of the immunization strategy at each scenario group of each decision period, that ensure a minimum value under all the future scenarios. Equations (56)–(57) are also the same as in model (DEM1), but also for the multistage environment.

Notice that model (DEM4) is an extension of the two-stage maxmin strategy given by model(DEM2) to the multistage scheme, where $\alpha = 0$, i.e., over the 100% of the scenarios. In Section 7.3 we introduce a more flexible scheme that allows an optimization of the VaR strategy given a selected class of potential (and variable) thresholds defined under different scenario groups and given probabilities of failure.

7.2 *Multistage stochastic dominance strategy*

The extension of the two-stage stochastic dominance strategy to a multi-stage scheme is based on the choice of a threshold, say ϕ^p from profile class $\mathcal{P}$ to satisfy given probabilities of failure, α^p, $p \in \mathcal{P}$. Moreover a $0-1$ variable $\nu^{\omega p}$ for each pair of threshold and scenario (since notice that the scenario groups of last stage are the scenarios) is introduced in order to optimize the stochastic dominance strategy. It is defined as follows,

$$\nu^{\omega p} = \begin{cases} 1 & \text{for } V_{t_k}^{\omega} < \phi^p \\ 0 & \text{otherwise,} \end{cases}$$

for $\omega \in \Omega, p \in \mathcal{P}$. The DEM to maximize the portfolio expected final value over the set of scenarios in a multistage environment is as follows,

$$\text{(DEM5)} \quad \max \sum_{g \in \mathcal{G}_{t_k}} w^{\mathbf{g}} V^{\mathbf{g}} = \sum_{\omega \in \Omega} w^{\omega} V_{t_k}^{\omega} \tag{61}$$

$$\text{s.t. constraints} \quad (49), (50), (52)\text{–}(54), (56)\text{–}(59) \quad \text{and}$$

$$V^{\mathbf{g}} + M\nu^{\mathbf{g}p} \geq \phi^p \quad \forall g \in \mathcal{G}_{t_k}, \quad p \in \mathcal{P} \tag{62}$$

$$\sum_{g \in \mathcal{G}_{t_k}} w^{\mathbf{g}} \nu^{\mathbf{g}p} \leq \alpha^p \quad \forall p \in \mathcal{P} \tag{63}$$

$$V^{\mathbf{g}} \geq 0 \quad \forall g \in \mathcal{G}_{t_k} \tag{64}$$

$$\nu^{\mathbf{g}p} \in \{0, 1\} \quad \forall g \in \mathcal{G}_{t_k}, \quad p \in \mathcal{P}. \tag{65}$$

Notice that model (DEM5) does not consider the maxmin strategy related constraints (51) and (55) from model (DEM4), since instead it considers the stochastic dominance constraints (62) and (63).

7.3 *Multistage VaR & stochastic dominance strategy*

Following the risk averse two-stage models introduced in Section 5, in this section we introduce an extension of the multistage VaR strategy. Notice that the multistage maxmin model (DEM4) provides a 0% VaR strategy over all the scenarios, by introducing an immunization set of constraints for each scenario group at each decision period.

Let us consider the class of threshold profiles $\mathcal{P}$ as the set of threshold profiles ϕ^p to be satisfied as in model (DEM5). In the most ambitious situation of modeling, we could choose this set to be included by the thresholds to satisfy under each scenario group at each decision stage, i.e., $\mathcal{P} = \{\phi^g, g \in \mathcal{G}_t, t = t_0, \cdots, t_{k-1}\}$. So, in this case, $|\mathcal{P}| = |\mathcal{G}_{t_0}| + |\mathcal{G}_{t_1}| + \cdots + |\mathcal{G}_{t_{k-1}}|$ and, then, the threshold, for $p \in \mathcal{P}$ is to be reached by V^p with probability of failure α^p. Then, the set of immunization constraints defined for each $p \in \mathcal{P}$ are as follows,

$$\sum_{i \in \mathcal{I}:t<t^i} v_{i,t(p)}^{\omega} z_i^p + M\nu^{\omega p} \geq V^p \quad \forall \omega \in \Omega_{\mathbf{g}}, \quad p = g, \quad g \in \mathcal{G}_t, \quad t = t_0, \ldots, t_{k-1},$$

where the $0-1$ variable $\nu^{\omega p}$ is defined as follows,

$$\nu^{\omega p} = \begin{cases} 1 & \text{for } \sum_{i \in \mathcal{I}:t<t^i} v_{i,t(p)}^{\omega} z_i^p < V^p \\ 0 & \text{otherwise.} \end{cases}$$

Moreover, the constraints

$$\sum_{\omega \in \Omega_p} w^{\omega} \nu^{\omega p} \leq \alpha^p \quad \forall p \in \mathcal{P}$$

forces the optimization with a given probability of failure, α^p. The new DEM is defined as follows,

$$\text{(DEM6)} \quad \max \sum_{p \in \mathcal{P}} \gamma^p V^p + \sum_{g \in \mathcal{G}_{t_k}} w^{\mathbf{g}} V^{\mathbf{g}} \tag{66}$$

$$\text{s.t. constraints} \quad (49), (50), (52)\text{–}(54), (56)\text{–}(60) \quad \text{and}$$

$$\sum_{i \in \mathcal{I}: t < t^i} v^{\omega}_{i,t(p)} z^p_i + M \nu^{\omega p} \geq V^p \quad \forall \omega \in \Omega_{\mathbf{g}} \quad p = g, \quad g \in \mathcal{G}_t, \quad t = t_0, \dots, t_{k-1} \tag{67}$$

$$\sum_{\omega \in \Omega_p} w^{\omega} \nu^{\omega p} \leq \alpha^p \quad \forall p \in \mathcal{P} \tag{68}$$

$$\nu^{\omega p} \in \{0, 1\} \quad \forall \omega \in \Omega_{\mathbf{g}} \quad p = g, \quad g \in \mathcal{G}_t, \quad t = t_0, \dots, t_{k-1} \tag{69}$$

$$V^p \geq \phi^p \quad p \in \mathcal{P}. \tag{70}$$

With this choice of class $\mathcal{P}$, there are $|\Omega| \cdot (|\mathcal{T}| - 1)$ $0-1$ variables and immunization constraints (67), and $|\mathcal{P}|$ constraints (68) to force the optimization with the different probabilities of failure. Notice that in this case the cardinality of class $\mathcal{P}$ can be very big and, so, the model solving computing time is unaffordable. However, the class $\mathcal{P}$ to consider can be reduced and, then, the computational effort for the solution of (DEM6), by selecting a subset of potential situations in a number which is not as ambitious as that given above. We consider that this reduction can be instance dependent. In the particular case of $\phi^p = 0$ and the probable $v^{\omega}_{i,t(p)} z^p_i \geq 0$, $\forall p \in \mathcal{P}$ (or for $|\mathcal{P}| = 0$), then model (DEM6) (66)–(70) is as model (DEM4) (48)–(60), i.e, the multistage maxmin strategy. Alternatively, model (DEM6) for $|\mathcal{P}| = 1$ is the so-called multistage VaR strategy.

Solution considerations: It is also well known that for a large set of scenarios and a large set of securities the kind of models as (DEM5) and (DEM6) become large-scale multistage mixed 0–1 problems. In this case, we have to consider alternative decomposition methods to the traditional plain using of the state-of-the-art optimization engine of choice. Our alternative consists of using an adaptation of our Cluster-based Benders Decomposition to the multistage environment and to integrate it in our Branch-and-Fix Coordination approach for solving large scale multi-stage stochastic mixed 0–1 models, that have been proved to produce good results in reasonable computing time, see Aranburu *et al.* (2011).

8 Case study

We proceed to illustrate the main strategies (see a summary in Table 6) introduced in the previous sections by applying them to an illustrative example, assuming that the planning horizon (18 months) is divided into three periods of equal length (6 months), and the portfolio rebalancing is only allowed at the beginning of each sub-interval, i.e., $\mathcal{T} = \{t_0 = 0, t_1 = 0.5, t_2 = 1, t_3 = 1.5\}$ and $k = 3$. We assume that at the beginning of the PH the current interest rate is 1.5% (compounded semiannually) and that interest rates may move upwards and downwards by 100 basic points, such that at future three outlooks are considered in the future, where $P(r_{t+1} = 0.5|r_t = 1.5) = 0.25$, $P(r_{t+1} = 1.5|r_t = 1.5) = 0.35$ and $P(r_{t+1} = 2.5|r_t = 1.5) = 0.4$. Also, the expected interest rate outstanding at the beginning of each sub-interval is the current interest rate. The initial budget is 1 million of monetary units. We assume that transaction cost incurred at each portfolio rearrangement are a percentage $\beta = 0.15\%$ of the volume trade at each period. Principal and coupon repayments do not generate any transaction cost although other assumptions could easily be implemented.

We set up a market comprising eight different bonds classes, four with a high credit rating (AAA) -government bonds, for instance- and the other four from a financial institution with a high credit risk (for example, B or CCC). Seven of the possible asset classes are shown in Tables 1, 2 and 3 plus the class of Default (D), which is shown in Table 1, now denoted by $\mathcal{I}_8$.

In order to analyze what an investor would do in different situations and different markets, we will consider three instances in the case study. In Instance 1, we consider bonds i for $i \in \mathcal{I}_1 = \{1, 2, 3, 4\}$, i.e., they are bonds in the first class (high credit rating, AAA), and bonds i for $i \in \mathcal{I}_7 = \{1', 2', 3', 4'\}$, i.e., bonds in the class CCC (very high credit risk). In Instance 2, instead, we consider bonds i in the first class (high credit rating, AAA), i.e., $i \in \mathcal{I}_1 = \{1, 2, 3, 4\}$, and bonds i for $i \in \mathcal{I}_6 = \{1', 2', 3', 4'\}$, i.e., bonds in the class B (quite high credit risk). In Instance 3 we consider bonds in the first class and bonds i for $i \in \mathcal{I}_5 = \{1', 2', 3', 4'\}$, i.e., bonds in the class BB (some credit risk). We also assume that the market interest rate in force is 1.5%, and that each bond will have half-yearly ($h = 0.5$) coupon payments according to its *real* interest rate level. Some of the characteristics of these two classes of assets are taken from Tables 3 and 4 and are depicted in Table 5.

8.1 *Deterministic LP model*

Let us consider the deterministic model (LP) (2)–(13), where the aim is to maximize the final value of the portfolio in Instance 1: with AAA and CCC bonds and Instance 2: with AAA and B bonds. In both instances, as a deterministic strategy, it does not take into account any future possibilities

Table 5 Case Study Data.

Class j	Credit rating	Default Prob: q	Recovery Rate: z	Annual Coupon: c
1	AAA	0.0000	0.694	1.5%
5	BB	0.0097	0.374	2.11857%
6	B	0.0493	0.374	4.77999%
7	CCC	0.2798	0.374	23.3719%

so none of the risks actually affect the optimal solution. It consists of investing, in the initial period $t_0 = 0$, all in the risky bond that matures on the PH, $i = 3'$. And then, reinvesting all the coupon payments in the same bond throughout all the decision stages $t_1 = 1$ and $t_2 = 1.5$. In this way we can receive all the coupon payments but avoiding transaction costs. In any case, it must be stressed that all the securities in which this strategy proposes to invest, in both instances, are risky securities. This is not surprising given that risky securities offer higher returns, and the objective function envisaged does not penalize risk. This problem may be due to the fact that the risk has not being suitably treated in a deterministic model.

8.2 *Two-stage scheme*

Let us consider a two-stage scheme which assumes that changes may happen at the beginning of the second decision stage, and will stay in that way until the end of the PH. It has its advantages and disadvantages. On the one hand, one might think that it is better to behave in this way and build another two-stage model at the second stage. But, on the other hand, this scheme can produce very myopic decisions, since it makes decisions while taking into account very little of the future possibilities.

We assume that default probabilities may change over time and can be different depending on the scenario. Recovery rates, however, will be constant for each class in our case study, i.e., $z_i = z_l$, $\forall i, l \in \mathcal{I}_j$, $j \in \mathcal{J}$. The recovery rate values for the government bonds, AAA, is taken from Table 5, as the recovery rate of the type term loans, is $z_1 = 0.694$. For the second classes in both instances, i.e., the financial bonds, the values are also taken from Table 5, they are $z_5 = 0.374$, $z_6 = 0.374$ and $z_7 = 0.374$, respectively.

In order to structure the possibilities in the second source of uncertainty, we know that it may change in different ways. Each one of these changes has a transition probability, which is the conditional probability, $p(S_2|S_1)$, of transiting to state S_2, given the state S_1. We will assume that each of the asset classes will have several possibilities for the next maturity period.

The AAA rated bonds may have two possibilities as outlooks. In particular, $p(AAA/AAA) = 0.9194$, and $p(AA/AAA) = 0.0806$. This means that a AAA rated bond (class $j = 1$) is much more likely to stay in the same rating than to change. But in unusual circumstances it would change, and would go down to another quite safe stage. Moreover, as shown in Table 3, the probabilities of default in these classes are $q_{1,t} = 0.00$, and $q_{2,t} = 0.02$, $\forall t = t_1, \ldots, t^i$, $i \in \mathcal{I}_j$, $j = 1, 2$.

The B rated bonds (class $j = 6$) may have three possibilities in the future. In particular, transition probabilities are, $p(BB/B) = 0.06535$, $p(B/B) = 0.8753$ and $p(D/B) = 0.0593$. As shown in Table 3, the probabilities of default in these classes are $q_{5,t} = 0.0097$, $q_{6,t} = 0.0493$ and $q_{8,t} = 1$, $\forall t = t_1, \ldots, t^i$, $i \in \mathcal{I}_j$, $j = 5, 6, 8$.

The CCC rated bonds (class $j = 7$) may have three possibilities. In particular, the transition probabilities are $p(B/CCC) = 0.1339$, $p(CCC/CCC) = 0.53395$ and $p(D/CCC) = 0.33215$. Notice that a CCC rated bond is much more likely to change into the worst stage a bond could reach, i.e., default, D. Moreover, as shown in Table 3, the probabilities of default in these classes are $q_{5,t} = 0.097$, $q_{6,t} = 0.0493$, $q_{7,t} = 0.2798$ and $q_{8,t} = 1$, $\forall t = t_1, \ldots, t^i$, $i \in \mathcal{I}_j$, $j = 6, 7, 8$.

We consider two different instances with similar shape. We have a safe asset and a risky one. In order to describe all the instances at the same time, we will define the generic states S_1 and S_2 that will be $S_1 = CCC$ and $S_2 = B$ for Instance 1, $S_1 = B$ and $S_2 = BB$ for Instance 2. Figure 3 depicts the 18 scenarios considered in the two-stage models, where we consider the class of bonds: AAA ($j = 1$) and S_1. In order to build the scenario tree we have considered three aspects in which the outcomes might change: $J^1 = \{1, 2\}$ (the possible rating of the first asset class), $J^j = \{j - 1, j, 8\}$ (the possible rating of the second asset class), and $r = \{0.5\%, 1.5\%, 2.5\%\}$ (the risk free interest rate level). So, we define the set of scenarios as the cartesian product of those three outcomes, $\Omega = J^1 \times J^j \times r$, that for simplicity we consider independent risks. Then, the probability w^ω of scenario ω is computed as the product of the probabilities of the three independent outcomes.

8.3 *Two-stage stochastic model (DEM1)*

Let us analyze the stochastic optimal solution of the two-stage model (DEM1) (17)–(28) for optimizing strategies under the two sources of

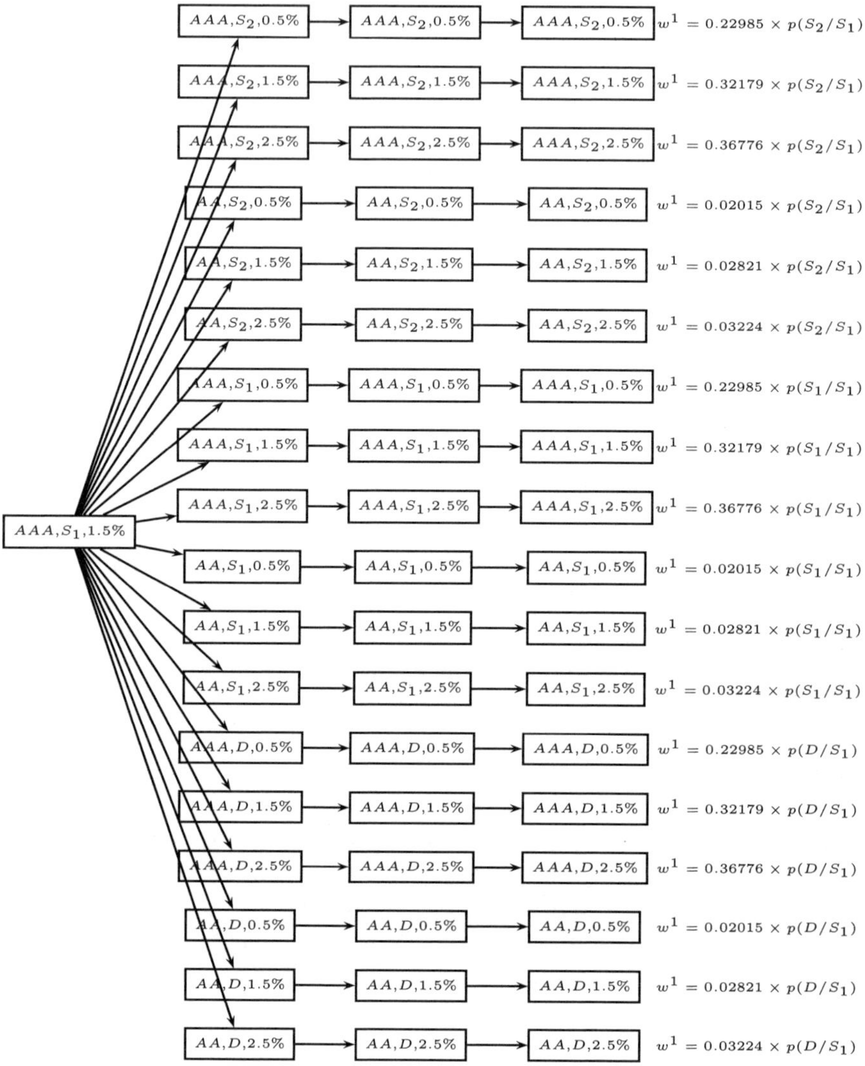

Figure 3 Two-stage scenario tree for the rating and interest rate variation. Instances 1 and 2

uncertainty and by considering the two-stage scenario tree information given in Figure 3, with AAA and CCC bonds, i.e., the Instance 1 of the case study. The aim of model (DEM1) is to maximize the expected final value of the portfolio, in this way, in a risk neutral environment. In this case, the optimal strategy at least deals with the risk of default at the initial period. As it is possible to have a defaulted bond in t_1 the optimal solution ensures a minimum wealth at the beginning (by investing everything in the sure asset $i = 1$ that matures in the next period $t_1 = 1$, thus avoiding all the transaction costs). After that, the strategy invests in the sure or the risky asset depending on the probability of default of the asset classes in the corresponding scenario. In those scenarios in which any of the risky bonds in the market may default, it invests everything in the sure bond that matures on the PH, $i = 3$. In the other case, this strategy invests everything in the risky asset that also matures at the PH, $i = 3'$.

The optimal solution of model (DEM1) for Instance 2 shows a similar strategy. In this case, the optimal solution is also to invest safe and short at the beginning and reinvest long in the next periods, investing in safe or risky depending on the risk of default of the classes in the corresponding scenario. The difference between both strategies comes from the fact that the risky asset pays different coupons. Notice, in any case, that the reinvestment of those payments is exactly the same.

8.4 *Mean-risk immunization model (DEM-MR)*

The results obtained by implementing the approach given in model (DEM-MR) (31)–(33) are the same as those obtained in the optimization of the two-stage stochastic model (DEM1).

The 0–1 variables reach their optimum value depending on the parameter ϕ. If ϕ is bigger than approximately the V-s involved, they all take value 0. In the other case, their optimum value is 1. This result is due to the aim of this approach, that is to ensure a minimum wealth that we would get in the model to immunize the portfolio. As in our specific case study we have modelized the default as a particular scenario, we are already dealing in a certain way with the risk of default and avoiding those decisions that do not exceed a minimum wealth.

In any case, independently of the value of parameter ϕ, the optimal strategy consists of investing safe and short at the beginning and reinvest long in the next periods, investing in safe or risky depending on the risk of default of the classes in the corresponding scenario.

8.5 *Two-stage maxmin immunization model (DEM2)*

Let us consider the optimal solution to the two-stage stochastic model (DEM2) (38)–(40) for optimizing immunization strategies under the two sources of uncertainty at the initial period $t_0 = 0$, for Instance 1. The strategy for the first period is observed to be very close to that proposed by Khang. Notice that, as it is an immunization strategy, the aim is to ensure a minimal value, so it advises to invest in $t_0 = 0$ in the safe assets in any of the possible scenarios. In addition, it invests in those safe assets that fit their duration as was proposed by Khan in his immunization theorem. But once it just immunizes the initial period, the optimal strategy for the other periods is very similar to the strategy that we obtained in the two-stage optimization model (DEM1), i.e., to invest in the bond that matures at the PH, which will or not be risky depending on the scenario.

The solution for Instance 2 is quite similar to that obtained for Instance 1. It also immunizes at period $t_0 = 0$ by equaling the maturity of the portfolio with its duration. Anyway, instead of taking two long term bonds, it invests something in the short term and adjusts the duration by investing more in the longest term. After that the strategy is quite similar to the previous one.

It is worth noting that when immunizing we only take into account non risky bonds. This is due to the fact that we are maximizing the expected value of the minimum wealth the portfolio would obtain, under any future realization of the uncertain variables, no matter whether they are very likely to happen or not. We would like to improve the performance of this model by adjusting it so as to maximize the expected value of the minimum wealth the portfolio would obtain under as many as the most probable scenarios.

8.6 *Two-stage VaR model (DEM3)*

When considering the two-stage scenario tree information given in Figure 3 for Instance 1, the "bad" scenarios are so probable that for any probability $\alpha \in (0\%, 10\%)$ we would obtain the same solutions optimizing model (DEM3) over the 100% and over the 90% of the scenarios, respectively. Notice that model (DEM3) always has to take into account at least one of the "bad" scenarios when immunizing. Moreover, when considering the scenario tree information given in Figure 3 for Instance 2 with probability $\alpha = 5\%$, the optimal strategy is also exactly the same as we had in model (DEM2). In this case, as the probabilities of the "bad" scenarios are higher than 5%, it is absolutely necessary to take into account one of them, at least, and as a result the decision is not to invest in risky bonds at the beginning.

The optimal strategy obtained by solving model (DEM3) model (41)–(45), with $\alpha = 6\%$ and the two-stage scenario tree information given in

Figure 3 (i.e., Instance 2), changes completely in comparison with the solution that has been obtained in model (DEM2). This is due to the fact that those scenarios in which default may happen are not sufficiently probable for the decision maker. So, the optimal strategy consists of investing everything in risky bonds that match the maturity and the duration of the portfolio. After the initial period $t_0 = 0$, the optimal solution depends on the scenario. In those "good" scenarios (in which there are no defaulted bonds) the portfolio is built up with risky bonds. It is worth pointing out that it is not exactly the strategy followed in scenarios $\omega = 4$, 5 and 6. The reason is that as they have such a small probability, they do not have a high weight in the objective function and then, they simply do whatever is the easiest. So, the optimal strategy is to invest in risky bonds for the "good" scenarios. The optimal solution for the "bad" scenarios, however, is to invest everything in safe assets.

8.7 *Multistage scheme*

Firstly, we introduce the main aspects for the construction of the multistage scenario tree adapted to our illustrative example, with market conditions similar to those in the two-stage environment. In the case of the interest rate changes, we consider the same outcomes but with different probability, such that $P(r_{t+1} = 0.5|r_t = 1.5) = 0.3$, $P(r_{t+1} = 1.5|r_t = 1.5) = 0.4$ and $P(r_{t+1} = 2.5|r_t = 1.5) = 0.3$.

Given the number of decision periods in the case study, the scenario tree has been designed as a three-stage one, still considering Instances 1 and 2, but a third instance is created by considering a new class of bonds, the speculative class $j = 5$, BB. These rated bonds may have three outlooks, where the transition probabilities are $p(BBB/BB) = 0.0633$, $p(BB/BB) = 0.9248$ and $p(D/BB) = 0.0119$. Moreover, as is shown in Table 3, the probabilities of default in these classes are $q_{4,t} = 0.0026$, $q_{5,t} = 0.0097$ and $q_{8,t} = 1$, $\forall t = t_1, \ldots, t^i$, $i \in \mathcal{I}_j$, $j = 4, 5, 8$.

We now consider three different instances with similar shape. We have a safe asset and a risky one. In order to describe all the instances at the same time, we will define generic states S_1 and S_2 that will be $S_1 = CCC$ and $S_2 = B$ (Instance 1), $S_1 = B$ and $S_2 = BB$ (Instance 2), and $S_1 = BB$ and $S_2 = BBB$ (Instance 3).

Figure 4 depicts the 27 scenarios considered for the three-stage and the three instances. For Instance 1 the classes of bonds: AAA ($j = 1$) and CCC ($j = 7$) are considered. In order to build a simple but

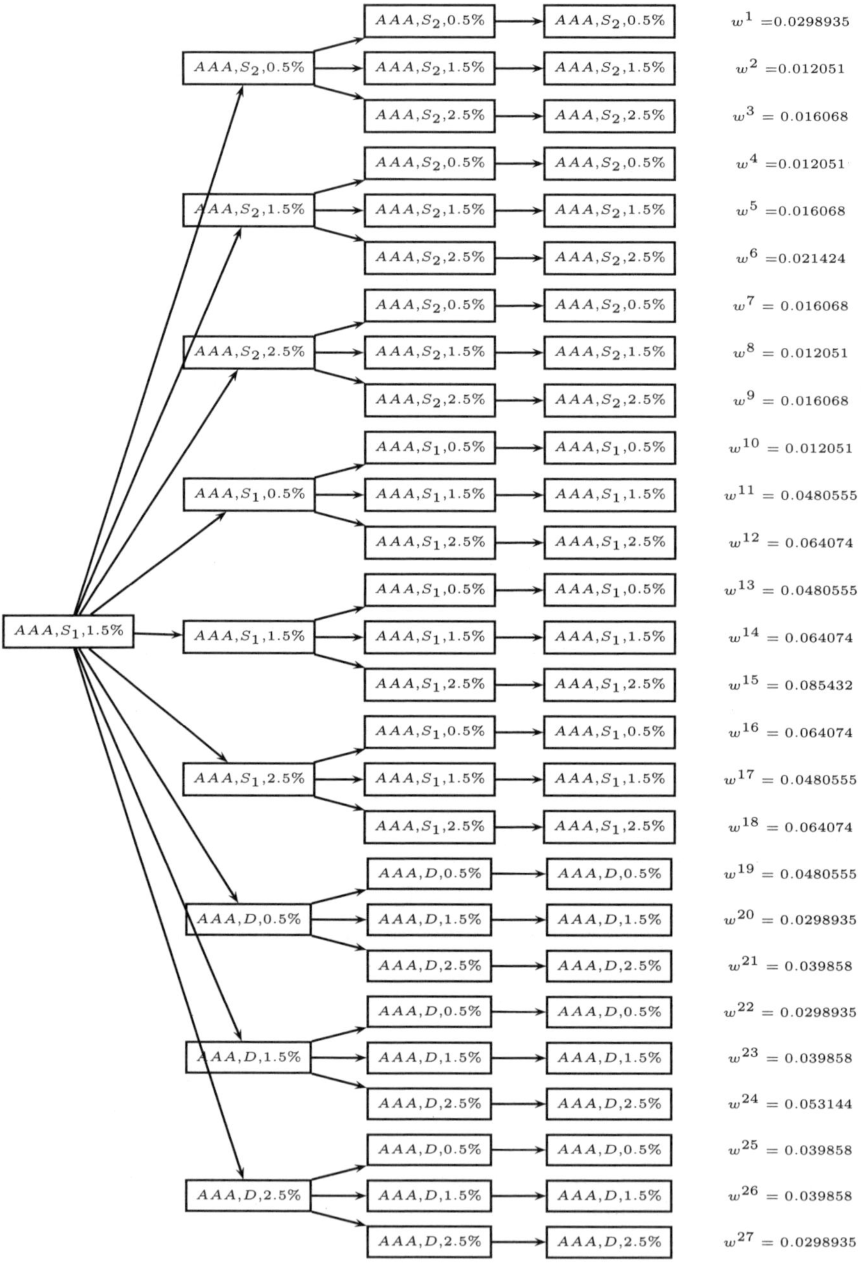

Figure 4 Three-stage scenario tree for rating and interest rate variation. Instances 1,2,3

representative three-stage scenario tree we have considered different aspects at the different stages. In stage $t_1 = 0.5$, the outcomes result from $J^7 \times r$, where $J^7 = \{6, 7, 8\}$ (the possible rating of the second asset class) and $r = \{0.5\%, 1.5\%, 2.5\%\}$ (the risk free interest rate level). So, a set of nine scenarios is defined as the cartesian product of those two sets of outcomes. In the next stage $t_2 = 1$, the scenarios are defined as the cartesian product given by ($\Omega = J^7 \times r \times J^1$), since for simplicity they are considered as independent risks, and where $J^1 = \{1, 2\}$ (the possible rating of the first asset class). Then, the probability w^ω of scenario ω is computed as the product of the probabilities of the three independent outcomes. For Instance 2 the classes of bonds: AAA ($j = 1$) and B ($j = 6$) are considered. In stage $t_1 = 0.5$, the outcomes result from $J^6 \times r$, where $J^6 = \{5, 6, 8\}$, and $r = \{0.5\%, 1.5\%, 2.5\%\}$. In the next stage $t_2 = 1$, the scenarios are defined as the cartesian product given by $\Omega = J^6 \times r \times J^1$, that for simplicity are considered as independent risks Finally, for Instance 3 the classes of bonds: AAA ($j = 1$) and BB ($j = 5$) are considered. Similarly as the above instances, in stage $t_1 = 0.5$, the outcomes result from $J^5 \times r$, where $J^5 = \{4, 5, 8\}$ and $r = \{0.5\%, 1.5\%, 2.5\%\}$. In the next stage $t_2 = 1$, the scenarios are defined as the cartesian product given by $\Omega = J^5 \times r \times J^1$.

8.8 *Multistage maxmin immunization model (DEM4)*

All of the optimal solutions to the multi-stage model (DEM4) (48)–(60) for Instances 1, 2 and 3, at the initial period t_0 are equal and consist of the same strategy, i.e., investing part of the portfolio in the safe asset that matures before the end of PH and the rest of the portfolio on the safe asset with the longest maturity. As the portfolio is made up almost half and half, it matches the duration with the PH, so it is following in some way the strategy proposed by Khang. It is also important to note that it invests in safe assets, thus avoiding the risk of default. So, it is dealing with both risks.

The optimal strategies for Instances 1, 2 and 3 in stage t_1 depend on the scenario groups in that period but in any case, they are very similar. In any of the scenarios in which default may not occur, the optimal solution consists of changing all the investments in risky bonds (since it does not consider the possibility of default in the next periods). However, if there is a probability of default, it invests in safe assets, reorganizing the portfolio depending on the interest rate. In those scenarios for which the risky asset improves or maintains its level, the strategy is very similar to that in the case of default, investing in risky assets instead of investing in those safe ones. This strategy is different depending on the current interest rate. If interest rates have fallen (they are in 0.5), it considers that they can maintain their level or change into 1.5 or 2.5, so they would be bigger in any case and thus, the optimal solution consists of investing everything short. If the interest rates maintain their level (1.5), they can go down or up and, thus, the optimal solution consists of matching the duration with the PH. Finally, if interest rates go up to 2.5, the possibilities considered in the scenario tree are to stay or to fall and thus, the optimal strategy is to invest long. The main difference between the strategies to follow in Instances 1 and 2 depends on the risky asset maintains its rating. So, for Instance 2, the optimal solution is also to invest short if the interest rate can only rise, matching the duration with the PH if the interest rate is still 1.5, and investing long if the interest rate can just fall. For Instance 1, the optimal strategy is always to invest short in those scenarios where the risky asset is still CCC. Notice that we can invest in risky asset and, as their reinvestments would give such big returns (real interest rates are around 20–22%), the different strategies barely affect the optimality of the strategy, so, anything that might be done is profitable.

As in the three-stage scenario tree considered, see Figure 4, there is no change after period t_2, so the optimal solution is to invest everything in the asset that matures in the PH but, sometimes it can leave something remaining for the other assets, thereby, avoiding transaction costs.

Anyway, in those scenarios with no default, it invests in risky assets and invests in safe assets, for the "bad" scenarios.

8.9 *Multistage stochastic dominance model (DEM5)*

Let us summarize some of the results obtained by solving the three-stage extension to the stochastic dominance strategy, model (DEM5) (61)–(65). We have considered two different situations: one in which we ask for a minimum wealth -always the same- that is higher than the one obtained with the recovery rate; and another in which we ask for two minimum values -one higher than we get with safe assets and another higher than the recovery rate. In both of these, the probability has been taken as $\alpha = 0.06$. After solving several situations, we can conclude that these strategies are very dependent on the values of the potential profiles, ϕ^p, $p \in \mathcal{P}$. It is very important to choose them in order to have the best results or those that best match the needs of the investor. This may not be so easy a priori, since the decision maker does not necessarily need to know what could be obtained in different situations at the beginning of the research.

Case 1: $|\mathcal{P}| = 1$, $\phi^1 = 5 \cdot 10^5$, $\alpha^1 = 0.06$

Let us summarize the optimal strategy corresponding to Instance 1. For Instances 2 and 3, the results are very similar. In this case, the optimal strategy at period t_0 consists of investing all the portfolio in the asset with the smallest maturity. So, it invests everything in the short term. It is also important to note that it invests in safe assets, thus, avoiding the risk of default. But it is not immunizing against interest rate changes, due to the fact that it is not matching the duration of the portfolio with the PH.

In the scenario groups where default may not occur in perido t_1, the optimal solution consists of changing all the investments in risky bonds (as they do not consider the possibility of default in the next periods). In the case of default, it would keep the safe assets instead, reorganizing them depending on the interest rate.

Case 2: $|\mathcal{P}| = 2$, $\phi^1 = 5 \cdot 10^5$, $\phi^2 = 1.05 \cdot 10^6$, $\alpha^1 = \alpha^2 = 0.06$

Let us summarize the optimal strategy corresponding to Instance 2. For Instances 1 and 3, the results are very similar. In this case, we analyze the optimal solution at period t_0. It consists of investing part of the portfolio in assets that mature before the end of PH and the rest in the assets with the longest maturity. The investment strategy shares between risky and safe assets in order to ensure, on the one hand, returns that are at least ϕ^1 in the most probable cases but, on the other hand, it ensures a minimum wealth (ϕ^2) in situation of default. As the portfolio is made up almost half and half, it matches the duration with the PH, so, it is following in some way the strategy proposed by Khang. It is also important to notice that it invests in risky assets since the failure scenarios are not considered sufficiently probable. The optimal strategy for each scenario group in period t_1 is very similar no matter if the risky asset changes its rating or not. The only difference is that in defaulted scenarios the optimal solution would be to invest safe, and risky in other cases. In any case, the strategy is different depending on the current interest rate. Both risks are dealt with.

It is worthy to point out that as the three-stage scenario tree does not consider any changes after period t_2, the optimal solution in both Cases 1 and 2 is to invest everything in the asset that matures in the PH but, sometimes, in order to avoid paying transaction costs, it can leave something remaining for other assets. Anyway, in those scenarios with no default, it invests in risky assets. On the other hand, it invests in safe assets for the "bad" scenarios.

8.10 *Multistage VaR & stochastic dominance model (DEM6)*

Finally, we summarize the results of the optimization of model (DEM6) (66)–(70).

Case 1: $\mathcal{P} = \{g \in \mathcal{G}_t : t = t_0, \cdots, t_{k-1}\}$ **and** $\phi^p = 0$ **for all** $p \in \mathcal{P}$

When implementing this model in Instance 1 with $\alpha \in (0, 0.10)$, the optimal strategies are exactly those obtained with model (DEM4). This is because default is so probable that it is important to take this into account when optimizing. That prevents it from changing behavior.

The same happens with Instance 2 if $\alpha \leq 0.5$. Even for Instance 2 with $\alpha = 0.6$ the optimal solution does not change. It is worth pointing out that, since the default scenarios in period t_0 do not reach in probability the value of α, the improvement that would achieve in V_0 is compensated by what it would lose in V_g for $g \in \mathcal{G}_t, t = t_1, \cdots, t_k$.

Let us summarize the optimal solutions for the different stages in Instance 3 with $\alpha = 0.05$. In this case, the optimal solution at period t_0 consists of matching the duration with the HP in order to avoid immunization risk, but investing in risky assets. It seems that default has such a small probability that the model does not take this into account when optimizing. At period t_1, it is clear that this is a much riskier strategy since in those failure scenarios we would get very little return. But, in terms of expected return, this is the best strategy due to the fact that failure has a such small probability.

Case 2: $\mathcal{P} = \{g \in \mathcal{G}_t : t = t_0, \cdots, t_{k-1}\}$ **and** $\phi^p = 500000$ **for all** $p \in \mathcal{P}$

In order to ensure a minimum wealth for even those failure scenarios, let us consider a threshold of 500000 monetary units for all potential profiles, $\mathcal{P}$. The optimal solution obtained for Instances 1 and 2, is the same as that obtained in Case 1. This happens since the expected wealth is higher than ϕ^p, $p \in \mathcal{P}$. For high values of this potential profile, for example 1.200.000, the problem becomes infeasible.

Finally, let us summarize the optimal solution for the different stages in Instance 3 with $\alpha = 0.05$. The optimal solution at period t_0 consists of investing part of the portfolio in assets that mature before the PH and the rest in the assets with the longest maturity. The investment strategy shares between risky and safe assets in order to ensure, on the one hand, as high as possible the minimum return over all the periods but, on the other hand, to ensure a minimum wealth (ϕ^p) in situation of default. As the portfolio is made up almost half and half, it matches the duration with the PH, so it is following in some way the strategy proposed by Khang. It is also important to note that it invests in risky assets since the failure scenarios are not considered to be that probable. The optimal solution at period t_1 consists of investing short if the interest rates are going to rise, long if they are going to fall, and following the strategy proposed by Khang when both possibilities exist. In any case, it invests in risky assets in no failure scenarios, and it invests in safe assets in scenarios of default.

It is important to point out that, with this strategy, we are maximizing over all the periods of time, but we will only get the final value if we maintain the strategy until the end of the PH.

9 Models comparison and conclusions

This paper proposes several stochastic models for selecting portfolios in a market in which there are transaction costs and bonds with different credit ratings. Uncertainty is introduced into the model through a scenario analysis scheme. In this case there are two parameters whose random behavior must be taken into account: trends in interest rates and the probabilities of default of the various institutions which issue the bonds. Different immunization strategies are considered. The validity of the proposed strategies is performed by using an illustrative case study. No definitive conclusions can be drawn from the case study (the aim of the paper has merely been to present the immunization strategies of choice), but the results that have been obtained seem to be reasonable. In any case, see in Aranburu (2011) the decisions offered by each of the tested strategies as well as more computational testing.

By considering the advantages and disadvantages of the different models tested in the case study, see Table 6, we can observe that the models that offer better immunization strategies are the models (DEM4), see Section 7.1 (100% multistage VaR model), (DEM5), see Section 7.2 (multistage stochastic dominance), and (DEM6), see Section 7.3 (the VaR & multistage stochastic dominance) for good choices of the potential profiles. Each of them has advantages and disadvantages, and the decision maker should choose between the three depending on its preferences. Model (DEM5) would be better for those investors who are only interested in the final value of the portfolio and they also know a priori which possible thresholds can be more interesting for their purposes. On the contrary, model (DEM6) would be better for those investors with no initial

Table 6 Comparing the Different Models as Tested in the Case Study.

Stages	Model	Strategy	Advantages	Disadvantages
	LP	• Invest everything in the risky asset that matures at the PH	• Easy	• Myopic • No risks
Two	**DEM1**	• $\underline{t_0}$: safe & short • $\underline{t > t_0}$: "Good" scen.: risky asset mat. PH "Bad" scen.: safe asset mat. PH	• Risk of default treated	• Myopic • Interest Rate risk not treated
	DEM-MR	• $\underline{t_0}$: safe & short • $\underline{t > t_0}$: "Good" scen.: risky asset mat. PH "Bad" scen.: safe asset mat. PH	• Risk of default treated	• Myopic • Interest Rate risk not treated
	DEM2	• $\underline{t_0}$: safe & Dur.=PH • $\underline{t > t_0}$: "Good" scen.: risky asset mat. PH "Bad" scen.: safe asset mat. PH	• Both risks treated	• Myopic • Risk adverseness not treated (too risk av.)
	DEM3	• $\underline{t_0}$: safe & Dur.=PH (Instance 1) risky & Dur.=PH (Instance 2) • $\underline{t > t_0}$: "Good" scen.: risky asset mat. PH "Bad" scen.: safe asset mat. PH	• Both risks treated • Risk adverse model	• Myopic • Maybe big lose in worst cases
Multi	**DEM4**	• $\underline{t_0}$: safe & Dur.=PH • $\underline{t > t_0}$: risky in "good" scen. & safe else r will rise: short r will fall: long r can fall or rise: Dur.=PH	• Not myopic • Both risks treated	• Risk adv. not treated (too risk av.)
	DEM5	$\boxed{\lvert\mathcal{P}\rvert = 1}$ well chosen • $\underline{t_0}$: safe & short • $\underline{t > t_0}$: risky in "good" scen. & safe else short, long or Dur.=PH depending on r	• Not myopic • Risk adverse model	• Interest Rate risk not treated • ϕ dependent • CNT

(*Continued*)

Table 6 *(Continued)*

Stages	Model	Strategy	Advantages	Disadvantages
		$\lvert\mathcal{P}\rvert = 2$ well chosen		
		• t_0: safe & short (Instance 1) share (safe-short) & short (Inst. 2, 3) • $t > t_0$: risky in "good" scen. & safe else short, long or Dur.=PH dep. on r	• Not myopic • Both risks • Risk adverse • Bad sit. covered	• ϕ dependent • CNT
	DEM6	$\phi = 0$		• Maybe big lose in worst cases • Max. $\forall V_t$, get just V_{t_k} • CNT
		• t_0: safe & short (Inst. 1, 2) risky & short (Inst. 3) • $t > t_0$: risky in "good" scen. & safe else short, long or Dur.=PH dep. on r	• Not myopic • Both risks • Risk adverse • Not ϕ dep.	
		$\phi \neq 0$	• Not myopic • Both risks • Risk adverse • Not so ϕ dep. • Bad sit. covered	• Max. $\forall V_t$, get just V_{t_k} • CNT
		• t_0: safe & short (Inst. 1, 2) share (safe-short) & short (Inst. 3) • $t > t_0$: risky in "good" scen. & safe else short, long or Dur.=PH dep. on r		

Notation: Dur.=PH means the strategy that matches the duration with the PH.
CNT means that it **can become** Computationally Not Treatable or very hard to solve.

expectations and are also interested in optimizing the portfolio all over the PH. It could be very interesting, for example, for those decision makers who would like to dissolve the portfolio at any time in order to use the new information in the market. Model (DEM4), instead, could be very interesting for a very risk averse investor. In any case, models (DEM5) and (DEM6) can become computationally non treatable for real markets with many possible future scenarios in case of using plain state-of-the-art optimization engines instead of using proven decomposition approaches, meanwhile (DEM4) can be more easily solved.

As a future work we are planning an extensive computational experience with large scale cases in fixed-income security portfolios, scenarios and profiles to test the validity of our stochastic mixed integer optimization algorithm approaches for problem solving in this type of financial application. The first piece of the research will consist of investigating the performance of the above models (and some strategies that do not require constraints with variables from different scenarios, such as the Scenarios Immunization, CVaR and some of its variants, Conditional expectation above VaR, conditional expectation shortfall on reaching the threshold and CVaR& Stochastic dominance) for bigger instances still by plain use of the MIP solver. Probably, this experiment may show that we need to use decomposition alternatives for real-life sized model solving. So, our second piece of research will consist on developing an extension of our Branch-and-Fix Coordination algorithm, see Escudero *et al.* (2012) to consider all the mentioned risk aversion strategies (beside the risk neutral one). Notice that some of the strategies require constraints whose involved variables belong to different scenarios, making more difficult the development of the mentioned extension.

References

Altman, E. I. and V.M. Kishore. Almost everything you wanted to know about recoveries on defaulted bonds. *Financial Analysts Journal* 52: pp. 57–64, 1996.

Aranburu, L. Stochastic decision models in discrete time. PhD thesis, Universidad del País Vasco, Bilbao (Vizcaya), Spain, 2011.

Aranburu, L., L.F. Escudero, A. Garín and G. Pérez. On solving two-stage stochastic linear problems by using a new approach, Cluster Benders Decomposition. *TOP*, 20: pp. 279–295, 2012.

Artzner, P., F. Delbaen, L. Eber and D. Health. Coherent measures of risks. *Mathematical Finance*, 9: pp. 203–228, 1999.

Beale, E.M.L. On minimizing a convex function subject to linear inequalities. *Journal of the Royal Statistical Society Ser B*, 17: pp. 173–184, 1955.

Benders, J. Partitioning procedures for solving mixed variables programming problems. *Numerische Mathematik*, 4: pp. 238–252, 1962.

Beraldi, P., A. Violi and F. De Simeone. A Decision Support System for strategic asset allocation. *Submitted for publication*, 2010.

Bertocchi, M. V. Moriggia and J. Dupačová. Horizon and stages in applications of stochastic programming in Finance. *Annals of Operations Research*, 142: pp. 63–78, 2006.

Bierwag, G.O. and C. Khang. An immunization strategy is a minimax strategy. *Journal of Finance*, 34: pp. 389–399, 1979.

Bierwag, G.O. Duration Analysis. *Ballinger Publishing Company*, 1987.

Birge, J.R. and F. Louveaux. Introduction to Stochastic Programming. *Springer*, 1997.

Bradley, S.P. and D.B. Crane. A dynamic model for bond portfolio management. *Management Science*, 19: pp. 139–151, 1972.

Charnes, A. and W.W. Cooper. Chanced-constrained programming. *Management Science*, 5: pp. 73–79, 1959.

Consigli, G., M. Bertocchi and M.A.H. Dempster (eds.). Stochastic optimization methods in Finance and Energy. *Springer*, to appear 2011.

Dantzig, G.B. Linear programming under uncertainty. *Management Science*, 1: pp. 197–206, 1955.

Dantzig, G.B. A proof of the equivalence of the programming problem and the game problem. In T.C. Koopmans (ed.), *Activity Analysis of Production and Allocation*, Yale University Press. 7th. edition, 1971.

Dentcheva, D. and A. Ruszczynski. Optimization with stochastic dominance constraints. *SIAM Journal on Optimization*, 14: pp. 548–566, 2003.

Dentcheva, D., R. Henrion and A. Ruszczynski. Stability and sensitivity of optimization problems for first-order stochastic dominance. *SIAM Journal on Optimization*, 18: pp. 322–337, 2007.

Dempster, M.A.H. (ed.) Stochastic programming. *Academic press*, 1980.

Dupačová, J., G. Consigli and S.W. Wallace. Scenario for multistage stochastic programs. *Annals of Operations Research*, 100: pp. 25–53, 2000.

Dupačová, J., N. Gröwe-Kuska and W. Römisch. Scenario reduction in stochastic programming. An approach using probability metrics. *Mathematical Programming Ser A*, 95: pp. 493–511, 2003.

Escudero, L.F., A. Garín, M. Merino and G. Pérez. A two-stage stochastic integer programming aproach as a mixture of Branch-and-Fix Coordination and Benders' decomposition schemes. *Annals of Operations Research*, 152: pp. 395–420, 2007.

Escudero, L.F., A. Garín, M. Merino and G. Pérez. On multistage integer programming model for incorporating logical constraints in assets and liabilities management under uncertainty. *Computational Management Science*, 6: pp. 307–327, 2009.

Escudero, L.F., A. Garín, M. Merino and G. Pérez. A general algorithm for solving two-stage stochastic mixed 0–1 first stage problems. *Computers & Operations Research*, 36: pp. 2590–2600, 2009.

Escudero, L.F., A. Garín, M. Merino and G. Pérez. On BFC-MSMIP: an exact branch-and-fix coordination approach for solving multistage stochastic mixed 0–1 problems. *TOP*, 17: pp. 96–122, 2009.

Escudero, L.F., A. Garín, M. Merino and G. Pérez. An exact algorithm for solving large-scale two-stage stochastic mixed integer problems: some theoretical and experimental aspects. *European Journal of Operational Research*, 204: pp. 105–116, 2010.

Escudero, L.F., A. Garín, M. Merino and G. Pérez. On BFC-MSMIP strategies for scenario cluster partitioning, and twin node family branching selection and bounding for multistage stochastic mixed integer programming. *Computers & Operations Research*, 37: pp. 738–752, 2010.

Escudero, L.F., A. Garín, M. Merino and G. Pérez. An algorithmic framework for solving large scale multi-stage stochastic mixed 0–1 problems with nonsymmetric scenario trees. *Computers & Operations Research*, 39: pp. 1133–1144, 2012.

Fong, H.G. and O. Vasicek. The tradeoff between return and risk in immunized portfolios. *Financial Analysts Journal*, 39: pp. 73–78, 1983.

Gaivoronski, A.A. and G. Pflug. Finding optimal portfolios with constraints on value-at-risk. In B. Green (ed.), *Proceedings of the Third International Stockholm Seminar on Risk Behaviour and Risk Management.* Stockholm University, 1999.

Gaivoronski, A.A. and G. Plug. Value-at-risk in portfolio optimization: properties and computational approach. *Journal of Risk*, 7: pp. 11–31, 2005.

Geyer, A. and W.T. Ziemba. The Innovest Austrian Pension Fund financial planning model InnoALM. *Operations Research*, 56: pp. 797–810, 2008.

Gollmer, R., F. Neiser and R. Schultz. Stochastic programs with first-order stochastic dominance constraints. *SIAM Journal on Optimization*, 19: pp. 552–571, 2008.

Gollmer, R., U. Gotzes and R. Schultz. A note on second-order stochastic dominance constraints induced by mixed-integer recourse. *Mathematical Programming Ser A*, 126: pp. 179–190, 2011.

Golub, B., M. Holmer, R. McKendall, L. Pohlman and S.A. Zenios. Stochastic management models for money management. *European Journal of Operational Research*, 85: pp. 282–296, 1995.

Heitsch, H. and W. Römisch. Scenario tree modeling for multistage stochastic programs . *Mathematical Programming Ser A*, 118: pp. 371–406, 2009.

Higle, J.L. and S. Sen. Stochastic decomposition. *Kluwer publishers*, 1996.

Hoyland, K. and S.W. Wallace. Generation scenario trees for multi-stage decision problems. *Management Science*, 47: pp. 295–307, 2001.

Hull, J. Options, Futures and Other Derivative Securities. *Pearson Prentice Hall*, 1989.

Kall, P. and S.W. Wallace. Stochastic programming. *Wiley*, 1994.

Khang, C. A dynamic global portfolio immunization strategy in the world of multiple interest rates changes: a dynamic immunization and minimax theorem. *Journal of Financial and Quantitative Analysis*, 18: pp. 355–363, 1983.

Kusy, M.I. and W.T. Ziemba. A bank Asset and Liability Management Model. *Operations Research*, 34: pp. 356–376, 1986.

Markowitz, H.M. Portfolio selection. *Journal of Finance*, 7: pp. 77–91, 1952.

Markowitz, H.M. Portfolio selection. *Wiley*, 1959.

Mulvey, J.M., R.J. Vanderbei and S.A. Zenios. Robust optimization of large scale systems. *Operations Research*, 43: pp. 264–281, 1995.

Ogryczak, W. and A. Ruszczyński. From stochastic dominance to mean-risk models: semideviations as risk measures. *European Journal of Operations Research*, 116: pp. 33–50, 1999.

Pflug, G. Scenario tree generation for multiperiod financial optimization by optimization discretization. *Mathematical Programming Ser B*, 89: pp. 251–271, 2001.

Rockafellar, R.T. and S. Uryasev. Optimization on conditional value-at-risk. *Journal of Risk*, 2: pp. 21–41, 2000.

Rockafellar, R.T. and S. Uryasev. Conditional value-at-risk for general loss distributions. *Journal of Banking & Finance*, 26: pp. 1443–1471, 2002.

Shapiro, A., D. Dentcheva and A. Ruszczyński. Lectures on stochastic programming modeling and theory. *SIAM-MPS*, 2011.

Schultz, R. and S. Tiedemann. Risk aversion via excess probabilities in stochastic programs with mixed-integer recourse. *SIAM Journal on Optimization*, 14: pp. 115–138, 2004.

Schultz, R. and S. Tiedemann. Conditional value-at-risk in stochastic programs with mixed integer recourse. *Mathematical Programming, Ser B*, 105: pp. 365–386, 2006.

Van Slyke, R. and R.J-B. Wets. L-shaped linear programs with application to optimal control and stochastic programming. *SIAM Journal of Applied Mathematics*, 17: pp. 638–663, 1969.

Wallace, S.W. and W.T. Ziemba (eds.). *Applications of Stochastic Programming.* MPS-SIAM-Series in Optimization, 2005.

Wets, R.J-B. Programming under uncertainty: The equivalent convex program. *SIAM Journal of Applied Mathematics*, 14: pp. 89–105, 1966.

Wets, R.J-B. and W.T. Ziemba (eds.). Stochastic programming. State of the Art 1998 (Special issue dedicated to the main lectures in the VIII International Conference on Stochastic Programming) *Annals of Operations Research*, 85, 1999.

Zenios, S.A., M. Holmer, R. McKendall and C. Vassiadou-Zeniou. Dynamic models for fixed-income portfolio management under uncertainty. *Journal of Economic Dynamics and Control*, 22: pp. 1517–1541, 1998.

Zenios, S.A. and W.T. Ziemba (eds.) *Handbook of Asset and Liability Modeling (Vol. 1 on Theory and methodology and Vol. 2 on Applications and Case Studies)*. North Holland-Elsevier, 2006.

Ziemba, W.T. *The stochastic programming approach to Asset Liability and Wealth Management.* Charlottesville, VA:AIMR, USA, 2003.

Ziemba, W.T. and J.M. Mulvey (eds.) *Worldwide Asset and Liability Modeling.* Cambridge University Press, 1998.

Chapter 9

Stochastic Programming and Optimization in Horserace Betting[1]

William T. Ziemba*

Summary

Racetrack betting is simply an application of portfolio theory. The racetrack offers many bets that involve the results of one to about ten horses. Each race is a special financial market with betting, then a race that takes one or a few minutes. Unlike the financial markets, one cannot stop the race when one is ahead, or have the market going almost 24/7. There is a well-defined end point. Like standard portfolio theory, the key issues are to get the means right. In this case, it is the probabilities of, say, two, three or more horses finishing first, second, third, etc., in a given order, and to bet well. For the latter, the Kelly capital growth criterion is widely used and that maximizes the expected logarithm of final wealth. Transaction and price pressure odds changes fit well into the stochastic programming models.

Professional syndicates or teams have been successful as hedge funds with gains approaching one billion over several years for the most successful. In the modern era, there are two features used extensively. First, there are rebates for large bettors of the track take similar to discounts at Costco. So instead of facing a 13–30% transaction cost, it's more like 10%. So to win, the bettors must make back this 10% disadvantage before profits ensue. And this is not easy as the markets are quite efficient. Also over half the betting is not recorded in the pools until the race is being run. This is because monies are bet near the start of the race and come from many off track sites which are combined with the on-track bets into the track pool. All this takes time. So estimates of future

*Alumni Professor of Financial Modeling & Stochastic Optimization (Emeritus), University of British Columbia, Vancouver, BC, Canada. ICMA Centre, University of Reading, UK.
Visiting Professor, Sabanci University, Istanbul,Turkey, Luiss — Guido Carlo University, Rome, Italy, Korea Advanced Institute for Science and Technology, University of Cyprus, wtzimi@mac.com.

[1]Some of the material in this paper is also being published simultaneously in the book Ziemba (2012).

prices are crucial. Secondly, betting exchanges such as Betfair in London allow short as well as standard long bets. This allows for more arbitrage and the ability to take advantage of known biases. I have been involved in this research since the late 1970s with six books and a number of articles.

In this paper I relate the theory, computations and examples of real races and experiences for various bets such as win, place and show, exactas, triactors, superfectas, super hi five, place pick all, double, pick 3, 4, 5 and 6. In the Halifax presentation I showed two of the greatest races ever by two undefeated female horses, one that was still then running (Zenyatta) and one retired (Personal Ensign) in 1988. The previous US undefeated horse was Colin in 1907! These and other great races can be seen free on the website chef-de-race.com.

1 Introduction

Racetrack betting is simply an example of portfolio analysis. The investment horizon is short with betting for a period and then a race for about 1–2 minutes. Some of the bets involve multiple horses in a given race while others involve multiple races. The wagers are basically of two types: high probability of winning low payoff bets and low probability high payoff bets. The latter can return a million dollars or more. Table 1 describes a number of the bets.

Examples of some of these follow. In all cases I made these actual bets. More detail on them with charts, etc. are in Ziemba (2012).

Investing in traditional financial markets has many parallels with racetrack and lottery betting and much of the analysis is similar. Behavioral anomalies such as the favorite-longshot bias are pervasive and also exist and are exploitable in the S&P500 and FTSE100 futures and equity puts and calls options markets. This is not discussed in this paper but I use this in personal and private managed accounts and in an offshore hedge

Table 1 Common US and Canadian Racetrack Wagers.

High probability low payoff	win
Low probability high payoff	pick 3–6
One horse is involved	win
Two horses are involved	place, exacta
Three horses are involved	show, triactor
Four horses are involved	superfecta
Two races are involved	double
Three races are involved	pick 3
Four races are involved	pick 4
Five races are involved	pick 5
Six races are involved	pick 6
N races are involved	place pick all

fund. Biases there favor buying high probability favorites and selling low probability longshots just like the high probability low payoff racing wagers. But in complex low probability high payoff exotic wagers such as the Pick 6, the bias reverses to overbet the favorite so one must include other value wagers in the betting program. Fundamental information such as breeding is important and is especially useful for the Kentucky Derby and Belmont Stakes where horses have never run that far before.

The idea is that more stamina is needed to win these races from the sires in the horses lineage. Since the horses have never run this distance before, a forecast of how they might do from their breeding is helpful. See Hausch, Bain and Ziemba (2006) and Gramm and Ziemba (2008, 2012) who study this by merging the odds (prices) with expert opinion (breeding measured by dosage).

A horse named *Stay Thirsty* in race 12 on Travers Day (see section 9) has a dosage profile of 4-6-16-0-0. That is 4 brilliant points (pure speed); 6 intermediate points, 16 classic points, zero solid and zero professional points. These are categories on the speed-stamina space. You can think of this as a discrete probability distribution. Now each chef-de-race stallion (that is a stallion which breeds consistent characteristics in their offspring) in the horse's pedigree counts: 16 for first generation sires, 8 each for the second generation, 4 each for the four third generation and finally 2 for the eight fourth generation sires. So each generation is equally important. If a chef is in two categories, the points are split. Some generations may have no chefs.

The dosage index is then

$$\mathrm{DI} = \frac{\text{Brilliant} + \text{Intermediate} + 1/2\ \text{Classic}}{\text{Solid} + \text{Professional} + 1/2\ \text{Classic}}$$

Despite its simplicity and crude weighting, the index does seem to work. An example of the pedigree and dosage of the 2005 Belmont Stakes winner, Afleet Alex, is in Tables 2 and 3. Parenthesis in Table 2 show categories for chef-de-race stallions: B = Brilliant, C = Classic.

A dual qualifier is a horse whose dosage index is 4.00 or lower, which is the limit suggested for maximum speed for Kentucky Derby winner, and within 10 pounds of the top 2 year old horse on the experimental free handicapping ratings. Gramm and Ziemba (2008, 2012), Hausch, Bain and Ziemba (2006) and the website chef-de-race.com show that such horses have superior performance in the Kentucky Derby and Belmont Stakes.

Table 2 Pedigree for the 2005 Belmont Stakes Winner Afleet Alex.

<table>
<tr><td rowspan="8">Northern Afleet</td><td rowspan="4">Afleet</td><td rowspan="2">Mr. Prospector (B/C)</td><td>Raisea Native (B)</td></tr>
<tr><td>Gold Digger</td></tr>
<tr><td rowspan="2">Polite Lady</td><td>Venetian Jester</td></tr>
<tr><td>Friendly Ways</td></tr>
<tr><td rowspan="4">Nuryette</td><td rowspan="2">Nureyev (C)</td><td>Northern Dancer (B/C)</td></tr>
<tr><td>Special</td></tr>
<tr><td rowspan="2">Stellarette</td><td>Tentam</td></tr>
<tr><td>Square Angel</td></tr>
<tr><td rowspan="8">Maggy Hawk</td><td rowspan="4">Hawkster</td><td rowspan="2">Silver Hawk</td><td>Roberto (C)</td></tr>
<tr><td>Gris Vitesse</td></tr>
<tr><td rowspan="2">Strait Lane</td><td>Chieftain</td></tr>
<tr><td>Level Sands</td></tr>
<tr><td rowspan="4">Qualique</td><td rowspan="2">Hawaii</td><td>Utrillo II</td></tr>
<tr><td>Ethane</td></tr>
<tr><td rowspan="2">Dorothy Gaylord</td><td>Sensitivo</td></tr>
<tr><td>Gaylord's Touch</td></tr>
</table>

Table 3 Dosage Index Calculation for 2005 Belmont Stakes Winner Afleet Alex.

Generation	Sire	Brilliant	Intermediate	Classic	Solid	Professional
1	Northern Afleet					
2	Afleet					
	Hawkster					
3	Mr. Prospector	2		2		
	Nureyev			4		
	Silver Hawk					
	Hawaii					
4	Raise a Native	2				
	Venetian Jester					
	Northern Dancer	1		1		
	Tentam					
	Roberto			2		
	Chieftain					
	Utrillo II					
	Sensitivo					
	Total	5	0	9	0	0

Note:Dosage Index $= (5 + 0 + 9/2)/(0 + 0 + 9/2) = 2.11$.

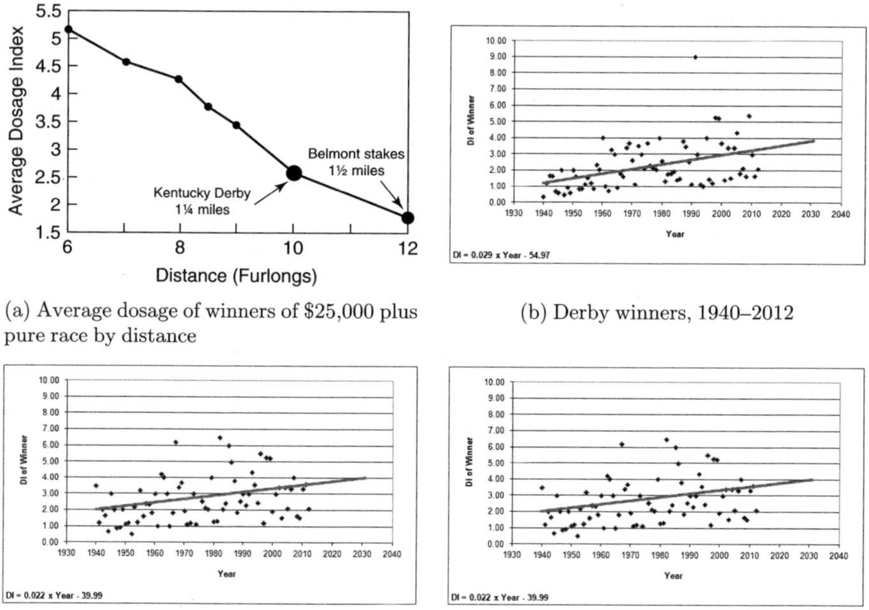

(a) Average dosage of winners of $25,000 plus pure race by distance

(b) Derby winners, 1940–2012

(c) Preakness winners, 1940–2012

(b) Belmont winners, 1940–2012

Figure 1 Dosage indices. Source: Steve Roman.

Figure 1a shows the average dosage index or speed over stamina for the average winner of the $1\frac{1}{4}$ mile Kentucky Derby and $1\frac{1}{2}$ mile Belmont stakes as well as a large number of high quality races at different distances. It was compiled years ago from race data of Steve Roman. Steve has an updated graph of this on the chef-de-race website which has similar conclusions. Observe that the winners of longer races have lower dosages which means more stamina and less speed.

Steve Roman has updated the graphs individually for the Kentucky Derby (Figure 1b), the Preakness (Figure 1c), and the Belmont (Figure 1d) up to the current races. Observe that the breed is moving more towards speed as the regression line is positive and a number of horses with dosage indices above the historical 4.0 cutoff have been winners. But, observe that in the long $1\frac{1}{2}$ mile Belmont all the recent winners have had low dosage. And 2012 was no exception as the Dual Qualifier Union Rags with a dosage index of 2.14 won the race and the favourite Dullahan with a 4.20 dosage finished seventh. There have been horses with dosage above 3.5 who won the Belmont Stakes such as Commendable (5.00 in 2000) and Sarava (4.50

in 2002). However, the last ten winners from 2012 back to 2003 had dosage indices of 2.14, 2.56, 1.75, 2.56, 2.43, 3.00, 3.00, 2.11, 1.77, and 1.88. Even the great Sunday Silence with a 3.8 dosage got crushed in the Belmont. It is not clear how useful Steve's regression line is here as there is a lot of noise in the system. It is believed that the breed has changed toward speed and that shows up more in the Kentucky Derby and less in the Belmont. Gramm and Ziemba (2008, 2012) study this as well.

Kelly and fractional Kelly betting is used extensively in racetrack betting. Full Kelly is the maximization of the expected logarithm of final wealth subject to constraints, see some specific formulations below. That means: take an expected utility approach with $u(w) = log\ w$. Log with very low Arrow-Pratt risk aversion $-u''(w)/u'(w) = 1/w \cong 0$ is very risky short term despite wonderful long term growth properties; see MacLean, Thorp and Ziemba (2011) for an extensive treatment of the key ideas and major papers. Fractional Kelly is simply the idea to blend cash with the Kelly strategy. This, under lognormal asset assumptions, amounts to a less risky negative power utility function rather than log which is the most risky utility function one would ever want to use.

Fractional Kelly leads usually to less growth and more security and a less violent wealth path. Half Kelly is a frequently used strategy. It has 75% of the full Kelly growth but the security, measured by the probability of breaking even rises from 87% with full Kelly to 95.4% with half Kelly. For lognormal assets this is $u(w) = -1/w$ and this is approximate otherwise. To show this visually, see Figure 2(a) for the Kentucky Derby from 1934 to 2005 and Figure 2(b) with the dosage filter to eliminate horses that cannot run $1\frac{1}{4}$ miles on the first Saturday in May of their three year old career. These use the Dr Z system discussed in section 4. The system that bets on

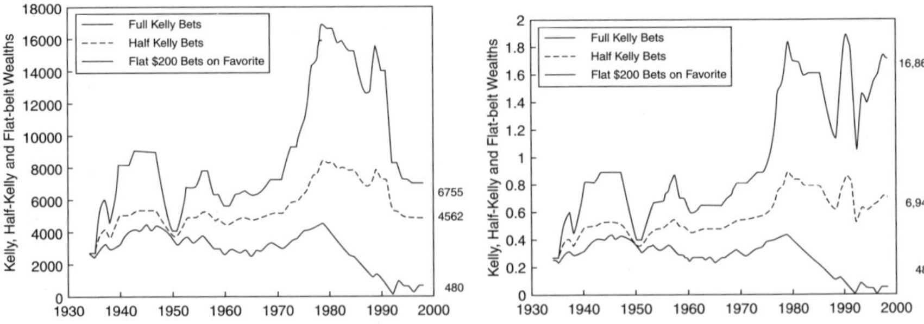

(a) Kelly, half Kelly and betting on the favorite

(b) same as (a) with dosage filter

Figure 2 Wealth History of Some Kentucky Derby bets, 1934–2005.

the favorite turns \$2500 into \$480 so is a loser; while the full and half Kelly Dr Z systems have gains.

In all cases the strategy to win is the same as in the financial markets:

1. get the mean right: thus one must have accurate probabilities of various outcomes;
2. use the actual odds and a betting model such as the Kelly criterion to optimize the bet sizes, that is, the allocations.

For situations with not many wagers, the Kelly capital growth maximizing expected logarithm or, its safer version, fractional Kelly, is useful as a decision tool, especially with many repeated bets. Then one has a stochastic program to maximize the expected utility using a logarithmic utility function of final wealth subject to various constraints. The Kelly strategy bets more on the attractive situations. In wagers where one makes hundreds of bets, it is often better to use a tree approach where many of the bets are of equal value. Besides being more convenient to make these multiple bets, this gets around integer problems as the wagers will be integers that can easily be bet. Whereas the Kelly optimization needs modifications to produce integer wagers. Also this approach can be computerized to print out the tickets — see the examples later in the paper, and the higher probability wagers can be bet more to approximate a Kelly strategy.

2 The importance of getting the mean right

Table 4 and Figure 3 show that getting the mean right is the most important aspect of any portfolio decision problem. Chopra and Ziemba (1993) discuss that and look at the effect of errors in means, variances and covariances using the cash equivalent of the approximate versus exact optimal solutions. Basically it is in the ratio 20:2:1 for errors in means,

Table 4 Average Ratio of CEL for Errors in Means, Variances and Covariances.

t Risk Tolerance	Errors in Means vs Covariances	Errors in Means vs Variances	Errors in Variances vs Covariances
25	5.38	3.22	1.67
50	22.50	10.98	2.05
75	56.84	21.42	2.68
	↓	↓	↓
	Error Mean	Error Var	Error Covar
	20	2	1

Source: Chopra and Ziemba (1993).

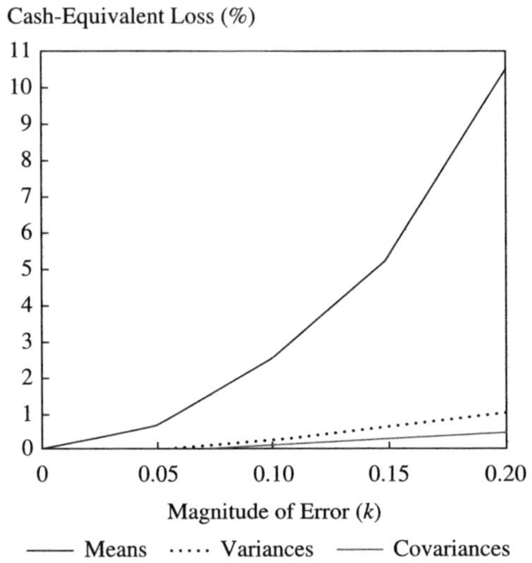

Figure 3 Typical Relative Importance of Errors in Means, Variances and Covariances in Terms of Certainty Equivalent.
Source: Chopra and Ziemba (1993).

variances and covariances in terms of error impact. We measure risk aversion by the Arrow-Pratt risk aversion index $R_A(w) = -u''(w)/u'(w)$, where primes denote differentiation of the utility of wealth function u. In investment practice, risk tolerance $t = 100/(0.5 * R_A)$ is typically used.

Referring to Table 4, we see that low risk aversion utility functions such as log with $R_A = 1/w \cong 0$, the effect of the errors is more like 100:3:1 so getting the mean right is even more important. And for horse racing, that is the probabilities for horses coming first, second, third, etc.

3 The favorite-longshot bias

The favorite-longshot bias is the tendency in horseracing, sports betting, and financial options for the most likely outcome to be underbet and the less likely outcomes overbet. So people tend to like junk and dislike the best possibilities. This bias has been well known to Irish and other bookmakers who actually create the bias with the bets they offer for the last 100+ years. Figure 4 shows the 1949 study by Griffin for 1386 races in 1947 for races at Churchill Downs, Belmont and Hialeah. In this graph, there are the number of entries, winners and winners times odds for every odds group. The axes show the odds, equivalent to the subjective probabilities, versus the actual number of winners, the objective probabilities.

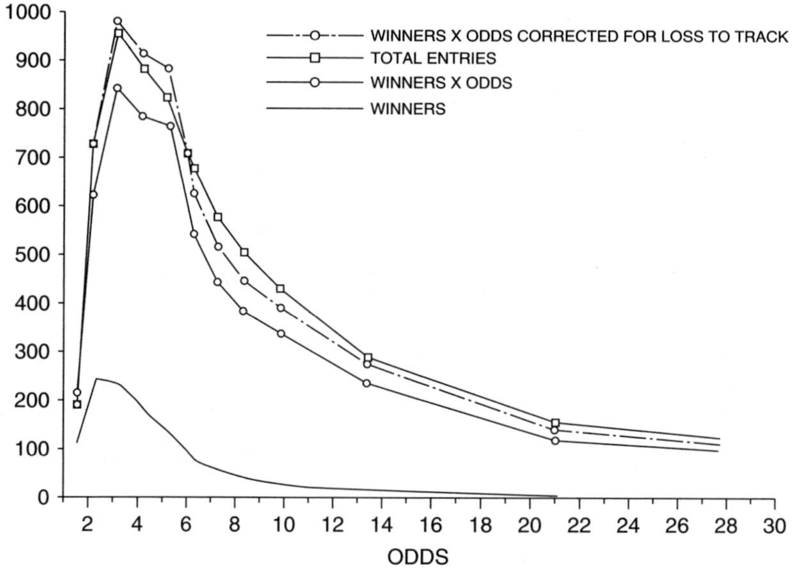

Figure 4 Griffith's 1949 Study on the Favorite-longshot Bias; see Hausch, Lo and Ziemba (2008) for the Reprinted Paper.

Figure 5 shows the effective track payback less breakage for various odds levels in California and New York, more than 300,000 races over various years and tracks, as of 1986 as reported in Ziemba and Hausch (1986).

There actually was a small profit, about 3%, in betting horses to win at US odds of 3–10 (UK odds of 1.30 or less), but at odds of 100-1, the fair odds are about 700-1 so that such bets were worth about 13.7 cents per dollar bet. The California and New York graphs differ slightly because of different track takes. There are approximately three piecewise linear segments, small profits on extreme favorites with favorites underbet and longshots overbet, more and more losses as the odds lengthen and extremely poor returns at high odds levels, like lotto tickets.

Since 1986 there have been a number of developments that have tended to influence the betting and have shifted these graphs, such as:

1. There are no longer separate pools for individual races at different racetracks. All the races now have pooled betting which now comes in late with about half the bets not recorded in the pools until the race is already running;
2. There are rebates where tracks send a signal with the results and the rebate shops and the track share the track commission between the

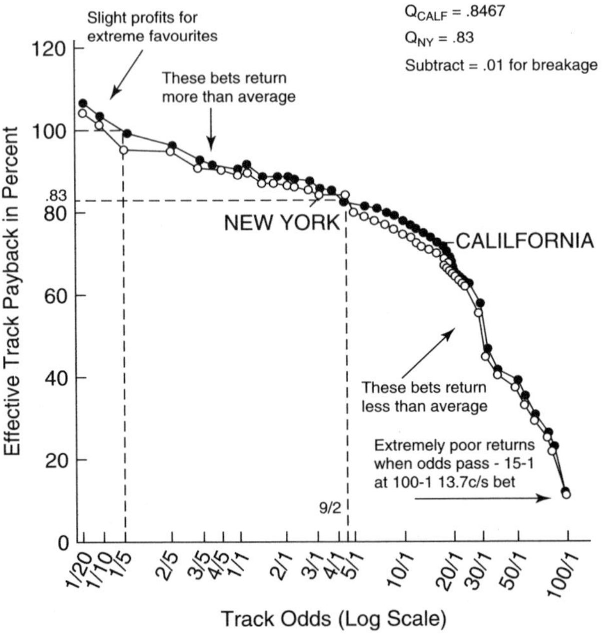

Figure 5 The Effective Track Payback Less Breakage for Various Odds Levels in California and New York (more than 300,000 Races over Various Years and Racetracks). Source: Ziemba and Hausch (1986).

rebate shop, the tracks and the bettors. So instead of facing 13–30% transaction costs, large bettors are actually charged about 10% net; and

3. Betting shops such as Betfair, offer long and short bets on racing and many other events such as political campaigns, etc. Online internet betting of this sort is legal in Canada, the UK and many other countries but it is not legal in the US.

Figure 6a–c look more closely at the extreme favorites in the US and the UK. Figure 6c shows that in Britain, bookies construct odds, creating the favorite-longshot bias to clear the market and equilibrate bettor demand.

Figure 7 shows that the bias curve may be different for different types of races. It shows the bias for the Kentucky Derby for 1903–1986. Other higher quality races like the Derby may well have flatter biases. See also Tompkins, Ziemba and Hodges (2008) who demonstrate similar biases in the S&P500 and FTSE100 index futures options. I use such ideas in an offshore hedge fund and in personal and private investment accounts.

Figure 8 and its data in Table 5 and Figures 9a-b show the bias based on more recent data. Observe that the favorites are no longer underbet

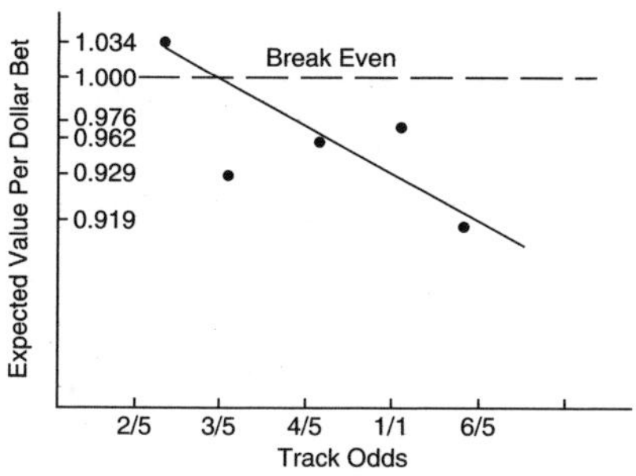

(a) Rates of return on extreme favorites in 1509 races run at various tracks during 1955-1962.
Source: Fabricand (1965, 1979)

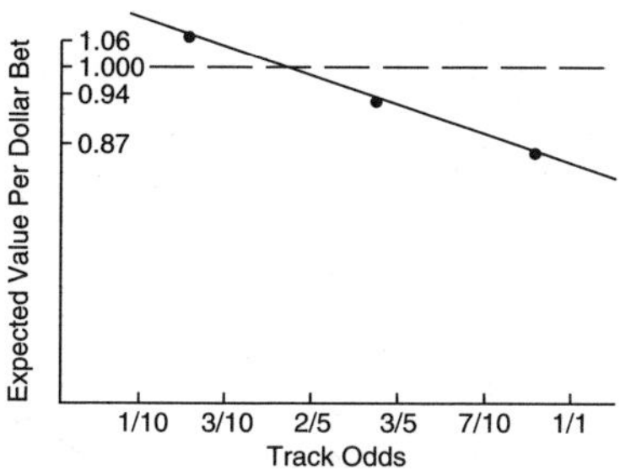

(b) Rates of return on extreme favorites in 233 races run in New York during 1980.
Source: Ziemba and Hausch (1984)

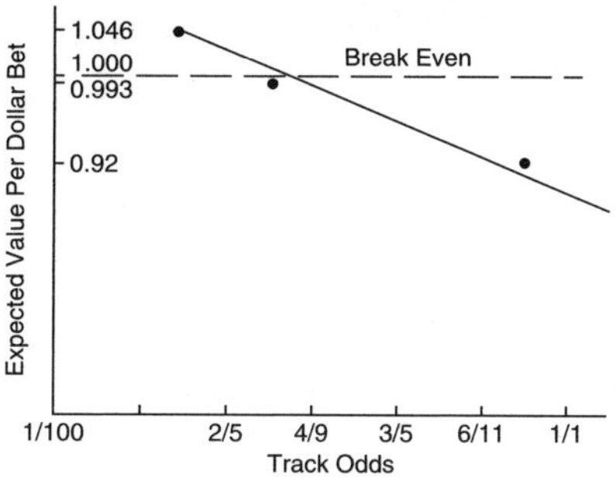

(c) Rates of return on extreme favorites in 1599 races run on British flat racing tracks in 1950, 1965 and 1973.
Source: Figgis (1974) and Lord Rothschild (1978)

Figure 6 Extreme Favorites, Small Profits: see Ziemba and Hausch (1986) for these References.

enough to turn a profit betting them and the flatness of the curve until you get to fairly long priced horses. You can still short longshots on Betfair and make a profit if you are careful. Additional discussion and results are in Hausch and Ziemba (2008).

This bias forms part of the Kahneman-Tversky (1979) prospect theory where low probability events are overestimated and high probability events are underestimated. This also forms a part of the behavioral finance literature. Thaler and Ziemba (1988) discuss reasons for the bias as do Ziemba and Hausch (1984). These include the fact that there are more bragging rights from picking longshots than from favorites: 50-1, *wow was I smart* while 2–5 is *an easy pick*. Transactions costs are another factor: bet \$50 to win \$10 is hardly worth the effort.

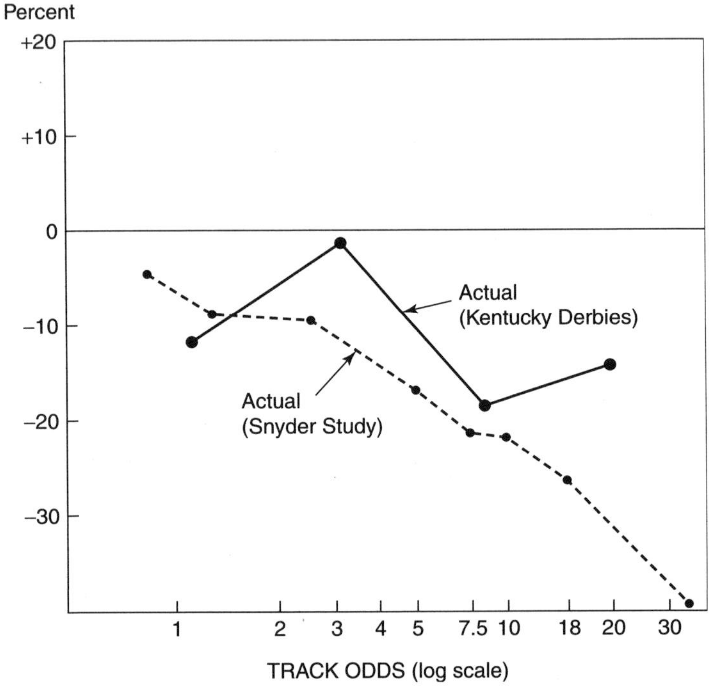

Figure 7 Expected Return Per Dollar bet with and without the Track Take Deducted for Different Odds Levels in the Kentucky Derby 1903–1986 and in 35,285 Races Run During 1947–1975, from Data in Snyder (1978).
Source: Ziemba and Hausch (1987).

4 Place and show optimization with transactions costs

The Dr Z system, co-developed with Donald Hausch with some early help from Mark Rubinstein, presents a winning method for betting on underpriced wagers. The idea of the system is simple: use the data from a simple market, in this case the win probabilities to fairly price bets in the more complex markets, such as place and show. For example, with ten horses, there are 720 possible finishes for show. Then one searches for mispriced place and show opportunities. This is a weak form violation of the efficient market hypothesis based solely on prices. How much to bet depends on how much the wager is out of whack and it is a good application of the Kelly betting system. The formulation below shows such an optimization. There is a lot of data here on all the horses and not much time at the track. So a simplified approach is suggested. Don and I solved thousands of such models with real data and estimated approximation regression equations

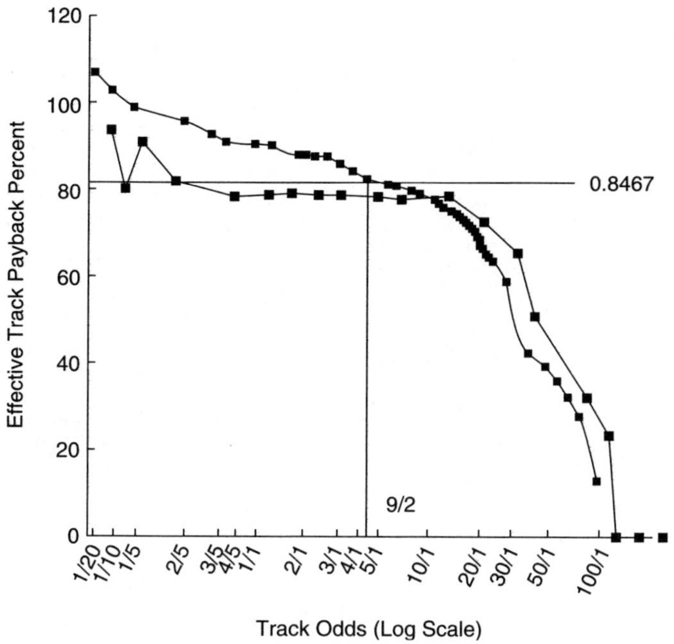

Figure 8 Effective Track Payback Less Breakage for Various Odds Levels in California. Source: Ziemba and Hausch [1986] and Hausch and Ziemba (2008).

that only involve four numbers, namely, the amounts bet to win in the total pool and the horse under consideration for a bet, plus the total place or show pool and the place or show bet on the horse under consideration.

These equations appear below. In our books Ziemba and Hausch (1984, 1986, 1987) and papers Hausch, Ziemba and Rubinstein (1981) and Hausch and Ziemba (1985), we study this in various ways, including different track takes, multiple bets for place and show on the same horse and how many can play the system before the edge is gone. This system revolutionized the way racetrack betting was perceived, viewing it as a financial market, not just a race. This led to pricing of wagers and the explosion of successful betting by syndicates in the US,Hong Kong and elsewhere; see, for example, my joint books referenced here and Hausch, Lo and Ziemba (1994, 2008) and Hausch and Ziemba (2008).

The effect of transactions costs which is called slippage in commodity trading is illustrated with the following place/show horseracing optimization formulation; see Hausch, Ziemba and Rubinstein (1981). Here q_i is the probability that i wins, and the Harville probability of an ij finish

Table 5 Data for the Effective Track Payback Less Breakage for Various Odds Levels in California.

Odds Level	Races	Avg Est Wins	Wins	Total Est Wins	Fav/Longshot	Odds Range
1	9	0.952381	8	8.571429	0.933333287	1/20 - 1/12
2	54	0.9090909	41	49.0909086	0.835185194	1/12 - 1/7.4
3	217	0.835003	163	181.195651	0.89958009	1/7.4 - 1/4.5
4	745	0.7660169	484	570.6825905	0.848107176	1/4.5 - 1/2.7
5	5990	0.6582732	3197	3943.056468	0.810792345	1/2.7 - 1/1.65
6	18250	0.5406803	8095	9867.415475	0.820376929	1/1.65 - 1/1
7	35987	0.4260281	12561	15331.47323	0.819295041	1/1 - 1.65/1
8	85716	0.3144652	22018	26954.69908	0.816852005	1.65/1 - 2.72/1
9	120483	0.2219781	21543	26744.58742	0.805508781	2.72/1 - 4.5/1
10	134990	0.14957	16371	20190.4543	0.810828709	4.5/1 - 7.4/1
11	136303	0.0966803	10827	13177.81493	0.821608139	7.4/1 - 12/1
12	113285	0.0614053	5504	6956.299411	0.791225287	12/1 - 20/1
13	89063	0.0380625	2543	3389.960438	0.750156247	20/1 - 33/1
14	63568	0.023571	976	1498.361328	0.651378264	33/1 - 54.6/1
15	35808	0.0146932	281	526.1341056	0.534084366	54.6/1 - 90/1
16	9868	0.009349	28	92.255932	0.303503519	90/1 - 148.4/1
17	647	0.0059442	1	3.8458974	0.260017337	148.4/1 - 244.7/1
18	25	0.003213	0	0.080325	0	244.7/1 - 403.4/1
19	7	0.0020153	0	0.0141071	0	403.4/1 - 665.1/1
20	3	0.0012285	0	0.0036855	0	665/1 - 1090.6/1

Source: Ziemba and Hausch [1986] and Hausch and Ziemba (2008).

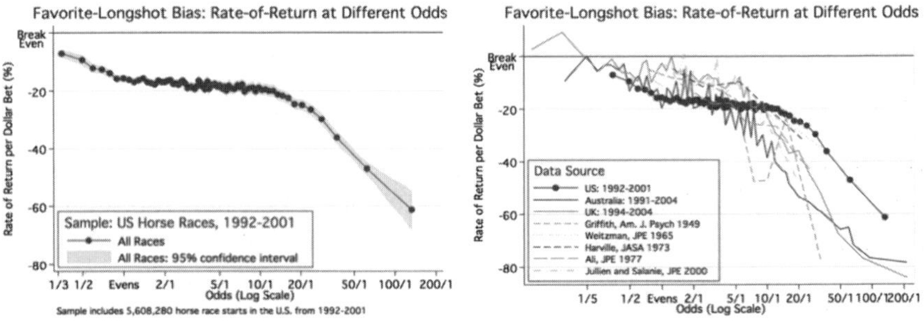

(a) Rates of return (b) Rates of return

Figure 9 Rate of Return at Different Odds.
Source: Snowberg and Wolfers (2008).

is $\frac{q_i q_j}{1-q_i}$, etc. That is, $q_j/(1-q_i)$ is the probability that j wins a race that does not contain i, that is, comes second to i. Q, the track payback, is about 0.82 (but is about 0.90 with professional rebates). The players' bets are to place p_j and show s_k for each of the about ten horses in the race out of the players' wealth w_0. The bets by the crowd are P_i with $\sum_{i=1}^{n} P_i = P$ and S_k with $\sum_{k=1}^{n} S_k = S$. The payoffs are computed so that for place, the first two finishers, say i and j, in either order share the net pool profits once each P_i and p_i bet's cost of say \$1 is returned. The show payoffs are computed similarly. The maximum expected utility model is

$$\max_{p_i,s_i} \sum_{i=1}^{n} \sum_{\substack{j=i \\ j\neq i}}^{n} \sum_{\substack{k=i \\ k\neq i,j}}^{n} \frac{q_i q_j q_k}{(1-q_i)(1-q_i-q_j)} \log \left[\begin{array}{c} \frac{Q\left(P+\sum_{l=1}^{n} p_l\right)-(p_i+p_j+P_{ij})}{2} \\ \times \left[\frac{p_i}{p_i+P_i} + \frac{p_j}{p_j+P_j}\right] \\ +\frac{Q\left(S+\sum_{l=1}^{n} s_l\right)-(s_i+s_j+s_k+S_{ijk})}{3} \\ \times \left[\frac{s_i}{s_i+S_i} + \frac{s_j}{s_j+S_j} + \frac{s_k}{s_k+S_k}\right] \\ +w_0 - \sum_{\substack{l=i \\ l\neq i,j,k}}^{n} s_l - \sum_{\substack{l=i \\ l\neq i,j}}^{n} p_l \end{array} \right]$$

$$\text{s.t.} \sum_{l=1}^{n} (p_l + s_l) \leqslant w_0, \quad p_l \geqslant 0, \quad s_l \geqslant 0, \quad l = 1, \ldots, n,$$

While the Harville formulas make sense, the data indicate that they are biased. But for place and show, the win favorite-longshot bias and the second and third finish bias tend to cancel so the corrected Harville formulas are not needed here. For other bets to correct for this, professional bettors adjust the Harville formulas, using, for example, discounted Harville formulas,[2] to lower the place and show probabilities for favorites and raise them for the longshots; see papers in Hausch, Lo and Ziemba (1994, 2008) and Hausch and Ziemba (2008) and the discussion below on place pick all.

Rebate is added to final wealth inside the large brackets by adding the rebate rate times all the bets,winners and losers.

This is a non-concave program but it seems to converge when nonlinear programming algorithms are used to solve such problems. But a simpler way is via expected value regression approximation equations using 1000s of sample calculations of the NLP model. These are

$$\text{Ex Place}_i = 0.319 + 0.559 \left(\frac{w_i/w}{p_i/p}\right)$$

$$\text{Ex Show}_i = 0.543 + 0.369 \left(\frac{w_i/w}{s_i/s}\right).$$

[2]The discounted probabilities come from

$$q_i^* = \frac{q_i^\alpha}{\sum_i^n q_i^\alpha}$$

for α about 0.81 then one uses the q_i^* in the second place position. For third one uses α^2 about 0.64. These empirical numbers vary over time and by track. This is more important for exacta pricing than place and show because for the latter the win bias from the favorite-longshot and the second and third biases tends to cancel. The favorite-longshot bias is the empirical observation that favorites are underbet and longshots overbet; see the discussion above.

The expected value (and optimal wager) are functions of only four numbers — the totals to win and place for the horse in question and the totals bet. These equations approximate the full optimized optimal growth model. See Hausch and Ziemba (1985) for more on this plus additional features. This is used in Dr Z calculators. See the discussion in Ziemba and Hausch (1986) and Ziemba (2011) for a discussion of typical use at the first Breeders' Cup in 1994.

An example is the 1983 Kentucky Derby.

	Totals	#8 Sunny's Halo	Expected Value Per Dollar Bet	Optimal Bet (W_0=1000)
Odds		5-2		
Win	3,143,669	745,524		
Show	1,099,990	179,758	1.14	52

Sunny's Halo won the race

Win	Place	Show
7.00	4.80	4.00

Π = $52

15 second bet!

Watch board in lineup while everyone is at the TV

Here, Sunny's Halo has about 1/6 of the show pool versus 1/4 of the win pool so the expected value is 1.14 and the optimal Kelly bet is 5.2% of one's wealth.

You might ask: does the system still work in 2012 and what is changed? The main new features are:

1. these days we bet at rebate shops by phone or electronically. The rebate is a sharing of the track take by the track, the rebater and the bettor. The effect is to take all bets from a track take of 13–30% for various bets to about 10%;
2. betting exchanges in the UK and elsewhere allow for short as well as long wagers; and
3. there is a lot of cross track and last minute betting and this takes time to be sent to the pools at the racetrack. Hence, about 50% of the wagers don't actually appear in the pools until after the horses are running. So one must estimate the final odds (probabilities).

Syndicates exist that break even on their wagers yet make millions on the rebate. Regarding the Dr Z system, John Swetye works with me and we wager with rebate searching for bets at 80 racetracks. Basically the system still works but the task is not easy. One successful six month period in 2004 with a \$5000 bankroll, the system lost 7%, received a 9% rebate. The total wagers were \$1.5 million giving a 2% or \$30,000 profit.

5 The place pick all

This is a less commonly used and known bet but it is available at Santa Anita, for example. The idea is to create a ticket with I horses over I races where you have either the winner or the second place horse in each race i, $i = 1, \ldots, I$. The number of races I varies from about 7 to 12.

The probability that a ticket with j the chosen horse in race i *wins* that race is the probability that j is first plus the probability that j is second, namely

$$p_{ij} + \sum_{k=1,\ldots,K_i} p_{ik} \frac{q_{ij}}{1 - q_{ik}} \text{ for } k \neq j$$

where

$$q_{ik} = \frac{p_{ik}^{\alpha}}{\sum p_{ik}^{\alpha}} \text{ and } \alpha \cong 0.81 \text{ and is track dependent.}$$

These are discounted Harville formulas, see papers in Hausch, Lo and Ziemba (1994, 2008) for more on this.

Then the chance that a given ticket with $i = 1, \ldots, I$ is a winner is

$$\hat{p}_{ij} = \prod_{i=1,\ldots,I} \left\{ p_{ij} + \sum_{k=1,\ldots,K_i} p_{ik} \frac{q_{ij}}{1 - q_{ik}} \right\}$$

6 Some stochastic programming formulations

There are basically two strategies for the optimization: Kelly expected log optimization and the tree tickets approach that approximates the Kelly strategy. To supplement the rest of the paper, a few formulations appear here.

The expected log problems are typically solved using CONOPT (Drud 1994) which has produced good results even though the problems have non-concavity in them. The bets must be computed very fast as the odds are changing. While this may not be general but because of an epsilon optimality convergence criteria, the MINOS code (Murtagh and Saunders 1998) may converge to a non-optimal strategy. Hence, CONOPT is safer.

In general, the Kelly Elog optimization given modern computing can be used for essentially all the bets, even possibly Hong Kong's triple trio, namely, getting the 1-2-3 finish in any order in three races with 14 horses in each race of which many horses are 200-1 but they can still finish 3rd. This has 48 million combinations. I focus here on the US bets and suggest that we focus on Elog Kelly optimization for high probability low payoff bets and the tickets approach for the low probability high payoff events which can closely approximate the Kelly strategy and yields easily implemented tickets that are integers.

The simplest bet is the exacta. To win you must get the winner plus the second place finisher. This uses elements of the place pick all formulation except it is just for one race and it is not first or second but first and second. First is easy, it is just p_i, the probability that i wins. Second uses the discounted Harville formula so it is $\frac{q_j}{1-q_i}$ where $q_i = \frac{p_i^\alpha}{\sum p_i^\alpha}$, where $\alpha \cong 0.81$ So the probability of an ij finish is $\frac{p_i q_j}{1-q_i}$. Let s_{ij} be our bet on an ij finish. So the Kelly optimization problem is

$$\max_{x \in K} E_w \log W = \sum_{i=1}^{I} \sum_{j \neq i}^{J} p_i \frac{q_j}{1-q_i}$$

$$\times \log \left\{ W_0 + r \sum_{i=1}^{I} \sum_{j \neq i}^{J} x_{ij} + Q \frac{\left(E + \sum_{i=1}^{I} \sum_{j \neq i}^{J} x_{ij}\right)}{E_{ij} + x_{ij}} \left(\frac{x_{ij}}{x_{ij} + E_{ij}} \right) - \sum_{i=1}^{I} \sum_{j \neq i}^{J} x_{ij} \right\}$$

where r is the rebate percent payable on all bets, losers and winners, E is the total exacta bet by the crowd, with the E_{ij} their bets on ij and Q is the track payback. The constraints can include a maximin bet on any combination ij as well as on each i and on the total bet. The wealth is final wealth plus rebate plus profits minus the bets and Exp is the expected value. Other high probability low payoff bets are similar.

Let's now consider a tickets model as well as an expected log model. I supervised an unpublished MSc thesis on this at the Oxford Math Department, see Assamoi (2010). Some of the theory is there but no calculations. Before we consider this, let us do it for the place pick all. I am not aware of any published research or calculations, real or simulated, on this bet.

Let x_{ij} be the bet that either the chosen horse j will win or come second in race i. The probability of winning is $\hat{p}_{ij}$ as given above.

So the Kelly formulation is

$$\max_{x \in K} Exp \log W = \sum_{i=1}^{I} \sum_{j \neq i}^{J} \hat{p}_{ij}$$

$$\times \log \left\{ W_0 + r \sum_{i=1}^{I} \sum_{j \neq i}^{J} x_{ij} + Q \frac{\left(PL + \sum_{i=1}^{I} \sum_{j \neq i}^{J} x_{ij}\right)}{E_{ij} + PL_{ij}} \left(\frac{x_{ij}}{x_{ij} + PL_{ij}} \right) - \sum_{i=1}^{I} \sum_{j \neq i}^{J} x_{ij} \right\}$$

which is very similar to the exacta formulation where PL is the place pick all pool with PL_{ij} bet on ij, r is the rebate percent and Q is the track payback percent.

The ticket formulation breaks the picks, the js, into categories I, II and III for each race j. I's are high value and high probability of winning horses. II's are major contenders and III's are longer odds horses who could upset the favorites. Suppose there are eight races so I = 8, where n_{ij} is the number of horses in category ij.

	1	2	3	4	5	6	7	8
I	N_{1I}	N_{2I}						
II	N_{1II}	N_{2II}						
III	N_{1III}	N_{2III}						

Given the probabilities and other factors, the horses are put into I, II or III or not considered in each race i. The score 8 tickets have all I's and they have the most money on them. There are $\prod_{i=1}^{8} N_{iI}$ of these. The score 9 have 7 ones and one II. There are 8 such tickets with lower bets. The score 10 tickets have 6 I's and 2 twos or 7 I's and one III with even lower bets. There are $\binom{8}{2}$ and eight of these with the lowest bets. One might go to score 11 and have bet sizes to approximate a Kelly strategy. As you can see, the number of tickets gets very large here, as does the cost. We use a computer program to generate these tickets. For a sample printout, see section 11.

7 The Pick3 and Pick4: Theory of pricing the bets

Example: the 9/11 Pick3: don't trust odds from newspaper stories

On September 11, 2011, the tenth anniversary of the attacks on New York and Washington, the first three races at Belmont had winners 9-1-1. The Pick3 paid \$18.60. The parlay on the three track takes versus one for the Pick3 paid $4.20 * 4.20/2 * 6.20/2 = \$30.5$, which is more than the Pick3 payoff without even factoring in the two extra track takes. So the 9-1-1 Pick3 was overbet just like popular numbers in lotteries are. The track take for the Pick3 is 26% and for win 16%.

So the fair value of the Pick3 with zero track take is \$18.60/0.74 = \$25.14 and for the parley

$$30.5 \left\{ \frac{1}{0.84} \right\}^3 = \$51.34,$$

so the numbers 9-1-1 were overbet. A newspaper article said the odds of such an outcome is a million to one. Actually by looking at the final odds on the charts for these three races, we can estimate that. Recall, the probability of winning is

$$\frac{Q + \Delta Q(odds)}{odds + 1}.$$

Referring to a table in Ziemba (2012) for the ΔQs and the payoffs below, the odds of 9-1-1 occurring are about one divided by the probability of this outcome which is

$$\left\{\frac{0.84 - .02315}{2.1}\right\}\left\{\frac{0.84 - .02315}{2.1}\right\}\left\{\frac{0.84 - .0345}{3.45}\right\}$$
$$= 0.3890 * 0.3890 * 0.2335 = 0.0353$$

since the payoffs for win were \$4.20, \$4.20 and \$6.90 for each \$2 bet. The odds against the 9-1-1 outcome then are found to be approximately $(1 - 0.0353)/0.0353 = 27.32$ to 1, not 1 million to 1.

8 The Pick4

This bet has no consolation prize so you need to win all four races to collect. The bet is usually offered twice daily, early and later with the late Pick4 covering the top races on the card. Like the other multiple race bets, it is hard to win as the probability that you win is $P^4 = P_1P_2P_3P_4$, namely, the product of the probabilities that you win each of the four races. For example, if each P_i is 1/2, so you have a 50% chance of winning each race, then $P^4 = \left\{\frac{1}{2}\right\}^4 = \frac{1}{16} = 6.25\%$ so once every 16 times you would win on average. And getting to 50% in each race is challenging. The bet is \$1 and sometimes 50 cents so you can take many multiple combinations to get P^4 to an acceptable level. As always, try to bet with an advantage so focus on horses of three types: good value and good chance to win and possible bombs to give a large payoff.

When facing the Pick4, determine a strategy to play the bet. If there are one to three standouts and some wide open races then you can have a simple ticket that singles the standouts and spreads in the contentious races. If the standouts are very strong, then singling make sense otherwise make the standout a I and have some backups as IIs.

I was at Santa Anita on Saturday April 9, 2011 for the Santa Anita Derby, a major prep race for the three year olds before the Kentucky Derby.

There was a standout, Premier Pegasus, who won the San Felipe $1\frac{1}{16}$ (the previous major race) by seven lengths, but in a soft time — he won so easily that his time (1:41:23) was not fast. He towered over the field but unfortunately had a minor injury and was scratched. So was Uncle Mo, another Kentucky Derby hopeful, who disappointed in the Wood Memorial at Aqueduct, finishing third at 1–10 win odds. But he was a dual qualifier and became the favorite. But he too was scratched. The rest of the field looked weak and it was hard to separate them. So spreading was suggested and I took seven horses in that 10th race — all except two very longshots. who were at over 40 to 1 odds. The other three races all had standouts and I singled all three. One was First Dude in race 8 who had never won a race but was second in last years Preakness and third in the Belmont and in the money in many Grade II races. His trainer, Bob Baffert, put him in a \$56k allowance against weaker competition to try to get that first win. He got it at 4-5, winning easily, paying \$3.60 for a \$2 win ticket.

It was good that I spread in the Santa Anita Derby race 10 as Midnight Interlude, trained by Bob Baffert at 13.90 to 1, was the winner, paying \$29.80 for a \$2 win ticket. Baffert has a knack of winning these top races.

Race 9's standout was Cambina, who also won at the essentially even money odds of 1.1 to 1 to pay \$4.20 for a \$2 win ticket.

Race 11's standout, Hey Maria, again at close to even money, won at 1.30 to 1 paying \$4.60 for a \$2 win ticket.

The Pick4 was a 50 cent bet but I bet \$10 on it costing $7 \cdot 1 \cdot 1 \cdot 1 \cdot \$10 = \$70$. The Pick4 paid \$99.20 for the 50 cents so I collected \$1995 which was a good gain. Like the Pick6 below, singling top standouts and spreading in tough races often is a very good strategy with low cost that can have a large payoff.

9 Example of Pick4 with embedded Pick3's and Doubles: Travers Day at Saratoga

Saturday, August 27, 2011 was the Travers Stakes at Saratoga. This is the mid summer Derby, the top race for three year old colts between the triple crown races in May and June and the Haskell at Monmouth Park on July 31st and the Breeders' Cup in November. This is a $1\frac{1}{4}$ mile race with a \$1 million purse.

A strategy for the \$1 million guaranteed Pick4 which was races 9–12 is to accumulate the picks of various handicappers and then cover the bets well. To be a Kelly bettor, you must bet more on the top favorites and less on the longer priced horses. In addition to the handicappers at Saratoga who discuss the races on TV for free and the ones with subscription services, there is betfair.com in London which provides bid-ask spreads on the races to give an idea of the final odds as they allow shorts as well as long bets. A drawback of betfair.com is that the spreads are wide but they close up near race time. Another drawback for US bettors is that you cannot use it in the US but Canadian and British bettors can use it.

The strategy is to establish bets on the Pick4 races 9–12 and have as backup the races 9–11 and 10–12 Pick3's and the races 9–10, 10–11 and 11-12 doubles. Thus if you get the P4 right you also win the two P3s and the three P2s. The strategy is to have multiple tickets, they can be through the score 8 system with I's, II's and III's or through straight tickets where you pick several horses in each race and box them all. That means that you win if any one of the horses in each race you selected wins, assuming you win all the races. These boxed tickets put more weight on the weaker horses than the multiple tickets approach. The multiple tickets approach is more complex but fully manageable. For a Pick5 or Pick6 we use a computer program to generate the tickets as there can be many of them. For a Pick4 there are 17 tickets if you do a score 8 down to II-II-II-II. Examples follow.

Both these approaches are discussed below.

Here are the races:

Race 9 was the 9th running of the Victory Ride, a Grade III, $100,000 race for three year old fillies. The past performances follow.

- Saratoga handicapper Mike Watermaker chooses one horse per race and here he chose #6 Hot Summer.
- The TV handicappers Dan Ulman and Mike Beer chose 2,5/6,8,9 and 6/5,8,9. This means double keying 2 and 5 as I's and 6,8,9 as II's. This is set up for the score tickets approach, see below.
- I use three additional services plus two of my colleagues have their picks, but one of them was rained out on that day by the great hurricane Irene. Our handicapper, who selects for superfecta as well so picks the best five horses, liked 2,5,6,8,9. One service ranked the horses 8-2-9-1-3, the other 2-9-1-8-1-3.
  The third, which are the Equiform pace numbers and final speed figures for the best final time in the last four races, #2 $74\frac{3}{4}$ twice; #5 74; #6 $73\frac{3}{4}$; #8 $71\frac{1}{4}$ and #9 76.

So the suggested picks were 2,5,6,9 because I skipped 8 so the ticket would not be too large. But in the multiple score tickets, 8 must be a II.

The result was 6-1-9. A \$16.40 win for a \$2 bet was a good start. Watermaker had it but as we will see that was it for his picks, the other three did not win. This is an important point — none of the handicappers will get it all right very often so it is important to use several outside handicappers plus your own analysis.

Race 10 was the 33rd running of the Ballerina, a Grade I race with a \$250,000 purse for three year old fillies. The past performances follow.

- Watermaker picked #7 Sassy's Image who was the favorite but did not win.
- Ulman and Beer chose 6/1, 2/3, 5, 7 and 6, 7/2. As we will see, #3 Hilda's Passion won and they basically threw her out because of the last race — a real clunker. But in the previous race to that, she ran a Beyer 107! Ranking her as a III meant that with a score 8 ticket 3 of the 8 spots are used up so one can still win with 2 II's and one I but this score 8 ticket will have the minimum bet \$1 or 50 cents depending on the track rules.
- Our handicapper liked 3, 6 and 7. One service ranked them 3-6-2-7-5, the other 1-7-5-2-6-4. Equiform's pace service got it right and as we will see, got the next two races correct as well for one third pick and three top ones. A very good performance. But that will not happen all the time. Equiform had #2 $77\frac{1}{4}$ and $77\frac{1}{2}$; #3 $79\frac{3}{4}$; #6 78 and $77\frac{1}{2}$ and #7 $77\frac{1}{2}$. So my picks were 2,3,6,7 and #3 Hilda's Passion was a short price and won paying \$7.20.

Race 11 was the 27th running of the Foxwoods King Bishop, a 7 furlong Grade I race with a purse of \$250,000. The race featured the return to the track of the two year old champion Uncle Mo who had been sidelined since being scratched just prior to the Kentucky Derby where he was favored to win. He had a variety of ailments but had good recent workouts. The big issue was, is this the old Uncle Mo? It was a tough challenge.

- Watermaker favored #3 Run Flat Out.
- The TV handicappers favored #1 Flashpoint with one double keying #4 Dominoes, the favorite in the Grade 2 Jim Dandy run on July 30 at $1\frac{1}{8}$ at Saratoga. He lost that race finishing third to Stay Thirsty, ran a Beyer 106 and was in Race 12 at the Travers. So Ulman and Beer had 1/2, 5, 7 and 1, 4/2, 6/7. Arguing that the return was too tough for Uncle Mo, they downgraded him to a II and a III, respectively.
- Our handicapper favored 4, 5 and 7. The two services had 1-2-7-4 and 1-7-5-2-6-4. The Equiform pace analysis had #1 76 (but double top); #2 79, #4 75, #5 75, and Uncle Mo #7 $78\frac{3}{4}$ but a long time ago. So this pointed to Caleb's Posse, #2.

I rather liked Caleb's Posse with the 79 at Saratoga in the Grade II Amsterdam on August 1 with a Beyer of 105 which was higher than Flashpoint's 104 at Monmouth. The 105 for Caleb's Posse was way above his previous high Beyer of 92 so that made it a risk. The only horse to win a Grade I was Uncle Mo who won two as a two year old, the Champaign and Breeders' Cup Juvenile, the latter with a 108 Beyer in the race he got the $78\frac{3}{4}$ speed figure.
My picks were 1,2 and 7.

The finish was 2-7-6-4 with Caleb's Posse nipping Uncle Mo at the wire. I had now won the first Pick3 there and the first two of the doubles.

Race 12, the final one in the Pick4 was the 142nd running of the Travers, a Grade I, $1 million purse race. This is the midsummer classic and one of the longest running top races in the US.

- The TV handicappers felt that the winner here would come out of either the Jim Dandy run at Saratoga or the Haskell run at Monmouth. So that favored #7 Coil, #10 Shackleford and #9 Stay Thirsty. They chose 7,9,10/3 and 10/6,9.
- Watermaker liked Shackleford, a front running horse who won the $1\frac{3}{16}$ mile Preakness and was ahead in the Kentucky Derby late in the race, but faded to fifth. His dosage profile was 6-13-9-0-2 for a dosage index of 3.62. This is under the usual Kentucky derby cutoff of 4.00 but is high. Stay Thirsty with 4-6-16-0-0 (2.25), Coil 2-0-9-1-0 (1.18) and Ruler on Ice, the winner of the Grade I Belmont Stakes in the slop with 6-1-9-0-0 (2.56) were more suited for this $1\frac{1}{4}$ mile race. But he still could win going wire to wire.
- Our handicapper favored 7,9,10 with 4 should there be rain and mud which there was not. The two handicapping services had the following rankings: 9-2-7-4-10-6 and 9-2-7-4-6-8-3-10. So they favored Stay Thirsty and did not like Coil much but gave him third and basically threw out Shackleford. They rather liked #2 Rattlesnake Bridge rated second and the mud runner Ruler on Ice, fourth with two wins on sloppy tracks and the rest mediocre. Rattlesnake did not look competitive to me with three Beyer's of 91, 90 and 91 with one win.
  Equiform's top picks were #6 $75\frac{3}{4}$; #7 $74\frac{1}{4}$, 74, 77; #9 $77\frac{1}{4}$ and #10 $76\frac{1}{2}$. So they suggested #9, Stay Thirsty.
  My picks were 6,7,9,10 on the bigger ticket and 7,9,10 on the smaller ticket.

The finish was 9-2-8-4.

My tickets were

2,5,⑥,9/2,③,6,7/1, ②,7/6,7,⑨,10	for $384 less rebate, and
2,5,9/③,6,7/1,②,7/7,⑨,10	for $162 where circles indicate the winners

So the big ticket won the Pick4, the two Pick3s and the three daily doubles. I made some win, place and show and superfecta bets. Some won, some lost.

The Pick4 paid $1453.00, the two Pick3s $471 and $195 and the three doubles $86.50, $58.00, and $48.80 per $2 bet. So there was a good gain.

The Pick3 tickets I played were

2,5,⑥,9/2,③,6,7/1,②,7	$96 for $2, to collect $86.50
2,③,6,7/1,②,7/6,7,⑨,10	$96.00 for $2, to collect $195
2,5,9/2,③,6,7/1,②,7	cost $72.00, a loser for $2
③,6,7/1,②,7/7,⑨,10	$54.00, a winner for $2, to collect $195

The doubles were

9/10	2,5,⑥,9/2,③,6,7	winner	$32 for $2 to collect $86.50
	2,5,9/③,6,7	loser	$18 for $2
10/11	2,③,6,7/1,②,7	winner	$24 for $2 to collect $58.00
	③,6,7/1,②,7	winner	$18 for $2 to collect $58.00
11/12	1,②,7/7,⑨,10	winner	$18 for $2 to collect $48.00 plus rebate on all the bets, winners and losers

I did not play a score ticket approach but it would have had all these winners. It would have cost more but could have returned more since scores 8,7,6,5 and 4 would likely have $20, $10, $5, $3, and $1 tickets. Rather than speculate, I will discuss the tickets approach in other situation where I actually made such bets. Rebate always help.

10 The Pick6: Theory of pricing the bets

2001 Breeders Cup insurance bets for SCA

In 2001, a racing colleague, Cary Fotias, creator of the Equiform ratings, and I were hired by SCA, a Dallas based sports insurance company, to help insure the Breeders' Cup which was at Belmont near New York City. We insured the $2 to $3 million part. So the insurance company would guarantee a pool of $3 million. For example, if $2.15 million was bet, they would be liable for

$850,000. We studied and proposed to bet a random amount if needed. The idea being to get to $3 million and return the insurance company's money by winning Pick 6s and Pick 5/6s. It was risky as September 11 had just occurred and all the Arab owners such as Sheik Mohammed of Dubai were not in attendance. Their horses and trainers were though. It turned out to be a glorious day so the crowd sent the Pick 6 pool well over $4 million. Our client, himself the world's most famous bridge player, Bob Hamman, said you two can just play about $25–30,000 of the tickets. So we had a $2000 ticket twice and what we call a gorilla ticket for $28,000. We had some 5/6's and got most of the money back. The Pick 6 paid $250,000. The race we lost was the sprint. Squirtle Squirt, which my handicapping colleague did not like at 9-1 beat the front running filly, Extra Heat at 14-1, who we had, and who had led all the way until the finish. So if she had won we would have had three about $450,000 Pick 6s plus more 5/6s. Squirtle Squirt had run at Belmont and had the very top jockey Jerry Bailey and was trained by the recently deceased legendary trainer Broadway Bobby Frankel. I should have overruled and included Extra Heat on the ticket, adding a bit of extra cost. Too bad. This was another example of a lot of operational risk involved in Pick6 tickets. It is very easy to have an error leading to a loss. Various models and research yields the horses to include. Then the optimization creates a good ticket. But it was fun! But the next week we won a similar case at Santa Anita, while guaranteeing a $1 million Pick 6, collecting $240,000 for the client and a nice bonus for us.

11 The one that got away: the hittable $2 million Pick 6 at the 2009 Breeders' Cup

The Breeders' Cup is now up to 14 major races over two days[3] and was held again at Santa Anita on Saturday, November 6 and 7, 2009. I went in 2008 and it is fun to see it live. This year on wide screen high definition TV it was wonderful to watch. Being at home, the handicapping and betting is a lot easier. There are many opinions. That's what makes a horse race. The spreads on Betfair are fairly tight and it is easy to bet from Canada and you frequently get better odds there than at the track. We don't actually bet at the track but bet through rebate shops that give back part of the track take. That's easy to do on the phone or by email. The rebaters take their cut and the track gets more easy, low expense business to up their revenues.

The big race was as usual the $5 million classic. It is no longer the world's richest race. The $10 million Dubai World Cup has that honor. But

[3]In 2011 there were 15 Breeders Cup races.

it is the most important race in the world and frequently determines the horse of the year. The two candidates for horse of the year are were both female. The 3-year-old Rachel Alexandra won all her 2009 races. She beat the top females in the top female races by 20 lengths. I was at Churchill Downs to see this in the Kentucky Oaks held on the first Friday in May, the day before the Kentucky Derby. In four races against males, she beat them handily. So she would normally be an almost sure bet for horse of the year. But Zenyatta, a five year old mare, had won all 13 of her races but always against females. She has a dynamite kick and just cruises by the other horses near the buzzer to win easily. Some of her races were in slow times (76 area on the Equiform scale that I follow) and some in fast times (81 area). To put this scale in perspective, the highest I ever saw was four 84's by Ghostzapper. One of my most treasured but small bets was on Ghostzapper's Breeders' Cup win. There was a top filly in that race, Azeri — I had watched her at Saratoga getting beat by females in a $1\frac{1}{4}$ race so the fact that she had numerous wins at short distances I was pretty sure she would not be in the top 4 in this male dominated $1\frac{1}{4}$ race. So a \$20 superfecta bet in 2004 boxing for \$5 Ghostzapper (1or 2) with Roses in May (1 or 2) with the two next leading horses, Pleasantly Perfect and Perfect Drift (3 or 4) and (3 or 4) came in to provide a \$5000 payoff. The big mistake was not betting more! Ghostzapper ran the $1\frac{1}{4}$ race in 1:59:02, faster than Secretariat's record setting Kentucky Derby 1:59:40. The 5-2 odds were quite generous given Ghostzapper's brilliant record.

The 13/13 of Zenyatta is historic since only Personal Ensign (13/13) in 1988 and Colin in 1907 had undefeated records in the US. Tesio, the great Italian trainer in the 1930s had the other three of the five undefeated horses since 1900, among horses with at least ten major races at major race tracks. Tesio was a great anomaly — a person looking at many many generations to bread cheap to cheap to get great champions. He did this without computers with a lot of help from his wife.

Currently, the way to check such matings is to go to Steve Roman who does this for ten generations back. See Steve's website, www.chef-de-race.com, for much valuable information.

The Europeans and many other handicappers were pushing for Rip Van Winkle, trained by Irish legend Aidan O'Brien, who was the pick since he had won his two races in 2009 when he did not face the superstar Sea the Stars. The three times Sea the Stars beat Rip Van Winkle he was close behind and well ahead of the competition. Unfortunately, the Sea the Stars connections preferred to hedge and cash in on this horse reputed to be the best in Europe in ten years by retiring him to stud duties. He had won the Epson Derby, the 2000 Guineas and the ARC. Perhaps they recalled the last "best horse of the decade", Dancing Brave, who "could not lose" but finished fourth in the classic. The polytrack at Santa Anita — not dirt or the European's grass — could have been a factor too.

Getting back to Zenyatta and Rip Van Winkle. The Betfair and track odds showed the local biases. You could get better odds on Zenyatta in Europe on Betfair and Rip Van Winkle in the US at the track. It was not possible to do an arbitrage here. I just concentrated on better odds on Zenyatta on Betfair. My assumption was despite the fact that her running times were not super outstanding she had the will to win. And indeed she did with Rip Van Winkle finishing out of the money. She could have been retired as the only undefeated mare who beat males in the toughest race in the world.

But let's discuss the Pick 6. In the Pick 6, to win you must have all six winners who share 75% of the net pool with the 5/6 sharing the remaining 25%. The Pick 6 was races 4–9. There were three standouts but they were definitely not certain winners. The other three races seemed wide open. So you could play the Pick 6 in the following very simple way: I thought about doing this but did not — it was a $2 million mistake. You have a single ticket with about 10 horses in the three wide open races and single the three standouts. That would cost about 10*10*10*1*1*1*$2=$2000, not a large Pick 6 ticket. You only win if all three standouts win and they did.

The payoffs for $2 win tickets were as follows

Race 4	Dancing in Silks	52.60	
Race 5	Value of York	63.20	
Race 6	Goldikova	4.80	the first standout
Race 7	Furthest Land	44.60	
Race 8	Conduit	3.80	the second standout
Race 9	Zenyatta	7.60	the third standout

Goldikova was the winner of the mile grass race in 2008 and was arguably better this year. Her connections were the same as those of Miesque, also a two time winner of this race. The second standout, Conduit, was also the winner last year on the $1\frac{1}{2}$ mile turf.

The Pick 6 paid $1,838,305.20 for one winning 6/6 ticket and the 3*9=27 Pick 5/6 tickets (of the 10 losers in the three wide open races) paid 27*$4822.40 for a total of $130,204.80 plus the 6% rebate on the $2000 of $120 for a grand total of $1,968,630.00. It is not quite $2 million but as Johnnie Hooker played by Robert Redford in the Sting said: "it's not enough but it's close". Of course, taxes would take 25% at the track and be sorted out later when filing and my winning would depress the Pick 6 and Pick 5/6 prizes.

The big question is would I have gotten these three winners from my 10 picks in these races? Of course, more than 10 was possible. So let's look at these three races. I use about 5-6 handicapping services plus my own analysis of the daily racing form and the Equiform pace numbers. So the idea is to handicap the handicappers. In many bets this is computerized.

Race 4. The Sprint, 6 furlongs Handicapper #1 had 3-1-5-6 (the winner, Dancing in Silks) -8-4 Handicapper #2 had 5-3-8-1-6. So at 12-1 that's one of the 10 for sure especially when you observe that its last race Beyer speed rating at 106 was the highest of any horse in the race and it was right there at Santa Anita. Three of the horses ran higher Beyers than #6 but not in their last race. Actually there were only nine horses in this race so likely I would have taken them all including the 20-1 shot #2 and the 30-1 shot #7, neither of which listed above. #2 had a Beyer of 110 in a Grade I at Santa Anita so must be used. #7 looked greatly outclassed.

Race 5 Juvenile 2-year-olds on polytrack with 13 horses The winner Vale of York (#7) was racing in the UK and in Italy. He always had short odds and had two wins in five races and was close in the other three races. The morning line odds were 20-1. Handicapper #1 had 5-13-4-9-10-8 (so no 7 but he was a foreign shipper so likely not rated by this US service). Handicapper #2 had 13-4-9-5-6. Only handicapper James Quinn with 13-8-11-5-7-6 had 7 anywhere. These are two-year olds so there is a lot of noise. There were four other longshot horses one at 20-1, one at 15-1, one at 30-1 and one at 50–1. So going with all 13 horses called all is suggested.

Race 7 Dirt Mile #2 Furthest Land won with 10 in the field. #1 had 4-3-7-1-9-2 (the winner). Handicapper #2 had 3-7-2-8. The pace numbers are competitive. So 2 at 20-1 must be used.

Summary: all three of the wide open races had winners that were competitive horses. So they would be on our ticket. But even if we bet on all horses, these three races, the ticket only costs 9*12*10*1*1*1=$2340. This ticket made a lot of sense so I should have played it.[4] It would have had 8+11+9=28 Pick 5/6s. Oh well, there is always next year.

---

[4]Another way to play this is to have three sets of tickets in which you assume that at least two of the three standouts will win. So you have $N_1^i, N_2^i, N_3^i, N_4^i, 1, 1$ combinations, $i = 1, 2, 3$ all at $2 each. So depending on the N_j^i you likely have a larger ticket than the three singles approach. You might win more than one Pick 6 and more Pick 5/6s, but you might miss the Pick 6 as well unless the tickets are well spread.

An example of the multiple ticket to approach to the Pick 6: Santa Anita, March 6, 2002 with $202,790 carryover from Sunday's wagers.

An example: Consider the Pick 6
Santa Anita, March 6, 2002 with $202,790 carryover from Sunday's wagers

Race	3	4	5	6	7	8
Win Payoff	3.60	4.60	12.00	9.20	18.00	7.20
					1-5 shot out of the money	

$$\text{Advantage} \equiv \prod_{i=1}^{6}(1+\text{edge}_i)$$

Pick 6 $86,347.60 – one large transaction cost

Parlay $15,452.40 - six races, six smaller transactions costs

Bet = 1922	1 Pick 6		Score 6	$4
Rebate = 154	11 Pick 5/6	692.00	7	$58
Net bet = 1768	Gross =	93,961.80	8	$376
			9	$1484
				1922

Score 9: you win if the score is 9 or less, here 3 I's, 2 II's, total 9 so we won.

Score 9: you win if the score is 9 or less, here 3 I's, 2 II's, total 9 so we won

Race	3	4	5	6	7	8
I	[1], 3	[3]	7	3	2	[2]
II	7	1,5	1,2,[5],6	4,7,8,[10]	1,[4]	5.9
III	5,6	2,4,8		1,2,5		1,6,7,8

8 was scratched

This $1922 ticket was actually 66 separate tickets.
The winning ticket will have one 6/6 winner and multiple 5/6 winners

			Morning Line
Race 7	1	Madame Pietra	3-1
	2	Love at Noon	3-5
	3	Harvest Girl	8-1
	4	Filigree	4-1
	5	Farah Love	12-1

- This was the behavioral key.
- Filigree, the third choice in the morning line went off at 8-1.
- The 3rd and 5th races back, he ran faster than the favorite Love at Noon ran in his last two races. So he had a chance to win and he did.
- Love at Noon went off at 1-5 and had most of the P5 money

In the following Equiform numbers, the top number is final speed number, other numbers are pace within the race.

EQUIFORM ®

7TH SAX MARCH 6 - 6 FURLONGS DIRT

MADAME PIETRA (118) 40

70	72½	67½	70½	64½	w76½	w71½	64½
	/////	/////					
80½	71½	73½	79	77½	78	77	67
SA6½	SA-6½	SA8	DMR6	SA6	SA6	DMR6	HOL6
70½	71	67½					
82	70½		81	83½	81	63	68½
JAN24	10/17	09/27	SEP02	03/16	JAN31	AUG05	05/28

LOVE AT NOON (118) 44

72½	73½	74½	75½	w78½	w^70½
71½	82½	75½	74½	85	70½
SA8½	SA7	HOL8½	CD8½	CD6½	SA6
69½	74½	72	72½	76½	
	84			85½	77½
01/20	DEC29	06/16	05/26	05/05	04/07

HARVEST GIRL (118) 19

w71½	71½	68½	--	70½	71	66½	66½	68½	w71½	68	69½	w68½	69½	w67½	61½
	/////		/////					/////		/////			/////		
80½	77	80½	71½	83½	81½	82½	79½	81	78½	75½	79½	77½	82	79½	69½
SA6	SA8	HOL6½	BM8½	FPX6½	SR6	SOL6	GG6	SA-6½	FPX6½	BMF8	SOL6	GG6	HOL5½	BM5½	BMF5½
	68½	69	65½	71½				69½	71½	68					
87		83		98	81½	82½	80½	84½	83½		78½	77½	75½	86½	80
02/14	01/05	11/21	10/13	09/14	08/06	JUL14	11/25	10/28	09/19	08/12	07/22	05/31	MAY18	09/12	08/15

FILIGREE (118) 40

71½	67½	74	71½	w74½	w71½	71½	71½	70½	70	70	70½	w73½	w70½	w70	68½
77½	72	82½	78½	75½	74½	75½	73½	70½	81½	61½	77	74	77½	78	73½
SA6½	SA5½	DMR6	DMR6½	HOL6½	HOL6	SA7	SA8½	SA6	SA5½	SA9	FPX6½	DMR6½	DMR6½	HOL6½	SA6
71½			70½	72½		67½	70			61½	68½	68	70	70	
77½	75½	85½	73½	73½	74½	75		67½	75½		74	68½	77½	79	75½
01/24	JAN02	09/05	07/23	06/27	05/28	04/16	03/03	01/31	JAN01	10/28	09/30	09/03	JUL31	05/26	04/16

FARAH LOVE (120) 26

73	w70½	w70½	67	66½	65	63½	67	62½	w66	69½	w66
/////						/////	/////				
78½	74½	79	72	72½	72	76½	76½	71½	76½	74½	67
SA-6½	SA6	SA6½	HOL6	SA6	SA6	SA-6½	HOL5½	HOL8½	SA6½	SA6½	HOL6
71½		69½				62½		65½	65½	69	
81½	77½	82½	72	68½	74½	80	66½		84	70	70½
02/07	01/17	01/05	DEC09	02/16	01/31	12/30	12/20	11/23	11/01	OCT11	07/23

Normally we approximate Kelly with more money on higher probability wagers, but in the following bets we made equal $2 bets.

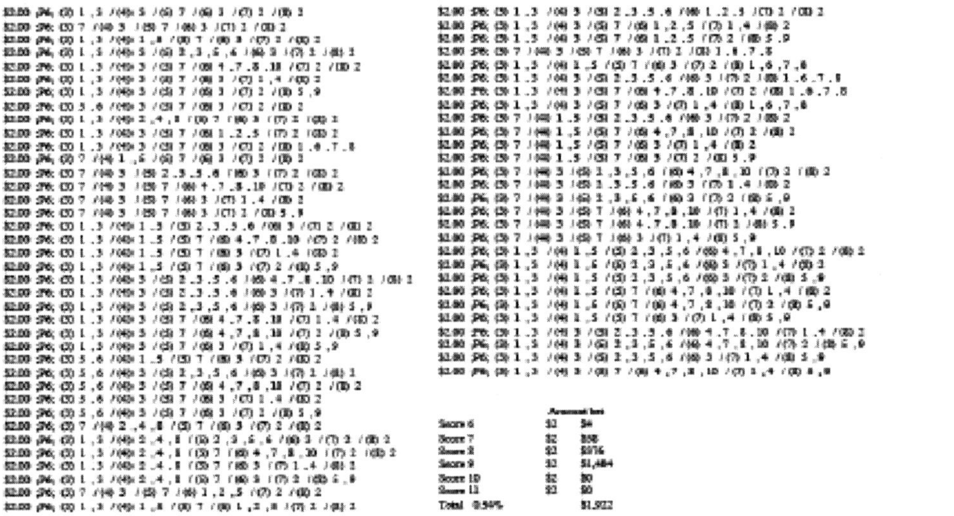

12 Professional racetrack betting syndicates

I had a hand with several of the major syndicate hedge fund teams through the *Beat the Racetrack* books Ziemba and Hausch (1984, 1986, 1987) and Hausch, Lo and Ziemba (1994, 2008) and other contacts. Hausch and I both talked to Bill Benter early in his Hong Kong career. He had started betting but had not put together a successful syndicate yet. So he quizzed us on the Dr Z system and other ideas in phone calls. We did help him a bit but as he said "we were academics spreading knowledge and he was a businessman so could not pay us". He did have other paid consultants on factor models and he pioneered successfully using 80+ factor models of two types:

1. predict the fair odds probabilities of various horses outcomes and compare these to the public's odds, and
2. bet most likely with the Kelly criterion.

I do not know if he picked up the Kelly from Ziemba and Vickson (1975) or from Ed Thorp's blackjack writings. Benter had been a blackjack player and Thorp introduced Kelly betting there as *Fortune's formula* so that may be where he learned it. A key early Thorp paper is in Ziemba and Vickson.

The second type of model is to include the track odds as one of the variables to get even better probability estimates.

Benter pioneered the use of such models. I had a bit of a hand in there as the major paper on this was published while I was the *Management*

Science departmental editor for finance and I processed and accepted it for publication. That's the Bolton and Chapman (1986) paper which along with the only paper Benter published are reprinted in Hausch, Lo and Ziemba. Chapman (1994) using Hong Kong data is in our book.

I met Benter in 1993 at the INFORMS meeting in Phoenix where I organized the finance sessions and helped on the racing sessions. I recall correcting Benter's (1994) paper in Hausch, Lo and Ziemba, which had one good new development. As discussed above, in the Dr Z method, the biases to win and being second and third tend to cancel so in the work I did with Donald Hausch, we did not need to make any changes except say that because of approximations, bets should not be made to place or show unless the expected value was significantly above break even. We suggested 1.10 for the best races at the best tracks and 1.14 and 1.18 for lesser races. This worked well for US place and show betting. Benter and others found that the Dr Z system did not really work well in Hong Kong as the biases there were different. Also, he discovered how to correct the second, third, etc biases through the discounted Harville formulations that are discussed above.

Victor Lo did his PhD thesis in Hong Kong, directed by statistician John Bacon-Shone on this problem and much of his research is in Hausch, Lo, Ziemba along with papers by others on this. Bacon-Shone has a joint paper in Hausch and Ziemba (2008) with the late Alan Woods who had his own small team in the Philippines after he left Benter.

Benter's real contribution is shown in Figure 10. Namely, he made it all work and in the process became a very rich man with total profits in the one billion area. His paper in Hausch, Lo and Ziemba plus the other papers made our book a cult item with originals selling for $2000 up to $12,000 on EBay and Amazon. Originals are still trading at high prices, about $600. I sold one for $1400 to one of the copycat syndicates in Australia who I was consulting for. Another syndicate wanted to buy up all the Hausch, Lo and Ziemba books and burn them keeping one for their research. I decided to make a second edition which was published with a new preface in 2008 along with the sports and lottos handbook (Hausch and Ziemba, 2008).

The gains in Hong Kong by Benter's team and others were in a market without rebates and high commissions. But they utilized several advantages.

1. Hong Kong Chinese bettors favor and dislike certain numbers from their culture, which makes horses with these numbers differ from the true odds.
2. All the horses are the same ones: mostly Australian geldings running in almost all the races on just two racetracks so prediction was easier than in the US.

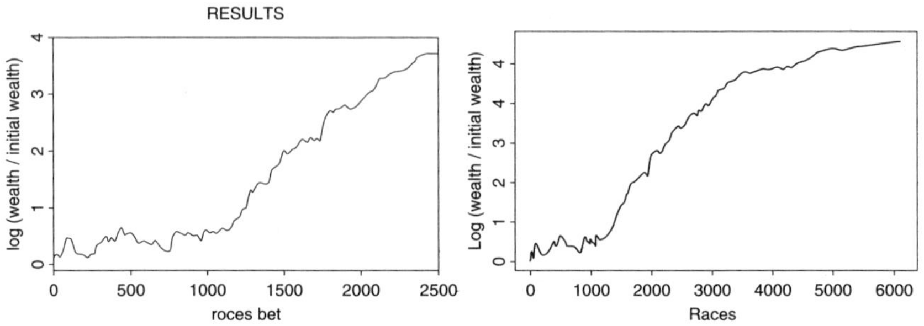

(a) to 1994 (b) to 2001

Figure 10 Benter's Hong Kong Racing Syndicate Returns.

3. Data feeds were every 12 seconds and later every minute giving access to pool odds which could be successfully used.
4. The market was deep with huge betting so the price impact was low and lastly, they could bet electronically into the pools.

Since the mainland takeover of Hong Kong in 1997, there have been some changes. But the syndicates continue and trade in many markets today in 2012 such as Japan and Korea as well as in the US, Canada and Europe. My personal experience consulting extensively for one other syndicate is that the setup cost for the research and computer implementation is a major time and financial undertaking. Like most markets it was easy earlier and much more difficult now. The syndicates with many workers and good experience have an edge on new ones, as I can report from our own activity on this.

13 Conclusion

Racetrack betting remains a very active set of markets. The basic betting problems are various versions of portfolio management. The problems are stochastic programs, usually one period but with non-concave objective functions because of the fractional functions inside the objective function. But the problems are easily solved and for many situations there are simplified strategies. The objective is usually the Kelly expected log criterion but in cases of low probability high payoff bets there can be hundreds or thousands of separate tickets and the bets must be integers. So a tickets network tree approach is useful and the Kelly strategy to bet more on the higher probability outcomes can be approximated.

The racetrack market is small compared to the financial markets such as currency and stock markets but there's enough there for a number of syndicates in the US, Australia and Hong Kong and elsewhere to make fifty to a hundred million dollars per year. It is not an easy market to enter at a high level as the setup costs are high and the competition fierce. Take a grass race at a mile with seven horses: one has not run in a year but did well then; one has only run on dirt; one has never run past 6 furlongs (three quarters of a mile); one was racing in France long distances $1\frac{1}{2}$ miles plus on grass losing consistently and the others have run similar distances on grass but not on this racetrack. Add in jockey and trainer changes and you see why "it is a supreme intellectual challenge" as argued by Andy Beyer, a noted racetrack writer. The models try to bypass this with probabilities and optimization.

References

Assamoi, K. V. Optimal investment strategies with Kelly capital growth criterion. Thesis: MSc in Mathematical Finance, Christ Church, University of Oxford, 2010.

Bacon-Shone, J. and A. Woods. Modeling money bet on horse races in Hong Kong. In D.B. Hausch and W.T. Ziemba (eds.), *Handbook of Sports and Lottery Markets*, pp. 17–25. North Holland, 2008.

Benter, W. Computer based horse race handicapping. In D. B. Hausch, V. Lo, and W. T. Ziemba (eds.), *Efficiency of Racetrack Betting Markets*, pp. 173–182. Academic Press, 1994.

Bolton, R. N. and R. G. Chapman. Searching for positive returns at the track: A multinomial logit for handicapping horse races. *Management Science 32*, 1040–1059, 1986.

Chapman, R. G. Still searching for positive returns at the track: empirical results from 2000 Hong Kong races. In D. B. Hausch, V. Lo, and W. T. Ziemba (eds.), *Efficiency of Racetrack Betting Markets*, pp. 173–181. Academic Press, 1994.

Chopra, V. K. and W. T. Ziemba. The effect of errors in mean, variance and co-variance estimates on optimal portfolio choice. *Journal of Portfolio Management, 19* : pp. 6–11, 1993.

Drud, A. S., CONOPT — A Large-Scale GRG code. *INFORMS Journal on Computing, 6*: pp. 207–216, 1994.

Gramm, M. and W. T. Ziemba. The dosage breeding theory for horse racing predictions. In D. B. Hausch and W. T. Ziemba (eds.), *Handbook of Sports and Lottery Markets*, pp. 307–340. North Holland, 2008.

Gramm, M. and W. T. Ziemba. Update on the dosage theory of horse race predictions for the triple crown races, mimeo, 2012.

Harville, D. A. Assigning Probabilities to the Outcomes of Multi-Entry Competitions. *Journal of the American Statistical Association, 68*: pp. 312–316, 1973.

Hausch, D. B., R. Bain, and W. T. Ziemba. An application of expert information to win betting on the Kentucky Derby, 1981–2001. *European Journal of Finance, 12*(4): pp. 283–302, 2006.

Hausch, D. B., V. Lo, and W.T. Ziemba (eds.). *Efficiency of Racetrack Betting Markets.* Academic Press, 1994.

Hausch, D. B., V. Lo, and W. T. Ziemba (eds.). *Efficiency of Racetrack Betting Markets* (2nd edition). Singapore: World Scientific, 2008.

Hausch, D. B. and W. T. Ziemba. Transaction costs, extent of inefficiencies, entries and multiple wagers in a racetrack betting model. *Management Science, 31*: pp. 381–394, 1985.

Hausch, D. B. and W. T. Ziemba (eds.). *Handbook of Sports and Lottery Markets.* North Holland, 2008.

Hausch, D. B., W. T. Ziemba, and M. E. Rubinstein. Efficiency of the market for racetrack betting. *Management Science* XXVII, pp. 1435–1452, 1981.

Kahneman, D. and A. Tversky. Prospect theory: an analysis of decisions under risk. *Econometrica, 47* (2): pp. 263–92, 1979.

Lane, D. and W. T. Ziemba. Arbitrage and risk arbitrage in team jai alai. In D. Hausch and W. T. Ziemba (eds.), *Handbook of Sports and Lottery Markets,* pp. 253–271. North Holland, 2008.

MacLean, L. C., E. O. Thorp, and W. T. Ziemba. *The Kelly Capital Growth Investment Criterion.* Singapore: World Scientific, 2011.

Murlagh, B. A. and M. A. Saunders. MINOS 5.5 User's Guide. Technical Report SOL 83-20R. Stanford University. Revised 1998.

Snowberg, E. and J. Wolfers. Examining explanations of a market anomaly: preferences or perceptions? In D.B. Hausch and W. T. Ziemba (eds.), *Handbook of Sports and Lottery Markets*, pp. 103–136. North Holland, 2008.

Thaler, R. H. and W. T. Ziemba. Anomalies: parimutuel betting markets: racetracks and lotteries. *Journal of Economic Perspectives, 2* : pp. 161–174, 1988.

Tompkins, R., W. T. Ziemba, and S. Hodges. The favorite-longshot bias in the S&P500 and FTSE100 index futures options: the return to bets and the cost of insurance. In D. B. Hausch and W. T. Ziemba (eds.), *Handbook of Sports and Lottery Markets*, pp. 161–180. North Holland, 2008.

Ziemba, W. T. Efficiency of racetrack, sports and lottery betting markets? In D. B. Hausch and W. T. Ziemba (eds.), *Handbook of Sports and Lottery Markets,* pp. 183–222. North Holland, 2008.

Ziemba, W. T. Place and show. *Wilmott Magazine* (November), pp. 32–42, 2011.

Ziemba, W. T. *Exotic Betting at the Racetrack.* In progress, 2012.

Ziemba, W. T. and D. B. Hausch. *Beat the Racetrack.* Harcourt, Brace and Jovanovich, 1984.

Ziemba, W. T. and D. B. Hausch. *Betting at the Racetrack.* Dr Z Investments, San Luis Obispo, CA, 1986.

Ziemba, W. T. and D. B. Hausch. *Dr Z's Beat the Racetrack.* William Morrow, 1987.

Ziemba, W. T. and R. G. Vickson (eds.). *Stochastic Optimization Models in Finance.* New York: Academic Press, 1975.

Ziemba, W. T. and R. G. Vickson (eds.). *Stochastic Optimization Models in Finance* (2nd edition). Singapore: World Scientific, 2008.

PART II

Papers in Production Planning and Logistics

Chapter 10

Multi-Stage Stochastic Programming for Natural Gas Infrastructure Design with a Production Perspective

Lars Hellemo*, Kjetil Midthun†, Asgeir Tomasgard** and Adrian Werner‡

Summary

We present a multi-stage stochastic model that analyzes investments in natural gas fields and infrastructure. New projects are evaluated together with existing infrastructure and planned expansions. Several uncertain factors both upstream and downstream such as reservoir volumes, the composition of the gas in new reservoirs, market demand and price levels can influence the optimal decisions. The model focuses also on the impact of the sequencing of field developments and new infrastructure on the expected security of supply. In order to analyze all these aspects in one model, we propose a novel approach to scenario trees, combining long-term and short-term uncertainty. Dimensionality and solution times of realistic investment cases from the Norwegian Continental Shelf are discussed.

1 Introduction

The pipeline transport system on the Norwegian Continental Shelf (NCS) is the world's largest subsea gas transport system with 7800 km of pipelines. When new projects such as platforms, pipelines, compressors or processing

*Department for Industrial Economics and Technology Management, Norwegian University of Science and Technology, Alfred Getz veg 3, NO–7491 Trondheim, Norway, lars.hellemo@iot.ntnu.no.
†Department for Applied Economics and Operations Research, SINTEF Technology and Society, NO–7465 Trondheim, Norway, kjetil.midthun@sintef.no.
**Department for Industrial Economics and Technology Management, Norwegian University of Science and Technology, Alfred Getz veg 3, NO–7491 Trondheim, Norway, asgeir.tomasgard@iot.ntnu.no, corresponding author.
‡Department for Applied Economics and Operations Research, SINTEF Technology and Society, NO–7465 Trondheim, Norway, adriantobias.werner@sintef.no.

facilities are considered, they should work well with existing and future infrastructure rather than be evaluated isolated from the total system. Current models for natural gas infrastructure development are often deterministic, where the distribution of uncertain parameters is replaced by mean values. There is, however, uncertainty in demand, prices, unplanned events, and gas quality and volumes from fields yet to be developed. Such uncertain parameters can be categorized with respect to the time horizon over which they vary. Some of the parameters vary from day to day, such as prices and demand in the markets, nominations in the long-term contracts, and unplanned events in the network. These variations will affect both the profitability of the operations in the natural gas value chain and the availability of network resources. Unplanned events can drastically reduce capacity in parts of the network and therefore cause large problems for the security of supply in the system. Examples of such unplanned events are production stops on fields, compressor failures, or reduced capacities of processing plants. By considering only average values for these parameters, important details may be lost through aggregation. For example, bottlenecks on the operational level during peak demand can be completely disguised when using only average demand values. On the long-term horizon there are also several uncertain parameters such as gas volumes in undeveloped reservoirs, the quality of the gas in such reservoirs, discoveries of new fields as well as long-term changes in price and demand levels.

The pipeline network exhibits system effects in the sense that the pressure and flow in one part of the network may influence the capacity in other parts of the network. To find an efficient routing of gas from production fields to markets we have to take into account relationships between pressure and flow and blending of gas with different quality. Consequently, these considerations are important both for evaluating how the proposed solutions work with existing infrastructure and for choosing solutions that are flexible enough to be reused for subsequent developments.

The infrastructure investment model we present here is part of the Ramona project[1] on production assurance and security of supply. We have developed tools for infrastructure design that value production assurance in addition to traditional profit or cost objectives. With the development of new gas fields with low quality gas, this task has become more challenging as the quality issues require more processing capacity or blending with good quality gas to reach target specifications on gas quality in addition to

[1]The Research Council of Norway, project number 175967.

volume requirements. With a sufficient number of scenarios we are able to use risk measures such as Conditional Value at Risk (CVaR) to ensure a high security of supply in the system.

Our main contribution is the development of a decision support system for infrastructure development that is very flexible with respect to the level of detail describing the physical processes in the natural gas value chain. The system is based on a multi-stage stochastic mixed-integer linear programming (MIP) model. This model can incorporate aspects such as pressure–flow relationship, multi-commodity flow, line pack, processing plants, and storages. At the same time we present a novel approach to analyzing both short-term and long-term uncertainty in the same model. We introduce a new scenario tree structure that combines strategic periods with operational periods, where the operational periods are representative for a given period. The structure allows a large number of such operational periods without causing more than a linear increase in the size of the scenario tree. We also present results from a realistic investment case on the NCS.

Paper structure We present an overview of the relevant literature in the following section. Uncertainty that needs to be incorporated in our stochastic MIP model is discussed in Section 3. In Section 4, we outline the model structure in general, while Sections 5 and 6 describe the main elements of the mathematical model formulation. Finally, we present a case study with preliminary computational results in Section 7. Section 8 concludes the paper. The notation will be introduced when needed in the mathematical formulation, while we refer to Appendix A for a full overview.

2 Literature review

In the literature, we find several examples of models that analyze different aspects of offshore petroleum investments. This is not surprising given the large risks and costs associated with such projects.

Strategic models There exist a number of deterministic investment models, and early overviews can be found in Sullivan (1988) and Haugland *et al.* (1988). Nygreen *et al.* (1998) present a multi-period MIP model used by the Norwegian Petroleum Directorate. The model is employed for the investment planning of fields in the North Sea which contain a mixture of oil and gas. In van den Heever and Grossmann (2001), a model for design and planning of offshore field infrastructure projects is presented. The model is a multi-period mixed-integer nonlinear programming model

(MINLP) and incorporates complex fiscal rules such as tariff, tax and royalty calculations. The net present value of projects is discussed in the light of these fiscal rules. There are also some models which incorporate uncertainty. Jørnsten (1992) presents an integer model for sequencing offshore oil and gas fields, where the objective is to maximize total economic benefit. Haugen (1996) develops a stochastic dynamic programming model to analyze a supplier's problem of scheduling fields and pipelines in order to be able to meet contractual agreements. The uncertainty in this model is in the resources (production profiles). In Jonsbråten (1998), a stochastic MIP model for optimal development of an oil field is presented. The objective of the model is to maximize the expected net present value of the oil field given uncertain future oil prices. Goel and Grossmann (2004) present a stochastic MIP model for the planning of offshore gas field developments. The expected net present value is maximized under uncertainty in reserves. A multi-stage investment model with decision-dependent uncertainty is presented by Tarhan *et al.* (2009) along with a branch-and-bound solution method for non-convex mixed-integer nonlinear sub-problems. They take into account nonlinear reservoir models and gradually revealed uncertainties concerning initial flow rate, recoverable volume and other characteristics of the reservoirs.

Operational models We use a value chain approach where we consider elements in the natural gas value chain from production field to market and optimize the resulting system. This approach has become even more valuable and important after the liberalization process, which meant an increase in flexibility for the participants in the value chain. In Ulstein *et al.* (2007), planning of offshore petroleum production is studied on a tactical level. The model has a value chain approach where production plans, network routing, processing of natural gas and sales in the markets are considered. In addition, multi-commodity flows and quality restrictions in the markets are taken into account. Pressure constraints in the network are, however, not included. The nonlinear splitting for chemical processing is linearized with binary variables. The resulting model is a mixed-integer programming model. Selot *et al.* (2008) presents an operational model for production and routing planning in the natural gas value chain. The authors combine a detailed infrastructure model with a complex contractual model but do not include a market for natural gas. The infrastructure model comprises nonlinear equations relating pressure and flow in wells and pipelines, multi-commodity flows and contractual agreements (delivery pressure and quality of the gas). The contractual model is based on a set of logical conditions for production sharing and customer requirements. The combined model is a mixed-integer nonlinear programming model. Li *et al.* (2011) present a global optimization approach to a stochastic pooling problem. They solve the integrated design and operations problem of industrial networks such as natural gas networks where the quality is stochastic.

The transportation of natural gas is one of the key elements when studying the natural gas industry and is paramount for analyzing the design of the network. Tomasgard *et al.* (2007) and Rømo *et al.* (2009) present a linearization of the Weymouth equation which enables the analysis of large networks and stochastic problems. There is a large number of publications with a technical approach to gas transportation. The models are detailed and accurate in their description of the physics of gas transportation, such as transient flow and interaction with compressors. A discussion of transient flows is given in Kelling *et al.* (2000), while the homepage of the Pipeline Simulation Interest Group (www.psig.org) gives a comprehensive overview on modeling, simulation and optimization of natural gas flows. Midthun *et al.* (2009) shows that it is difficult, if not impossible, to determine appropriate static capacities in a natural gas network. The system effects are discussed and a framework for economic analysis in natural gas networks is provided. Problems related to transport booking are studied by, e.g., Kalashnikov *et al.* (2010) who model the transport of gas by a transport company and a pipeline operating company as a stochastic bi-level problem.

Strategic models with operational variability In order to ensure that short-term fluctuations and peak demand situations can be accounted for in the system design, we include a representation of operational flexibility in a strategic investment model. Not much work can be found which focuses on both aspects at the same time. Schütz *et al.* (2009) include short-term variations in a strategic model for the Norwegian meat industry. De Jonghe *et al.* (2011) use an equilibrium model to study generation expansion. They integrate the short-term demand response in their strategic model and discuss the effects on flexibility of the generation capacity. The approach, however, considers only a one-period static model. Sönmez *et al.* (2011) analyze technology choice in LNG transport and discuss the impact of using a stochastic model for LNG throughput. They show that operational flexibility is important in order to cope with short-term variations and that is has a significant impact on profitability.

3 Uncertainty

3.1 *Uncertainty in the context of our model*

The investment problem we address contains uncertainty both on the long-term and the short-term horizons. In the *long term*, there is considerable uncertainty regarding future energy prices and demand. It is driven by the uncertainty about the future mix of energy sources, technology shifts, and imports from other geographical regions. Other important aspects are the gas composition and reservoir levels of new fields yet to start production. The uncertainty in gas composition poses challenges related to keeping the delivered gas quality within specifications, while the volume uncertainty will influence optimal pipeline designs. Also the gas quality specifications may not be perfectly known, for example, the maximum allowed content of contaminants such as CO_2 might be lowered in the future.

In the *short term* there is uncertainty in the prices in each market. These prices have a seasonal profile, in addition to high variability and large spikes. Also the demand in the long-term contracts is uncertain in the short term as the customers have the flexibility to nominate volumes within certain levels. In addition, parts of the production or transport capacity may be reduced during an unforeseen *event*. Such events can last much shorter than an operational period but may block network elements completely. We model this by stating how much of the network element's capacity is available over the operational period.

It is important to deal with the uncertainty in an appropriate way to ensure a robust and flexible infrastructure design allowing profitable operations under various, also adverse, conditions. For the investment analysis we need an optimization horizon of several years, typically between 20 and 50. The operational analysis, on the other hand, requires a much finer time resolution, such as days or hours. Combining the long time horizon, the fine time resolution and the uncertainty in both the long and the short term into a common paradigm, the size of a traditional scenario tree will explode. It is, therefore, worthwhile to investigate alternative approaches. One way of handling the effects of the short-term uncertainty is to iteratively run the investment model and a short-term model that incorporates the operational aspects (see, for instance, Myklebust (2010)). This approach poses, however, large challenges with respect to coordination and convergence of the solution. We propose a different approach that allows us to analyze long time horizons consisting of several years while still taking into account the effects of operational decisions and uncertainty in the same model. The following section describes our approach in more detail.

3.2 *The multi-horizon scenario tree*

We have developed a new scenario tree structure to represent both short-term and long-term uncertainty in a unified framework. To the best of our knowledge, such a framework has not been presented before. The approach includes key aspects of operational variability and uncertainty while keeping the size of the scenario tree manageable. Our approach is based on multi-horizon scenario trees, where the long-term and short-term horizons are handled differently but in a unified framework. The strategic nodes span long time periods such as years, while the operational nodes typically represent days. All decisions related to investments and network design are made in the strategic nodes, while all decisions related to operating the network are made in the operational nodes.

The operational nodes represent sub-trees in the scenario tree. In these sub-trees, the leaf nodes are always operational nodes while the root node will be an investment node. At each investment node we can then add several sub-trees. This way we can test the infrastructure in the connected investment node on an operational level. This connection is illustrated in Figure 1, where the strategic node is given as a square while the operational nodes are oval. The figure also shows that we distinguish between two different types of operational scenarios. The first type is included in the objective function of our model and provides us with the costs and revenues from operating the network. The benefits from adding these nodes are the greater detail we can add to the calculation of the profits with a finer time resolution as well as the possibility of adding several scenarios (using the expected value in the objective function). In addition, we have a second type of operational nodes (on the far right in Figure 1) that are used to find the security of supply (or production assurance) in the network. These operational nodes represent extreme scenarios that include events with negative impact on operations that may occur with a small probability as well as scenarios representing the normal operations with an associated

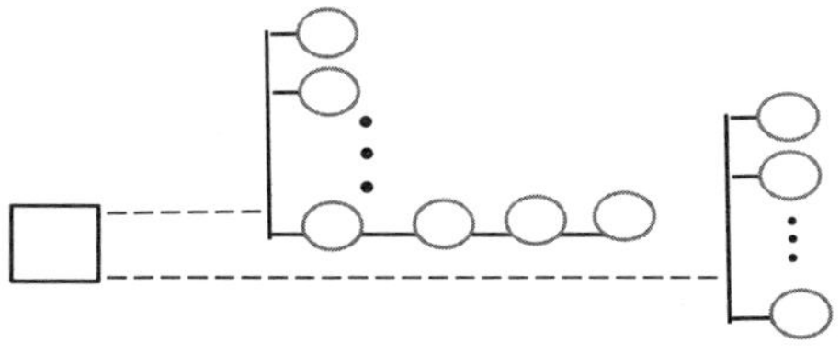

Figure 1 Coupling of Strategic Nodes (Square) and the Two Different Groups of Operational Subtrees (Ovals).

high probability. These scenarios are included to enforce the satisfaction of production assurance targets given the occurrence of such events. The results from these extreme scenarios do not contribute to the objective function in terms of revenue or operational expenses. Rather, production assurance targets may be enforced through hard constraints or violations penalized in the objective function. This way it is possible to influence the infrastructure solutions proposed by the model to be more robust against failures, also in situations that may not directly reduce overall profitability by much due to the short duration of the events.

The branching in our scenario tree is performed on the strategic nodes only, as illustrated in Figure 2. That is, the only connection between the operational nodes and the next strategic node is through the expected profits from operations (in addition to the expected production which will change the reservoir levels). This way, we can incorporate operational details and short-term uncertainty allowing us to thoroughly test design decisions, and at the same time use traditional scenarios expressing the long-term uncertainty, thus analyzing long time horizons without an explosion in problem size. This corresponds to a contingent scenario analysis

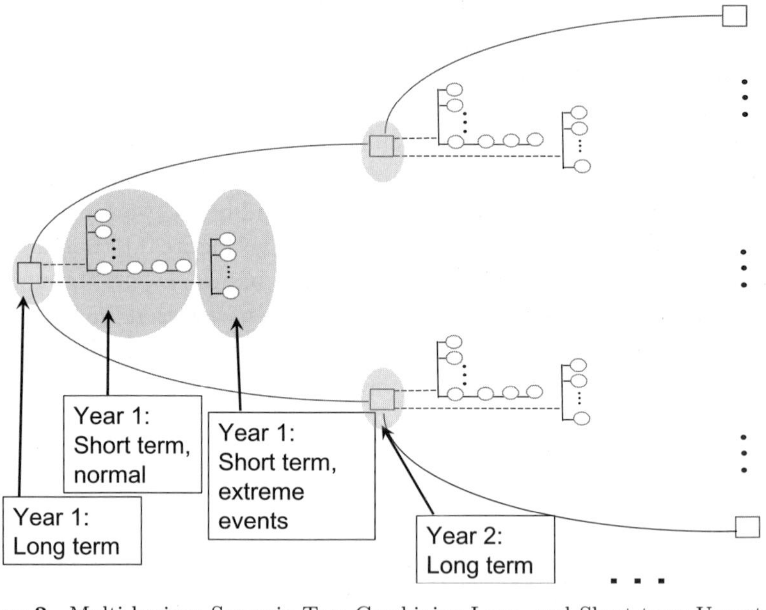

Figure 2 Multi-horizon Scenario Tree Combining Long- and Short-term Uncertainty for Investment and Operational Decision Nodes. The First Group of Operational Time Periods Denotes the Representative Normal Days, While the Second Group Represents Events.

for the operational problem for each strategic node and is, in general, a relaxation from the real information structure which would include branching in the main tree also at the operational level. Such a relaxation allows us to take into account the operational variability to some degree without suffering from the immense growth of the scenario tree needed for a model with frequent branching in a short-term horizon included in a strategic model. Observe that, if both the long-term uncertainty is independent of previous short-term realizations and the strategic decisions are independent of previous short-term decisions, the scenario tree structure actually reflects the exact information structure.

We denote the strategic scenarios of the tree by $s \in \mathcal{S}$ and the operational scenarios by $o \in \mathcal{S}^{Op}$ with the corresponding root node $R(o) \in \mathcal{S}$. The set $\mathcal{S}_{ts}^{OpSub}$ denotes all operational scenarios in the sub-tree associated with a strategic node in time period $t \in \mathcal{T}$ and scenario $s \in \mathcal{S}$. The operational (leaf) nodes $o \in \mathcal{S}^{Op}$ have no descendant nodes in $\mathcal{S}$, but all have a root node $R(o) \in \mathcal{S}$. We sometimes refer to o as the operational scenario contingent on $R(o)$. To be able to explore effects of time bindings (for example, when studying line pack or storages), we may consider groups $g \in \{1, \ldots, G\}$ consisting of several consecutive operational nodes. As discussed earlier in this section, we can distinguish two disjoint subsets: The set $\mathcal{G}_t^{norm} \subseteq \{1, \ldots, G\}$ defines all such groups representing normal operations in strategic period t while the set $\mathcal{G}_t^{extreme} \subseteq \{1, \ldots, G\}$ denotes all groups of nodes representing extreme situations in that period. Note that such an aggregation into groups of potentially varying length requires a scaling factor γ_g to include operational terms into the objective function in the right order of magnitude.

Note that, although not illustrated above, the approach can easily accommodate also more complex tree structures such as multistage operational sub-trees, branching at operational nodes within sub-trees. This is conceivable in particular for the sub-trees representing normal operating conditions.

4 Model structure

4.1 *Time structure, investment and operational decisions*

We assume that investment decisions are made at the beginning of an investment period $t \in \mathcal{T}$ while operational decisions are made at the operational time periods $d \in \mathcal{D}_t$ associated with the investment period t. Obviously, the sets of investment and operational periods are disjoint. In the remainder of this paper, we denote investment time periods often as "years" and operational time periods as "days". This is done purely for reasons of notational clarity. However, an investment time period may also cover, for example, several calendar years and varying time granularity can easily be taken into account.

4.2 *Modularity*

We employ a two-layered model design: A fully working *core model* comprises basic formulations of all necessary aspects while a number of separate *extension modules* are designed to improve functionality and/or level of detail and to allow for more sophisticated modeling of selected areas.

This modular approach enhances the flexibility and generality of the mathematical model because it allows conducting various analyses with different focus without compromising the underlying model and its implementation. It enables also a stepwise extension of the model as new functionality can be added as required. The level of detail can be chosen in a trade off between richness in modeling and computational complexity.

The flexibility in modeling will in some cases lead to a less tight formulation than what might be possible for a more specialized model, and we rely on the effectiveness of the presolve step of available commercial MILP solvers.

Core model

In the *core model*, the objective is to maximize net present value including cash flow from operations and penalties for insufficient production assurance. Main constraints included in the core model are the following:

- Investment decision modeling
- Demand satisfaction in the market nodes
- Mass balances in all network nodes
- Flow capacity limits in pipelines
- Production capacity limits
- Reservoir modeling

The main decisions concern the single project investments (network elements and alternatives: nodes and pipelines) and the flow through each of the elements invested in.

Extension modules

At present, the model contains the following modules:

Production assurance modeling. We penalize deviations from production assurance requirements in the objective function. The extension adds two sets of constraints. The first set ensures that the probability of days with non-satisfactory production assurance is not too large. The second set limits the expected value of such underperformance by applying risk measures such as Conditional Value at Risk (Rockafellar and Uryasev (2000)).

Pressure dynamics. This module allows to model pipelines with pressure dynamics. It expands on the core functionality of maximum flow constraints. We include upper and lower limits on pressures in each node and in pipelines, relationships between pressures into and out of a node as well as in ingoing and outgoing pipelines, and contracted pressure in market (customer) nodes. These features are described in Subsection 6.1.

Reservoir modeling. The core model uses a simple tank implementation for each reservoir where a certain maximum volume is given which is exhausted when the accumulated production has reached this level. This is replaced in the more advanced module by an approach similar to a production profile varying over time. The single production volumes are limited with respect to how much has been produced so far. Details on this module are given in Subsection 6.2.

Multi-component flow modeling. The extension module introduces variables and constraints for modeling heterogeneous flows, see Subsection 6.3. In most cases, this module will considerably increase the computational complexity.

Relations between projects. This module introduces constraints which specify that, from a given group of projects, at most one alternative (or a certain minimum/maximum number) can be started. Constraints may ensure the startup or shutdown of a certain number of projects from one group before the startup/shutdown of projects from another group or vice versa. Another relation controls the replacement of one (system of) project(s) by another. Mutually exclusive (systems of) projects may also be specified.

Processing nodes. The module adds the option to extract some components at dedicated processing plants in the network and to sell them in separate markets.

As the needs of the analysts develop, additional modules may be included, for example modules for line pack, storage or bidirectional pipelines.

5 Core model

5.1 *Objective function*

The main objective in the model is to maximize the expected net present value NPV, arising from infrastructure investments and their operation, taken over the values in the single long-term scenarios $s \in \mathcal{S}$. The parameter $Prob_{ts}$ denotes the absolute probability of strategic tree node at time period $t \in \mathcal{T}$ in scenario $s \in \mathcal{S}$. Included in the revenue, Rev_{ts}, are both potential salvage value from obsolete infrastructure and income from gas sales while costs $Cost_{ts}$ comprise both investment costs and operational expenses at each time period $t \in \mathcal{T}$.

$$NPV = \sum_{s \in \mathcal{S}} \sum_{t \in \mathcal{T}} Prob_{ts} \delta_t (Rev_{ts} - Cost_{ts}) \tag{1}$$

The single objective function terms are discussed in more detail in Subsection 5.4 below.

Cash flows are discounted in each investment period $t \in \mathcal{T}$ at a discount rate σ, using the discount factor δ_t,

$$\delta_t = \frac{1}{(1+\sigma)^t}$$

5.2 *Investment constraints*

All investment decisions are characterized by certain common properties such as startup, shutdown, capacities (possibly varying over time), and costs. We model these properties by means of a common class, denoted as *projects*, describing the common structure of all network elements such as fields, pipelines, or nodes. For the considered gas transport network, projects comprise pipelines, production nodes (fields), transport nodes (junctions, processing plants, etc.) and consumption nodes (markets). Projects can be either existing or new projects. Summarizing, a project $p \in \mathcal{P}$ is any option to invest in a network element. The *binary* variables $start_{pts}$, $stop_{pts}$, and prd_{pts} describe decisions about start, stop, and production, respectively, of the project under scenario s at time t:

The sets $\mathcal{I}(p)$ and $\mathcal{O}(p)$ denote the projects which are located directly upstream and downstream of a project p. For network nodes p, these sets

$\mathcal{I}(p), \mathcal{O}(p)$ mean the pipelines leading to and from p. For pipelines p, they denote the input and outlet nodes.

The startup date of *existing* projects $p \in \mathcal{P}^{ex}$ is set to the start of the optimization horizon, $start_{p1s} = 1$. Each *new* project $p \in \mathcal{P}^{New}$ can start up at most once during the optimization horizon and has a certain time window $Start_p^e, \ldots, Start_p^l \in \mathcal{T}$ for startup,

$$\sum_{t=Start_p^e \ldots Start_p^l} start_{pts} \leq 1$$

Each project can be shut down at most once during the optimization horizon,

$$\sum_{t=Start_p^e}^{T} stop_{pts} \leq 1$$

Production can only take place if a project has started up and has not been shut down yet. If there is a delay Δ_p between the time when deciding to start a *new* project $p \in \mathcal{P}^{New}$ and the time it is available for production, startup decisions after $t - \Delta_p$ do not need to be taken into account (but shut-down decisions should).

$$\forall t \in \{Start_p^e + \Delta_p \ldots T\} : \; prd_{pts} \leq \sum_{\tau=Start_p^e}^{t-\Delta_p} start_{p\tau s} - \sum_{\tau=Start_p^e}^{t} stop_{p\tau s}$$

To avoid an undue postponement of a project shutdown to reduce the net present value of shutdown costs, the following inequality enforces the immediate stop of a project after production halts.

$$\forall t \in \{Start_p^e, \ldots, T\} : \; prd_{pts} - prd_{pts} \leq stop_{pts}$$

The binary variable $stprd_{pt\tau s}$ indicates whether a project p which started at $\tau \in \mathcal{T}$ is producing at $t \in \mathcal{T}$.

$$prd_{pts} = \sum_{\substack{\tau \in \mathcal{T} \\ \tau \leq t-\Delta_p}} stprd_{pt\tau s}$$

$$\forall \tau \in \{Start_p^e, \ldots, t - \Delta_p\} : \; stprd_{pt\tau s} \leq start_{p\tau s}$$

Time series of parameters — profiles All projects can be described by parameters that change over time. For example, production capacity may be low at the beginning due to necessary adjustments, high during peak production and low again at the end when reservoirs are exhausted. We call these series of values "profiles" and distinguish two types: Floating profiles depend on the start date of a project and are relative to this start date. They are denoted by Π_{float}. An example is a capacity profile. Fixed profiles refer to calendar dates and are independent of project start. They are denoted by Π_{fix}. Examples are market demand or prices. See Appendix A.5 for a complete list of profiles used in the model.

5.3 *Operational constraints*

Mass balances We have modeled the flow *through* a network element in a unified way although it may have slightly different meanings for different network elements in real life. For example, the flow through production nodes $p \in \mathcal{P}^{Prod}$ is the total amount produced in that node in the considered period, the flow through consumption nodes $p \in \mathcal{P}^{Cons}$ is the total amount sold in that node in that period.

The following constraints ensure the mass balances in the network for each project p in each operational scenario $o \in \mathcal{S}_{ts}^{OpSub}$ and period $d \in \mathcal{D}_t$ in each strategic scenario $s \in \mathcal{S}$ and period $t \in \mathcal{T}$. For production nodes $p \in \mathcal{P}^{Prod}$, the amount produced must equal the outflow into the network (2a), the amount sold in a market node $p \in \mathcal{P}^{Cons}$ must equal the inflow from the network (2b) and the flow into a transport node $p \in \mathcal{P}^{Transp}$ must equal the flow out of it (2c). For pipelines $p \in \mathcal{P}^{Pipe}$, no explicit equations are necessary as they have exactly one direct upstream and downstream project. Note that the flow through a transport node $p \in \mathcal{P}^{Transp}$ is defined only implicitly through the mass balance (2c). It is specified in equation (2d) in order to state, e.g., the lower and upper flow bounds (3).

$$flow_{ptdo} = \sum_{j \in \mathcal{O}(p)} flow_{jtdo} \tag{2a}$$

$$flow_{ptdo} = \sum_{j \in \mathcal{I}(p)} flow_{jtdo} \tag{2b}$$

$$\sum_{i \in \mathcal{I}(p)} flow_{itdo} = \sum_{j \in \mathcal{O}(p)} flow_{jtdo} \tag{2c}$$

$$flow_{ptdo} = \sum_{j \in \mathcal{I}(p)} flow_{jtdo} = \sum_{j \in \mathcal{O}(p)} flow_{jtdo} \tag{2d}$$

Flow bounds Capacity bounds on the flow through a project depend on this project's age and are given through floating profiles. These profiles indicate the physically available capacity and are defined on a strategic level. For projects $p \in \mathcal{P}$ with startup decision made at τ, flow in the operational periods $d \in \mathcal{D}_t$ is bounded by the value of the floating profiles at a position relative to the project start, taking into account possible delays Δ_p. In addition, a stochastic parameter $CapPerc_{ptdo}$ denotes the percentage of that capacity which is actually available in the operational period $d \in \mathcal{D}_t$ under scenario $o \in \mathcal{S}^{Op}$. This allows us to model unforeseen events affecting network capacity, and the input parameters' consistency between reduced upper and lower bounds must be ensured.

$$flow_{ptdo} \leq CapPerc_{ptdo} \sum_{\substack{\tau \in \mathcal{T} \\ \tau \leq t}} \Pi^{MaxFlow}_{pt-\tau-\Delta_p+1} \cdot stprd_{pt\tau s} \tag{3a}$$

$$flow_{ptdo} \geq \sum_{\substack{\tau \in \mathcal{T} \\ \tau \leq t}} \Pi^{MinFlow}_{pt-\tau-\Delta_p+1} \cdot stprd_{pt\tau s} \tag{3b}$$

The amount sold at a consumption node $p \in \mathcal{P}^{Cons}$ cannot exceed the contracted volume in this node during that time period and scenario. In the presence of production assurance constraints, the sales volume bound Vol_{ptdo} is interpreted as contracted volume, in the absence of such constraints as possible sales in spot markets.

$$flow_{ptdo} \leq Vol_{ptdo}$$

Reservoir constraints In the core model, reservoir volumes are modeled through a simple tank implementation: the parameter $ResMax_p$ states an upper bound on the totally available volume in the reservoir determining the production capacity at field $p \in \mathcal{P}^{Prod}$ for operational scenarios $o \in \mathcal{S}^{Op}$.

$$\sum_{t \in \mathcal{T}} \sum_{g \in \mathcal{G}_t^{norm}} \gamma_g \sum_{d=d_g^0}^{d_g^L} flow_{ptdo} \leq ResMax_p \tag{4}$$

Additionally, there is a given yearly production rate on the field which should not be exceeded in any year $t \in \mathcal{T}$,

$$\sum_{g \in \mathcal{G}_t^{norm}} \gamma_g \sum_{d=d_g^0}^{d_g^L} flow_{ptdo} \leq ResProdRate_{pt}$$

5.4 *Objective function terms*

Investment costs

Costs Projects have costs that, like production characteristics, depend on the age of the project while other costs depend on calendar dates. It is therefore reasonable to operate with both floating and fixed profiles.

Costs incurred by investment decisions are startup, fixed investment and removal costs,

$$Cost_{ts}^{Inv} = \sum_{p \in \mathcal{P}} \left(Cost_{pts}^{invst} + Cost_{pts}^{invfix} + Cost_{pts}^{invrem} \right)$$

Startup costs may depend on the startup point of time; for example, it may be cheaper to start an extension project early rather than to wait until the additional capacity is needed.

$$Cost_{pts}^{invst} = \Pi_{pt}^{invst} \cdot start_{pts}$$

Fixed investment costs may occur at investment periods during the project's lifetime and depend on the age of the project,

$$Cost_{pts}^{invfix} = \sum_{\tau=1}^{t} \Pi_{pt-\tau+1}^{invfixC} \cdot stprd_{pt\tau s}$$

Removal costs are incurred by the stop of a project p and may depend on both the age of the project and the point of time, i.e. they have a fixed and a floating component.

$$Cost_{pt}^{invrem} = \Pi_{fix,pt}^{invremC} \cdot stop_{pts} + \sum_{\tau=1}^{t} \Pi_{float,pt-\tau+1}^{invremC} \cdot stop_{p\tau s}$$

In addition to costs associated with stopping we allow a positive salvage value where applicable.

Operational expenditure and revenue

Costs Total operational costs during investment period $t \in \mathcal{T}$ under scenario $s \in \mathcal{S}$ are composed of operational expenditure, fixed and variable costs arising for projects $p \in \mathcal{P}$ in all operational periods (days) $d \in \mathcal{D}_t$ in this investment period. These costs are representative costs, as only a few operational periods are considered, and they are scaled up to a yearly basis.

$$Cost_{ts}^{Op} = \sum_{p \in \mathcal{P}} \left(Cost_{pts}^{opex} + \sum_{o \in \mathcal{S}_{ts}^{OpSub}} \sum_{g \in \mathcal{G}_t^{norm}} \gamma_g \sum_{d=d_g^0}^{d_g^L} Prob_{do} \left(Cost_{ptdo}^{opfix} + Cost_{ptdo}^{opvar} \right) \right)$$

Operational expenditure (opex) is independent of the production volume or flow but can vary over the lifetime of the project and depends, hence, on the project's age. It arises only when the project is actually active (i.e. projects which have started up but are not producing yet do not generate opex).

$$Cost_{pts}^{opex} = \sum_{\tau=1}^{t} \Pi_{pt-\tau-\Delta_p+1}^{opex} \cdot stprd_{pt\tau s}$$

Fixed operational costs depend neither on the age of a project nor on the flow but on the operational date $d \in \mathcal{D}_t$ and the associated investment period t,

$$Cost_{ptdo}^{opfix} = \Pi_{ptd}^{opfix} \cdot prd_{ptR(o)}$$

Variable operational costs depend on the flow through that project and are independent of the project age; they may also include costs of production factors or similar,

$$Cost_{ptdo}^{opvar} = \Pi_{ptd}^{opvar} \cdot flow_{ptdo}$$

Revenue Revenue from operations comes from sales at market nodes at operational periods $d \in \mathcal{D}_t$ under scenarios $o \in \mathcal{S}_{ts}^{OpSub}$ illustrating normal operating conditions. This revenue is considered representative revenue, and the scaling factor γ_g is applied. For the core model we assume there is no sale of by-products at any node $p \in \mathcal{P}^{Prod} \cup \mathcal{P}^{Transp} \cup \mathcal{P}^{Cons}$. Sales price profiles at the single market nodes $p \in \mathcal{P}^{Cons}$ depend on the point of time and may be stochastic,

$$Rev_{ts}^{Op} = \sum_{o \in \mathcal{S}_{ts}^{OpSub}} \sum_{g \in \mathcal{G}_t^{norm}} \gamma_g \sum_{d=d_g^0}^{d_g^L} Prob_{do} \sum_{p \in \mathcal{P}^{Cons}} Price_{ptdo}^{Spot} \cdot flow_{ptdo}$$

Production assurance Production assurance is studied in terms of *deliverability*, referring to deliveries at market nodes in each operational scenario $o \in \mathcal{S}_{ts}^{OpSub}$ and period $d \in \mathcal{D}_t$ in strategic scenario $s \in \mathcal{S}$ and period $t \in \mathcal{T}$. It is defined as *(contracted volumes — deviations)/contracted volumes* where *deviations* means the difference between contracted and delivered volumes. The measure is aggregated over all markets $p \in \mathcal{P}^{Cons}$,

$$Del_{tdo} = \frac{\sum_{p \in \mathcal{P}^{Cons}} flow_{ptdo}}{\sum_{p \in \mathcal{P}^{Cons}} Vol_{ptdo}}$$

For the core model, we achieve sufficient production assurance by penalizing negative deviations from the target production assurance $PATarget_t$. The actual production assurance is realized through the investment and all operational decisions in a given investment period $t \in \mathcal{T}$, scenario $s \in \mathcal{S}$, and group $g \in \{1, \ldots, G\}$,

$$Dev_{gs} = \max \left\{ 0, PATarget_t - \sum_{o \in \mathcal{S}_{ts}^{OpSub}} \sum_{d=d_g^0}^{d_g^L} Prob_{do} \frac{Del_{tdo}}{d_g^L - d_g^0 + 1} \right\}$$

We then include penalty terms

$$\delta_t Pen_t \sum_{g \in \{1,\ldots,G\}} Dev_{gs}$$

for time period $t \in \mathcal{T}$ and scenario $s \in \mathcal{S}$ into the objective function (1).

6 Selected extension modules

As discussed in Section 4.2, extension modules provide more sophisticated modeling than the core model and can be switched on as required for various analyses. In the following, we describe some relevant modules in more detail, in particular, advanced pressure modeling, reservoir modeling, and multi-component flow/quality modeling.

6.1 *Pressure dynamics*

For this module it is convenient to consider a set $\mathcal{P}^{Nodes}$ comprising all network elements other than pipelines. The current model does not include a detailed compressor model. Instead we allow pressure increase within capacity bounds for projects representing a compressor.

Consider a pipeline $p \in \mathcal{P}^{Pipe}$ connecting network nodes i and j which have the respective pressures $pres_{itd}$ and $pres_{jtd}$. Simple flow constraints as in the core model are used for short pipelines where the pressure drop is not important for the flow calculation,

$$pres_{ptdo}^{out} = pres_{ptdo}^{in}$$

For all other pipelines, a linearization of the Weymouth equation (Gas Processors Suppliers Association (1998)) gives an upper bound on the flow $flow_{ptdo}$ coming from $i \in \mathcal{I}(p)$ and going out through $j \in \mathcal{O}(p)$ at each linearization breakpoint $b \in \mathcal{B} = \{1, \ldots, B\}$.

$$flow_{ptdo} \leq K_p^W \frac{P_{ib}}{\sqrt{P_{ib}^2 - P_{jb}^2}} pres_{itdo} - K_p^W \frac{P_{jb}}{\sqrt{P_{ib}^2 - P_{jb}^2}} pres_{jtdo}$$

The parameter K_p^W denotes the pipeline specific Weymouth constant while P_{ib} and P_{jb} are approximations of the pressures at the in- and outlet points of pipeline p at the breakpoint b.

Additionally, the pressure $pres_{itdo}$ in the node i needs to be within some minimum and maximum bounds,

$$\underline{P_i} \leq pres_{itdo} \leq \overline{P_i}$$

Such bounds are defined, e.g., through design parameters of the network and contractual agreements and may also have a time index, $\underline{P_{it}}$, $\overline{P_{it}}$, $t \in \mathcal{T}$. See Tomasgard *et al.* (2007) for further details.

6.2 *Reservoir modeling*

While the basic reservoir modeling in constraint (4) only restricts the *total* or aggregated volume produced on a field, the advanced reservoir constraints set upper bounds on the production volumes based on how much has already been produced from that field. Observe that these constraints do not apply for groups $g \in \mathcal{G}_t^{extreme}$ of operational periods testing the viability of decisions during more extreme scenarios.

The production rate at each time period t should be at most a certain percentage of the volume still available in the reservoir. This describes a floating profile of yearly production rates. Over time there may be several such percentages, resulting in a complex profile structure. A procedure is employed to derive a set of M linear constraints for projects $p \in \mathcal{P}^{Prod}$,

$$\sum_{g \in \mathcal{G}_t^{norm}} \gamma_g \sum_{d=d_g^0}^{d_g^L} flow_{ptdo} \leq Prod_{pm}^{Abs} + Prod_{pm}^{Rel} \cdot flow_{pts}^A$$

These constraints describe the production profile as a convex set where the production capacity at any time t depends on the accumulated production flow^A_{pts} until t which is given by

$$\mathit{flow}^A_{pts} = \sum_{\tau=1}^{t} \sum_{g \in \mathcal{G}^{norm}_t} \gamma_g \sum_{o \in \mathcal{S}^{OpSub}_{ts}} Prob_{do} \sum_{d=d^0_g}^{d^L_g} \mathit{flow}_{ptdo}$$

The accumulated production must not exceed the total volume to be produced from the reservoir,

$$\mathit{flow}^A_{pts} \leq MaxProd_p$$

The absolute and relative coefficients $Prod^{Abs}_{pm}$ and $Prod^{Rel}_{pm}$ $(m = 1, \ldots, M)$ and the limits $MaxProd_p$ are found based on user specified yearly production rates $ProdRate_{pt}$.

6.3 *Quality/multi-component flow modeling*

The fractions of the component flows going out of a split node $p \in \mathcal{P}^{Transp}$ into the different downstream pipelines $p_1, p_2 \in \mathcal{O}(p)$ must be equal for all components c_1, $c_2 \in \mathcal{C}$ in the flow:

$$\frac{\mathit{flow}^C_{p_1 c_1}}{\mathit{flow}^C_{p_2 c_1}} = \frac{\mathit{flow}^C_{p_1 c_2}}{\mathit{flow}^C_{p_2 c_2}}$$

This results in a bilinear expression which is approximated through a linearization using pre-defined split options $\alpha_{p_1 h}$, $h = 1, \ldots, H$, to find the share of the flow through a pipeline p_1. These pre-defined split options can be found based on, for example, the split percentages from a single-component run of the model or a fixed set of split options (Tomasgard *et al.*, 2007).

The following constraints define the flow $\mathit{flow}^C_{p_1 ctdo}$ of component $c \in \mathcal{C}$ through this pipeline at day $d \in \mathcal{D}_t$, year $t \in \mathcal{T}$, operational scenario $o \in \mathcal{S}^{OpSub}_{ts}$:

$$\forall\, h \in \{1, \ldots, H\} : \; \mathit{flow}^S_{p_1 chtdo} \leq \alpha_{p_1 htdo} \cdot \mathit{flow}^C_{pctdo}$$

$$\mathit{flow}^C_{p_1 ctdo} = \sum_{h \in \{1,\ldots,H\}} \mathit{flow}^S_{p_1 chtdo}$$

$$\sum_{h \in \{1,\ldots,H\}} \mathit{flow}^S_{p_1 chtdo} \leq UB_{p_1 tdo} \cdot \lambda_{p_1 htdo}$$

$$\alpha_{p_1 htdo} \cdot \mathit{flow}^C_{pctdo} \leq \mathit{flow}^S_{p_1 ctdo} + UB_{p_1 tdo} \cdot (1 - \lambda_{p_1 htdo})$$

With an upper bound

$$UB_{p_1tdo} = CapPerc_{p_1tdo} \sum_{\substack{\tau \in \mathcal{T} \\ \tau \leq t}} \Pi^{MaxFlow}_{p_1 t-\tau-\Delta_{p_1}+1}$$

on the capacity of pipeline p_1, the auxiliary binary decision variables λ_{p_1htdo} for the choice of split options $h \in \{1, \dots, H\}$ and the continuous variable $flow^S_{p_1chtdo}$ denoting the component flow of component $c \in \mathcal{C}$ through p_1 if option h would be chosen. Note that, for each split node p, these constraints need to be defined only for all but one downstream pipelines $\mathcal{O}(p)$.

The volume of each component $c \in \mathcal{C}$ in the flow from a production node $p \in \mathcal{P}^{Prod}$ is given through

$$flow^C_{pctdo} = flow_{ptdo} \cdot CompProd_{pctdo}$$

$CompProd_{pctdo}$ defines the ratio of component c contained in the flow produced at p.

The aggregated flow through the other nodes is the sum of the component flows,

$$flow_{ptdo} = \sum_{c \in \mathcal{C}} flow^C_{pctdo}$$

Component-wise mass balances for all $c \in \mathcal{C}$ replace the aggregated mass balances (2a)–(2c):

$$\forall p \in \mathcal{P}^{Prod}: \quad flow^C_{pctdo} = \sum_{p_1 \in \mathcal{O}(p)} flow^C_{p_1ctdo}$$

$$\forall p \in \mathcal{P}^{Cons}: \quad flow^C_{pctdo} = \sum_{p_1 \in \mathcal{I}(p)} flow^C_{p_1ctdo}$$

$$\forall p \in \mathcal{P}^{Transp}: \quad \sum_{p_1 \in \mathcal{I}(p)} flow^C_{p_1ctdo} = \sum_{p_1 \in \mathcal{O}(p)} flow^C_{p_1ctdo}$$

These constraints connect the multi-component module to all other model functionality. Additional constraints may reflect quality requirements at market nodes.

7 Case study

We present a case study based on a realistic set of investment opportunities from the NCS. The actual investment projects are confidential, and we altered the names and the geographical locations of the network nodes. The data used in our analysis is partly real and partly synthetic. Most of the upstream data is real while the market data (prices and demand) is synthetic, but based on an analysis of historical data. We consider uncertainty both in volumes in the reservoirs for not yet developed fields and in the quality of the gas.

The analyzed network structure is shown in Figure 3. The case includes existing fields, markets and pipelines connecting them. Candidate new investment projects comprise branch-offs and compressors within the main transport network and the development of new fields and pipelines connecting them to the existing network. The existing infrastructure is shown in black, while candidate projects are shown in light gray color. Production fields are indicated by squares, consumption nodes by triangles and nodes within the network by circles. Arcs in the network indicate the pipelines in the system while arrows show the direction of flow.

As can be seen from the figure, the investment possibilities include relatively isolated new parts of the network as well as extensions and changes to the existing network. The system perspective taken in our analysis is very important when there are close links between the existing system and the new investment possibilities. Even though the links between the new parts of the system and the existing ones may seem weak on a first inspection, the consequences of ignoring them may be large due to network effects on pressure or quality. For example, projects 180 and 181 (just above the center of the figure) are parts of two alternative branch-offs that would make the existing link between projects 83 and 84 obsolete. The branch-off is located at different places in the two cases, resulting in different pressure parameters. Examples of new fields are projects 201 through 203, while projects 204, 227, and 205 represent examples of pipeline alternatives with small, medium and large capacity, respectively, and corresponding costs.

Our test instance is a two-stage stochastic program with 11 years time horizon. The stochastic parameters are price and demand at the market nodes as well as the volumes and quality available from new fields. There are 240 projects in total, of which 182 are already existing infrastructure, and 58 represent investment opportunities. The two-stage model with 27 scenarios generates a model instance with 27.390.199 rows, 372.832 columns and 38.082.317 nonzero variables. Gurobi 4.6.0 reduces this to 599.988 rows, 110.872 columns and 2.094.249 nonzero variables during the presolve step. This presolved model instance has 20.441 continuous and 90.431 binary variables. It is solved to optimality in approximately 3 days on a six-core AMD Opteron processor 2431 with 24Gb memory. The computer was running Linux 2.6.18 (Rocks 5.3).

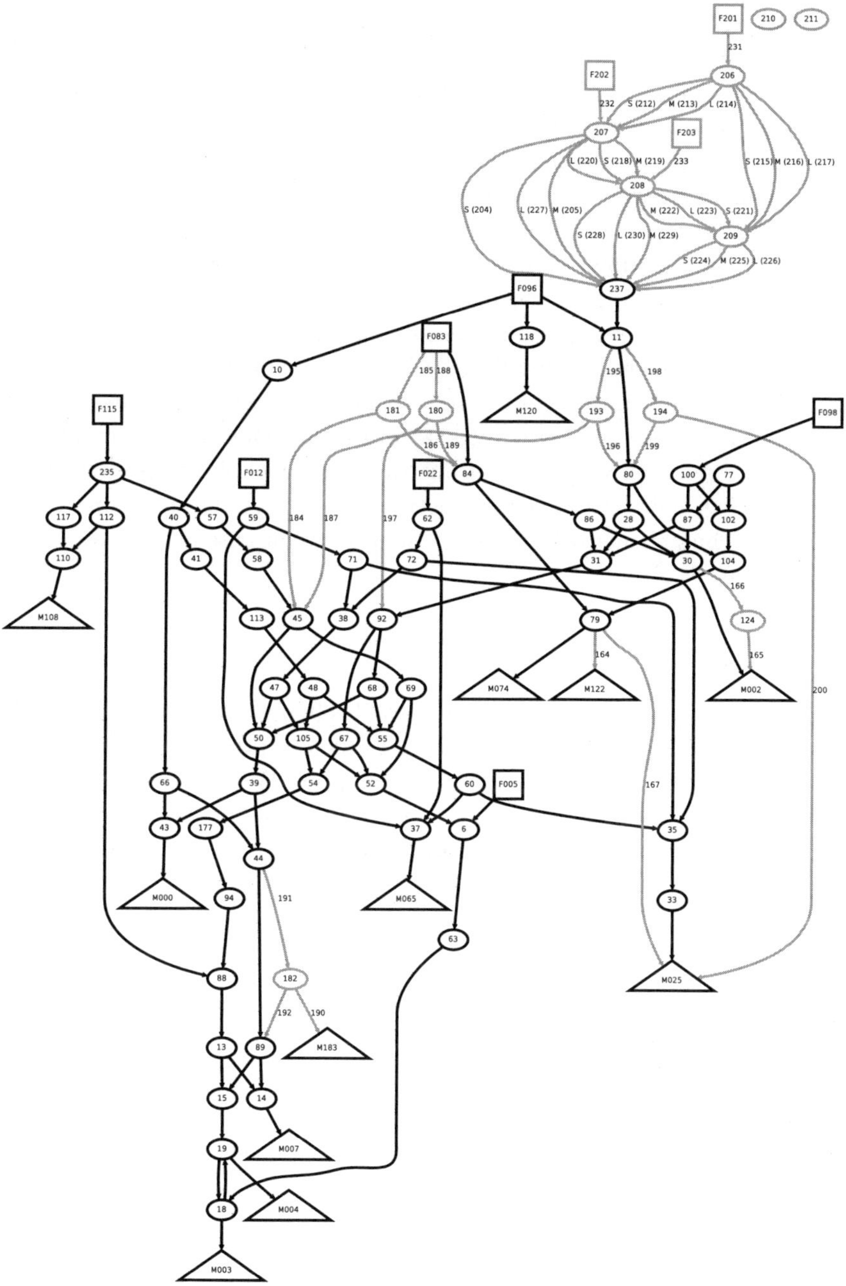

Figure 3 The Network Structure Considered in the Test Instance.

8 Conclusions

We have developed a multi-stage stochastic optimization model for analyzing natural gas infrastructure investment decisions in an integrated approach. New investment and shut-down options, including their proper timing, are studied in the light of the existing network structure, thus taking into account system effects as well as future production and development plans. Evaluating production assurance alongside common objectives such as profit maximization leads to a more robust and flexible infrastructure design. The inclusion of aspects such as pressure–flow relationships, multi-commodity flows or line-pack allows assessing the impact of the investment decisions on daily operations. We present a novel approach to analyze both long- and short-term uncertainty in a unified framework. For this purpose we introduce a new scenario tree structure, allowing a large number of operational periods without causing the typical explosion in size often observed with more traditional structures. We also discuss an application of the modeling framework to a realistic investment case on the Norwegian Continental Shelf.

Acknowledgments

The model presented in this paper has been developed as part of the Ramona project (The Research Council of Norway, project number 175967) on production assurance and security of supply. We thank our industrial partners Statoil and Gassco for many inspiring discussions and support as well as providing the initial layout of the case study.

References

De Jonghe, C., B. Hobbs, and R. Belmans. Integrating short-term demand response into long-term investment planning. Cambridge working papers in economics 1132, Faculty of Economics, University of Cambridge, 2011.

Goel, V. and I. Grossmann. A stochastic programming approach to planning of offshore gas field developments under uncertainty in reserves. *Computers and Chemical Engineering*, 28(8): pp. 1409–1429, 2004.

Gas Processors Suppliers Association. *Engineering Data Book*. Tulsa, 11th edition, 1998.

Haugen, K. A stochastic dynamic programming model for scheduling of offshore petroleum fields with resource uncertainty. *European Journal of Operational Research*, 88(1): pp. 88–100, 1996.

Haugland, D., Å. Hallefjord, and H. Asheim. Models for petroleum field exploitation. *European Journal of Operational Research*, 37: pp. 58–72, 1988.

Jonsbråten, T. Oil field optimization under price uncertainty. *Journal of the Operational Research Society*, 49(8): pp. 811–818, 1998.

Jørnsten, K. O. Sequencing offshore oil and gas fields under uncertainty. *European Journal of Operational Research*, 58: pp. 191–201, 1992.

Kalashnikov, V., G. Pérez-Valdés, A. Tomasgard, and N. Kalashnykova. Natural gas cash-out problem: Bilevel stochastic optimization approach. *European Journal of Operational Research*, 206(1): pp. 18–33, 2010.

Kelling, C., K. Reith, and E. Sekirnjak. A practical approach to transient optimization for gas networks. Technical report, Pipeline Simulation Interest Group (PSIG), 2000.

Li, X., A. Tomasgard, and P. Barton. Decomposition strategy for the stochastic pooling problem. *Journal of Global Optimization*, pp. 1–26, 2011.

Midthun, K. T., M. Bjørndal, and A. Tomasgard. Modeling optimal economic dispatch and system effects in natural gas networks. *Energy Journal*, 30: pp. 155–180, 2009.

Myklebust, J. *Techno-economic Modelling of Value Chains Based on Natural Gas — with Consideration of CO2 Emissions.* PhD thesis, Department of Industrial Economics and Technology Management, Norwegian University of Science and Technology, 2010.

Nygreen, B., M. Christiansen, K. Haugen, T. Bjørkvoll, and Ø. Kristiansen. Modelling Norwegian petroleum production and transportation. *Annals of Operations Research*, 82: pp. 251–267, 1998.

Rockafellar, R. and S. Uryasev. Optimization of Conditional Value-at-Risk. *Journal of Risk*, 2: pp. 21–42, 2000.

Rømo, F., A. Tomasgard, L. Hellemo, M. Fodstad, B. Eidesen, and B. Pedersen. Optimizing the Norwegian natural gas production and transport. *Interfaces*, 39(1): pp. 46–56, 2009.

Schütz, P., A. Tomasgard, and S. Ahmed. Supply chain design under uncertainty using sample average approximation and dual decomposition. *European Journal of Operational Research*, pp. 409–419, 2009.

Selot, A., L. Kuok, M. Robinson, T. Mason, and P. Barton. A short-term operational planning model for natural gas production systems. *AIChE Journal*, 54(2): pp. 495–515, 2008.

Sönmez, E., S. Kekre, A. Scheller-Wolf, and N. Secomandi. Strategic analysis of technology and capacity investments in the liquefied natural gas industry. Working paper, Carnegie Mellon Tepper School of Business, 2011.

Sullivan, J. The application of mathematical programming methods to oil and gas field development planning. *Mathematical Programming*, 42: pp. 189–200, 1988.

Tarhan, B., I. Grossmann, and V. Goel. Stochastic programming approach for the planning of offshore oil or gas field infrastructure under decision-dependent uncertainty. *Industrial & Engineering Chemistry Research*, 48(6): pp. 3078–3097, 2009.

Tomasgard, A., F. Rømo, M. Fodstad, and K. Midthun. Optimization models for the natural gas value chain. In G. Hasle, K.-A. Lie, and E. Quak, (eds.),

Geometric Modelling, Numerical Simulation and Optimization, pp. 521–558. Springer Verlag, 2007.

Ulstein, N. L., B. Nygreen, and J.R. Sagli. Tactical planning of offshore petroleum production. *European Journal of Operational Research*, 176(1): pp. 550–564, January 2007.

van den Heever, S. A. and I.E. Grossmann. A Lagrangean decomposition heuristic for the design and planning of offshore hydrocarbon field infrastructures with complex economic objectives. *Industrial & Engineering Chemistry Research*, 40: pp. 2857–2875, 2001.

A Notation

A.1 *Sets and indices*

Notation	Description
$b \in \mathcal{B} = \{1, \ldots, B\}$	break points for Weymouth linearization
$c \in \mathcal{C} = \{1, \ldots, C\}$	components in the flow
$d \in \mathcal{D}_t = \{1, \ldots, D_t\}$	operational periods associated with investment period t
d_g^0	first operational period in group g
d_g^L	last operational period in group g
$h \in \{1, \ldots, H\}$	split options for multi-component flow
$\mathcal{I}(p) \subset \mathcal{P}$	direct upstream ("input") projects of p
$g \in \{1, \ldots, G\}$	groups of consecutive operational periods
$\mathcal{G}_t^{norm} \subseteq \{1, \ldots, G\}$	sets of operational periods representing normal operation
$\mathcal{G}_t^{extreme} \subseteq \{1, \ldots, G\}$	sets of operational periods representing extreme situations
$n_0 \in \mathcal{S}$	root node in the scenario tree
$\mathcal{O}(p) \subset \mathcal{P}$	direct downstream ("output") projects of p
$i,\ j,\ p \in \mathcal{P}$	all projects
$\mathcal{P}^{Cons}$	consumption projects (markets etc.)
$\mathcal{P}^{ex}$	existing projects
$\mathcal{P}^{new}$	new projects

(*Continued*)

(*Continued*)

Notation	Description
$\mathcal{P}^{Nodes}$	network nodes (all projects except pipelines)
$\mathcal{P}^{Pipe}$	pipeline projects
$\mathcal{P}^{Prod}$	production projects (fields etc.)
$\mathcal{P}^{Transp}$	transport nodes
$s \in \mathcal{S} = \{1, \ldots, S\}$	investment scenarios
$o \in \mathcal{S}^{Op}$	all operational scenarios
$o \in \mathcal{S}^{OpSub}_{ts}$	all operational scenarios associated with strategic tree node determined through year t, investment scenario s
$t \in \mathcal{T} = \{1, \ldots, T\}$	investment time periods

A.2 *Decision variables*

Notation	Description
Strategic level:	
$prd_{pts} \in \{0, 1\}$	project p producing at t under scenario s
$start_{pts} \in \{0, 1\}$	project p started at $t \in \mathcal{T}$ under scenario s
$stop_{pts} \in \{0, 1\}$	p stopped/removed at t under scenario s
$stprd_{pt\tau s} \in \{0, 1\}$	project p started at τ and producing at t
Operational level:	
$flow_{ptdo}$	flow through project p
$flow^{A}_{pts}$	accumulated flow through production node p until (including) period t
$flow^{C}_{pcto}$	volume flow of component c through project p
$flow^{S}_{pchtdo}$	volume flow of component c through pipeline p if split option h is chosen
$\lambda_{phtdo} \in \{0, 1\}$	split option h chosen for flow
$pres_{ptdo}$	pressure in project p (in- and output pressures modeled through pressures in in- and output projects)

A.3 *Variables/functions*

Notation	Description
$Cost_{ts}$	all costs in period $t \in \mathcal{T}$, scenario $s \in \mathcal{S}$
$Cost_{ts}^{Inv}$	investment related costs
$Cost_{pts}^{invfix}$	fixed investment costs project p
$Cost_{pts}^{invrem}$	stop/removal costs project p
$Cost_{pts}^{invst}$	startup/investment costs project p
$Cost_{ts}^{Op}$	expected operations related costs
$Cost_{pts}^{opex}$	operational expenditure of project p
$Cost_{ptdo}^{opfix}$	fixed operational costs of project p
$Cost_{ptdo}^{opvar}$	variable operational costs of project p
Del_{tdo}	production assurance measure: deliverability
Rev_{ts}	all revenue in period $t \in \mathcal{T}$, scenario $s \in \mathcal{S}$
Rev_{ts}^{Op}	expected operations related revenue

A.4 *Parameters/constants*

Notation	Description
$CapPerc_{ptdo}$	percentage of flow capacity available for usage
$CompProd_{pctdo}$	fraction of component c in flow from production node p
δ_t	discount factor in (investment) period $t \in \mathcal{T}$
Δ_p	time between decision to invest and availability for production
γ_g	factor to scale operational cash flow/volumes to yearly basis
K_p^W	Weymouth constant for pipeline p
P_{ib}	approximation of pressure at in-/outlet point i of a pipeline, breakpoint b
$\underline{P}_i$, $\overline{P}_i$	lower/upper bounds for pressure in node i

(*Continued*)

(*Continued*)

Notation	Description
$PATarget_t$	performance target for production assurance
Pen_t	penalty for insufficient production assurance
$Price^{Spot}_{ptdo}$	sales price at market node/project p
$ResMax_p$	total volume to be produced from reservoir p
$Prod^{Abs}_{pm}$	absolute coefficient for linearization of reservoir profile
$Prod^{Rel}_{pm}$	relative coefficient for linearization of reservoir profile
$Prob_{do}$	absolute probability of scenario tree node representing operational period $d \in \mathcal{D}_t$, scenario $o \in \mathcal{S}^{Op}$
$Prob_{ts}$	absolute probability of scenario tree node representing strategic period $t \in \mathcal{T}$, scenario $s \in \mathcal{S}$
$R(n) \in \mathcal{S}$	root node of operational node $o \in \mathcal{S}^{Op}$
σ	discount rate
$Start^e_p$	earliest startup date for (new) project p
$Start^l_p$	latest startup date for (new) project p
UB_{ptdo}	upper bound on capacity of pipeline p
Vol_{ptdo}	contracted delivery volume in market node $p \in \mathcal{P}^{Cons}$

A.5 *Profiles*

Note that all strategic profiles have "yearly" (investment time unit) resolution while operational profiles have a "daily" resolution. This holds also for the floating operational profiles based on the age of the project although this age is expressed in investment time units.

Notation	Description
Investment level, fixed (time dependent) — at $t \in \mathcal{T}$ for project $p \in \mathcal{P}$:	
$\Pi^{invremC}_{fix,pt}$	removal costs (per year)
Π^{invst}_{pt}	startup/investment costs (per year)

(*Continued*)

(*Continued*)

Notation	Description
Investment level, floating (age dependent) — at $t \in \mathcal{T}$ for project $p \in \mathcal{P}$ started at $\tau \in \mathcal{T}$:	
$\Pi^{invfixC}_{pt-\tau+1}$	fixed investment costs
$\Pi^{invremC}_{float,pt-\tau+1}$	removal costs
Operational level, fixed (time dependent) — at $d \in D_t$, $t \in \mathcal{T}$ for project $p \in \mathcal{P}$:	
Π^{opfix}_{ptd}	constant operational costs (per day)
Π^{opvar}_{ptd}	variable operational costs (per unit of flow)
Operational level, floating (age dependent) — at $t \in \mathcal{T}$ for project $p \in \mathcal{P}$ started at $\tau \in \mathcal{T}$:	
$\Pi^{MaxFlow}_{pt-\tau+1}$	upper bound on daily flow
$\Pi^{MinFlow}_{pt-\tau+1}$	lower bound on daily flow
$\Pi^{opex}_{pt-\tau+1}$	operational expenditure (per year — averaged to daily levels)

Chapter 11

A Stochastic Programming Model for Optimizing the Production of Farmed Atlantic Salmon

Martin B. Hæreid*, Peter Schütz†, Asgeir Tomasgard‡

Summary

We present a multi-stage stochastic programming formulation for the long-term production planning problem of a salmon farmer. Uncertainty in prices, growth and mortality is included in the model formulation, but we mainly focus on modeling the growth and development of the biomass in this paper. The planning horizon is 4–5 years, spanning approximately 3 life cycles of releasing smolt and slaughtering salmon. The objective is to maximize the expected profits from farming salmon, with the main decisions being both the timing and the quantity of releasing smolt and selling salmon. This chapter focuses on the description of the stochastic model and important factors for the modelling. Further work will focus more on computations and analysis.

1 Introduction

Farming of Atlantic salmon has grown during the last decades to become one of Norway's most important industries. However, even as the salmon aquaculture industry has matured, it is still characterized by large variations in supply and varying margins, largely due to the biological

*Department of Industrial Economics and Technology Management, Norwegian University of Science and Technology, 7491 Trondheim, Norway, martin@hareid.com.
†Department of Applied Economics, SINTEF Technology & Society, 7465 Trondheim, Norway, peter.schutz@sintef.no.
‡Department of Industrial Economics and Technology Management, Norwegian University of Science and Technology, 7491 Trondheim, Norway, asgeir.tomasgard@iot.ntnu.no, corresponding author.

nature of salmon farming and the resulting risks. Dealing with these risks and understanding the uncertainty present in the production system, is an essential part of successful planning in salmon farming.

In the ocean, the salmon are subject to varying growth conditions, potential disease outbreaks, and escape, all of which introduce uncertainty in the production process. Then, when the salmon are ready to be slaughtered and sold, the salmon farmer faces a volatile salmon price. The objective of this article is to study and describe the uncertainty that is relevant when planning harvesting and sales, and to present stochastic programming as a tool for dealing with this uncertainty. A multistage stochastic model is proposed that considers uncertainty in both salmon growth and price development. The aim of the model is to provide salmon producers with a tool that can aid them in making profitable decisions regarding harvesting and future sales.

The importance of production planning for salmon farming is also reflected in the amount of research literature dedicated to the topic. A good overview over the literature regarding models for aquaculture and fish farming can be found in Øveraas and Rynning-Tønnesen (2011). Early optimization models mainly focused on timing harvesting decisions (see e.g. Lillestøl, 1986; Bjørndal, 1988). These models have later been extended to also include the feeding schedule in the harvesting decision (Arnason, 1992) and by introducing piecewise-continuous and weight-dependent prices (Mistiaen and Strand, 1998).

Forsberg (1999) formulates a deterministic linear multi-period integer programming model for harvesting different fish cohorts in order to maximize overall profits. Growth is modeled by feed type, fish size and water temperature, and the cohort is modeled with a size-range to represent the dispersal within the same fish cohort. Harvesting is modeled with time windows, biomass, size-range of harvested fish, limitations for harvest operations and recovery time between harvest operations. The sales revenue as well as the costs regarding feed, transportation and slaughtering are given as input parameters. However, the model does not consider certain important production restrictions like Maximum Allowable Biomass (MAB)-limits, access to well boats, etc. In addition, the production cycle before harvesting is not modeled, disregarding uncertainty earlier in the production cycle. In 2006, Forsberg and Guttormsen state that fish farmers have traditionally focused on either production planning or price forecasts, and conclude that the two topics must be pulled together in order to make good harvesting decisions. Prices are closely related to optimal decisions, making it a major source of uncertainty in production planning models.

Restrictions governing the amount of biomass per location and region were introduced in Norway in 2005. Stikholmen (2010) is one of few works that consider MAB-restrictions. The author analyzes the effect the introduction of MAB-limits had on the efficiency of the aquaculture industry. The results however, are inconclusive. Langan and Toftøy (2011) develop an optimization model focusing on long term smolt ordering plans, analyzing two different objective functions: maximizing profit and maximizing MAB utilization of the salt-water facilities. Their work shows that minimizing deviation from the MAB quota does not maximize the biomass output. Hæreid (2011) builds a stochastic model for planning the sales of slaughtered salmon given uncertain future market prices. The main decisions over a one-year time horizon are which contracts the salmon farmer should enter and how much salmon should be traded in the spot marked.

In Section 2, we first describe the production process and the related uncertainty, before we discuss the market for salmon in Section 3. The biomass growth model is described in more detail in Section 4. The multi-stage stochastic programming model is given in Section 5. We conclude in Section 6.

2 Salmon Farming

Producers in general face two main types of risk: production risk and price risk. Production risk refers to the effects of uncertainty in the output of production, while price risk describes the effects of uncertainty in the revenue obtained from the quantity produced (Sandmo, 1971; Just and Pope, 1978). In salmon aquaculture, both of these risks are present, production risk through the uncertainty in biomass development, and price risk, through the uncertainty associated with entering contracts and trading fish in the spot market. In a long term perspective, additional uncertainty is also present, mainly regarding the strategic conditions under which salmon producers operate. Note that this section focuses primarily on explaining and measuring the sources of uncertainty, without a detailed discussion of the resulting risk. For a further discussion regarding the risks in salmon farming, the reader is referred to Bergfjord (2009), Kumbhakar and Tveterås (2003), and Tveterås (1999).

2.1 *Production in seawater*

After 6 to 18 months in freshwater, salmon undergo a process called smoltification. Smoltification enables the fish to survive in seawater, facilitating the transition from land based production in fresh water to marine farms along the coast. Salmon are at this stage referred to as smolt, and usually weigh from 60 to 150 grams at the time of release. Once in seawater, salmon are farmed in net cages containing anywhere from approximately 30,000 to 200,000 fish, depending on size and layout of the production facility (Cermaq ASA, 2012). Net cages are rectangular or round, and are made of either plastic or steel. Rectangular cages most commonly range in size from 10 by 10 to 40 by 40 meters, while circular cages can be anywhere from 40 to 200 meters in circumference (AKVA group, 2010).

Feeding the salmon is normally done using automatic feeding systems that follow advanced veterinary plans ensuring optimal conditions for fish health and growth. As the fish grow, sorting may be required, either because only fish of a certain size are to be harvested, or because lowering size variations in a cage increases feed intake and prevents harmful hierarchies from developing among the fish (NFR, 2005). Sorting can be performed using well boats or on-site sorting equipment.

After 12 to 18 months in the sea, the salmon are ready to be slaughtered. Fish can either be slaughtered immediately on-site, or transported with a well boat to the nearest slaughtering facility. Here, the fish are gutted, cleansed, sorted, filleted and packed. The salmon are either sold as whole fish or processed further and sold as value added products. Once a site is harvested, the location must be fallowed for at least 2 months before it is ready to be used by the next generation of salmon (FHL, 2011).

2.2 *Uncertainty in production*

The are two main uncertain factors in biomass development. First, there is the uncertainty associated with salmon growth rates. This includes both the uncertainty in overall growth due to uncertain growth conditions, and the uncertainty in the growth distribution within a biomass caused by varying growth rates among fish of the same weight. Second, the loss of fish, either due to mortality or escape, can drastically reduce biomass. Due to the presence of this uncertainty combined with the salmon's high sensitivity to its marine environment, salmon aquaculture can be considered as more risky than traditional land-based livestock production (Tveterås, 1999).

2.2.1 *Biomass development*

Salmon health is affected by many biological factors. Among the most important ones are oxygen concentration, salinity, pH-value, ammonia and carbon dioxide content, lighting conditions, feed, sea temperature. Of these factors, lighting and feed can to a large degree be controlled by the producer, while the first four factors mentioned are usually within acceptable values (Agnalt, Bakketeig, Haug, Knutsen and Opstad, 2009). This leaves temperature as the main source of the uncertainty in determining growth rates. Marine farms are typically located in sheltered surroundings in fjords or in coastal areas, where the variations in water temperature can be significant. Since salmon are cold blooded, and therefore extremely sensitive to changes in temperature, these variations in temperature are important to understand.

Figure 1 displays the monthly mean temperature for the same salmon farm for the years 1998 to 2006, showing considerable variations in temperature from the seasonal pattern. Each year is represented by a colored line; the scale is given in the left vertical axis. Also shown is the in-sample relative standard deviation, shown as the dashed line, with the scale given by the right vertical axis. Notice that the uncertainty is especially large from June to September. The effect of these temperature variations on salmon growth are illustrated in Figure 2, which shows the monthly growth for a 5 kg salmon. In addition to the uncertainty discussed above, it

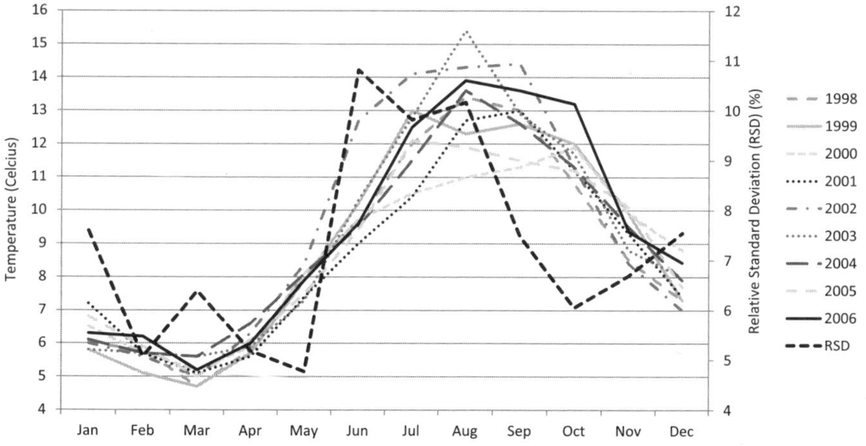

Figure 1 Monthly Mean Temperature and the In-Sample Relative Standard Deviation for a Salmon Farm in Sør-Trøndelag, 1998 to 2006, Data from Marine Harvest.

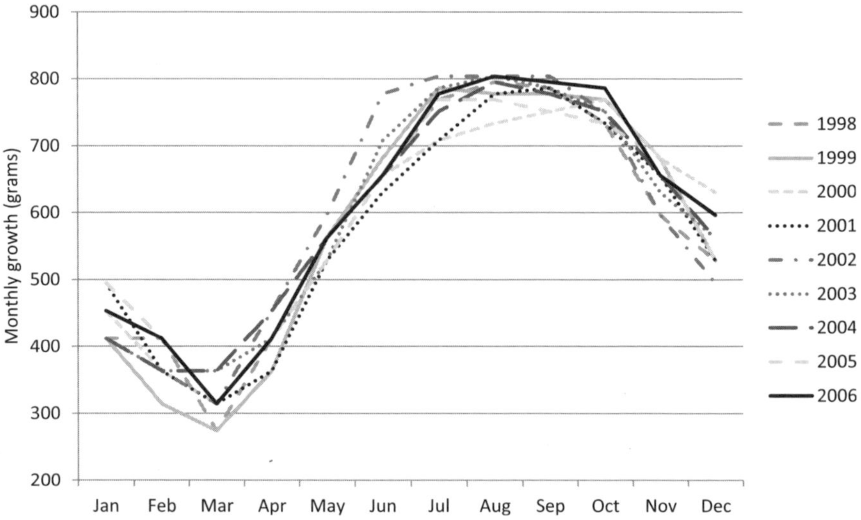

Figure 2 Monthly Growth for a 5 kg Salmon in a Salmon Farm in Sør-Trøndelag, 1998 to 2006 (Skretting, 2011).

is also worth mentioning that the growth rates among salmon of the same size also varies.

2.2.2 *Loss in production*

Among the main reasons for losses occurring while the salmon are in the ocean, are mortality of young fish and disease outbreaks. Approximately 80% of fish mortality occurs before the salmon are 0.5 kg, due to either deformities, injuries from transportation or release, or fish not coping with the transition to seawater. In addition to causing mortality, disease affects biomass development by causing reduced appetite and situations where premature slaughtering is necessary. Therefore, much effort is put into minimizing the concentration of bacteria, viruses, and toxic algae. Some of the most important diseases include Infectious Pancreatic Necrosis (IPN), Pancreas Disease (PD), Heart and Skeletal Muscle Inflammation (HSMI), Infectious Salmon Anaemia (ISA), Salmonid Rickettsial Septicaemia (SRS), Gill Disease (GD), and sea lice (Marine Harvest, 2010).

Another reason for losses in the production process is the escape of farmed salmon. The main reasons for this are winter storms, propeller damage, and wear and tear on equipment. Fish escaping through holes in nets account for approximately 65% of the reported cases (AKVA group,

2010). In recent years, better management of these problems has led to a reduced number of escaped salmon, which contrasts with the increased number of salmon produced. Large escapes (>10,000 fish) account for only 19% of all incidents, but amount to 91% of the number of escaped fish (Jensen, 2010). This means that even though the number for escaped salmon is small in comparison with losses due to mortality, the consequence of an escape can be far greater for the unlucky producer.

2.2.3 *Consequences of the uncertainty in biomass development*

If not handled carefully, the uncertainties in biomass development discussed above can have costly consequences. Higher growth than expected may necessitate unplanned harvesting in order to satisfy MAB restrictions, resulting in biomass which must be sold in the spot market. This increases the exposure to the price risk discussed in the following sections. Another possible consequence of the salmon growing faster than expected is that the salmon reach the delivery weight earlier than planned. This might require renegotiation of contract terms or the purchase of a replacement product in the spot market. In cases where the biomass development is lower than expected, a producer may be forced to purchase fish in the spot market in order to fulfill contract commitments. This can be costly since the producer might not be able to buy a replacement product in time due to limited availability, or since a premium above the spot price must be paid when buying the product from a competing producer. If the product is not delivered, non-refundable costs (transportation, booked slaughtering capacity, etc.) and costs associated with loss of customer goodwill may apply. In addition, since poor growth conditions and disease outbreaks often affect the entire industry, chances are high that the shortage applies to the majority of producers. The result may be a decrease in market supply combined with increase in producer demand (to fulfill contracts), thus putting pressure on prices and making purchasing fish in the spot market undesirable.

2.3 *The Norwegian regulatory framework*

Environmental challenges involved with farming salmon in seawater have necessitated stringent governmental regulation. This was most recently illustrated by the weak regulation in Chile, and the vast problems and decline in supply that followed (Barrionuevo, 2008). The main tool for ensuring sustainable development in Norway has been licensing and has been practiced since 1973. Today's regulatory framework was established in 2005, and is described in Laksetildelingsforskriften (FKD, 2004). In 2010, a total of approximately 105 companies were operating 960 licenses in Norway (Sandberg, 2010). Licenses are awarded by the Ministry of Fisheries, while governing is left to the Directorate of Fisheries. In most of Norway, the biomass limit per license is normally 780 tons live weight, while in the northernmost parts, the limit is 900 tons. A license is valid within a specified region, and the sum of licenses in a region gives the region's Maximum Allowable Biomass (MAB). Within a region, one license can be utilized in up to four sites, while two licenses together can be utilized in up to six sites. In addition, each site has a site specific MAB-limit, which can be exploited using one or more licenses.

The most recent development in the Norwegian regulatory framework is related to lice zone regulations (FKD, 2009). This regulation gives The Norwegian Food Safety Authority the authority to establish regions in which fallowing must be conducted simultaneously for all facilities. The purpose is to avoid contamination between different locations and generations of fish.

3 The Market for Atlantic Salmon

Understanding the dynamics of the market for salmon is essential for any company involved in the salmon industry. According to surveys done by Bergfjord (2009), Norwegian fish farming companies consider future salmon prices to be the most important source of risk. This is not surprising, considering the turbulent economic environment that has characterized the industry during its rapid development over last decades. A substantial variation in margins has led to frequent bankruptcies, often associated with periods of persistently low salmon prices (Bergfjord, 2009). Figure 3 shows the yearly average operating margin (operating result/operating income) for the period 1986 to 2009. Even as the industry has matured, future salmon prices remain hard to foresee. This is largely because of the unpredictable supply resulting from the biological nature of salmon farming, and the almost perfect negative correlation between the supply of salmon and the salmon market price (Marine Harvest, 2010).

3.1 *Supply*

In line with economic theory, supply and demand determine the price in a competitive market. In the salmon industry, the volatility in this price can

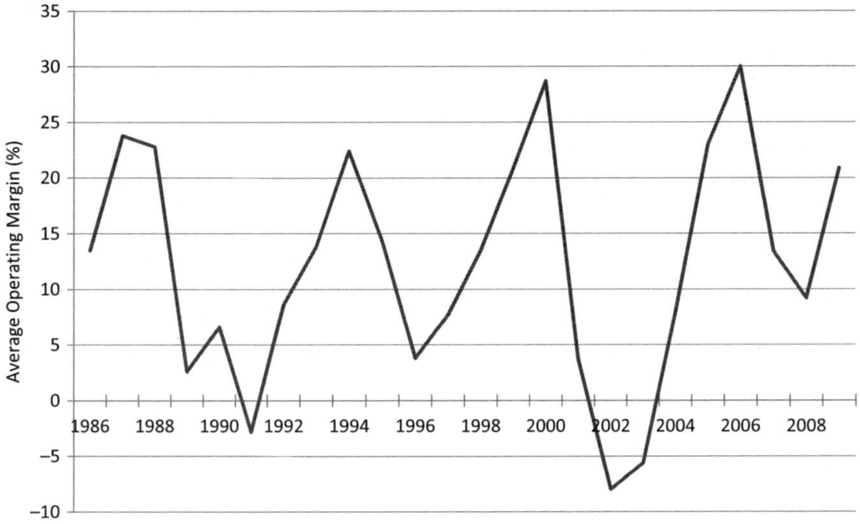

Figure 3 Average Operating Margin, 1986 to 2009 (Directorate of Fisheries, 2009)

in part be explained by looking at the price elasticity of supply. As most of the farmed salmon is marketed fresh and is thereby perishable, salmon produced in one period has to be consumed in the same period. In the short term, the production level is difficult and expensive to adjust, as the planning/production cycle is approximately three years long. This, combined with government regulations limiting capacity, results in a short run price supply elasticity that is close to zero (Andersen, Roll and Tveterås 2008).[1] At the same time, short term supply is greatly affected by uncertain biological factors and the resulting development in biomass, as discussed above. The result is a short term supply that is both inelastic and uncertain, hereby introducing the possibilities for both periods of over and undersupply, causing prices to fluctuate. In addition, even though a variety of sizes and quality creates different product segments, salmon can be considered a relatively homogeneous product. This, along with the development of international trade, has created a worldwide market where the price in a region is not only sensitive to changes in local supply, but also governed by changes in the supply in other countries. This was recently illustrated by the outbreak of the ISA virus in Chile in 2007, where

[1]The study showed that a 1% increase in the sales price only induced about a 0.05% increase in supply.

the reduction in supply created an imbalance between global supply and demand.

In the long run, the situation changes substantially as supply becomes elastic. Asche, Salvanes and Steen (1997) estimate a long-run supply elasticity of about 1.5 using annual aggregated data. This result is confirmed by Asche, Kumbhakar and Tveterås (2007), deriving the results from a cost function. Andersen *et al.* (2008) argue that the difference in price responsiveness of salmon supply in the short and long run plays a large part in explaining the observed cyclical profitability in the salmon farming industry. The reader is referred to Andersen *et al.* (2008) for a more detailed discussion on the reasons and consequences of the elasticity of supply being larger in the long run than in the short run.

3.2 *Price*

Resulting from the preceding discussion above is a salmon price that exhibits significant volatility. Figure 4 shows the weekly price of Atlantic salmon for the period 1995 to 2010. The prices shown in the figure are for fresh, head-on, gutted, superior salmon, packaged and delivered from slaughtering, also referred to as HOG (head-on gutted). The average price

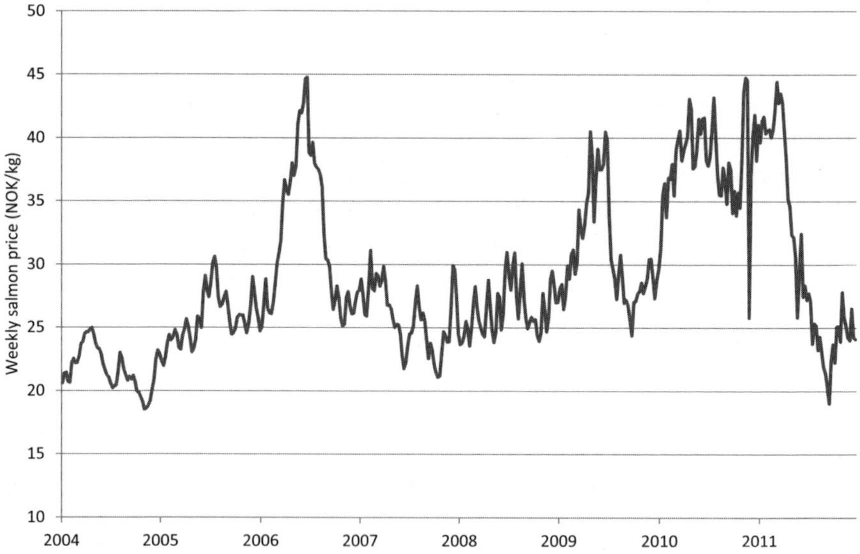

Figure 4 Weekly Price of Atlantic Salmon, 2004 to 2011 (Fish Pool, 2012)

for Norwegian whole salmon during the last decade has been approximately 25 NOK/kg HOG, with peaks at 45 NOK/kg and lows at 15 NOK/kg. A study by Alnæs and Skagen (2009) revealed that the annualized volatility at times during this period approached 50%. For a discussion of the characteristics of salmon price volatility, see Oglend and Sikveland (2008). In addition, Asche and Guttormsen (2001) provide insight into how the biological production cycle can cause patterns in the relative prices between the different sizes throughout the year.

For salmon farmers wishing to secure future volumes and prices, the traditional method has been to enter into long term contracts with processing companies, wholesalers or supermarkets. Now, with the recent establishment of Fish Pool, financial tools present new possibilities for reducing the exposure to price risk. Fish Pool is the first and only organized marketplace for trading derivatives with salmon as the underlying asset. Trading financial salmon contracts provides salmon farmers with an alternative to entering physical contracts, with the additional benefits that a purely financial market entails. With the longest contracts stretching nearly three years, it is theoretically possible for salmon farmers to secure all or part of their earnings at a contractual price. Still, Fish Pool is in an early stage of development, and investigations carried out by Alnæs and Skagen (2009) show that limited liquidity prevents the futures, forwards, and options available from being fully efficient, thereby limiting the possibilities for creating a perfect hedge.

3.3 *Other sources of uncertainty*

In addition to the risk associated with uncertain future prices and biomass development, the exploratory study done by Bergfjord (2009) shows that Norwegian fish farming companies consider institutional risk factors to be of great importance. In the study, market regulation, area access, changes to the licensing system, environmental regulations, taxation, certification systems, and animal health regulations are mentioned as the most important institutional risk factors. Governmental restrictions, like quotas and tolls set by the EU on Norwegian salmon, are almost always enforced after a prior warning, and will thereby give producers the chance to adapt. But in the long term, these changes can in any case be vital to overall yield in the industry. This is clearly illustrated by the modernization of the industry in Norway, which has undergone dramatic structural and legislative changes since companies in 1991 for the first time were allowed to own more than one license. The introduction of maximum allowable biomass in 2005, along with the more recent lice zone restrictions, have been important steps in bringing the industry closer towards a viable state of operation. It is predicted that the strict governmental involvement will continue, and that the coming years will without doubt bring additional regulations that again will change the conditions under which salmon farmers operate.

Other sources of uncertainty worth mentioning are the price of feed, future exchange rates, health concerns, and changing natural conditions. As the salmon industry continues to globalize, currency exchange rates become ever more important, especially for export dependent countries like Norway. In regards to changing natural conditions, there is an uncertainty associated with the effects of a general increase in sea temperature in Norway due to global warming. Salmon growth rates are highest when the temperature is between 13 and 17 degrees Celsius, which for Norway implies that a limited increase in sea temperature will have a positive effect on productivity. But if the temperature increases to the point where it approaches the limits of what is suitable for the salmon, mortality rates will increase, and the salmon industry might have to move farther north to ensure suitable conditions (Lorentzen, 2010).

4 Biomass Growth Model

The basis for keeping track of biomass development in the model is the definition of a set $\mathcal{F}$ of fish classes f. Each fish belongs to a fish class f with an expected weight V_f, and as the fish grows, it advances from one fish class to another.

The growth in kilograms in period t for fish of fish class f at location i is given by σ_{fit}. The weight at the end of a period t is then calculated by adding the growth during t, σ_{fit}, to the weight at the beginning of the period, V_f. The new value $(V_f + \sigma_{fit})$ will fall between the upper and the lower limit of one of the fish classes f:

$$V_{\underline{f}} \leq (V_f + \sigma_{frs}^t) \leq V_{\overline{f}}. \tag{1}$$

Here, $V_{\underline{f}}$ is the expected weight of a fish in fish class $\underline{f}$, the fish class with defined weight closest to $(V_f + \sigma_{frs}^t)$ from below. Correspondingly, $V_{\overline{f}}$ is the expected weight of a fish in fish class $\overline{f}$, the fish class with defined weight closest to $(V_f + \sigma_{frs}^t)$ from above.

Based on the value $(V_f + \sigma_{fit})$, the model will redistribute all fish among the defined fish classes reflecting the growth in biomass during the past period. This is done by distributing the fish among the two classes $\underline{f}$ and $\overline{f}$, defined by equation 1. We calculate the share $\delta_{rf\underline{f}t}$ of fish belonging to class f in period t moving to class $\underline{f}$ in period $t+1$ as

$$\delta_{rf\underline{f}t} = \frac{(V_f + \sigma^t_{frs}) - V_{\underline{f}}}{V_{\overline{f}} - V_{\underline{f}}}. \tag{2}$$

The share $\delta_{rf\overline{f}t}$ of fish being redistributed to class $\overline{f}$ is correspondingly given as

$$\delta_{rf\overline{f}t} = \frac{V_{\overline{f}} - (V_f + \sigma^t_{frs})}{V_{\overline{f}} - V_{\underline{f}}}. \tag{3}$$

Note that $\delta_{rf\underline{f}t} + \delta_{rf\overline{f}t} = 1$. Let m_{fit} be the number of fish in fish class f at the beginning of period t in location i. We allow m_{fit} to take on real values due to the high number of fish. At the beginning of the next period, $t+1$, the number of fish m_{fit} is then distributed among fish classes $\underline{f}$ and $\overline{f}$ according to $m_{\underline{f}it+1} = \delta_{rf\underline{f}t} \cdot m_{fit}$ and $m_{\overline{f}it+1} = \delta_{rf\overline{f}t} \cdot m_{fit}$.

In reality, the biomass development is less homogeneous than in the presented growth model. Fish should therefore be distributed among several fish classes. However, despite the distribution of fish being slightly distorted, the average weight from the model corresponds well to the observed biomass growth. Due to the regulations regarding the Maximum Allowable Biomass (MAB), it is more important to accurately model the total amount of biomass in the system rather than the correct distribution among the different fish classes. The need for good biomass information together with the focus on production planning decisions justifies in our opinion the use of this simple growth model. If sales planning decisions become more important, a more sophisticated model that keeps better track of the distribution among different fish classes is preferable.

Skretting (2011) provide relative daily growth rates based on water temperature. They use a total of 34 weight classes, ranging from 30 gr to 5.25 kg, to discretize the weight distribution. However, in order to apply the above growth model, we require a combination of both sufficiently large time resolution and sufficiently detailed partitioning of fish weight. The reason for this is that in case the growth of fish in fish class f during period t, σ_{fit}, is such that $\underline{f} = f$ for $\delta_{rf\underline{f}t}$, a share of the fish will remain in f. If this repeats for every period, some of the fish will never grow, a situation which clearly is not realistic. To prevent this from occurring, $\underline{f}$ must in most periods be larger than f in order to advance all fish to a new class. This is ensured by having sufficiently long periods combined with a small enough difference in weight from one fish class to the next. A longer period duration implies a larger σ_{fit}, while the more detailed the partitioning, the less the fish need to grow in order to advance to the next class.

We therefore discretize the weight distribution by means of 82 fish classes, with average weights ranging from 30 gr to 8 kg. Interpolating on the growth rates given by Skretting (2011), we calculate an absolute growth σ_f in biomass per month for each fish class. Based on the historic water temperatures (see Figure 1), we generate scenarios for future water temperatures for the different locations and time periods. Using the growth model, these temperatures are then translated into growth rates σ^s_{fit}. With this information, we can calculate the shares $\delta^s_{rf\underline{f}t}$ and $\delta^s_{rf\overline{f}t}$, that will be part of the scenario tree for the stochastic programming model.

5 Model Formulation

In this section, we present a multi-stage stochastic programming formulation for the long-term production planning problem of a salmon farmer. The objective is to maximize the expected profit of the produced biomass over a planning horizon of 4–5 years. The model focuses on the seawater phase of the value chain and the main decisions are the timing as well as the quantity of releasing smolt, slaughtering, and selling salmon.

5.1 *Model properties*

Accurately modeling the situation faced by a salmon producer who seeks to optimize profits requires an overwhelming degree of detail. Therefore, assumptions and simplifications are essential in making the task feasible and developing a computationally tractable model.

The industry classifies fish using a set of characteristics such as weight, age, vaccines, feed type, quality, etc. The model presented in this paper includes only weight as this is by far the most important parameter, both for producers and for customers (however, other characteristics can be easily included). For producers, knowing the weight of the fish is necessary in order to be able to decide the type and amount of feed, perform growth estimates, and calculate biomass accounts. From a customer perspective, fish of different weights are suitable for different uses, and are thereby viewed as different products. To establish a connection between the way producers and customers deal with weight, we introduce two classifications: salmon belongs both to a fish class $f \in \mathcal{F}$ for production and to a sales class $j \in \mathcal{J}$ for the market. The different sets of fish and sales classes represent a disjoint discretization of fish weight.

End-of-horizon conditions are modeled by utilizing business rules that restrict when harvesting is permitted in each location. Depending on the length of the planning period, some fish may not be ready for harvesting until after the final period. For example, with a planning horizon of twelve months, this would apply to both fish that are released just before and during the planning period. Without specifying the periods for which harvesting is allowed for each location, sub-optimal harvesting decisions will be taken for the final periods in order to maximize the profits of the planning period. The restriction in harvesting periods will be modeled by properly defining harvesting variables rather than by adding constraints to the model formulation.

5.2 *Notation*

Let us first introduce the following notation for our problem formulation:

- Sets

$\mathcal{F}$	Set of fish classes
$\mathcal{J}$	Set of sales classes
$\mathcal{F}(j)$	Set of fish classes f in sales class j, $\mathcal{F}(j) \subset \mathcal{F}$
$\mathcal{I}(r)$	Set of locations in region r
$\mathcal{R}$	Set of regions
$\mathcal{S}$	Set of scenarios
$\mathcal{T}$	Set of periods
$\mathcal{N}$	Set of event nodes in the scenario tree

- Indices and superscripts

$f, \hat{f}$	Fish class indices
j	Sales class index
i	Location index
r	Region index
s	Scenario superscript
$t, \hat{t}$	Time period indices
n	Event node index

- Parameters, constants, and coefficients

K_f	Cost per period for keeping one kilogram of biomass belonging to fish class f
m^s_{fi0}	Initial number of fish in fish class f in location i, equal for all scenarios s (period $t = 0$ is not part of the set of periods $\mathcal{T}$)
MAB_{it}	Maximum Allowable Biomass in kilograms for location i in period t ($MAB_{it} = 0$ for fallowing periods)
MAB_r	Maximum Allowable Biomass in kilograms for region r
Q_f	Processing ratio (loss factor due to gutting) for fish class f
$\overline{S}_r$	Slaughtering capacity per period in region r
T_f	Production lead time from ordering smolt of fish class f until delivery
U^s_{ft}	Available number of smolt belonging to fish class f in period t in scenario s
V_f	Average weight in kilograms of a fish belonging to fish class f
α^s_{jt}	Price per kilogram for sales class j in period t in scenario s
β^s_{jt}	Price per kilogram for spot purchases of sales class j in period t in scenario s
γ^s_{ft}	Price per kilogram for procuring smolt belonging to fish class f in period t in scenario s
$\delta^s_{\hat{f}fit}$	The share of fish class $\hat{f}$ that has grown to become part of fish class f due to the growth at location i in period t in scenario s ($\delta^s_{\hat{f}fi0} = 1$)
ϵ^s_{fit}	The survival rate for a fish in fish class f at location i in period t in scenario s ($\epsilon^s_{fi0} = 1$)
η^s_{ft}	Deviation from the ordered amount of smolt belonging to fish class f in period t in scenario s
ι^s_{jt}	Minimum demand in kilograms for sales class j in period t in scenario s
κ^s_{fit}	Cost per kilogram for deploying smolt belonging to fish class f at location i in period t in scenario s
μ^s_{jt}	Maximum demand in kilograms for sales class j in period t in scenario s
σ^s_{fit}	The growth in kilograms in period t for fish class f at location i in scenario s
π^s	Probability of scenario s.

- Decision variables

m^s_{fit} Number of fish in fish class f in location i in region r at the beginning of period t after harvesting and release in scenario s

u^s_{ft} Number of smolt of fish class f ordered in period t in scenario s

v^s_{fit} Number of smolt of fish class f released at location i in period t in scenario s

w^s_{fit} Number of fish in fish class f harvested from location i in period t in scenario s

x^s_{jt} Number of kilograms of sales class j sold in period t in scenario s

y^s_{jt} Number of kilograms of sales class j purchased spot in period t in scenario s

5.3 *Stochastic variables and time structure*

The time horizon of the model is 4–5 years, spanning approximately 3 life cycles in order to ensure that biomass constraints are satisfied and that smolt production is planned in time. Time resolution in this model is weeks or months, and it is a considered a tactical model where the important issues are timing of biomass decisions. The number of stages should be large enough to capture the dynamics of seasons and the biomass cycles in the different locations in the model. Uncertain parameters are the growth of fish σ^s_{fit}, the demand for fish ι^s_{jt}, the survival rate ϵ^s_{fit}, the price α^s_{jt} and smolt deliveries η^s_{ft}. The outcomes of these variables are represented as scenarios in an event tree. The objective is weighted with the probability of scenario s, π_s.

5.4 *Mathematical formulation*

We now give a multi-stage stochastic programming formulation of our production planning problem under uncertainty. First, we discuss the objective function before we explain the different constraints. The model is formulated as deterministic equivalent problem, including a set of non-anticipativity constraints.

5.4.1 *Objective function*

The model's objective is to maximize the expected profits of the salmon farmer. The profits are here defined as the revenue from selling slaughtered salmon minus the costs of procuring salmon in the spot market, the costs of procuring smolt, and the costs of keeping biomass at sea.

$$\max \sum_{s \in \mathcal{S}} \pi_s \left(\sum_{t \in \mathcal{T}} \left(\sum_{j \in \mathcal{J}} \alpha^s_{jt} x^s_{jt} - \sum_{j \in \mathcal{J}} \beta^s_{jt} y^s_{jt} - \sum_{f \in \mathcal{F}} \gamma^s_{ft} V_f u^s_{ft} \right.\right.$$

$$\left.\left. - \sum_{f \in \mathcal{F}} \sum_{i \in \bigcup_{r \in \mathcal{R}} \mathcal{I}(r)} \kappa^s_{fit} V_f v^s_{fit} - \sum_{f \in \mathcal{F}} \sum_{i \in \bigcup_{r \in \mathcal{R}} \mathcal{I}(r)} K_f V_f m^s_{fit} \right) \right) \quad (4)$$

The first term of the objective function (4) corresponds to the income from selling salmon, whereas the second term denotes the cost of procuring additional salmon in case demand exceeds the available biomass. The third and the fourth term represent the cost of procuring and deploying smolt. The last term of the objective function is the inventory holding cost or the cost of keeping the biomass at sea.

5.4.2 *Constraints*

We start by presenting the constraints that are included in the model. Note that the same decisions are taken in each time period, irrespective of the stage. At the end of this section, we will also discuss the issue of initial and end-of-horizon conditions.

Biomass Development Constraint (5) is a mass balance constraint for the different fish classes. For any given period t, the amount of fish at sea belonging to fish class f plus the amount harvested from this fish class has to be equal to the amount of fish that has grown into fish class f at the end of the previous time period plus the number of fish that has been released into f in period t.

$$m^s_{fit} + w^s_{fit} = \sum_{\hat{f} \le f} (\epsilon^s_{\hat{f}it-1} \delta^s_{\hat{f}fit-1} m^s_{\hat{f}it-1}) + v^s_{fit},$$

$$f \in \mathcal{F}, i \in \bigcup_{r \in \mathcal{R}} \mathcal{I}(r), t \in \mathcal{T}, s \in \mathcal{S} \quad (5)$$

Smolt Release Ordering and releasing smolt are important decisions when planning production for 4–5 years. With constraint (6), we enforce that the smolt order for fish class f does not exceed the production capacity for this fish class. Using constraint (7), we distribute the available amount of smolt among the different locations. This constraint also includes η^s_{ft}, that captures the uncertainty to the smolt production.

$$u^s_{ft} \leq U^s_{ft}, \quad f \in \mathcal{F}, t \in \mathcal{T}, s \in \mathcal{S}, \tag{6}$$

$$\sum_{i \in \bigcup_{r \in \mathcal{R}} \mathcal{I}(r)} v^s_{fit} \leq u^s_{ft-T_f} + \eta^s_{ft}, \quad f \in \mathcal{F}, t \in \mathcal{T}, s \in \mathcal{S} \tag{7}$$

Slaughtering The following constraint (8) ensures that the producer does not harvest more than what the slaughtering facilities can manage. Slaughtering capacity is aggregated on a regional level and we assume that all locations within region r must use slaughtering capacity in the given region.

$$\sum_{i \in \mathcal{I}(r)} \sum_{f \in \mathcal{F}} w^s_{fit} \leq \overline{S}_r, \quad r \in \mathcal{R}, t \in \mathcal{T}, s \in \mathcal{S} \tag{8}$$

Sales and Demand Equation (9) ensures that the amount of fish sold in the market is either harvested or procured in the spot market.

$$x^s_{jt} = \sum_{f \in \mathcal{F}(j)} \sum_{i \in \bigcup_{r \in \mathcal{R}} \mathcal{I}(r)} Q_f V_f w^s_{fit} + y^s_{jt}, \quad j \in \mathcal{J}, t \in \mathcal{T}, s \in \mathcal{S} \tag{9}$$

This equation also

- links the production fish classes f to the corresponding sales class j,
- determines the amount supplied by each in order to satisfy demand,
- enforces procuring salmon in the spot market if the amount to be delivered exceeds the available harvested salmon, or if it is not profitable to harvest the available fish.

Total sales of each sales class j are constrained by lower and upper limits on demand, see inequalities (10) and (11). The lower limit is influenced by contract sales that are agreed upon ahead of the delivery period. The upper limit on sales can be interpreted as both the total size of marked for this particular fish class, but also as the sales department's ability to push fish into the market. Note that we assume that the price is independent of the amount of fish sold.

$$x^s_{jt} \geq \iota^s_{jt}, \quad j \in \mathcal{J}, t \in \mathcal{T}, s \in \mathcal{S}, \tag{10}$$

$$x^s_{jt} \leq \mu^s_{jt}, \quad j \in \mathcal{J}, t \in \mathcal{T}, s \in \mathcal{S} \tag{11}$$

Maximum Allowable Biomass Constraints (12) and (13) represent the regulatory restrictions regarding the maximum amount of biomass, both at a given location and within the region. The constraints enforce that the biomass at the end of period t does not exceed the limits defined in the regulatory framework.

$$\sum_{f\in\mathcal{F}} \epsilon^s_{fit}(V_f + \sigma^s_{fit})m^s_{fit} \leq MAB_{it}, \quad i \in \bigcup_{r\in\mathcal{R}} \mathcal{I}(r), t \in \mathcal{T}, s \in \mathcal{S}, \tag{12}$$

$$\sum_{i\in\mathcal{I}(r)} \sum_{f\in\mathcal{F}} \epsilon^s_{fit}(V_f + \sigma^s_{fit})m^s_{fit} \leq MAB_r, \quad r \in \mathcal{R}, t \in \mathcal{T}, s \in \mathcal{S} \tag{13}$$

We also use constraint (12) to enforce fallowing requirements. Fallowing requirements are exogenously given and can, for example, be due to lice zone regulations covering multiple location within a region or the company's smolt release schedule for a given location in combination with the fallowing requirements ahead of the release. For fallowing periods, MAB_{it} is set equal to zero.

Non-anticipativity and Non-negativity The structure of the scenario tree and the relationship between stages, periods and scenarios is enforced by non-anticipativity constraints (14) (see Rockafellar and Wets, 1991) for each event node in the scenario tree where uncertainty is resolved. Let the set of scenarios passing through event node n in a scenario tree be given by $S(n)$. Then the constraints can be expressed as

$$\begin{aligned} &\frac{1}{|S(n)|} \sum_{s' \in S(n)} (m^{s'}_{fit}, u^{s'}_{ft}, v^{s'}_{fit}, w^{s'}_{fit}, x^{s'}_{jt}, y^{s'}_{jt}) \\ &= (m^{s}_{fit}, u^{s}_{ft}, v^{s}_{fit}, w^{s}_{fit}, x^{s}_{jt}, y^{s}_{jt}), \\ &\qquad f \in \mathcal{F}, j \in \mathcal{J}, i \in \bigcup_{r\in\mathcal{R}} \mathcal{I}(r), n \in \mathcal{N}, t \in \mathcal{T}(n), s \in \mathcal{S}(n) \end{aligned} \tag{14}$$

The final constraints (15) are the non-negativity constraints; the indices are omitted.

$$m, w, x, y, z \geq 0. \tag{15}$$

Initial and end-of-horizon conditions For reasons of simplicity, only the for parameter the initial biomass m^s_{fi0} is defined above and included in the formulation. Another important parameter that should be properly initialized is the amount of smolt that were ordered before the start of the planning horizon. A simple way of resolving this issue is by incorporating the existing orders into η^s_{ft} and using it as upper bound on the smolt deployment for $t \leq T_f$.

In order to avoid emptying all locations before the end of the planning horizon, the state of the production system in the final period has to be carefully defined. There are several ways to ensure that production is ongoing also after the model's planning horizon. On way of securing future production is to require a minimum total amount of biomass in the system. This can easily be supplemented by minimum requirements for each fish class. Another way of avoiding the end-of-horizon problem is to value the biomass still in the ocean based on future sales prices and the time it takes until the salmon can be sold.

6 Conclusions

We have presented a multi-stage stochastic programming model for making biomass decisions for farmed Atlantic salmon. Typical uncertain parameters are demand, price, smolt deliveries, survival rates and growth rates. Decisions on biomass are constrained by limits on maximum allowable biomass, and the key to a profitable business is to plan the release of smolt to the marine farms, the feeding and management of the salmon in the farms and the timing of slaughtering. Uncertainty plays a major role in order to stay within the biomass limits and in order to meet demand. The work in this book chapter will be followed up with analysis on real data for Norwegian salmon farms. The main focus in the further work will be to study how uncertainty affects decisions and how the use of a stochastic optimization model will influence profit as compared to todays routines based on deterministic optimization and simulation.

References

Agnalt, A.-L., I. Bakketeig, T. Haug, J. A. Knutsen and I. Opstad (eds.). *Coast and Aquaculture 2009*, Institute of Marine Research, Bergen, Norway, 2009. (in Norwegian).

AKVA group. Cage farming aquaculture, AKVA group, Bryne, Norway, 2010.

Alnæs, A. B. and M. R. Skagen. Risk Management in the Market for Salmon — a study of Fish Pool's derivatives, Master's thesis, Department of Finance and Management Science, Norwegian School of Economics and Business Administration, Bergen, Norway, 2009. (in Norwegian).

Andersen, T. B., K. H. Roll and S. Tveterås. The price responsiveness of salmon supply in the short and long run, *Marine Resource Economics*, 23(4): pp. 425–437. 2008.

Arnason, R. Optimal feeding schedules and harvesting time in aquaculture, *Marine Resource Economics*, 7(1): pp. 15–35. 1992.

Asche, F. and A. G. Guttormsen. Patterns in the relative price for different sizes of farmed fish, *Marine Resource Economics*, 16(3): pp. 235–247, 2001.

Asche, F., S. C. Kumbhakar and R. Tveterås. Testing cost versus profit function, *Applied Economics Letters*, 14(10): pp. 715–718, 2007.

Asche, F., K. G. Salvanes and F. Steen. Market delineation and demand structure, *American Journal of Agricultural Economics*, 79(1): pp. 139–150, 1997.

Barrionuevo, A. Salmon virus indicts Chile's fishing methods, New York Times, 27 March 2008.

Bergfjord, O. J. Risk perception and risk management in Norwegian aquaculture, *Journal of Risk Research*, 12(1): pp. 91–104, 2009.

Bjørndal, T. Optimal harvesting of farmed fish, *Marine Resource Economics*, 5(2): pp. 139–159, 1988.

Cermaq ASA. Cermaq fish farming. last accessed: 27. January 2012. URL: *http://www.cermaq.com/portal/wps/wcm/connect/cermaqen/home/fishfarming/*, 2012.

Directorate of Fisheries. Time series production of edible fish — socioeconomic perspective — finished. Directorate of Fisheries, Bergen, Norway 2009. (in Norwegian). URL: *http://www.fiskeridir.no/statistikk/akvakultur/loennsomhet/matfiskproduksjon-laks-og-regnbueoerret/tidsserier-matfiskproduksjon-samfunnsoekonomisk-perspektiv-avsluttet.*

FHL. Norwegian Aquaculture, Fiskeri- og havbruksnæringens landsforening (FHL) and Eksportutvalget for fisk (EFF), Oslo, Norway, 2011. (in Norwegian).

Fish Pool. List of spot prices. last accessed: 8. February 2012. URL: *http://fishpool.eu/iframe.aspx?iframe=extranett/spothistory.asp*, 2012.

FKD. Regulations on the granting, amendment and withdrawal of licenses for salmon, trout and rainbow trout, Department of Aquaculture, Seafood and Markets, Ministry of Fisheries and Coastal Affairs 2004. (in Norwegian). URL: *http://www.lovdata.no/cgi-wift/ldles?doc=/sf/sf/sf-20041222-1798.html.*

FKD. Regulations for control of lice in fish farms, Department of Aquaculture, Seafood and Markets, Ministry of Fisheries and Coastal Affairs 2009. (in Norwegian). URL: *http://www.lovdata.no/cgi-wift/ldles?doc=/sf/sf/sf-20090818-1095.html.*

Forsberg, O. I. Optimal harvesting of farmed atlantic salmon at two cohort management strategies and different harvest operation restrictions, *Aquaculture Economics & Management*, 3(2): pp. 143–158, 1999.

Forsberg, O. I. and A. G. Guttormsen. The value of information in salmon farming: Harvesting the right fish at the right time, *Aquaculture Economics & Management*, 10(3): pp. 183–200, 2006.

Hæreid, M. B. Allocating sales in the farming of Atlantic salmon, Master's thesis, Department of Industrial Economics and Technology Management, Norwegian University of Science and Technology, Trondheim, Norway, 2011.

Jensen, Ø. Escapes of fishes from Norwegian sea-cage aquaculture: causes, consequences and prevention, *Aquaculture Environment Interactions*, 1(1): pp. 73–80, 2010.

Just, R. E. and R. D. Pope. Stochastic specification of production functions and economic implications, *Journal of Econometrics*, 7(1): pp. 67–86, 1978.

Kumbhakar, S. C. and R. Tveterås. Risk preferences, production risk and firm heterogeneity, *The Scandinavian Journal of Economics*, 105(2): pp. 275–293, 2003.

Langan, T. B. and T. Toftøy. Optimizing production in salmon farming — planning under uncertainty, Master's thesis, Department of Industrial Economics and Technology Management, Norwegian University of Science and Technology, Trondheim, Norway, 2011. (in Norwegian).

Lillestøl, J. On the problem of optimal timing of slaughtering in fish farming, *Modeling, Identification and Control*, 7(4): pp. 199–207, 1986.

Lorentzen, T. Increased sea temperature affects the farming of salmon, *Klima: Norsk magasin for klimaforskning*, 2: pp. 31–33, 2010. (in Norwegian).

Marine Harvest. Salmon farming industry handbook 2010, Marine Harvest AS, Bergen, Norway, 2010.

Mistiaen, J. A. and I. E. Strand. Optimal feeding and harvest time for fish with weight-dependent prices, *Marine Resource Economics*, 13(4): pp. 231–246, 1998.

NFR. Research needs within animal welfare in Norway, Report from the steering committee, Norsk forskningsråd, Oslo, Norway, 2005. (in Norwegian).

Oglend, A. and M. Sikveland. The behaviour of salmon price volatility, *Marine Resource Economics*, 23(4): pp. 507–526, 2008.

Øveraas, S. T. and C. Rynning-Tønnesen. Production optimization in the salmon farming industry — smolt ordering under uncertainty, Project work, Department of Industrial Economics and Technology Management, Norwegian University of Science and Technology, Trondheim, Norway, 2011.

Rockafellar, R. T. and R. J.-B. Wets. Scenarios and policy aggregation in optimizatio n under uncertainty, *Mathematics of Operations Research*, 16(1): pp. 119–147, 1991.

Sandberg, P. Key figures from aquaculture industry 2009, Directorate of Fisheries, Bergen, Norway, 2010.

Sandmo, A. On the theory of the competitive firm under price uncertainty, *American Economic Review*, 61(1): pp. 65–81, 1971.

Skretting. The Norwegian feed catalog 2011, Skretting AS, Stavanger, Norway, 2011. (in Norwegian).

Stikholmen, N.-A. The development of productivity in salmon farming — a study of the period from 2001 to 2008 using DEA and the Malmquist index, Master's thesis, Tromsø University Business School, Tromsø, Norway, 2010. (in Norwegian).

Tveterås, R. Production risk and productivity growth: some findings for Norwegian salmon aquaculture, *Journal of Productivity Analysis*, 12(2): pp. 161–179, 1999.

Chapter 12

Prioritizing Network Interdiction of Nuclear Smuggling

Dennis P. Michalopoulos*, David P. Morton† and J. Wesley Barnes‡

Summary

We develop a stochastic network interdiction model for prioritizing locations for installing radiation detectors along a nation's border. In this one-country model, we characterize the smuggler population by a set of possible threat scenarios, where the identity of the smuggler is unknown at the time we install detectors. Detector performance depends on the threat scenario, as well as a number of additional factors such as terrestrial background radiation, geometric attenuation, and exposure time. Furthermore, the budget for installing detectors is unknown at the time the installation plan must be proposed. We model the budget as having a known probability distribution, and consequently, the solution to the problem is a rank-ordered priority list of installation locations, where one or more locations are assigned to each priority level. Upon its realization, we exhaust the budget by installing detectors at locations ranked from highest to lowest priority. The identity of the smuggler is subsequently revealed. Having full knowledge of the interdictor's actions, the smuggler then selects an origin-destination path, which maximizes his evasion probability. Modeling the problem as a bilevel stochastic mixed-integer program, we present methods for strengthening the resulting formulation, exact and heuristic solution algorithms, and computational results. We also introduce a performance measure that quantifies the value of our prioritization model.

*Graduate Program in Operations Research and Industrial Engineering, The University of Texas at Austin, Austin, TX 78712, USA, dennis@videologygroup.com.
†Graduate Program in Operations Research and Industrial Engineering, The University of Texas at Austin, Austin, TX 78712, USA, morton@mail.utexas.edu.
‡Graduate Program in Operations Research and Industrial Engineering, The University of Texas at Austin, Austin, TX 78712, USA, wbarnes@mail.utexas.edu.

1 Introduction

Models for interdicting nuclear smuggling are motivated by documented incidents of illicit trafficking of nuclear material and by repeated attempts Of terrorist groups to obtain a nuclear weapon. An International Atomic Energy Agency (IAEA) database on illicit trafficking reports over 1700 confirmed incidents from 1993–2009. About 20% of those involved "unauthorized possession and related criminal activities" and 15 involved highly-enriched uranium or plutonium (International Atomic Energy Agency, 2011).

Weapons-grade material has been seized by authorities in Russia, Germany, the Czech Republic, Lithuania, Bulgaria, Kyrgyzstan, Georgia, Greece, and France. In the majority of these cases, the material was found to have originated in Russia or other parts of the Former Soviet Union (FSU). This points to the vulnerability of our "first line of defense," which deals with nuclear material protection, control, and accountability (US Department of Energy, 2001; US General Accounting Office, 2001). While these efforts are critically important, they are by themselves insufficient, in part, because a complete inventory of the nuclear material that existed in Russia at the beginning of the 1990s is unrealistic.

Al Qaeda operatives have expressed a desire to carryout a nuclear attack on the US and Europe, and Al Qaeda has repeatedly attempted to obtain nuclear material, technology, and expertise. Osama bin Laden said that acquiring a nuclear weapon or other weapon of mass destruction is a "religious duty." A radical Saudi cleric issued a *fatwa* authorizing use of nuclear weapons against American civilians. For further discussion of terrorist aims to acquire nuclear weapons, see, for example, Bunn (2010).

The threat that a terrorist could attack the US with a nuclear bomb may manifest itself in one of several ways. For instance, a terrorist may: (i) be supplied with nuclear material, or a weapon, via a rogue nation; (ii) purchase nuclear material smuggled out of the FSU; or, (iii) deploy a radiological dispersal device (RDD), comprised of more easily obtained radioactive material. The manner in which the Soviet Union dissolved, the pursuit of nuclear technologies by rogue nations, and the pursuit of nuclear energy by an increasing number of nations, have arguably increased the likelihood such threats could be realized. However, for that to occur, nuclear or other radioactive material must be smuggled across a transportation network. This provides an important opportunity to interdict the illicit material and smugglers (Levi, 2007).

The US Domestic Nuclear Detection Office (DNDO) is part of the Department of Homeland Security, and oversees the Global Nuclear Detection Architecture. As of FY 2007, this involved 74 federal programs and a budget of \$2.8 billion, which includes \$1.1 billion for combating nuclear smuggling internationally, \$220 million for supporting detection at the US border, and \$900 million for security and detection activities within the US (US General Accounting Office, 2009). One of the key international programs is called the Second Line of Defense (SLD), which since 1998, has worked to install and maintain radiation detectors at border crossings in Russia, countries of the FSU, and other key countries (US General Accounting Office, 2006). Domestic programs overseen by the DNDO carry out similar efforts for US border crossings.

There is a modest but growing literature in the operations research community related to detecting smuggled nuclear material. There are studies related to multi-layered, multi-detector, protocols at a single port (Boros *et al.*, 2009; Elsayed *et al.*, 2009; Gaukler *et al.*, 2011; McLay *et al.*, 2011; Wein *et al.*, 2007; Wein *et al.*, 2006), analyses involving detection activities within, and around, a city (Chang *et al.*, 2009; McLay *et al.*, 2011; Wein and Atkinson, 2007), and work that involves interdicting nuclear smugglers on a larger transportation network (Dimitrov *et al.*, 2011; Morton *et al.*, 2007; Pan and Morton, 2008). Further related issues involve understanding the role of deterrence (Bier and Haphuriwat, 2011; Merrick and McLay, 2010), understanding the role of secrecy and deception (Brown *et al.*, 2009; Salmerón, 2011), and seeking interdiction plans that are robust to a range of models of smuggler behavior (Bayrak and Bailey, 2008; Gutfraind *et al.*, 2011; Sullivan *et al.*, 2011).

The models of Dimitrov *et al.* (2009), Morton *et al.* (2007) and Pan and Morton (2008) are stochastic Stackelberg games involving a leader (an interdictor) and a follower (a nuclear smuggler). The interdictor installs radiation detectors on a transportation network knowing the smuggler's origin-destination pair, and other characteristics we describe below, only through a probability mass function. After the detectors have been installed, a specific smuggler is revealed, who selects a path to maximize the probability of evading detection. The interdictor's goal is to minimize the evasion probability, subject to a budget constraint, which limits the number of detectors that can be installed. In this chapter, we extend these models by assuming that the budget is uncertain at the time we must commit to a *prioritization* of the locations for detector installation. Like the work of Dimitrov *et al.* (2011) and Morton *et al.* (2007), we restrict attention to models in which the smuggler can encounter at most one detector on his origin-destination path. We call this a one-country model because this situation arises when all origins, or all destinations, are in one country, and we can only install detectors at checkpoints on the border of that country.

Under budget uncertainty, the interdictor postulates a set of budget scenarios with an associated probability mass function. The optimal solution to the prioritized model is a rank-ordered list of detector installation locations. After the budget is revealed, detectors are installed in order from highest to lowest priority until the budget is exhausted. In lieu of assuming a fixed budget, the prioritization model provides a well-hedged approach to decision making when resources that impact those decisions are not known with certainty. A second motivation for a prioritized solution is that detector installations — and more generally, other plans for hardening infrastructure — occur over time as resources become available. When a budget increment becomes available, we cannot reposition detectors, which have already been installed. In this second setting the probability mass function associated with the budget scenarios represents the relative likelihood of a smuggling attempt when the system is in the state associated with each budget increment. While the idea of a time-phased deployment of detectors applies, to keep the discussion concrete we restrict our lexicon to the former notion of having a probabilistic budget.

To illustrate our main idea, consider two fixed budget levels, $b' < b$. It is likely that the optimal detector installation plan under b' is not a subset of the optimal plan under b. Suppose that we determine the optimal installation plan assuming budget level b, and then begin installing detectors at appropriate locations. Subsequently, we learn that our budget has been reduced to b'; i.e., our funds are insufficient to complete the original installation plan. The subset of detectors already installed may have little overlap with the optimal plan under b'. Thus, the resulting set of installations is inferior to the optimal set under this reduced budget.

Alternatively, suppose that we are given b as our installation budget for this year, and we are told that next year, there is the possibility of receiving additional funds, $b''-b$. Restated, in the first scenario, our total interdiction budget is b, and in the second scenario, our total interdiction budget is $b'' > b$. A myopic approach to solving this problem is to: (i) determine the optimal installation plan under b; (ii) install detectors at the appropriate locations; and, (iii) if additional funds $(b'' - b)$ are realized the following year, determine additional installation locations given that the first set is fixed. Once again, the first installation plan, assuming budget b, may have little overlap with the optimal plan under b'', but given that the first set of detectors has already been deployed, we are forced to implement a potentially inferior overall interdiction plan.

A better approach to dealing with the situation just posed is to consider two budget scenarios with realizations b and b'', where each scenario is weighted according to the likelihood of its occurrence. With the optimal prioritized solution in hand, we install detectors at the highest-priority locations that are funded under b, and if we obtain additional funds the following year, we install detectors at the remaining highest-priority locations until the total budget, b'', is consumed.

As a final motivating example, consider the following three interdiction plans superimposed on a map of Russia for budgets of $b = 7$, 8, and 9 detectors. These plans are optimal solutions to a instances of a model that we describe in Section 3, assuming a single budget scenario, known with certainty. In Figure 1a, we have labeled one of the locations as "Checkpt. 1." When we view the interdiction plan in Figure 1b for $b = 8$, we see that Checkpt. 1 is no longer part of the solution. Instead, the solution now includes Checkpt. 2 and Checkpt. 3 on the eastern side of Russia. Finally, when $b = 9$, Checkpt. 1 rejoins the solution and is included in the interdiction plan with Checkpt. 2 and 3. Clearly certain smugglers have a similar preference for Checkpts. 2 and 3, and thus when $b = 7$, the optimal

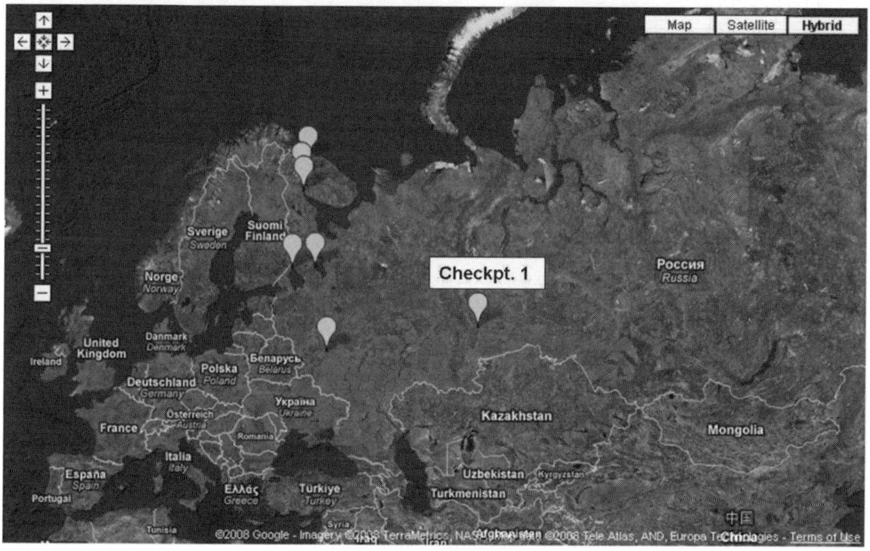

(a) $b = 7$

Figure 1 Installation plan for Russian customs checkpoints when $b = 7–9$, where detector installation locations are indicated by icons.

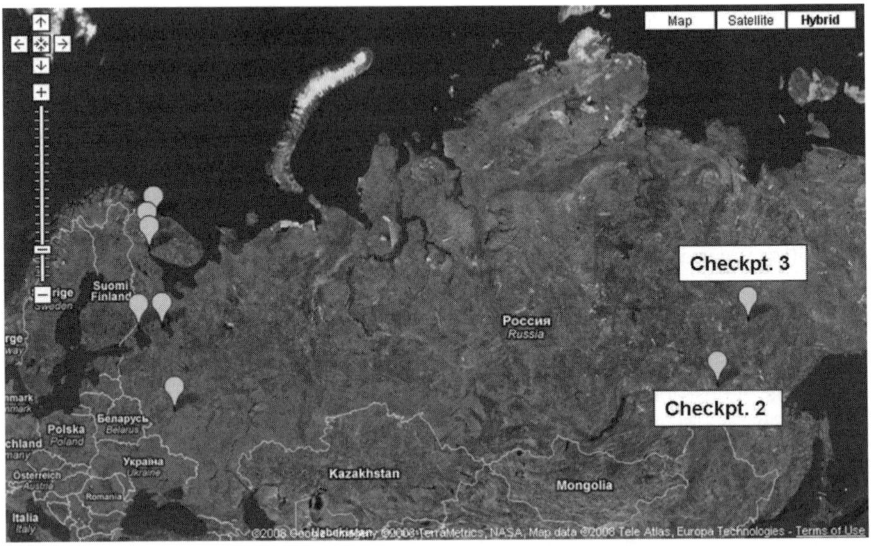

(b) $b = 8$

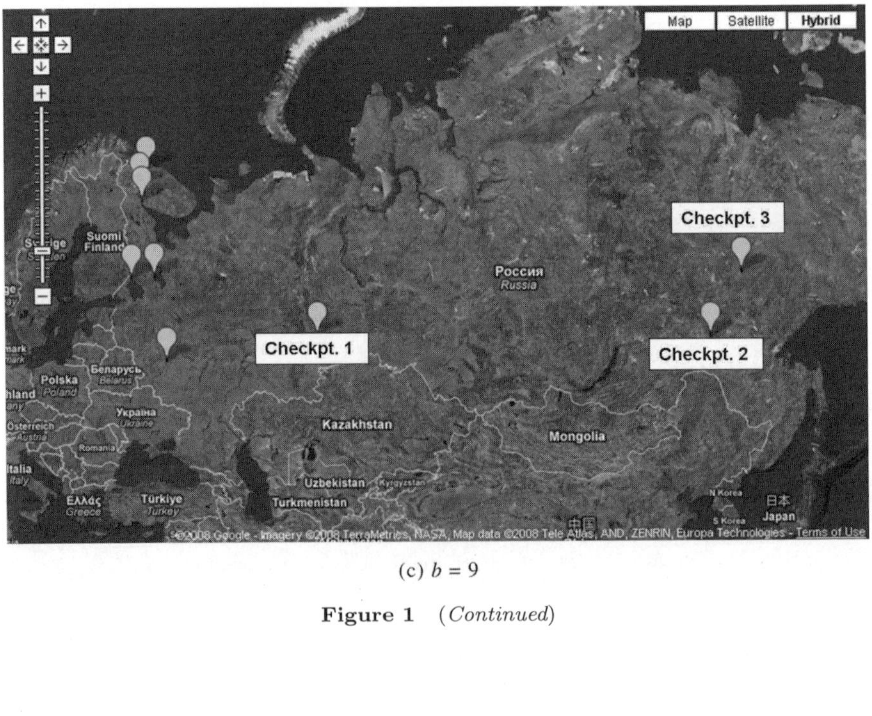

(c) $b = 9$

Figure 1 (*Continued*)

plan instead includes the choice of Checkpt. 1. With the additional budget provided under $b = 8$, the optimal plan discards Checkpt. 1 in favor of covering both Checkpts. 2 and 3, since this results in a greater decrease in smuggler evasion probability. When the budget value increases once again ($b = 9$), we have sufficient resources to cover all three checkpoints, as reflected in Figure 1c. These types of changes can be disconcerting to decision makers when b is uncertain, and in this setting, our prioritization model can provide a more robust approach for choosing a set of detector installation locations.

Some attention has been devoted to the idea of prioritization in the literature. Mettu and Plaxton (2003) and Plaxton (2006) consider a k-median problem with an uncertain budget. They seek a prioritized list of locations, which minimizes the competitive ratio, i.e., the worst-case ratio, over all values of k, of the total distance customers must travel to their closest facility to the minimum distance when k is known. Lin *et al.* (2006) carry out further related work on other combinatorial optimization problems. Dean *et al.* (2008) consider adaptive and non-adaptive versions of the stochastic knapsack problem. Both models seek to maximize expected reward, assume that the size of each item is a random variable, and that the knapsack's residual capacity is realized only after selecting an item. See Dean *et al.* (2005) for related work on more general stochastic packing problems. Allen *et al.* (2003) and Seref *et al.* (2009) consider incremental optimization problems, where a prespecified initial solution is given and then iteratively modified towards a specified goal.

The notion of solutions which are naturally nested at particular budget increments is studied by Witzgall and Saunders (1988) and Hochbaum (2009) in the context of the selection problem of Balinski (1970) and Rhys (1970). Nehme and Morton (2010) study the same issue of nestedness for models with submodularity of the benefit and cost functions. In this work, the budget increments at which solutions are nested is an *output* of a parametric optimization model. In contrast, in the model we describe here, these budget increments are instead specified by the decision maker; i.e., are input to the prioritization model. Koc *et al.* (2008) propose the use of prioritization modeling for project portfolio selection under budgetary uncertainty. The prioritization scheme in Koc *et al.* (2008) is the same that we consider here, but the underlying application differs.

The remainder of this chapter is organized as follows. Section 2 presents our approach to modeling smuggler threat scenarios, detection events, and estimating detection probabilities. In Section 3, we formulate our prioritized bipartite stochastic network interdiction problem (PrBiSNIP). In Section 4, we introduce a performance measure for quantifying the inherent value of our model. Section 5 discusses three methods for tightening the formulation. In Section 6, we present computational results related to two different bipartite smuggling networks: One for securing the Russian border, and one for securing the northern and southern borders of the contiguous United States. Section 7 concludes.

2 Computing detection probabilities

We briefly review how we model threat scenarios and compute associated detection probabilities. For further details see the discussion in Dimitrov *et al.* (2008, 2011) and Thoreson and Schneider (2010). We specify a threat scenario via the smuggler's origin-destination (O-D) pair as well as: (i) the type of nuclear material being smuggled, e.g., weapons-grade plutonium, highly-enriched uranium, and spent nuclear fuel; (ii) the material's mass of the material; (iii) shielding, e.g., lead and borated-polyethylene; and, (iv) the transporting vehicle. Two smugglers may have some overlapping characteristics (e.g., O-D pair and mass of nuclear material), but they may differ with respect to additional threat scenario parameters (e.g., shielding).

For a given threat scenario and for a given location and type of detector, the detection probability (DP) can be derived as a function of the threat characteristics, the survey strategy of the personnel operating the detector, and the detector alarm algorithm. DPs can be estimated by simulating detector response with Monte Carlo N-Particle eXtended (MCNPX), which is a stochastic radiation transport code for modeling the interaction of radiation with its surrounding environment. Using MCNPX, alarm algorithms are applied to the results to translate count rate probability density functions into detection probabilities.

Accurate computation of DPs is complicated by a number of factors. Naturally occurring radioactive material (NORM) and other authorized radionuclide-bearing cargoes can induce false, i.e., "nuisance" alarms. NORM-bearing materials include fertilizer, cat litter, granite, and marble. Compared to background, radiation from most NORM increases detector count rates but does not dramatically alter spectra. Proper consideration of NORM can aid in mitigating nuisance alarms during the alarm algorithm design process. However, educated smugglers can also use NORM-bearing payloads to conceal nuclear material. Spectra for selected NORM can be obtained from published data sources such as Ely *et al.* (2006), which provides spectra for tile and fertilizer.

Nuisance alarms occur when NORM causes the count rate to exceed the alarm algorithm's threshold, triggering the alarm. Note that these are different from statistical false alarms, which occur when the threshold is exceeded in the absence of any local radioactive source, NORM or otherwise. A specific alarm threshold is typically selected to control the false alarm probability (FAP), which implies that alarm thresholds (and hence FAPs) are parameters that are under our control. The FAP also depends on interrogation time, which is a primary concern for portal detectors, where many short measurements generate a spatial spectrum profile.

Vehicle-specific characteristics are also important factors when considering FAP. Setting aside shielding by cargo, vehicle profiles must be considered because they shield background radiation. This baseline depression due to the vehicle profile can be approximately compensated for in real-time detection. An alarm algorithm can adjust thresholds based on baseline depression by vehicle type predicted using a particle transport model or empirical data. An extensive library of depressions by vehicle type and detector location can be found in Lo Presti *et al.* (2006). Ely *et al.* (2005) describe a second method which bins detector signals into energy windows, which is viable since most baseline depression does not strongly affect count ratios between energy bins.

The detection probabilities we compute via MCNPX serve as input for the prioritization models that we describe in the next section. In Section 5 we return to threat scenarios and show that if smugglers under multiple threat scenarios have identical rank-ordered preferences for the checkpoints they can traverse, then these threat scenarios can be aggregated to achieve an equivalent but smaller optimization model. This can greatly reduce the requisite scale of our prioritization model instances.

3 Prioritized bipartite stochastic network interdiction

In this section, we introduce our interdiction model (PrBiSNIP), whose goal is to minimize the probability a nuclear smuggler can evade detection on a bipartite transportation network. The assumptions of our model are presented in Section 3.1, followed by its formulation in Section 3.2.

3.1 *Model assumptions*

In PrBiSNIP, we assume that the characteristics of the smuggler threat scenario we will face are not known *a priori*, but rather are governed by a known probability mass function (PMF). Smuggler evasion events on each arc are mutually independent, and the interdiction budget is assumed to be a random variable, independent of smuggler identity. For smuggler ω, the evasion probability for traveling through checkpoint k undetected, is given by q_k^ω when a detector is installed, and $p_k^\omega > q_k^\omega$ otherwise.

We assume that the interdictor and the smuggler have identical perceptions; i.e., they "agree" on all indigenous evasion probabilities at intermediate points throughout the network, as well as the evasion probability at each checkpoint, both with and without a detector installed. In the one-country interdiction model we consider, each smuggler can encounter at most one detector on an O-D path. As a result, the transportation network may be transformed to a network with bipartite structure, where one set of nodes consists of smuggler O-D pairs, and the other set of nodes consists of border checkpoints through which smugglers may choose to travel. An edge connects an O-D node to a checkpoint node, if a smuggler with this O-D pair can choose to traverse the associated checkpoint. We assume that each smuggler selects a path to maximize the probability of evading detection from his origin to his destination, knowing which subset of checkpoints have detectors. A known PMF is assumed to govern the threat scenarios.

3.2 *Formulation*

The timing of decisions, and observations of the stochastic parameters in PrBiSNIP, follows a three-stage process: First, the interdictor prioritizes the checkpoints, assigning them to a set of predetermined priority levels before the interdiction budget is realized. Second, after the budget is revealed, the interdictor installs detectors at locations ranging from highest to lowest priority until the budget is exhausted, where we assume unit installation costs. Third, the smuggler's threat scenario is revealed, and having full knowledge of the interdictor's actions, the smuggler travels from his origin, s^ω, through a checkpoint k, and then to his destination, t^ω. The smuggler selects this path to maximize his evasion probability. In a preprocessing step, we can compute a path for each O-D pair via each intermediate checkpoint, and hence the smuggler's path selection reduces to a choice of k. Modeled as a stochastic mixed-integer program (MIP), the formulation of our problem is given as follows.

Sets, Indices, and Parameters

$\omega \in \Omega$	Sample space of smuggler threat scenarios
$k \in K$	Candidate locations for installing detectors(checkpoints)
$K^\omega \subseteq K$	Subset of checkpoints smuggler ω may choose to traverse
$\beta \in B$	Sample space of budget scenarios

$l \in L$	Priority levels
p_k^ω	Smuggler ω's evasion probability at $k \in K$ if no detector is installed
q_k^ω	Smuggler ω's evasion probability at $k \in K$ if a detector is installed
(s^ω, t^ω)	Smuggler ω's O-D pair
γ_k^ω	Product of ω's evasion probabilities from s^ω to k and k to t^ω, $k \in K^\omega$
n_l	Number of checkpoints assigned to priority level l (budget increment)
I_l^β	Equals 1 if budget scenario β funds priority level l, 0 otherwise
ϕ^ω	PMF on Ω
ψ^β	PMF on B

Decision Variables

x_{kl} Equals 1 if checkpoint $k \in K$ is assigned to priority level l, 0 otherwise

Formulation

Under budget scenario β, smuggler ω's decision problem is given by

$$h(x, (s^\omega, t^\omega), I^\beta) = \max_{k \in K^\omega} \left\{ \gamma_k^\omega p_k^\omega (1 - x_k^\beta), \gamma_k^\omega q_k^\omega x_k^\beta \right\}, \tag{1}$$

where $x_k^\beta = \sum_{l \in L} I_l^\beta x_{kl}$, $x = (x_{kl})_{k \in K, l \in L} \in X$, and

$$X = \left\{ x : \sum_{k \in K} x_{kl} = n_l, \sum_{l \in L} x_{kl} \leq 1, \; x_{kl} \in \{0, 1\}, \, k \in K, \, l \in L \right\}.$$

The constraints in X state that: (i) exactly n_l checkpoints are assigned to each priority level l; (ii) any checkpoint can be assigned to at most one l; and, (iii) interdiction choices are modeled as binary decision variables.

Linearizing h as defined by equation (1), PrBiSNIP can be formulated as the following stochastic linear MIP:

$$z^* = \min_{x, \theta} \sum_{\omega \in \Omega} \sum_{\beta \in B} \phi^\omega \psi^\beta \theta^{\omega\beta} \tag{2a}$$

$$\text{s.t.} \sum_{k \in K} x_{kl} = n_l, \; l \in L \tag{2b}$$

$$\sum_{l \in L} x_{kl} \leq 1, \; k \in K \tag{2c}$$

$$x_k^\beta = \sum_{l \in L} I_l^\beta x_{kl},\ k \in K,\ \beta \in B \tag{2d}$$

$$\theta^{\omega\beta} \geq \gamma_k^\omega p_k^\omega (1 - x_k^\beta),\ k \in K^\omega,\ \beta \in B,\ \omega \in \Omega \tag{2e}$$

$$\theta^{\omega\beta} \geq \gamma_k^\omega q_k^\omega x_k^\beta,\ k \in K^\omega,\ \beta \in B,\ \omega \in \Omega \tag{2f}$$

$$x_{kl} \in \{0, 1\},\ k \in K,\ l \in L. \tag{2g}$$

The objective function in (2a) is the evasion probability, formed by a weighted sum of evasion probabilities conditional on ω and β. Constraints (2b), (2c), and (2g) encode the requirements for a valid priority list. Constraint (2d) defines the value x_k^β, which equals 1 if checkpoint k has a detector installed under budget scenario β, and 0 otherwise. Constraints (2e) and (2f), coupled with the minimization in (2a), linearize the "max" in equation (1), ensuring $\theta^{\omega\beta} = h(x, (s^\omega, t^\omega), I^\beta)$.

Model (2) can be simplified. Since $p_k^\omega > q_k^\omega$ and x_k^β is binary, the right-hand side of (2f) can be replaced by $\gamma_k^\omega q_k^\omega$, which yields $\theta^{\omega\beta} \geq \gamma_k^\omega q_k^\omega$, $k \in K^\omega$, $\beta \in B$, $\omega \in \Omega$. Now, let $\bar{q}^\omega = \max_{k \in K^\omega} \gamma_k^\omega q_k^\omega$, $\omega \in \Omega$, and note that $\bar{q}^\omega$ denotes the maximum evasion probability for ω if *every* location $k \in K^\omega$ receives a detector, i.e.,

$$\theta^{\omega\beta} \geq \bar{q}^\omega,\ \omega \in \Omega,\ \beta \in B \tag{3}$$

can replace constraint (2f). If we let $r_k^\omega = (\gamma_k^\omega p_k^\omega - \bar{q}^\omega)^+ \equiv \max\{\gamma_k^\omega p_k^\omega - \bar{q}^\omega, 0\}$, and $\bar{\theta}^{\omega\beta} = \theta^{\omega\beta} - \bar{q}^\omega$, then model (2) can be transformed into the following stochastic MIP, where $\bar{\theta}^{\omega\beta}$ has simple lower bounds of zero:

$$\bar{z}^* = \min_{x, \bar{\theta}} \sum_{\omega \in \Omega} \sum_{\beta \in B} \phi^\omega \psi^\beta \bar{\theta}^{\omega\beta} \tag{4a}$$

$$\text{s.t.} \sum_{k \in K} x_{kl} = n_l,\ l \in L \tag{4b}$$

$$\sum_{l \in L} x_{kl} \leq 1,\ k \in K \tag{4c}$$

$$x_k^\beta = \sum_{l \in L} I_l^\beta x_{kl},\ k \in K,\ \beta \in B \tag{4d}$$

$$\bar{\theta}^{\omega\beta} \geq r_k^\omega (1 - x_k^\beta),\ k \in K^\omega,\ \beta \in B,\ \omega \in \Omega \tag{4e}$$

$$x_{kl} \in \{0, 1\},\ k \in K,\ l \in L. \tag{4f}$$

The above transformation eliminates a total of $|B| \sum_{\omega \in \Omega} |K^\omega|$ structural constraints from (2f), and replaces them with simple lower bounds, $\bar{\theta}^{\omega\beta} \geq 0$, $\omega \in \Omega$, $\beta \in B$, which are absorbed in (4e). The linear programming (LP) relaxation of model (4) is at least as strong as that of model (2). The optimal values of models (2) and (4) are related by $z^* = \bar{z}^* + \sum_{\omega\in\Omega} \phi^\omega \bar{q}^\omega$.

Model (4) has decision variables: x_{kl}, $k \in K$, $l \in L$; x_k^β, $k \in K$, $\beta \in B$; and, $\bar{\theta}^{\omega\beta}$, $\omega \in \Omega$, $\beta \in B$. When we focus on the priority list, we refer to $(x_{kl})_{k\in K, l\in L}$ as the solution of model (4). Other times, including the next section, we focus on which checkpoint receives a detector installation under a specific budget scenario, and in this case, we refer to $x^\beta = (x_k)_{k\in K}$ as the solution under budget scenario β. Which notion we refer to as a solution will be clear from context.

4 The value of prioritization

As indicated in Section 1, optimal interdiction plans derived by assuming deterministic budgets can vary significantly under different values of the budget. As such, assuming a fixed budget in the face of uncertainty can produce a poor solution once the true budget is revealed. Prioritization deals effectively with this problem, but at a cost of additional model complexity. Thus, we seek to measure the benefit, or "value," one gains from solving the prioritization model. Our approach is in the spirit of Birge's (1982) "value of the stochastic solution." The typical benchmark is to compare the solution obtained by solving a stochastic program with that obtained when replacing the stochastic parameters with their expectations, but such an approach does not yield a reasonable priority list in our setting and we need another benchmark. The benchmark method we employ for obtaining a prioritized list is based on a greedy myopic procedure alluded to in one of our examples in Section 1.

Our myopic prioritized solution is constructed as follows. Let budget scenarios $\beta_1, \beta_2, \beta_3, \ldots$ be such that $b^{\beta_1} < b^{\beta_2} < b^{\beta_3} < \cdots$. First, for budget scenario β_1, we solve the fixed budget problem, i.e., BiSNIP under budget value b^{β_1}. (This is equivalent to solving PrBiSNIP with a single budget scenario given by b^{β_1}.) Let $z_G^{\beta_1}$ and $x_G^{\beta_1} = (x_k^{\beta_1})_{k\in K}$ denote the optimal value and optimal solution for locating detectors, respectively, under this fixed budget. Next, after fixing each $x_k^\beta = 1$ for locations that received detectors under $x_G^{\beta_1}$, we proceed to budget scenario β_2, and then determine the optimal value, $z_G^{\beta_2}$, and solution under b^{β_2}, given that the detectors located under scenario β_1 cannot be relocated. We proceed to budget scenario β_3, following the same type of procedure of fixing checkpoints for receiving detectors that was used for β_2. The process continues for each subsequent budget scenario until the scenario with the largest budget realization has been considered.

Note that the myopic approach represents how a decision maker who does not account for budgetary uncertainty would augment an existing interdiction plan as the future unfolds and further funds become available. For each budget scenario, the procedure selects a subset of checkpoints without considering the possibility of obtaining additional funds for locating detectors.

Now, let z^* denote the optimal value of PrBiSNIP, as indicated in model (2), and let z_G^β denote the myopic value for each $\beta \in B$. Then

$$VoP(B) = \sum_{\beta \in B} \psi^\beta z_G^\beta - z^*, \tag{5}$$

quantifies the *value of prioritization* via the difference between the value obtained from the nested myopic solutions, x_G^β, $\beta \in B$, and z^*, the optimal value of model PrBiSNIP (2).

5 Improving the formulation of PrBiSNIP

Computational experiments using a branch-and-bound algorithm (B&B) indicate that PrBiSNIP has a particularly weak LP relaxation. Even for modest-size problem instances, model (4) requires significant computational effort to solve to (near-) optimality. In this section, we begin by showing that the number of checkpoints that smuggler ω might access for each $\beta \in B$ is a function of the budget realization, b^β, and we use this fact to tighten the formulation. Then, we describe conditions under which threat scenarios can be aggregated while maintaining an equivalent optimization model. Finally, we present a row generation algorithm that further tightens the formulation prior to employing B&B.

5.1 *Restricting the size of K^ω*

In constraint (4e), for each $\beta \in B$, we consider all possible checkpoints that smuggler ω can choose to travel through. However, we can restrict the size of K^ω, $\omega \in \Omega$, depending on the value of $b^\beta = \sum_{l \in L} I_l^\beta n_l$, i.e., on the value of the budget realization under scenario β. For a fixed $\omega \in \Omega$, let $k_1, \ldots, k_{|K^\omega|}$ correspond to the ordering

$$r_{k_1}^\omega \geq r_{k_2}^\omega \geq \cdots \geq r_{k_{|K^\omega|}}^\omega. \tag{6}$$

We define $K^{\omega\beta} \subseteq K^\omega$ as $K^{\omega\beta} = \{k_1, k_2, \ldots, k_m, k_{m+1}\}$, where $m = \min\{b^\beta, |K^\omega|\}$. Given (6), the definitions of k_{m+1} and $r_{k_{m+1}}$ are clear when $m < |K^\omega|$. If $m = |K^\omega|$ then $r_{m+1}^\omega \equiv 0$, and in this case k_{m+1} indexes an artificial checkpoint with an evasion probability of zero. The index ordering $k_1, \ldots, k_{|K^\omega|}$ depends on ω and the value of m depends on both ω and β, but we suppress these dependencies to simplify notation. When the values of r_k^ω lead to ties in the ordering condition (6), those ties can be broken arbitrarily.

The intuition behind our construction is that if $m < |K^\omega|$ then under budget scenario β, smuggler ω will never choose to travel through any $k \notin K^{\omega\beta}$. This result comes from the fact that we can interdict at most b^β of smuggler ω's checkpoints, and the assumption that smuggler ω always selects a checkpoint with maximum evasion probability in the residual network. Using these facts, it follows that

$$\bar{\theta}^{\omega\beta} \geq r_k^\omega - (r_k^\omega - r_{k_{m+1}}^\omega)x_k^\beta,\ k \in K^{\omega\beta},\ \omega \in \Omega,\ \beta \in B \tag{7}$$

valid for model (4) and dominates constraint (4e).

To verify this claim, we first show that for each (ω, β), it suffices to consider only $k \in K^{\omega\beta}$. For any $\beta \in B$, the result is trivial for $\omega \in \Omega$ such that $m = |K^\omega|$, since this implies that $K^{\omega\beta} = K^\omega \cup \{k_{m+1}\}$ and $r_{k_{m+1}}^\omega = 0$. Thus, we restrict our attention to the case when $m < |K^\omega|$. Constraint (4e) implies that $\bar{\theta}^{\omega\beta} = \max_{k \in K^\omega}\{r_k^\omega : x_k^\beta = 0\}$, $\beta \in B$, and since we can interdict at most $m = b^\beta$ checkpoints, it follows that $\bar{\theta}^{\omega\beta} \geq r_{k_{m+1}}^\omega$ is valid for (4).

Next, we show that (7) is valid and dominates (4e) by considering what happens when (i) $x_k^\beta = 0$, and (ii) $x_k^\beta = 1$, $k \in K^{\omega\beta}$. If $x_k^\beta = 0$, then (7) reduces to $\theta^{\omega\beta} \geq r_k^\omega$, i.e., (4e) and (7) are equivalent in case (i). If $x_k^\beta = 1$, then (7) reduces to $\theta^{\omega\beta} \geq r_{k_{m+1}}^\omega$, and (4e) reduces to $\theta^{\omega\beta} \geq 0$. Since $\theta^{\omega\beta} \geq r_{k_{m+1}}^\omega \geq 0$, it follows that (7) dominates (4e), which verifies the claim.

We can once again reformulate PrBiSNIP through a transformation similar to the one described in Section 3. In particular, let $\tilde{\theta}^{\omega\beta} = \bar{\theta}^{\omega\beta} - r_{k_{m+1}}^\omega$ and $\tilde{r}_k^{\omega\beta} = r_k^\omega - r_{k_{m+1}}^\omega$, where we indicate the dependence of $\tilde{r}_k^{\omega\beta}$ on both ω and β because m depends on both these indices. Substituting $\tilde{\theta}^{\omega\beta}$ and constraint (7) into model (4) yields the following formulation:

$$\min_{x,\tilde{\theta}} \sum_{\omega \in \Omega} \sum_{\beta \in B} \phi^\omega \psi^\beta \tilde{\theta}^{\omega\beta} \tag{8a}$$

$$\text{s.t.} \sum_{k \in K} x_{kl} = n_l,\ l \in L \tag{8b}$$

$$\sum_{l \in L} x_{kl} \leq 1,\ k \in K \tag{8c}$$

$$\tilde{\theta}^{\omega\beta} \geq \tilde{r}_k^{\omega\beta}\left(1 - \sum_{l \in L} I_l^\beta x_{kl}\right), \; k \in K^{\omega\beta}, \, \beta \in B, \, \omega \in \Omega \tag{8d}$$

$$x_{kl} \in \{0, 1\}, \, k \in K, \, l \in L. \tag{8e}$$

Models (4) and (8) are equivalent. The advantage of the reformulation in model (8) is that its LP relaxation is at least as strong as that of model (4), and as our analysis suggests, it can in fact be stronger. For simplicity, we have eliminated variables x_k^β in presenting model (8). That said, we still employ these variables when convenient in the discussion below.

5.2 *Threat scenario aggregation*

We return to the notation of model (4). Consider two smugglers indexed by $\omega_1, \omega_2 \in \Omega$, and for notational simplicity assume these two smugglers have the same set of checkpoints, denoted K. Finally, suppose that for these two smugglers we have

$$r_{k_1}^\omega \geq r_{k_2}^\omega \geq \cdots \geq r_{k_{|K|}}^\omega, \quad \text{for } \omega = \omega_1 \quad \text{and} \quad \omega = \omega_2 \tag{9}$$

for $k_1, k_2, \ldots, k_{|K|} \in K$. Restated, both smugglers, while they may have different evasion probabilities at some or all checkpoints, can rank-order the checkpoints in an identical manner.

The motivation for considering this situation in the context of the PrBiSNIP model can arise as follows. Smugglers ω_1 and ω_2 could be identical in every way, including their O-D pair and the type of material they carry, but the shielding and/or mass of the material that the two smugglers carry may differ. As a result, these two smugglers will be equally vulnerable to detection through indigenous law enforcement while traveling: from their origin to checkpoint k; from checkpoint k to their destination; and, through checkpoint k if no detector is installed. In terms of the notation we introduce in model (2) we have $\gamma_k^{\omega_1} = \gamma_k^{\omega_2}$ and $p_k^{\omega_1} = p_k^{\omega_2}$ for all $k \in K$. The two smugglers are distinguished only by the evasion probability at a checkpoint k if a detector is installed at that checkpoint. For example, other factors being equal, a smuggler transporting material with greater mass will have a smaller evasion probability because he will provide a larger "signal" to a detector. Then, given the definitions of $\bar{q}^\omega$ and $r_k^\omega = (\gamma_k^\omega p_k^\omega - \bar{q}^\omega)^+$ from Section 3, we see that this leads to smugglers ω_1 and ω_2 rank-ordering their checkpoints in an identical manner.

In the setting above, there may be fewer positive values of, say, $r_k^{\omega_1}$ than of $r_k^{\omega_2}$, but they will satisfy the requisite ordering condition (9).

The same result can arise under other conditions, e.g., when two smugglers have different O-D pairs in close geographic proximity to one another. Specifically, since $r_k^\omega = (\gamma_k^\omega p_k^\omega - \bar{q}^\omega)^+$, two smugglers with different but identically ordered values of $\gamma_k^\omega p_k^\omega$ can be aggregated into a single scenario.

If we fix an interdiction plan, $x = (x_{kl})_{k \in K, l \in L}, \in X$, then for ω_1 and ω_2 satisfying the ordering condition (9), we have $k^* \in \text{argmax}_{k \in K} r_k^{\omega_1}(1 - x_k^\beta)$, and $k^* \in \text{argmax}_{k \in K} r_k^{\omega_2}(1 - x_k^\beta)$, for the same checkpoint k^*, which implies that $\bar{\theta}^{\omega_1 \beta} = r_{k^*}^{\omega_1}$, and $\bar{\theta}^{\omega_2 \beta} = r_{k^*}^{\omega_2}$. The contribution of ω_1 and ω_2 to the objective function is then given by $\psi^\beta(\phi^{\omega_1} r_{k^*}^{\omega_1} + \phi^{\omega_2} r_{k^*}^{\omega_2})$. Of course, x is not known ahead of time, but we can replace ω_1 and ω_2 by a single scenario, say $\bar{\omega}$. In doing so, the objective function coefficient of $\bar{\theta}^{\bar{\omega}\beta}$ is $\psi^\beta(\phi^{\omega_1} + \phi^{\omega_2})$, and the evasion probability of each $k \in K$ equals

$$\psi^\beta \left(\frac{\phi^{\omega_1} r_k^{\omega_1} + \phi^{\omega_2} r_k^{\omega_1}}{\phi^{\omega_1} + \phi^{\omega_2}} \right).$$

Extending these ideas to $n \geq 2$ smuggler scenarios yields the following proposition.

Proposition 1. *Consider model (4) and let $x \in X$. Let $\Omega_n, n \in N$, partition Ω such that every smuggler $\omega \in \Omega_n$ rank-orders his checkpoints in the same fashion, i.e., for each $n \in N$, $\exists\, k_1, k_2, \ldots, k_{|K|}$ such that $r_{k_1}^\omega \geq r_{k_2}^\omega \geq \cdots \geq r_{k_{|K|}}^\omega$, $\forall \omega \in \Omega_n$. Define ω_n such that $\phi^{\omega_n} = \sum_{\omega \in \Omega_n} \phi^\omega$ and $r_k^{\omega_n} = \sum_{\omega \in \Omega_n} \phi^\omega r_k^\omega / \phi^{\omega_n}$. If $\bar{\theta}^{\omega_n \beta} = \max_{k \in K} r_k^{\omega_n}(1 - x_k^\beta)$ then $\phi^{\omega_n} \psi^\beta \bar{\theta}^{\omega_n \beta} = \psi^\beta \sum_{\omega \in \Omega_n} \phi^\omega \bar{\theta}^{\omega \beta}$.*

Proof. Every $\omega \in \Omega_n$ rank-orders their checkpoints the same. So, $\exists\, k^*$ such that $r_{k^*}^\omega = \max_{k \in K} r_k^\omega (1 - x_k^\beta)$, $\forall \omega \in \Omega_n$. This same k^* maximizes $\psi^\beta \sum_{\omega \in \Omega_n} \phi^\omega r_k^\omega (1 - x_k^\beta)$. This implies that

$$\begin{aligned}
\phi^{\omega_n} \psi^\beta \bar{\theta}^{\omega_n \beta} &= \max_{k \in K} \phi^{\omega_n} \psi^\beta r_k^{\omega_n} (1 - x_k^\beta) \\
&= \max_{k \in K} \psi^\beta \sum_{\omega \in \Omega_n} \phi^\omega r_k^\omega (1 - x_k^\beta) \\
&= \psi^\beta \sum_{\omega \in \Omega_n} \phi^\omega \max_{k \in K} r_k^\omega (1 - x_k^\beta) \\
&= \psi^\beta \sum_{\omega \in \Omega_n} \phi^\omega r_{k^*}^\omega \\
&= \psi^\beta \sum_{\omega \in \Omega_n} \phi^\omega \bar{\theta}^{\omega \beta}.
\end{aligned}$$

■

Under the conditions stated in Proposition 1, and using the notation defined in the same proposition, we can reformulate model (4) as

$$\min_{x,\bar{\theta}} \sum_{n\in N}\sum_{\beta\in B} \phi^{\omega_n}\psi^{\beta}\bar{\theta}^{\omega_n\beta} \tag{10a}$$

$$\text{s.t.} \sum_{k\in K} x_{kl} = n_l,\ l \in L \tag{10b}$$

$$\sum_{l\in L} x_{kl} \le 1,\ k \in K \tag{10c}$$

$$\bar{\theta}^{\omega_n\beta} \ge r_k^{\omega_n}(1 - \sum_{l\in L} I_l^{\beta} x_{kl}),\ k \in K^{\omega_n},\ \beta \in B,\ n \in N \tag{10d}$$

$$x_{kl} \in \{0,1\},\ k \in K,\ l \in L. \tag{10e}$$

For each $\beta \in B$ in model (10), $r_k^{\omega_n}$ and $\bar{\theta}^{\omega_n\beta}$ still represent conditional evasion probabilities, but are now conditioned on the event Ω_n. Thus, $\phi^{\omega_n} = P(\Omega_n)$ denotes the probability of realizing a threat scenario in Ω_n. Extending the notation we introduce above, we use K^{ω_n} to denote the checkpoints that smugglers $\omega \in \Omega_n$ can traverse.

5.3 *Row generation algorithm*

We again return to the notation of model (4), and in this section, we present a class of valid inequalities for tightening the LP relaxation of PrBiSNIP. They are a straightforward generalization of the step inequalities derived to tighten the formulation of BiSNIP (Morton *et al.*, 2007), i.e., SNIP on a bipartite network with a fixed budget level. The inequalities are also of the form of the star inequalities of Atamturk *et al.* (2000). These *step inequalities* can be identified on an as-needed basis by iteratively solving a shortest path *separation problem.* Let $T^{\omega} = \{k_1, k_2, \ldots, k_n\} \subseteq K^{\omega}$ satisfy the ordering condition $r_{k_1}^{\omega} > r_{k_2}^{\omega} > \cdots > r_{k_n}^{\omega} > r_{k_{n+1}}^{\omega} \equiv 0$, where we require strict inequalities. Then the n-step inequality is given by

$$\bar{\theta}^{\omega\beta} \ge r_{k_1}^{\omega} - (r_{k_1}^{\omega} - r_{k_2}^{\omega})x_{k_1}^{\beta} - \cdots - (r_{k_{n-1}}^{\omega} - r_{k_n}^{\omega})x_{k_{n-1}}^{\beta} - (r_{k_n}^{\omega} - r_{k_{n+1}}^{\omega})x_{k_n}^{\beta}. \tag{11}$$

Step inequalities exploit the structure of PrBiSNIP through the rank-ordering that smugglers assign to their sets of accessible checkpoints. To see this, consider the two-step inequality $\bar{\theta}^{\omega\beta} \ge r_{k_1}^{\omega} - (r_{k_1}^{\omega} - r_{k_2}^{\omega})x_{k_1}^{\beta} - (r_{k_2}^{\omega} - 0)x_{k_2}^{\beta}$. If $x_{k_1}^{\beta} = x_{k_2}^{\beta} = 0$, then smuggler ω selects k_1, and $\bar{\theta}^{\omega\beta} = r_{k_1}^{\omega}$ by the two-step inequality. If $x_{k_1}^{\beta} = 1$, then $\bar{\theta}^{\omega\beta}$ "steps down" to k_2 and reduces to

$\bar{\theta}^{\omega\beta} \geq r^{\omega}_{k_2}\left(1 - x^{\beta}_{k_2}\right)$, which is already an existing constraint in model (4), as well as reformulated models (8) and (10).

In general, when $n \geq 2$, the step inequalities are not redundant when x takes on fractional values in the linear programming relaxation of (4), leading to x^{β}_{k}-values that are fractional. The value of the step inequality is that it removes fractional solutions that are otherwise feasible to the LP relaxation of PrBiSNIP. In the case of the two-step inequality, this can be seen by continuously increasing $x^{\beta}_{k_1}$ from 0 to 1, and then continuously increasing $x^{\beta}_{k_2}$ from 0 to 1.

Since there are an exponential number of step inequalities, adding all of them to the formulation of PrBiSNIP is not a practical solution approach. Our procedure for iteratively identifying a most violated inequality is based on a shortest path separation problem. The procedure takes as input $(x^{lp}, \bar{\theta}^{lp})$, the solution to the LP relaxation of PrBiSNIP, and maximizes the right-hand side of (11) by solving a shortest path problem on a directed acyclic network. The directed network has nodes $k \in K^{\omega}$ plus one additional node labeled $k_{|K^{\omega}|+1}$. We define a directed arc in the network from k_j to k_i with length $(r^{\omega}_{k_i} - r^{\omega}_{k_j})x^{\beta,lp}_{k_i}$ for each pair of nodes satisfying $r^{\omega}_{k_i} > r^{\omega}_{k_j}$, where $x^{\beta,lp}_{k} = \sum_{l \in L} I^{\beta}_{l} x^{lp}_{kl}$, and $r^{\omega}_{k_{|K^{\omega}|+1}} \equiv 0$. We then find a shortest path from node $k_{|K^{\omega}|+1}$ to node k_1, and denote the set of nodes in this shortest path by T^*.

We now present our row generation algorithm, which we describe in the context of model (4). However, we note that the same procedure applies to models (8) and (10), which we later consider in our computational results.

PrBiSNIP Row Generation Algorithm

Input: Optimality tolerance $\varepsilon > 0$
Output: ε-optimal priority list of interdiction locations, $x^* = (x^*_{kl})_{k \in K,\, l \in L}$

Step 1: Solve the LP relaxation of (4) to obtain $(x^{lp}, \bar{\theta}^{lp})$.

Step 2:
for each (ω, β) **do**

Step 2a: Let T^* denote an optimal solution to

$$v^{\omega\beta} = \min_{T \subseteq K^{\omega}} \sum_{k_i \in T} (r^{\omega}_{k_i} - r^{\omega}_{k_{i+1}}) x^{\beta,lp}_{k_i}, \tag{12}$$

where $x_k^{\beta lp} = \sum_{l \in L} I_l^{\beta} x_{kl}^{lp}$.

Step 2b: If $\bar{\theta}^{\omega\beta,lp} < r_{k_1}^{\omega} - v^{\omega\beta}$, add the step inequality defined by T^* to (4).

end for

Step 3: If a step inequality was added for at least one $(\omega, \beta) \in \Omega \times B$, go to Step 1.

Step 4: Solve model (4), with the added step inequalities, to ε-optimality via B&B.

Any step inequality defined by $T \subseteq K^{\omega}$ which does not have $k_1 \in T$ satisfying $r_{k_1}^{\omega} = \max_{k \in K^{\omega}}\{r_k^{\omega}\}$ is dominated by a step inequality with $k_1 \in T$. Thus, when we solve the separation problem (12) in Step 2a, we are maximizing the right-hand side of inequality (11) over all $T \subseteq K^{\omega}$. The condition $\bar{\theta}^{\omega\beta,lp} < r_{k_1}^{\omega} - v^{\omega\beta}$ implies that we have found a $T^* \subseteq K^{\omega}$ for which inequality (11) is violated, and hence we add that inequality to model (4). After no more violated step inequalities are found, we solve the problem to ε-optimality by applying B&B to the tightened formulation.

We close this section with the following proposition, which provides conditions under which a step inequality for pair (ω, β) is valid for additional budget scenarios.

Proposition 2. *Assume that budget scenarios β' and β'' satisfy $I^{\beta'} \leq I^{\beta''}$, where I^{β} denotes the vector $I^{\beta} = (I_l^{\beta})_{l \in L}$ which appears in constraint (4d). Furthermore, assume that T^* defines a most-violated step inequality for budget scenario β'', i.e., $\bar{\theta}^{\omega\beta'',lp} < r_{k_1}^{\omega} - v^{\omega\beta''}$. If $\bar{\theta}^{\omega\beta',lp} < r_{k_1}^{\omega} - v^{\omega\beta''}$, then T^* also identifies a violated step inequality for β'.*

Proof. We have $x_k^{\beta,lp} = \sum_{l \in L} I_l^{\beta} x_{kl}^{lp}$, $k \in K$, $\beta \in B$. By assumption we have $I^{\beta'} \leq I^{\beta''}$, which implies $\sum_{l \in L} I_l^{\beta'} x_{kl}^{lp} \leq \sum_{l \in L} I_l^{\beta''} x_{kl}^{lp}$, and hence $x_k^{\beta',lp} \leq x_k^{\beta'',lp}$. This implies that $v^{\omega\beta'} \leq v^{\omega\beta''} \Rightarrow r_{k_1}^{\omega} - v^{\omega\beta'} \geq r_{k_1}^{\omega} - v^{\omega\beta''}$, and so $\bar{\theta}^{\omega\beta',lp} < r_{k_1}^{\omega} - v^{\omega\beta''} \leq r_{k_1}^{\omega} - v^{\omega\beta'}$. It then follows that T^* defines a violated step inequality for β'. ■

The value of Proposition 2 is that the identification of a most-violated step inequality for budget scenario β may imply a violated inequality for any scenario that funds fewer priority levels. The hope is that the early identification of violated inequalities will reduce the total number of separation problems, which need to be solved.

6 Computational experiments and results

In this section, we present computational results for two different one-country smuggling networks, one for bordering crossings leaving Russia and the other for border crossings entering the US. The former models are somewhat larger, and so in Section 6.1, we consider two different Russian model instances, and discuss results related to the computational methods presented in Section 5. The latter models have geographic structures that lead to interesting structural results, and so in Section 6.2, we discuss the value of prioritization (Section 4) within the context of a model designed to secure the northern and southern borders of the contiguous US. Finally, in Section 6.3, we return to the larger Russian models. For these models we illustrate and present results for a simple heuristic that generates granular priority lists ($n_l = 1, l \in L$) from "coarse" lists ($n_l \geq 2, l \in L$).

6.1 *Russian model*

Tables 1 and 2 show computational results for two Russian model instances: PrBiSNIP-R1 and PrBiSNIP-R2. These models differ only in that PrBiSNIP-R1 uses detectors with an alarm algorithm that accounts for the depression of background radiation due to the transporting vehicle while PrBiSNIP-R1 latter does not. (See the discussion in Section 2.) Both instances consist of 1320 scenarios and 265 checkpoints, resulting in $265\times|L|$ binary decision variables. For each variant of our algorithm, we report total computation time (in seconds), and the number of generated step inequalities. Both problem instances are solved with a code implemented in a C++ programming environment on a 3.73 GHz Dell Xeon dual-processor machine with 8 GB of memory, using CPLEX version 10.1 with an absolute MIP solution tolerance of $\varepsilon = 0.0001$.

The results show that aggregating smuggler scenarios (Section 5.2) reduces computational effort more significantly than restricting their checkpoint sets (Section 5.1). We see that scenario aggregation can drastically reduce the number of step inequalities required to tighten the formulation. This outcome is not surprising, since the aggregation reduces the number of $(\omega, \beta) \in \Omega \times B$ combinations for the algorithm to consider. However, the value of restricting smuggler checkpoint sets is evident when implementing both preprocessing schemes together. The algorithm exhibits the best performance when using both methods, as indicated by the reported solution times and number of generated step inequalities in Tables 1 and 2.

Table 1 Row generation results for PrBiSNIP-R1, $n_l = 1, l \in L$, where $\times$ indicates that we did not obtain a solution with absolute error within $\varepsilon = 0.0001$ of optimal within two hours. The columns under "Rest. Checkpts.", "Scen. Agg.", and "Both" are, in turn, the results obtained by restricting checkpoint sets based on budget scenarios (Section 5.1), threat scenario aggregation (Section 5.2), and both. For each problem instance, the term "cpu" denotes the time required to execute the algorithm of Section 5.3, and "no. $\geq$" denotes the number of generated step inequalities.

	Rest. Checkpts.		Scen. Agg.		Both	
$\|B\|$	cpu	no. $\geq$	cpu	no. $\geq$	cpu	no. $\geq$
5	27.53	1745	7.08	252	0.01	18
6	70.77	2365	11.94	316	0.02	26
7	137.87	3165	38.61	401	0.07	37
8	226.25	4645	46.99	461	0.07	56
9	358.09	5900	77.89	641	0.10	72
10	611.01	8771	154.10	923	0.16	115
11	1222.45	11497	536.41	1341	0.35	151
12	3017.01	15592	1710.92	1599	1.15	200
13	$\times$	$\times$	5698.31	1937	6.23	268
14	$\times$	$\times$	$\times$	$\times$	5.69	313
15	$\times$	$\times$	$\times$	$\times$	14.70	380
16	$\times$	$\times$	$\times$	$\times$	55.83	452
17	$\times$	$\times$	$\times$	$\times$	246.54	524
18	$\times$	$\times$	$\times$	$\times$	252.61	611
19	$\times$	$\times$	$\times$	$\times$	1701.24	718
20	$\times$	$\times$	$\times$	$\times$	$\times$	$\times$

6.2 *US model*

Figure 2 shows the 136 motor-crossing checkpoints we consider in the US model instances, and groups them into four clusters of checkpoints. Here, we consider 140 origin-destination threat scenarios, with half originating in Canada and the other half in Mexico, and we denote the model instance PrBiSNIP-US.

Similar to the example presented in Section 1, checkpoints in each cluster will enter, leave, and possibly reenter the solution under different budget levels, when we optimize under deterministic budget forecasts. For certain budget values, b, significant changes in the optimal solution occur when there are just enough detectors to interdict an entire cluster of checkpoints. As shown in Figure 3, when the budget increases from $b = 10$ to $b = 11$, we have the ability to interdict all checkpoints on the US-Mexico border east of Big Bend. In Figures 4a and 4b, when the budget increases

Table 2 Row generation results for PrBiSNIP-R2, $n_l = 1, l \in L$. The table is to be read using the same conventions as in Table 1.

	Rest. Checkpts.		Scen. Agg.		Both	
$\|B\|$	cpu	no. $\geq$	cpu	no. $\geq$	cpu	no. $\geq$
5	36.58	1126	5.80	226	0.09	20
6	40.55	1621	11.12	266	0.06	32
7	60.19	2186	20.96	324	0.10	45
8	129.09	2767	28.28	366	0.13	59
9	166.40	3352	43.05	465	0.12	73
10	324.36	4775	120.18	714	0.27	108
11	311.09	5738	230.84	1101	0.44	142
12	803.65	7411	593.30	1396	0.90	179
13	4369.96	9363	3045.01	1746	2.27	221
14	×	×	×	×	8.50	256
15	×	×	×	×	20.78	342
16	×	×	×	×	35.99	425
17	×	×	×	×	46.72	476
18	×	×	×	×	143.18	569
19	×	×	×	×	680.88	679
20	×	×	×	×	804.42	760

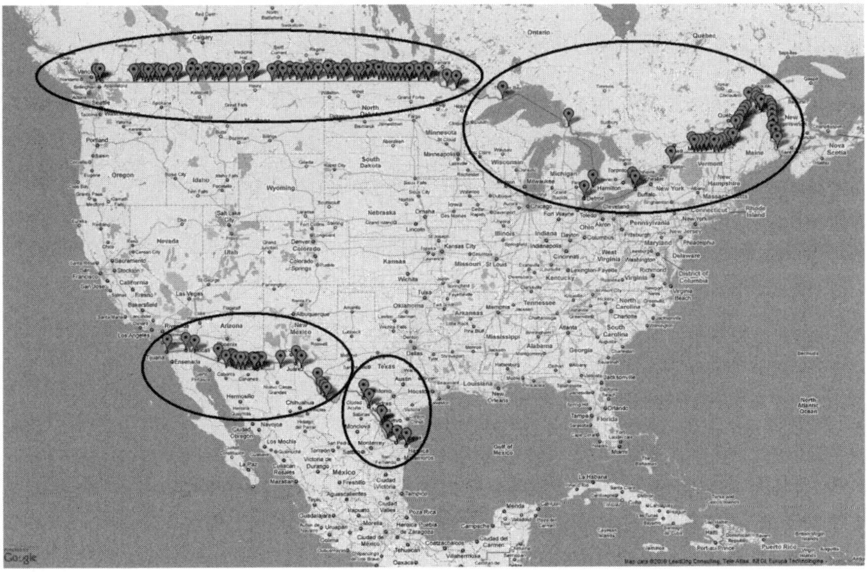

Figure 2 136 motor-crossing checkpoints on the US border grouped into four clusters.

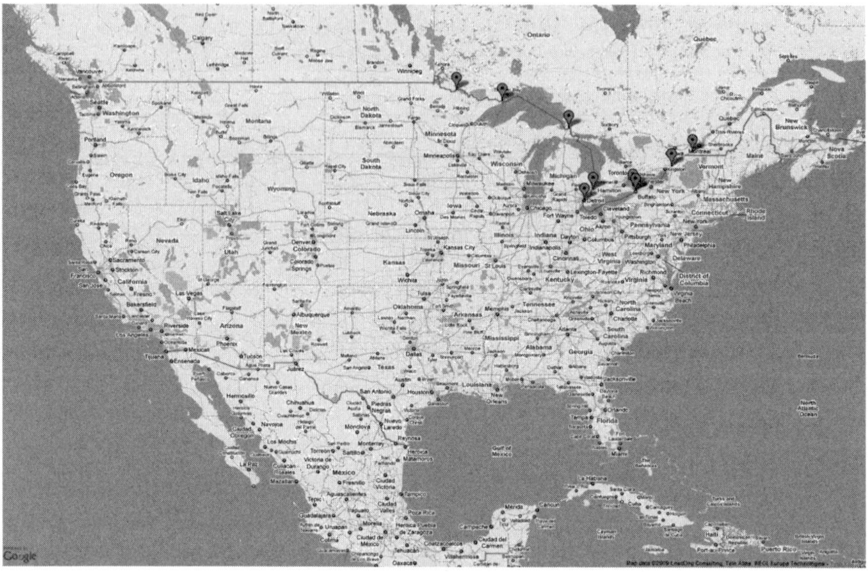

(a) $b = 10$

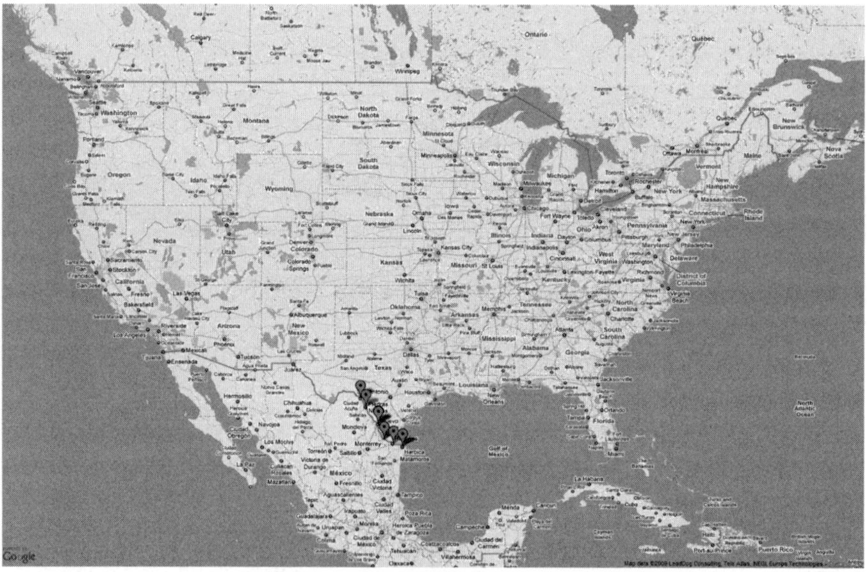

(b) $b = 11$

Figure 3 Part (a) of the figure shows the optimal solution to the US model instance with a budget to install detectors at $b = 10$ locations. Part (b) of the figure is identical but for $b = 11$. Note that the full number of checkpoints are not visible in the map due to their close proximity.

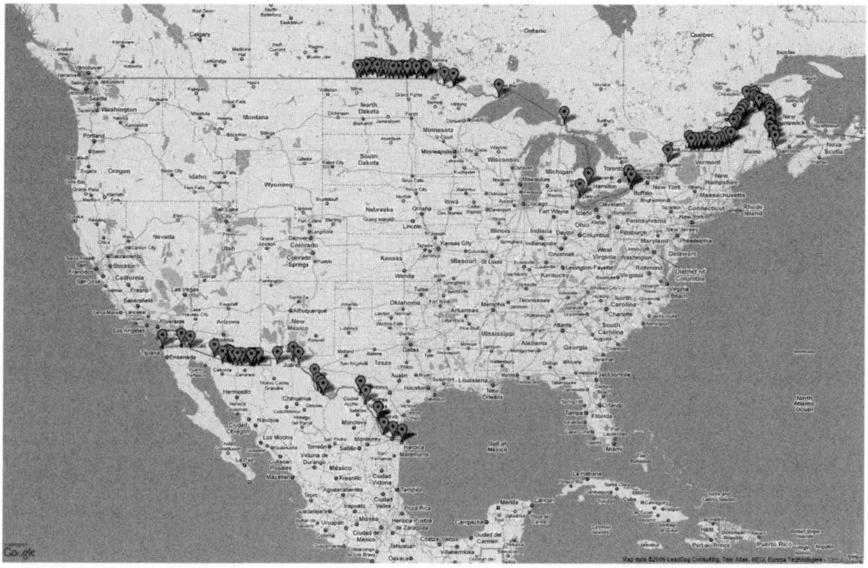

(a) $b = 96$

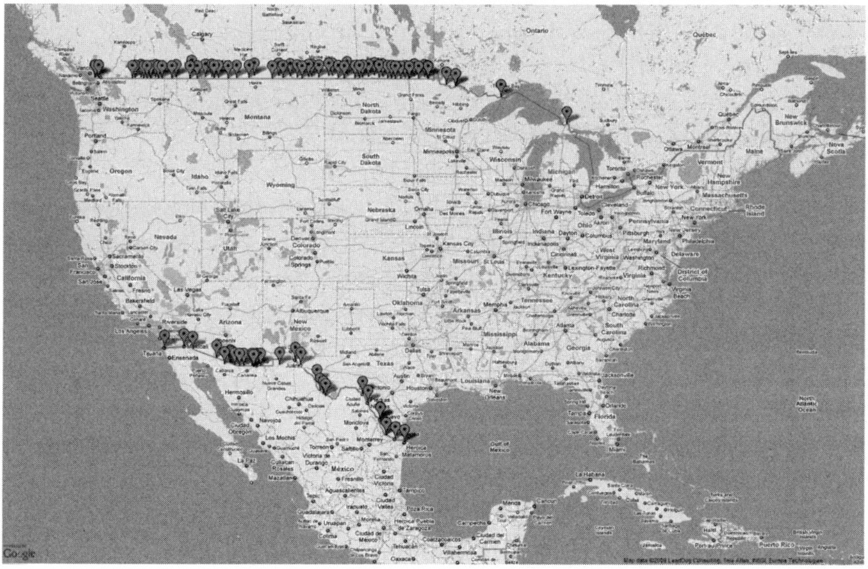

(b) $b = 97$

Figure 4 Part (a) of the figure shows the optimal solution to the US model instance with a budget to install detectors at $b = 96$ locations. Part (b) of the figure shows the optimal solution for $b = 97$.

to $b = 96$ to $b = 97$, we can interdict all checkpoints from Mexico and all checkpoints from Canada, west of Lake Huron.

Given this type of geographic structure in the solution, we can construct budget scenarios where solutions are naturally nested. For example, consider the instance shown in Figure 5, where the prioritized and greedy myopic solutions (Section 4) coincide. Here, we see that β_1 funds 11 checkpoints, which is exactly enough to secure the US-Mexico border east of Big Bend. Budget scenario β_2 funds an additional 23 checkpoints, i.e., 34 in total, which is exactly enough to secure the remainder of the US-Mexico border to the west. Each budget scenario allows for the interdiction of an additional cluster of checkpoints. Thus, when $B = \{\beta_1, \beta_2\}$ with $b^{\beta_1} = 11$ and $b^{\beta_2} = 34$, and $\psi^{\beta_1} = \psi^{\beta_2} = 0.5$, the prioritized and greedy solutions coincide, which implies that $VoP(B) = 0$. In this case, there is no value in dealing with the additional complexity of the prioritization model, since the greedy myopic procedure discussed in Section 4 yields the same solution.

Table 3 displays the value of prioritization under different sets of equally-likely budget scenarios for PrBiSNIP-US. In each instance, we report the number of budget scenarios, "$|B|$", the largest realization of the budget, "$b^{\max}$", and "$VoP(B)\%$", i.e., the percent gap between the values of the greedy and the optimal prioritized solutions, relative to the optimal value z^*. Here, all MIPs are solved within an absolute tolerance of $\varepsilon = 0.001$. In all problem instances, $n_l = 5$ for all $l \in L$. For example, when $|B| = 4$, we have model (4), with budget realizations of $b^{\beta_1} = 5$, $b^{\beta_2} = 10$, $b^{\beta_3} = 15$, and $b^{\beta_4} = 20$; probability mass function $\psi^{\beta_1} = \psi^{\beta_2} = \psi^{\beta_3} = \psi^{\beta_4} = \frac{1}{4}$; and, as shown in Table 3, the prioritized solution has an evasion probability that is smaller by 1.84% over that obtained by the myopic heuristic that we describe in Section 4. As the table indicates, the value of the prioritized solution grows as the magnitude of the maximum budget realization and the number of budget scenarios, grows. When we have 10 equally-likely budget realizations of $5, 10, \ldots, 50$, the value of the prioritized solution is 24% better than that of the nested myopic solution. This suggests that in some settings, solving the optimal prioritization model can yield substantial improvements over what is arguably a very natural heuristic scheme.

6.3 *Heuristic generation of granular priority lists*

In this section, we consider the heuristic generation of "granular" priority lists ($n_l = 1, l \in L$) from "coarse" priority lists ($n_l \geq 2$ for some $l \in L$). The motivation here is the following: A decision maker might be uncomfortable with placing multiple checkpoints on each priority level, but obtaining

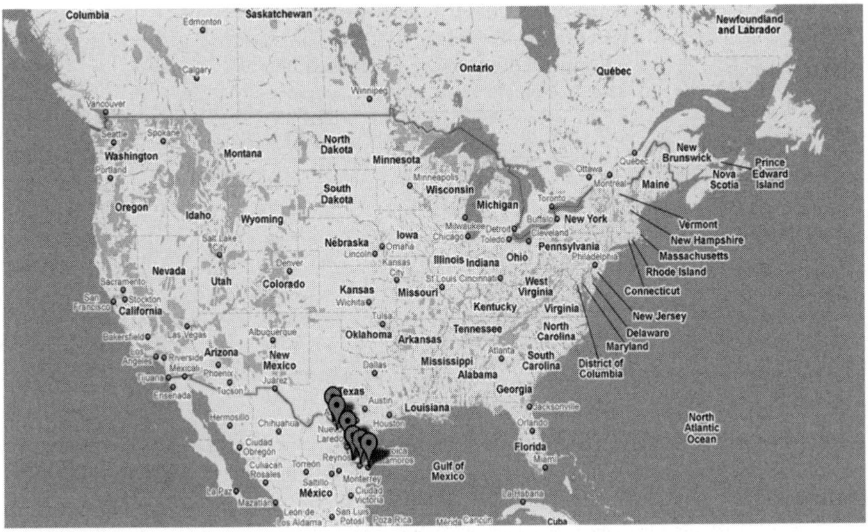

(a) $b^{\beta_1} = n_{l_1} = 11$

(b) b^{β_2} =34, n_{l_2} = 23

Figure 5 An example with two equally likely budget scenarios where the prioritized and greedy solutions are equal. Part (a) of the figure shows the optimal solution under $b^{\beta_1} = 11$ (i.e., $n_{l_1} = 11$) locations. Part (b) of the figure shows the solution for $b^{\beta_2} = 34$ ($n_{l_2} = 23$), under scenario β_2.

Table 3 The value of prioritization (Section 4) for PrBiSNIP-US, expressed as a percentage of z^*, under different sets of equally likely budget scenarios.

$\|B\|$	$b^{\max}$	$VoP(B)\%$
1	5	0.00
2	10	0.00
3	15	1.02
4	20	1.84
5	25	2.34
6	30	2.88
7	35	9.57
8	40	15.31
9	45	20.31
10	50	24.02

granular priority lists can be computationally expensive for large $|B|$. For example, finding the optimal priority list for $|B| = 20, n_l = 1, l \in L$, which contains 20 checkpoints, tends to be more computationally expensive than finding the optimal list for $|B| = 5, n_l = 4, l \in L$, which also contains 20 checkpoints.

We present a heuristic approach to generating a granular priority list from a coarse list. To illustrate, consider an instance where $n_l = 2, l \in L = \{l_1, l_2, l_3\}$, $B = \{\beta_1, \beta_2, \beta_3\}$, and the optimal priority list is, notionally, given by

$$x^* = (\underbrace{k_1, k_2}_{l_1} \vdots \underbrace{k_3, k_4}_{l_2} \vdots \underbrace{k_5, k_6}_{l_3}).$$

First, we map the coarse list defined by x^* to a granular list with one checkpoint at each priority level, i.e., $n_l = 1, l \in L = \{l_1, l_2, \ldots, l_6\}$, $B = \{\beta_1, \beta_2, \ldots, \beta_6\}$. Next, a heuristic local neighborhood search — generated by permuting the components of x^* within each original priority level, one level at a time — is performed. Restated, we first evaluate both granular solutions generated by interchanging k_1 and k_2. For example, we can generate two new solutions by interchanging the priority levels of k_1 and k_2, which yields

$$\hat{x}^1 = (\underbrace{k_1}_{l_1}, \underbrace{k_2}_{l_2} \vdots \underbrace{k_3}_{l_3}, \underbrace{k_4}_{l_4} \vdots \underbrace{k_5}_{l_5}, \underbrace{k_6}_{l_6}),$$

$$\hat{x}^2 = (\underbrace{k_2}_{l_1}, \underbrace{k_1}_{l_2} \vdots \underbrace{k_3}_{l_3}, \underbrace{k_4}_{l_4} \vdots \underbrace{k_5}_{l_5}, \underbrace{k_6}_{l_6}),$$

and

$$\hat{z}^i = \sum_{\omega \in \Omega} \sum_{\beta \in B} \phi^\omega \psi^\beta \max_{k \in K^\omega} \left\{ r_k^\omega \left(1 - \sum_{l \in L} I_l^\beta \hat{x}_{kl}^i \right) \right\}, \; i = 1, 2.$$

Selecting the better of these two solutions, we proceed by fixing k_1 and k_2 at their current levels and carrying out the same procedure with k_3 and k_4. Finally, after k_1, k_2, k_3, and k_4 are fixed, we proceed by evaluating the two granular solutions generated by interchanging k_5 and k_6. If we let L^0 denote the original (coarse) set of priority levels, then the size of the solution neighborhood is equal to $\sum_{l \in L^0} \binom{n_l}{2}$.

During the heuristic procedure, we track the best heuristic solution and objective function value, denoting them as $\bar{x}$ and $\bar{z}$, respectively. Thus, for any solution, $\hat{x}^i$, if $\hat{z}^i < \bar{z}$, we let $\bar{x} \leftarrow \hat{x}^i$, and $\bar{z} \leftarrow \hat{z}^i$. Clearly, $\bar{z}$ yields an upper bound on the optimal value of the granular list model, and we immediately obtain a lower bound via the optimal value of the coarse list model. Thus, if z^* denotes the value of the optimal coarse list, then the difference between $\bar{z}$ and z^* provides an estimate of the gap between $\bar{x}$ and the optimal granular list.

Table 4 shows the heuristic results for different coarse list structures. Here, we assume that $n_l = n, \forall l \in L$. In each case, we indicate $|B|$ and n for the coarse list problem, and the resulting number of budget scenarios in the granular list problem. For each coarse list instance, we apply both smuggler scenario aggregation and restricted smuggler checkpoints, and solve the problem using the algorithm of Section 5.3. We report the total computation time, i.e., the total time required to solve the coarse list problem and carry out the heuristic and the estimate of the relative optimality gap.

The best results are observed when the heuristic is applied to coarse list instances where $n_l = 2, l \in L$. In many cases, the approximate granular solution is near-optimal, and requires far less computational effort than the optimal solution. Our procedure is able to quickly find approximate solutions to instances from Tables 1 and 2 that could not be solved to optimality in under two hours, e.g., $|B| = 20, n_l = 1, l \in L$.

7 Conclusion

We have presented a stochastic network interdiction model for installing radiation detectors along a nation's border. We consider a variant of an interdiction model in which the budget is unknown at the time an installation plan is created, and hence the solution takes on the form of a rank-ordered priority list of candidate locations. The resulting stochastic

Table 4 Heuristic computational results for PrBiSNIP-R1 and PrBiSNIP-R2. Here, we consider coarse lists resulting from different combinations of $5 \leq |B| \leq 10$, and $2 \leq n_l \leq 5$, $l \in L$. For each instance, the term "cpu" denotes the total time (seconds) required to solve the coarse list problem plus the time to execute the heuristic. The term "rel. gap (%)" denotes the relative gap between the optimal coarse list and heuristic objective function values.

Coarse ($n > 1$)		Granular ($n = 1$)	PrBiSNIP-R1		PrBiSNIP-R2	
$\|B\|$	n	$\|B\|$	cpu	rel. gap (%)	cpu	rel. gap (%)
5	2	10	0.60	0.92	0.66	0.92
5	3	15	1.26	1.51	1.81	1.66
5	4	20	4.53	1.73	4.07	1.83
5	5	25	220.94	2.11	35.64	2.32
6	2	12	1.03	0.85	1.33	0.81
6	3	18	3.90	1.32	3.86	1.53
6	4	24	69.10	1.55	15.32	1.69
6	5	30	373.21	2.12	182.48	2.10
7	2	14	1.46	0.77	2.07	0.81
7	3	21	19.06	1.24	9.99	1.44
7	4	28	4714.62	1.43	39.65	1.67
7	5	35	457.28	2.00	×	×
8	2	16	2.53	0.72	2.98	0.77
8	3	24	1065.14	1.15	35.42	1.33
8	4	32	5237.97	1.60	737.91	1.55
8	5	40	681.96	1.91	×	×
9	2	18	8.38	0.68	4.81	0.73
9	3	27	×	×	654.39	1.25
9	4	36	2132.86	1.55	×	×
9	5	45	1570.83	1.88	6423.94	2.04
10	2	20	104.27	0.64	11.25	0.70
10	3	30	×	×	2506.57	1.22
10	4	40	2558.40	1.48	×	×
10	5	50	×	×	×	×

MIP formulation can be tightened by aggregating smuggler threat scenarios, restricting smuggler border checkpoint choices within each budget scenario, and carrying out a separation procedure that iteratively generates members of a class of valid inequalities. Computational results show that the best results are obtained when all three procedures are implemented together. Additionally, we presented a heuristic approach for transforming "coarse" priority lists with two or more checkpoints on each level into "granular" lists with one checkpoint on each level. The heuristic is able to find near-optimal solutions to problem instances that could not be solved to optimality in under two hours by CPLEX, even with our algorithmic enhancements. The resulting upper and lower bounds suggest that the heuristic might aid in solving larger granular list problems via exact methods. Finally, we report results that measure the value of solutions from the optimal prioritization model over a natural myopic heuristic scheme for forming a prioritized list. We show that this value can be significant, and we give associated insight on the geographic structure of solutions for placing detectors for a US model instance.

Acknowledgements

This work has been supported by the National Science Foundation through grants CMMI-0653916 and CMMI-0800676, the Defense Threat Reduction Agency through grant HDTRA1-08-1-0029, and the US Department of Homeland Security under Grant Award Number 2008-DN-077-ARI021-05. The views and conclusions contained in this document are those of the authors and should not be interpreted as necessarily representing the official policies, either expressed or implied, of the US Department of Homeland Security.

References

Allen, D., I. Ismail, J. Kennington and E. Olinick. An incremental procedure for improving path assignment in a telecommunications network. *Journal of Heuristics*, 9: pp. 375–399, 2003.

Atamtürk, A., G. L. Nemhauser and M. W. P. Savelsbergh. The mixed vertex packing problem. *Mathematical Programming*, 89: pp. 35–53, 2000.

Balinski, M. L. Selection problem. *Management Science*, 17: pp. 230–231, 1970.

Bayrak, H. and M. D. Bailey. Shortest path network interdiction with asymmetric information. *Networks*, 52: pp. 133–140, 2008.

Bier, V. M. and N. Haphuriwat. Analytical method to identify the number of containers to inspect at U.S. ports to deter terrorist attacks. *Annals of Operations Research*, 187: pp. 137–158, 2011.

Birge, J. R. The value of the stochastic solution in stochastic linear programs with fixed recourse. *Mathematical Programming*, 24: pp. 314–325, 1982.

Boros, E., L. Fedzhora, P. B. Kantor, K. Saeger and P. Stroud. A large-scale linear programming model for finding optimal container inspection strategies. *Naval Research Logistics*, 56: pp. 404–420, 2009.

Brown, G. G., W. M. Carlyle, R. Harney, E. Skroch and R. K. Wood. Interdicting a nuclear-weapons project. *Operations Research*, 57: pp. 866–877, 2009.

Bunn, M. *Securing the Bomb 2010*. Project on Managing the Atom, Belfer Center for Science and International Affairs, Harvard Kennedy School and Nuclear Threat Initiative, Cambridge, MA and Washington, DC, April 2010.

Chang, J., M. Xie and F. Roberts. Design and deployment of a mobile sensor network for the surveillance of nuclear materials in metropolitan areas.

In *Proceedings of 15th International Conference on Reliability and Quality of Design (ISSAT09)*, 2009.

Dean, B., M. X. Goemans and J. Vondrák. Adaptivity and approximation for stochastic packing problems. In *Proceedings of the 16th ACM-SIAM Symposium on Discrete Algorithms*, pp. 395–404. Vancouver, British Columbia, 2005.

Dean, B., M. X. Goemans and J. Vondrák. Approximating the stochastic knapsack problem: the benefit of adaptivity. *Mathematics of Operations Research*, 33: pp. 945–964, 2008.

Dimitrov, N. B., M. A. Gonzalez, D. P. Michalopoulos, D. P. Morton, M. V. Nehme, E. Popova, E. A. Schneider and G. G. Thoreson. Interdiction modeling for smuggled nuclear material. In *Proceedings of the 49th Annual Meeting of the Institute of Nuclear Materials Management*, 2008.

Dimitrov, N. B., D. P. Michalopoulos, D. P. Morton, M. V. Nehme, F. Pan, E. Popova, E. A. Schneider and G. G. Thoreson. Network deployment of radiation detectors with physics-based detection probability calculations. *Annals of Operations Research*, 187: pp. 207–228, 2011.

Elsayed, E. A., C. M. Young, M. Xie, H. Zhang and Yada Zhu. Port-of-entry inspection: Sensor deployment policy optimization. *IEEE Transactions on Automation Science and Engineering*, 6: pp. 265–276, 2009.

Ely, J. and R. Kouzes. Spies, lies and nuclear threats. Technical report, Health Physics Society, Homeland Security Committee, McLean, VA, 2005.

Ely, J., R. Kouzes, J. Schweppe, E. Siciliano, D. Strachan and D. Weier. The use of energy windowing to discriminate SNM from NORM in radiation portal monitors. *Nuclear Instruments and Methods in Physics Research A*, 560: pp. 373–387, 2006.

Gaukler, G. M., C. Li, R. Cannaday, S. S. Chirayath and Y. Ding. Detecting nuclear materials smuggling: using radiography to improve container inspection policies. *Annals of Operations Research*, 187: pp. 65–87, 2011.

Gutfraind, A., A. Hagberg, D. Izraelevitz and F. Pan. Interdiction of a Markovian evader. In *Operations Research, Computing and Homeland Defense*, pp. 3–15. INFORMS, Hanover, Maryland, 2011.

Hochbaum, D. S. Dynamic evolution of economically preferred facilities. *European Journal of Operational Research*, 193: pp. 649–659, 2009.

International Atomic Energy Agency. Illicit Trafficking Database, Fact Sheet: January 1993-December 2009. http://www-ns.iaea.org/security/itdb.asp. Accessed 5 December 2011.

Koc, A., D. Morton, E. Popova, S. Hess, E. Kee and D. Richards. Optimizing project prioritization under budget uncertainty. In *Proceedings of the 16th International Conference on Nuclear Engineering, Orlando, Florida, May 11–15*, 2008.

Levi, M. *On Nuclear Terrorism*. Harvard University Press, 2007.

Lin, G., C. Nagarajan, R. Rajaraman and D. P. Williamson. A general approach for incremental approximation and hierarchical clustering. In *Proceedings of the 17th ACM-SIAM Symposium on Discrete Algorithms*, pp. 1147–1156. Miami, Florida, 2006.

Lo Presti, C., D. Weier, R. Kouzes and J. Schweppe. Calculating gamma-ray signatures from aged mixtures of heavy nuclides. *Nuclear Instruments and Methods in Physics Research A*, 562: pp. 281–297, 2006.

McLay, L. A., J. D. Lloyd and E. Niman. Interdicting nuclear material on cargo containers using knapsack problem models. *Annals of Operations Research*, 187: pp. 185–205, 2011.

McLay, L. A., J. D. Lloyd and E. Niman. Nuclear threat detection with mobile distributed sensor networks. *Annals of Operations Research*, 187: pp. 45–63, 2011.

Merrick, J. R. W. and L. A. McLay. Is screening cargo containers for smuggled nuclear threats worthwhile? *Decision Analysis*, 7: pp. 155–171, 2010.

Mettu, R. R. and C. G. Plaxton. The online median problem. *SIAM Journal on Computing*, 32: pp. 816–832, 2003.

Morton, D. P., F. Pan and K. J. Saeger. Models for nuclear smuggling interdiction. *IIE Transactions*, 39: pp. 3–14, 2007.

Nehme M. V. and D. P. Morton. Efficient nested solutions of the bipartite network interdiction. In *Proceedings of the IIE Annual Conference*. Cancun, Mexico, 2010.

Pan, F. and D. P. Morton. Minimizing a stochastic maximum-reliability path. *Networks*, 52: pp. 111–119, 2008.

Plaxton, C. G. Approximation algorithms for hierarchical location problems. *Journal of Computer and System Sciences*, 72: pp. 425–443, 2006.

Rhys, J. M. W. Selection problem of shared fixed costs and network flows. *Management Science*, 17: pp. 200–207, 1970.

Salmerón, J. Deception tactics for network interdiction: A multiobjective approach. *Networks*, 2011. DOI: 10.1002/net.20458.

Seref, O., R. K. Ahuja and J. B. Orlin. Incremental network optimization: theory and algorithms. *Operations Research*, 57: pp. 586–594, 2009.

Sullivan, K. M., D. P. Morton, F. Pan and J. C. Smith. Interdicting stochastic evasion paths with asymmetric information on bipartite networks, 2011. working paper.

Thoreson, G. G. and E. A. Schneider. Efficient calculation of detection probabilities. *Nuclear Instruments and Methods in Physics Research A*, 615: pp. 313–325, 2010.

US Department of Energy. MPC&A program strategic plan, July 2001. Washington DC.

US General Accounting Office. Nuclear nonproliferation: Security of Russia's nuclear material improving; further enhancements needed, 2001. GAO-01-312.

US General Accounting Office. Combatting nuclear smuggling: Corruption, maintenance, and coordination problems challenge US efforts to provide radiation detection equipment to other countries, 2006. GAO-06-311.

US General Accounting Office. Nuclear detection: Domestic Nuclear Detection Office should improve planning to better address gaps and vulnerabilities, 2009. GAO-09-257.

Wein, L. M., Y. Liu, Z. Cao and S. E. Flynn. The optimal spatiotemporal deployment of radiation portal monitors can improve nuclear detection at overseas ports. *Science and Global Security*, 15: pp. 211–233, 2007.

Wein, L. M., A. H. Wilkins, M. Baveja and S. E. Flynn. Preventing the importation of illicit nuclear materials in shipping containers. *Risk Analysis*, 26: pp. 1377–1393, 2006.

Wein, L. M. and M. P. Atkinson. The last line of defense: Designing radiation detection-interdiction systems to protect cities from a nuclear terrorist attack. *IEEE Transactions on Nuclear Science*, 54(3): pp. 654–669, 2007.

Witzgall, C. J. and P. B. Saunders. Electronic mail and the "locator's" dilemma. In R. D. Ringeisen and F. S. Roberts, editors, *Applications of Discrete Mathematics*, pp. 65–84. SIAM, Philadelphia, 1988.

Zhuang, J. and V. M. Bier. Secrecy and deception at equilibrium, with applications to anti-terrorism resource allocation. *Defence and Peace Economics*, 22: pp. 43–61, 2011.

Chapter 13

Sawmill Production Planning Under Uncertainty: Modelling and Solution Approaches

Masoumeh Kazemi Zanjani*, Mustapha Nourelfath† and Daoud Ait-Kadi‡

Summary

We investigate the operational level production planning problem in the sawing units of sawmills under the uncertainty in the quality of materials and demand. In order to widen the application of the production planning model in different types of sawmills, we also take into account setup constraints. We propose two-stage stochastic programming, robust optimization, and multi-stage stochastic programming models to formulate different aspects of this problem. As demand and yield own different uncertain natures, they are modeled separately and then integrated. Demand uncertainty is considered as a dynamic stochastic data process during the planning horizon which is modeled as a scenario tree. The uncertain yield is modeled as scenarios with a stationary probability distribution during the planning horizon. Yield scenarios are then integrated into each node of demand scenario tree, constituting a hybrid scenario tree. We also propose two solution strategies to find good solutions with an acceptable gap to the optimal solution, while taking into account setup constraints in the production planning model. The first strategy is based on the progressive hedging algorithm (PHA), while the second strategy is a successive approximation heuristic which solves the problem by considering

*Department of Mechanical and Industrial Engineering, Concordia University, 1515 St. Catherine St. West, EV4.243, Montreal(QC),Canada, H3H 1M8, kazemi@encs.concordia.ca, corresponding author.

†Interuniversity Research Centre on Enterprise Networks, Logistics and Transportation (*CIRRELT*), Department of Mechanical Engineering, Université Laval, Québec, Canada, mustapha.nourelfath@gmc.ulaval.ca.

‡Interuniversity Research Centre on Enterprise Networks, Logistics and Transportation (*CIRRELT*), Department of Mechanical Engineering, Université Laval, Québec, Canada, daoud.aitkadi@gmc.ulaval.ca.

only a subset of scenarios which is updated at each iteration. We conduct a case study with respect to a realistic scale prototype sawmill. Computational experiments highlights the effectiveness of the proposed tools for production planning under uncertainty in the sawing units of sawmills.

1 Introduction

Production planning in the sawing units of sawmills which can be classified as divergent-type production systems is a challenging issue. More precisely, in such production systems, several products can be produced after processing a common material. In the majority of cases, the type of yielded products and their volume are variable, due to the variable characteristics of inputs. Moreover, some by-products might also be produced for which the demand is almost zero. The variable mix of products, in addition to the existence of by-products makes production planning more difficult. Moreover, there are other sources of uncertainties, namely demand, that sawmills as many manufacturing environments are faced with, which complicate the production planning and control.

The goal of this work is to address the operational level production planning problem for the sawing units of sawmills. In sawing units, different classes of raw materials (logs) are sawn into different pieces of lumber by means of different cutting patterns. However, due to the non-homogeneity in the characteristics of logs, the quantities of lumber sawn by different cutting patterns (the process yields) are random variables. The non-stationary and uncertain behaviour of the lumber demand in the market during the planning horizon is another uncertain factor that is considered in this study. Furthermore, in many sawmills, regarding the technological constraints for changing machine and equipments setups, only one process family comprised of processes with similar setup requirements can be executed in each period. The operational level production planning in sawing units is to determine the optimal quantity of log consumption from different classes and the selection of appropriate cutting patterns to fit against product demands. The part of the demand that cannot be fulfilled on time due to machine capacities, log inventory, and random yields will be postponed to the following periods by considering a backorder cost. The objective is to minimize log consumption cost, as well as product inventory/backorder costs.

The described production planning problem can be formulated as a linear program (LP) or a mixed-integer program (MIP) (by considering set-up constraints). The uncertainty in the yields of cutting patterns and the demand can be represented as random coefficients in the constraint matrix and the random right-hand-side vector in the mentioned models, respectively.

The existing approaches in the literature for sawmill production planning can be classified into two groups: (1) Deterministic optimization models where the average process yields and demand are taken into account (Gaudeault *et al.*, 2011); (2) combined optimization type solutions linked to real-time simulation subsystems (Mendoza *et al.*, 1991; Maness and Adams, 1991; Maness and Norton, 2002). Implementing the production plan proposed by a deterministic sawmill production planning model results in extra inventory of products with lower quality and price, and simultaneous backorder of products with higher quality and price. On the other hand, in the second type of sawmill production planning approaches, the stochastic characteristics of logs are taken into account by assuming that all the input logs are scanned through an X-ray scanner before planning. However, such approaches involve the following limitations to be implemented in many sawmills: logs needed for the next planning horizon are not always available in sawmills to be scanned before planning. Furthermore, to implement this method, the logs should be processed in the production line in the same order they have been simulated, which is not an easy practice. Finally, scanning logs before planning is a time consuming process in high capacity sawmills, and delays the planning process significantly.

It has been shown in the literature that in mathematical programming models which include random parameters in their right-hand–side and/or technological coefficients, the stochastic programming approach (Kall and Wallace, 1994; Birge and Louveaux, 1997) results in higher quality solutions compared with mean-value deterministic models. Stochastic programming and robust optimization (Mulvey *et al.*, 1995) has been applied in several production planning contexts under uncertain environments (see for example, Escudero *et al.* (1993); Bakir and Byrne (1998); Leung *et al.* (2006, 2007); Kazemi Zanjani *et al.*, 2010a, b, c, 2011).

In this paper, we first address the sawmill production planning problem by only taking into account the random quality of materials. We propose a two-stage stochastic program with recourse to address this problem. We also propose an efficient approach to model the random yields, which results in yield scenarios very close to those that might be observed in real sawmills. In order to validate the proposed stochastic sawmill production planning tool and to compare it with the existing deterministic model, we implement the plans proposed by each model through Monte Carlo simulation under circumstances close to real sawmills. The stochastic and deterministic sawmill production planning models are compared in terms of the realized backorder size and the precision of their proposed plans. The proposed two-stage stochastic model is applied to a prototype medium capacity sawmill in Quebec (Canada). Our computational results indicate that the proposed stochastic model results in a more realistic production plan than a deterministic model, thus it serves as a viable tool for sawmill planning by considering random characteristics of logs. Next, we investigate the robustness in sawmill production planning while considering the random process yields. In other words, we consider a service sensitive company that wants to establish a reputation for always meeting a target customer service level, where we define the service level as the proportion of the customer demand that can be fulfilled on time. Thus, the need for robustness has been mainly recognized in terms of determining a robust customer service level by minimizing the products' backorder size variability in the presence of different scenarios for random yields. The robustness in the products' inventory size is also considered in this problem. We propose two robust optimization (RO) models (Mulvey *et al.*, 1995) with two alternative variability measures which are inspired from Takriti and Shabbir (2004) and Shabbir and Sahinidis (1998). The proposed robust optimization (RO) models are applied to a realistic-scale prototype sawmill. A comparison between the backorder/inventory size variability in the two-stage stochastic model and the two robust optimization models is provided. Finally, the tradeoff between the backorder/inventory size variability and the expected total cost in the two RO models is discussed and a decision framework to select among them is proposed.

As the third step, we address the production planning problem in sawing units by taking into account simultaneously the random yield and demand. We formulate this problem as a multi-stage stochastic program with recourse. As demand uncertainty originates from market conditions and yield uncertainty is due to non-homogeneity in the quality of raw materials, they are modeled separately and independently. We assume that the uncertain demand evolves as a discrete time stochastic process during the planning horizon with a finite support, thus it is modeled as a scenario tree. The uncertain yields are modeled as scenarios with stationary probability distribution during the planning horizon. Finally, yield scenarios are integrated into the demand scenario tree, forming a hybrid scenario tree with two types of branches. The goal of the multi-stage stochastic model is to determine implementable plans for production that takes into account the possible demand and yield scenarios, to provide recourse actions in the future, and to minimize the expected cost of raw material consumption, holding inventory, and backorders. The proposed approach is applied to a prototype sawmill. Numerical results indicate that the solution to the multi-stage model is far superior to the optimal solution of the mean-value deterministic and the two-stage stochastic models. Furthermore, it is shown that the significance of using multi-stage stochastic programming is increased as the variability of random demand is augmented in the scenario tree.

Finally, in order to widen the application of our proposed multi-stage stochastic production planning model in sawmills, where changing machines and equipments setups during the production shift is not possible, we included setup constraints in the model. The resulting production planning model is a multi-stage stochastic mixed-integer programming (MS-MIP) model lacking any special structure. As the development of efficient decomposition and cutting plane algorithms to obtain a good solution in a reasonable amount of time is not straightforward in this model, we propose two heuristic solution strategies to find good solutions with a small gap to the optimal solution.

The first solution strategy is a scenario decomposition approach that decomposes the stochastic model for the possible scenarios of the random events. Each scenario sub-problem then becomes a deterministic problem that can either be solved directly by a commercial solver (e.g. CPLEX MIP) or an efficient heuristic algorithm. Following the original decomposition scheme proposed by Rockafellar and Wets (1991), we address the issue of using the local information yielded by the sub-problems as global solutions so as to guide the overall algorithm toward a unique solution.

The second solution strategy is inspired from the Scenario Updating (SU) method proposed by Lulli and Sen (2006). It is based on solving instances of the problem, which contain only a subset of scenarios in the scenario tree. The subset of scenarios is updated by adding those scenarios that imply either certain degradation or an improvement of the objective function value. We improve the SU algorithm in the following directions. The first improvement is motivated by the fact that, in our experiments, the scenario selection rules proposed in Lulli and Sen (2006) for updating the scenario tree did not converge fast. We propose a new scenario selection rule and we combine it with some of the most appropriate rules in the SU method in order to increase the convergence rate of the algorithm as well as the quality of the approximate solution. Furthermore, the proposed lower bound on the objective function value of a stochastic MIP model in Lulli and Sen (2006) is more appropriate for a multi-stage stochastic MIP with partial recourse. As the problem we are addressing is a multi-stage stochastic MIP with complete recourse, we consider only the proposed upper bound to evaluate different scenarios to be added into the current scenario tree.

The remainder of this paper is organized as follows. In the next section, we discuss briefly production planning in the sawing units of sawmills. In Section 3, we propose a two-stage stochastic linear program and two robust optimization models for sawmill production planning under the uncertainty of process yields. In Section 4, the multi-stage stochastic program proposed to address sawmill production planning under yield and demand uncertainty is provided. In Section 5, a model and two solution strategies for sawmill production planning under yield and demand uncertainty while considering setup constraints are presented. Our concluding remarks are given in Section 6.

2 Production planning in the sawing units of sawmill

2.1 *Sawing process in sawmills*

There are a number of processes that occur at a sawmill: log sorting, sawing, drying, planing and grading (finishing). The raw materials in sawmills are the logs which are transported from different districts of forests after the felled trees have been bucked. Logs are classified in sawmills according to various attributes, namely: diameter class, species, length, taper, etc. Logs are broken down into different dimensions of lumber by means of different cutting patterns. From each log, several pieces of sawn lumber are produced depending on the cutting pattern. Figure 1 illustrates the

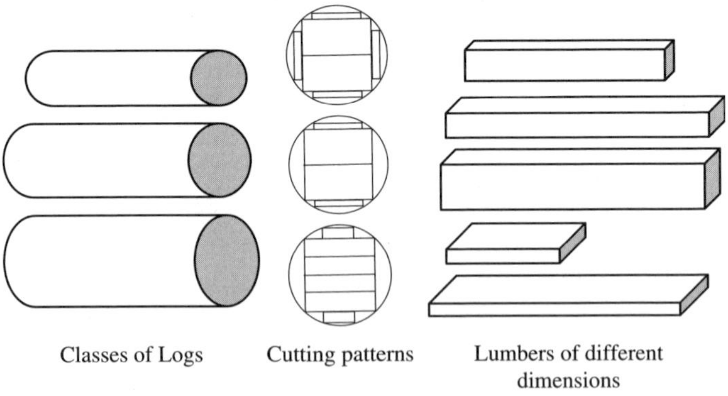

Figure 1 Sawing Process in Sawmills.

sawing process in the sawing units of sawmills. The lumber quality (grade) as well as its quantity yielded by each cutting pattern depends on the quality and characteristics of input logs. The mix of lumber types as well as by-products (e.g., chips, sawdust, etc.) that should be produced after sawing each individual log is determined by means of an X-ray scanner and a log sawing optimizer. More precisely, at the beginning of the sawing line, the logs are passed through an X-ray scanner. The log characteristics (scanning results) are then transferred into a log sawing optimization software. The optimizer, in turn, with the objective of maximizing the total volume/value, determines the optimal mix of lumbers and by-products. It should be mentioned that at the beginning of each production shift (period) the desired types of cutting patterns in addition to the updated market prices of different lumber types are entered into the optimizer. Having determined the optimal product mix for each log, the logs are then passed through a set of machines. Due to technological limitations in most sawmills, the optimizer parameters as well as machine setups can only be adjusted at the beginning of each production shift. As a consequence, only compatible processes (a combination of log diameters and cutting patterns) which share the same machine setups and parameters can be run in each period.

Despite the classification of logs in sawmills, variety of characteristics (in terms of diameter, number of knots, internal defects, etc.) might be observed in different logs of the same class. As it is not possible in many sawmills to scan the logs before planning, the exact yields of cutting patterns for different log classes cannot be determined *a priori*. Uncertainty of the demand in the lumber market is another important parameter that should be taken into account in sawmill production planning.

Production planning in sawing units at the operational level consists of the selection of the optimal mix of log classes and corresponding cutting patterns in each period (day) in the planning horizon in order to fulfill the lumber demand, considering log inventory, and machine capacities. The objective is to minimize the material consumption plus product backorder and inventory costs.

2.2 *A deterministic LP model for sawmill production planning*

Consider a sawmill with a set of products (lumber types) P, a set of classes of raw materials (logs) C, a set of production processes A, a set of resources (machines) R, and a planning horizon consisting of T periods.

For modeling simplicity, we define a process as a combination of a log class and a cutting pattern. To state the deterministic linear programming model for this production planning problem, the following notation is used:

2.2.1 *Notation*

Indices

p product (lumber type)
t period
c raw material (log) class
a production process
r resource (machine)

Parameters

h_{pt} inventory holding cost per unit of product p in period t
b_{pt} backorder cost per unit of product p in period t
m_{ct} raw material cost per unit of class c in period t
I_{c0} the inventory of raw material of class c at the beginning of the planning horizon
I_{p0} the inventory of product p at the beginning of the planning horizon
s_{ct} the quantity of material of class c supplied at the beginning of period t
d_{pt} demand of product p by the end of period t
ϕ_{ac} the units of raw material class c consumed by process a (consumption factor)
ρ_{ap} the units of product p produced by process a (yield of process a)
δ_{ar} the capacity consumption of resource r by process a
M_{rt} the capacity of resource r in period t

Decision variables

X_{at} the number of times each process a should be run in period t (production plan)
I_{ct} inventory size of raw material of class c by the end of period t
I_{pt} inventory size of product p by the end of period t
B_{pt} backorder size of product p by the end of period t

2.2.2 The deterministic LP model

$$\text{Minimize} \quad Z = \sum_{p\in \mathrm{P}} \sum_{t=1}^{T} (h_{pt} I_{pt} + b_{pt} B_{pt}) + \sum_{c\in \mathrm{C}} \sum_{t=1}^{T} \sum_{a\in A} m_{ct} \phi_{ac} X_{at} \tag{1}$$

Subject to
Material inventory constraint

$$I_{ct} = I_{ct-1} + s_{ct} - \sum_{a\in A} \phi_{ac} X_{at}, \quad t = 1,\ldots,T, \quad c \in \mathrm{C}, \tag{2}$$

Product inventory constraint

$$I_{p1} - B_{p1} = I_{p0} + \sum_{a\in \mathrm{A}} \rho_{ap} X_{a1} - d_{p1},$$

$$I_{pt} - B_{pt} = I_{pt-1} - B_{pt-1} + \sum_{a\in \mathrm{A}} \rho_{ap} X_{at} - d_{pt}, \quad t = 2,\ldots,T, \quad p \in \mathrm{P}, \tag{3}$$

Production capacity constraint

$$\sum_{a\in \mathrm{A}} \delta_{ar} X_{at} \le M_{rt}, \quad t = 1,\ldots,T, \quad r \in \mathrm{R}, \tag{4}$$

Non-negativity of all variables

$$X_{at} \ge 0, I_{ct} \ge 0, I_{pt} \ge 0, B_{pt} \ge 0,$$
$$t = 1,\ldots,T, \quad p \in \mathrm{P}, \quad c \in \mathrm{C}, \quad a \in A. \tag{5}$$

The objective function (1) minimizes the total inventory and backorder costs for all products and raw material cost for all classes in the planning horizon. Constraint (2) corresponds to the raw material inventory balance. Constraint (3) indicates the product inventory balance. Constraint (4) requires that the total production do not exceed the available production capacity.

3 Production planning in sawing units by taking into account the random yield

In this section, we propose a two-stage stochastic program with recourse for operational level production planning in the sawing units of sawmills by taking into account random process yields. In the following we first present our proposed approach for modeling random yields. Then, we provide the two-stage stochastic sawmill production planning model.

Table 1 Example of Scenarios for Yields of a Process in Sawmills.

	Products					
Scenarios	**P1**	**P2**	**P3**	**P4**	**P5**	**P6**
1	1	0	1	0	1	1
2	2	1	1	0	1	0
3	1	0	0	1	1	1
4	2	0	0	1	0	1

3.1 *Modelling the random process yields*

We model the random yield as a number of scenarios. A scenario for the yields of process a (combination of a log class c and a cutting pattern s) in a sawmill is defined as possible quantities of lumber that can be produced by cutting pattern s after sawing each log of class c. As an example of the uncertain yields in sawmills, consider the cutting pattern s that can produce six products (P1–P6) after sawing the logs of class c. Table 1 represents four scenarios among all possible scenarios for the yields of this process.

In this work, we assume that all the logs that will be processed in the next planning horizon are supplied from the same type of forest. Hence, a stationary probability distribution can be considered for the quality of logs and random yields during the planning horizon. In sawmills thousands of logs in each class are sawn in each period. As a consequence, instead of considering the yields of individual logs as scenarios, we propose to consider the average yield of a batch of logs as a scenario in the stochastic model. In fact, the probability that the yield parameter be equal to the yield of each individual log is very small (by the law of large numbers (LLN) in statistics). Thus, more probable and realistic scenarios would be those defined as the average yields of a number of logs. Such scenarios with their probability distributions in sawmills can be determined as follows.

(1) Take a sample of logs in each log class (e.g. 3000) and let them be processed by each cutting pattern. Compute the average yield for the sample.
(2) Repeat step 1 for a number of replications (e.g. 30).
(3) By the Central Limit Theorem (CLT) in statistics, the average yield has an approximate normal distribution. Thus, based on the average yields computed for each replication in step 2, estimate the mean and variance of the normal distribution corresponding to the average yields of each process.

It should be noted that the implementation of step 1 in this approach is very difficult in sawmills. In fact, the high production speed in sawmills makes it almost impossible to track the logs through the production line and to observe the result of sawing individual logs. As a more feasible alternative, we used the set of yield scenarios generated by a log sawing simulator (Optitek). Optitek was developed by FPInnovations, a research company for Canada's solid wood products industry. It was developed based on the characteristics of a large sample of logs in different log classes, as well as sawing rules available in Quebec sawmills. The inputs to this simulator include the characteristics of logs in each class (obtained by X-ray scanning), and the desired cutting pattern. Regarding the characteristics of input logs and the sawing rules corresponding to the requested cutting pattern, the simulator generates different quantity of lumber yields for each log. Thus, in order to implement step 1 in the proposed scenario generation approach, a sample (e.g. 3000) of yields is randomly selected among the set of scenarios already generated by Optitek, and the average yield for the sample is computed, accordingly.

3.2 *A two-stage stochastic model with recourse for sawmill production planning*

To include the random nature of process yields in sawmill production planning, we expand the model (1)–(5) to a two-stage stochastic linear program with recourse. We represent the random yield vector by $\underset{\sim}{\xi}$, where $\underset{\sim}{\xi} = \{\rho_{ap} | a \in \mathrm{A}, p \in \mathrm{P}\}$. We also represent each realization (scenario) of random process yields by ρ_{ap} $(\underset{\sim}{\xi}^i)$. We denote the total number of yield scenarios by N, and the probability of each scenario i by p^i, respectively. In two-stage stochastic programming, we explicitly classify the decision variables according to whether they are implemented before or after an outcome of the random variable is observed. In this production planning problem, in the first-stage (planning stage) the decision maker does not have any information about the process yields due to lack of complete information on the characteristic of raw materials (logs). However, the production plan should be determined before the complete information is available. Thus, the first-stage decision variable is the production plan (X_{at}). In the second stage (plan implementation stage), the logs are sawn and their realized yields are available. At this stage the product inventory or backorder sizes, which are called the second-stage decisions (recourse actions), can be calculated. The objective of the two-stage stochastic sawmill production planning model is to minimize the material consumption cost, plus the expected inventory and backorder costs (recourse costs) for all yield scenarios. The resulting deterministic equivalent of the two-stage stochastic model can be formulated as follows.

$$\text{Minimize} \quad Z = \sum_{c\in \mathrm{C}}\sum_{t=1}^{T}\sum_{a\in A} m_{ct}\phi_{ac}X_{at} + \sum_{i=1}^{N}\sum_{p\in \mathrm{P}}\sum_{t=1}^{T} p^i[h_{pt}I^i_{pt} + b_{pt}B^i_{pt}] \tag{6}$$

Subject to

$$I_{ct} = I_{ct-1} + s_{ct} - \sum_{a\in A}\phi_{ac}X_{at}, \quad t = 1,\ldots,T, \quad c \in \mathrm{C}, \tag{7}$$

$$I^i_{p1} - B^i_{p1} = I_{p0} + \sum_{a\in \mathrm{A}} \rho_{ap}(\underset{\sim}{\xi}^i)X_{a1} - d_{p1},$$

$$I^i_{pt} - B^i_{pt} = I^i_{pt-1} - B^i_{pt-1} + \sum_{a\in \mathrm{A}} \rho_{ap}(\underset{\sim}{\xi}^i)X_{at} - d_{pt},$$

$$t = 2,\ldots,T, \quad p \in \mathrm{P}, \quad i = 1,\ldots,N, \tag{8}$$

$$\sum_{a\in \mathrm{A}} \delta_{ar}X_{at} \le M_{rt}, \quad t = 1,2,\ldots,T, r \in \mathrm{R}, \tag{9}$$

$$X_{at} \ge 0, I_{ct} \ge 0, I^i_{pt} \ge 0, \quad B^i_{pt} \ge 0,$$

$$c \in C, \quad p \in P, \quad t = 1,\ldots,T, \quad a \in A, \quad i = 1,\ldots,N. \tag{10}$$

In the two-stage stochastic program (6)–(10), I^i_{pt} and B^i_{pt} denote the inventory and backorder sizes of product p in period t under the yield scenario i, respectively.

3.3 *Validation of the stochastic sawmill production planning model by Monte Carlo simulation*

In this subsection, we propose an approach to compare the plans proposed by the stochastic and deterministic sawmill production planning models. This approach is based on implementing the plans of the mentioned models under conditions very close to real sawmills through Monte Carlo simulation. We assume that the company is very service sensitive, i.e., the realized total backorder size after implementation of the production plan is more crucial than the realized inventory size. Thus, the following key indicators of performance are considered to compare the deterministic and stochastic models:

(1) Backorder gap (BO gap): the gap between the realized total backorder size of the deterministic and the stochastic models' plans, after implementing the mentioned plans.
(2) Plan precision: the gap between the planned total backorder size determined by the production planning model and the realized total backorder size, after implementing the model's plan. This indicator evaluates also the extent to which the yield scenarios considered in the stochastic model are close to the scenarios that can be observed in the real production process.

In order to compute the realized total backorder size after implementing the plans, we propose to use Monte Carlo simulation. The main objective of this simulation is to implement the production plans virtually, by considering the yield scenarios that might be realized during the plan implementation in real sawmills. Hence, the following features are considered for the simulator:

(1) To get the production plans proposed by the deterministic and stochastic models as well as the products demands, as the inputs.
(2) To simulate the production plan implementation based on the received production plan as follows:
    (2.1) To determine a sample size equal to the number of times each process should be run in each period (production plan).
    (2.2) To take randomly a sample of scenarios (with the size determined in 2.1) for the yields of each process, from the set of available scenarios for the yields of that process. It should be mentioned that this step is equivalent to selecting a random sample of logs from the inventory of logs in each class to be sawn by each cutting pattern, while implementing the production plan in sawmills.
(3) To compute the total production size of each product at the end of each period, after simulating the plan implementation for that period (step 2).
(4) To compute the backorder or inventory size of each product in each period based on the total production size of that product (computed in step 3) and its demand for that period.

Figure 2 illustrates the main features of the simulator which is designed to simulate the plan implementation in sawmills.

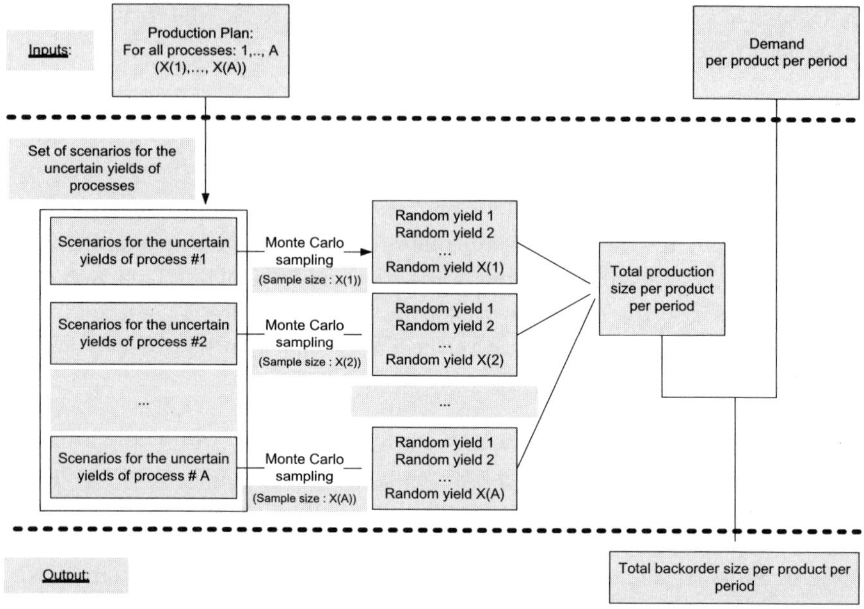

Figure 2 Simulation of the Production Plans Implementation in Sawmills.

3.4 *Application of the proposed stochastic production planning model for a prototype sawmill*

In the following, we describe the numerical experiments using the proposed two-stage stochastic production planning model in a prototype medium capacity sawmill. We first describe the characteristics of the test problem; then we compare the stochastic and mean-value deterministic models' plans by the proposed Monte Carlo simulation approach for different demand levels.

3.4.1 *Prototype sawmill*

The prototype sawmill is a typical medium capacity softwood sawmill located in Quebec (Canada). The sawmill focuses on sawing high-grade products to the domestic markets as well as export products to the USA. It is assumed that the input bucked logs into the sawmill are categorized into three classes. Five different cutting patterns are available. The sawmill produces 27 products of custom sizes (e.g. $2(\text{in}) \times 4(\text{in})$, $2(\text{in}) \times 6(\text{in})$ lumber) in four lengths. In other words, there are 15 processes able to produce 27 products with random yields. We consider two bottleneck machines: Trimmer and Bull. The planning horizon consists of 30 periods (days). It would be worth mentioning that the data used in this example are based on data gathered from different sawmills in Quebec (Canada). As the list of custom sizes, machine parameters and prices are proprietary, they are not reported in this paper.

3.4.2 *Comparison between the stochastic and deterministic sawmill production planning models*

Four different demand levels (D1, D2, D3, D4) are considered, where D2 $= 2 \times$ D1, D3 $= 3 \times$ D1, D4 $= 4 \times$ D1. For each demand level, 60 demand scenarios are generated randomly which are distinguished by the distribution of total demand between different products. Hence, a total of 240 (4×60) test problems are solved by the deterministic LP and stochastic models. The plan implementation simulation is run for 1000 replications. The expected total backorder size computed in 1000 replications is used to compute the key indicators of performance for the test problems. Figures 3 and 4 compare the backorder gap (BO gap) and the plan precision, respectively, computed for the 60 test problems, corresponding to the 60 demand scenarios, in each of the 4 demand levels. It is worth mentioning that the values of BO gap and plan precision are computed as follows:

BO gap $= 100 \times (BO_D - BO_s)/BO_D$,
Plan precision $= 100 \times (BO_{sim} - BO_{plan})/BO_{sim}$,

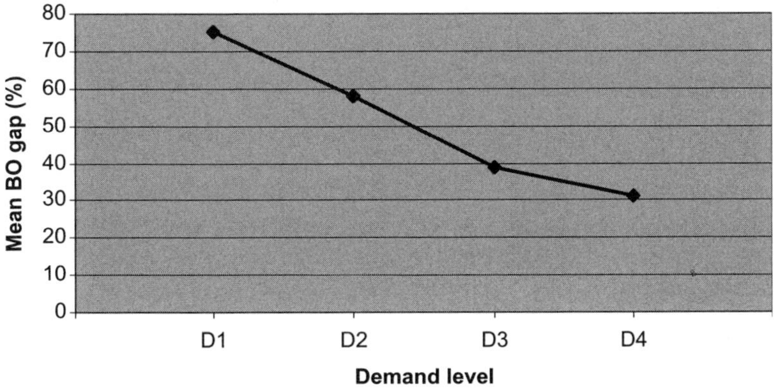

Figure 3 Backorder Gap of the Stochastic and Deterministic Sawmill Production Planning Models.

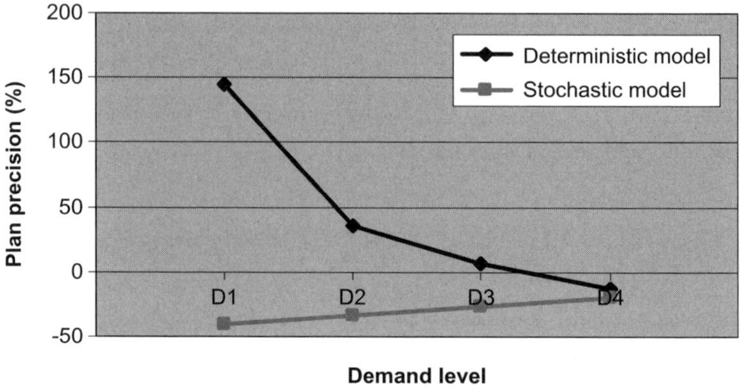

Figure 4 Plan Precision Comparison of the Deterministic and Stochastic Sawmill Production Planning Models.

where BO_D and BO_S denote the realized total backorder size of the deterministic and stochastic models after their implementation (through Monte Carlo simulation), respectively; BO_{sim} and BO_{plan} denote the realized total backorder size after plan implementation and the total backorder size determined by the production planning model, respectively.

As can be observed in Figure 3, the production plan proposed by the stochastic model results in smaller realized backorder size (after implementing the plan) than the deterministic model plan, for the four demand levels. However, the gap between the total backorder size of the stochastic and the deterministic models' plans decreases, as the demand is increased. This should come as no surprise. As we mentioned before, the sawmill example is a medium capacity sawmill, where thousands of logs are sawn in each period in the planning horizon. By the law of large numbers (LLN) in statistics, as the demand is increased, the average yields of each process in each period (observed through Monte Carlo sampling in the plan implementation simulator) will be closer to their expected values considered in the deterministic model.

Figure 4 compares the average precision of plans proposed by the stochastic and deterministic models, for the four demand levels. As can be observed in Figure 4, the precision of the production plan proposed by the stochastic model is higher than the deterministic one, for all demand levels. By increasing the demand, the average yields (observed after implementing the plan through Monte Carlo simulation) get closer to the yield scenarios considered in the stochastic model. Hence the precision of the plans proposed by the stochastic model improves for larger volumes of demand. For the lower demand levels, the stochastic model proposes a relatively pessimistic plan. On the other hand, the deterministic model provides an optimistic plan, since it does not take into account different yield scenarios that might be observed in sawmills. However, as the demand increases, the average yields of each process in each period observed through Monte Carlo simulation get closer to average yields which are used in the deterministic model (LLN). Thus, the precision of the plan proposed by the deterministic model increases, as the demand is increased.

Regarding the simulation results, it is clear that the two-stage stochastic model provides a more realistic production plan in sawmills, in terms of the realized backorder size, than the mean value deterministic model. The deterministic model provides an optimistic plan, since it considers the deterministic yields (expected values). On the other hand, as the stochastic model considers yield scenarios which are close to those that might be observed in real sawmills and finds a production plan with minimum expected backorder and inventory size for all the yield scenarios, the production plan provided by this model is more realistic. Finally, regarding the high precision of the production plan proposed by the stochastic model, it can be concluded that the proposed approach for modeling the random yields (see Subsection 3.1) generates high quality scenarios which are very close to those that might be observed in real sawmills.

3.5 *Taking into account service robustness in the stochastic sawmill production planning model*

It is important to note that the stochastic programming approach focuses on optimizing the expected performance (e.g. cost) over a range of possible scenarios for the random parameters. We can expect that the system would behave optimally in the *mean* sense. However, the system might perform poorly at a particular realization of scenarios such as the worst-case scenario. More precisely, unacceptable inventory and backorder size for some scenarios might be observed by implementing the solution of two-stage stochastic model (6)–(10). To handle the trade-off associated with the expected cost and its variability in stochastic programs, Mulvey *et al.* (1995) proposed the concept of robust optimization. In this subsection, two robust optimization (RO) models are proposed for sawmill production planning while considering random yields. We consider a service sensitive company that wants to establish a reputation for always meeting customer service level. We also define the customer service level as the proportion of the customer demand that can be fulfilled on time, and we use the expected backorder size as a measure for evaluating the service level. Thus, the need for robustness has been mainly recognized in terms of determining a robust customer service level by minimizing the products backorder size variability in the presence of different scenarios for random yields. The robustness in the products inventory size is also considered in this problem. Two alternative variability measures are used in the robust optimization model which can be selected depending on risk aversion level of decision maker about backorder/inventory size variability and the total cost. The first variability measure is the upper partial moment of order 2 (UPM-2) (used also in Takriti and Shabbir, 2004), and the second one is the upper partial variance (UPV) which is the quadratic version of upper partial mean (UPM) of Shabbir and Sahinidis (1998).

3.5.1 *Robust optimization models for sawmill production planning*

To state the robust optimization model for this production planning problem, the following notation is used in addition to the quantities introduced in 2.2.1 and 3.2:

3.5.1.1 Notations

Parameters

λ Goal programming parameter ($\lambda \geq 0$)
R^* Target inventory/backorder cost

Decision variables

Δ_+^i The variability measure of inventory and backorder cost for scenario i

3.5.1.2 The robust optimization models

$$\begin{aligned}\text{Minimize } Z = &\sum_{c\in \mathrm{C}}\sum_{t=1}^{T}\sum_{a\in A} m_{ct}\phi_{ac}X_{at} + \sum_{i=1}^{N}\sum_{p\in \mathrm{P}}\sum_{t=1}^{T} p^i[h_{pt}I_{pt}^i + b_{pt}B_{pt}^i]\\ &+\lambda\sum_{i=1}^{N} p^i\Delta_+^{i^2}\end{aligned} \tag{11}$$

Subject to

Material inventory constraint

$$I_{ct} = I_{ct-1} + s_{ct} - \sum_{a\in A}\phi_{ac}X_{at}, \quad t = 1, \ldots, T, \quad c \in \mathrm{C}, \tag{12}$$

Production capacity constraint

$$\sum_{a\in \mathrm{A}} \delta_{ar} X_{at} \leq M_{rt}, \quad t = 1, 2, \ldots, T, \quad r \in \mathrm{R}, \tag{13}$$

Product inventory constraint

$$\begin{aligned} I^i_{p1} - B^i_{p1} &= I_{p0} + \sum_{a\in \mathrm{A}} \rho^i_{ap} X_{a1} - d_{p1}, \\ I^i_{pt} - B^i_{pt} &= I^i_{pt-1} - B^i_{pt-1} + \sum_{a\in \mathrm{A}} \rho^i_{ap} X_{at} - d_{pt}, \\ &\qquad t = 2, \ldots, T, \quad p \in \mathrm{P}, \quad i = 1, \ldots, N, \end{aligned} \tag{14}$$

Recourse cost variability

$$\begin{aligned} \Delta^i_+ &\geq \sum_{p\in \mathrm{P}} \sum_{t=1}^{T} (h_{pt} I^i_{pt} + b_{pt} B^i_{pt}) - \sum_{i'=1}^{N} \sum_{p\in \mathrm{P}} \sum_{t=1}^{T} p^{i'} (h_{pt} I^{i'}_{pt} + b_{pt} B^{i'}_{pt}), \\ &\qquad i = 1, \ldots, N, \quad \text{(RO-UPV)} \\ \Delta^i_+ &\geq \sum_{p\in \mathrm{P}} \sum_{t=1}^{T} (h_{pt} I^i_{pt} + b_{pt} B^i_{pt}) - R^*, \\ &\qquad i = 1, \ldots, N, \quad \text{(RO-(UPM-2))} \end{aligned} \tag{15}$$

Non-negativity of all variables

$$\begin{aligned} &X_{at} \geq 0, \quad I_{ct} \geq 0, \quad I^i_{pt} \geq 0, \quad B^i_{pt} \geq 0, \quad \Delta^i_+ \geq 0, \\ &c \in C, \quad p \in P, \quad t = 1, \ldots, T, \quad a \in A, \quad i = 1, \ldots, N. \end{aligned} \tag{16}$$

The objective function (11) is to minimize the raw material consumption cost, the expected inventory and backorder costs, in addition to inventory and backorder cost variability for all scenarios in the planning horizon. The inventory and backorder costs are computed by multiplying the inventory and backorder unit cost by the inventory and backorder size, respectively. λ is the goal programming parameter that models the trade-off between the expectation and variability of the recourse cost in the objective function. For $\lambda = 0$, model (11)–(16) would be the two-stage stochastic model in Subsection 3.2. Constraints (15) compute the inventory and backorder cost variability for each scenario. Depending on the type of variability measure that is used in the RO model, the mentioned cost variability is defined as follows. In the RO-UPV model it is defined as the difference between the total inventory and backorder cost of each scenario and the expected inventory and backorder cost, while in the RO-(UPM-2) model it denotes the difference between the total inventory and backorder cost of each scenario and a target inventory and backorder cost (R^*).

3.5.2 *Application of robust optimization approach for the sawmill example*

In this section, we report the results of the implementation of the two robust optimization models for the prototype sawmill described in Subsection 3.4.1.

3.5.2.1 RO-(UPM-2) model results

Remember from Subsection 3.5.1 that the RO-(UPM-2) model requires a target recourse cost R^*. It should be noted that the target cost can be determined based on the desired service level. In this sawmill example, we provide the target cost as a percentage of the optimal expected backorder and inventory cost when $\lambda = 0$ (the two-stage stochastic program). We consider a range of values for parameter λ to generate a range of optimal solutions. Figure 5 illustrates the tradeoff between the backorder/inventory cost variability and raw material cost for different values of λ for each R^*. As expected, for a given value of R^*, increasing λ reduces the backorder/inventory cost (size) variability. Thus, we can expect more control on the excess of each scenario's backorder/inventory cost over the target cost (R^*) as well as decreased expected backorder cost (size), although at the expense of increased raw material cost and the expected inventory cost (size). In other words, by enforcing the backorder/inventory cost variability measure in the objective function of model (11)–(16), the

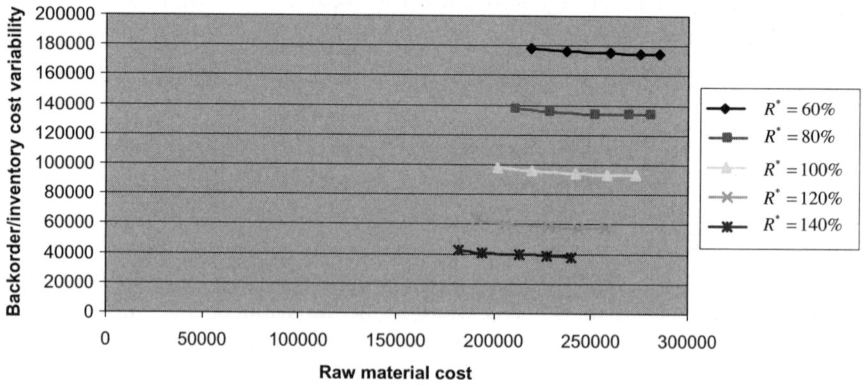

Figure 5 Raw Material Cost and Backorder/Inventory Cost Variability Tradeoff in RO-(UPM-2) Model for Different Values of λ and R^*.

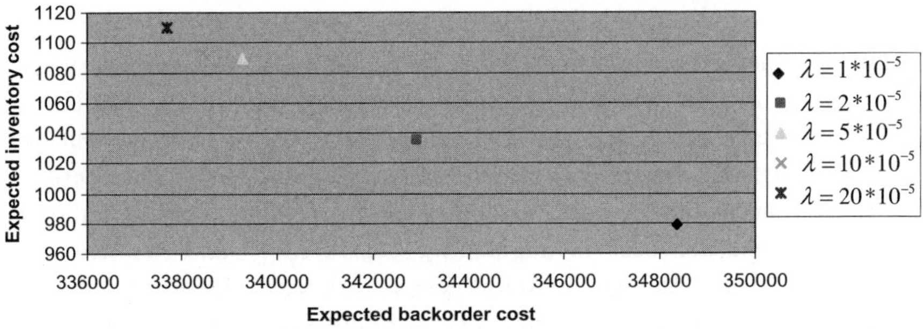

Figure 6 Tradeoff between the Expected Backorder and Inventory Costs for Different Values of $\lambda(R^* = 100\%)$ in RO-(UPM-2).

production level and consequently raw material consumption is increased in order to minimize the excess of backorder/inventory cost of all scenarios over the target cost. Furthermore, the increased inventory cost (size) is also the result of increasing the production level and raw material consumption.

Figure 6 illustrates the tradeoff between the expected backorder and inventory cost by enforcing the robustness parameter in the RO-(UPM-2) model for $R^* = 100\%$. As can be observed from the results presented in Figure 5, decreasing the target recourse cost R^* in this example does not necessarily decrease the variability measure. This implies that the control on the excess of the backorder/inventory cost of scenarios over a target cost might be limited depending on the yield scenarios as well as problem constraints (i.e. raw material inventory and machine capacity constraints). In other words, by imposing a target cost on the variability measure in the RO model it might not be feasible to achieve a plan with small recourse cost variability. In this example, for target costs R^* below or equal to the expected recourse cost of the two-stage stochastic model, the recourse cost variability can be decreased into a limited value by enforcing the value of λ. On the other hand, for higher values of R^* (120% and 140%) more robust production plans with less variable backorder/inventory cost (size) can be achieved at the expense of lower service level (higher expected backorder size). From the above discussions, it can be concluded that, if the decision maker wishes to use the RO-(UPM-2) model to obtain a robust production plan, he/she should choose a value of λ which reflects his/her risk aversion about backorder/inventory cost (size) variability as well as increased raw material consumption and expected inventory cost (size). Moreover, it might not be feasible to achieve a completely robust production plan by considering any desirable service level (target cost R^*), depending on the yield scenarios and problem constraints.

3.5.2.2 RO-(UPV) model results

Figure 7 illustrates the trade-off between the backorder/inventory cost variability and the expected backorder/inventory cost in the RO-(UPV) and the 2-stage stochastic models. In the RO-(UPV) model, as it can be observed from Figure 7, by increasing the value of the parameter λ, the backorder/inventory cost variability decreases significantly, while the expected backorder cost is augmented considerably and the expected inventory cost and the raw material cost is decreased. In other words, by enforcing the backorder/inventory cost variability measure in the objective function of model (11)–(16), a higher expected backorder/inventory cost is determined by the model to minimize the excess of backorder/inventory cost of all scenarios over the expected backorder/inventory cost. Thus, the expected backorder size is increased and consequently production levels and raw material consumption are decreased. Furthermore, the decreased expected inventory cost is also the result of decreasing the production level and raw material consumption.

From the above discussions, it can be concluded that if the decision maker wishes to have a robust production plan by using the RO-UPV model, he/she should choose a value of λ which reflects his/her risk aversion about backorder size variability as well as increased expected backorder cost (size). Since the customer service level is defined in this work as the proportion of customer demand that can be fulfilled, the increased expected backorder cost (size) leads to decreased customer service level.

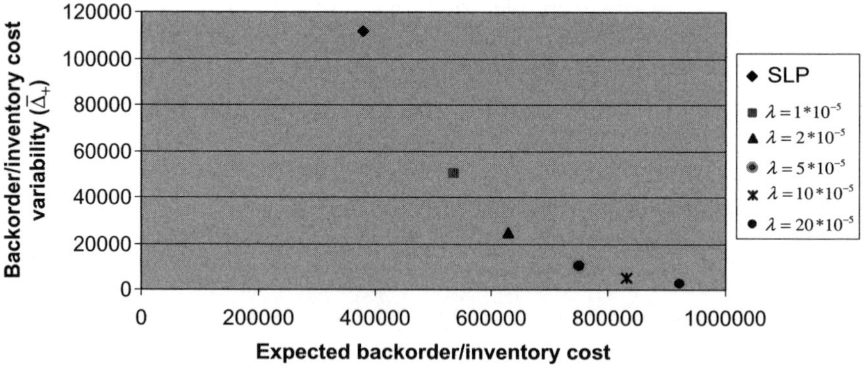

Figure 7 Expected Backorder/Inventory Cost and Backorder/Inventory Cost Variability Tradeoff in RO-(UPV) Model for Different Values of λ.

3.5.2.3 Performance comparison between RO-(UPM-2) and RO-UPV models

Figure 8 illustrates the difference between the robustness of optimal solutions in the RO-(UPM-2) and RO-UPV models for different values of R^* and λ. In Figure 8, as the target cost R^* increases (the service level decreases) in the RO-(UPM-2) model, the robustness of plans proposed by this model approaches those the of RO-UPV model. However, the more robust solution of the RO-UPV model might exhibit larger expected backorder costs (lower customer service level) compared with those of the RO-(UPM-2) model. The comparison between the total costs of both models is presented in Figure 9.

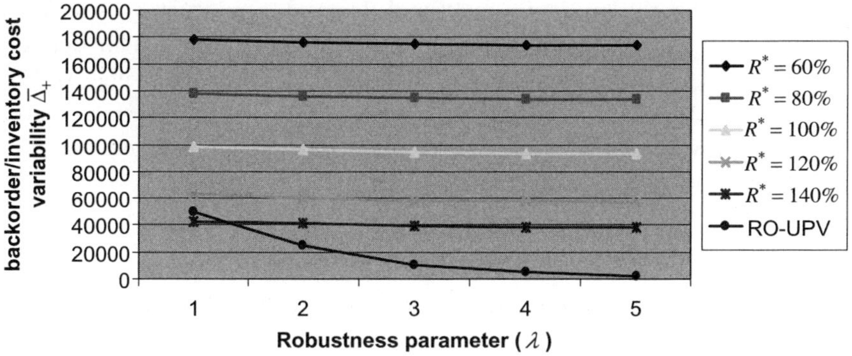

Figure 8 Comparison Between the Performance of Two Robust Optimization Models in Controlling Recourse Cost Variability.

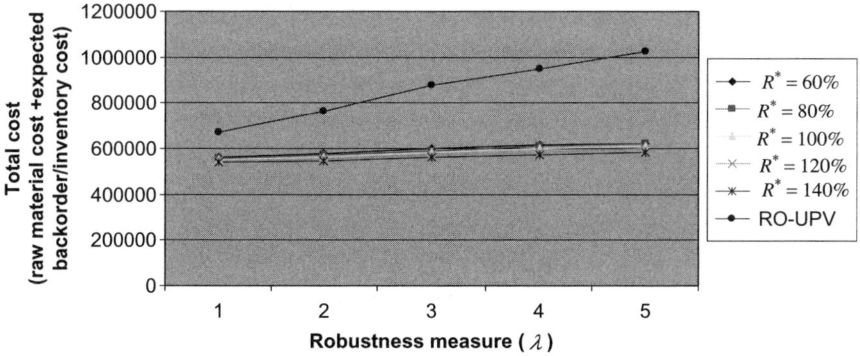

Figure 9 Comparison between the Total Cost Resulting from Two Robust Optimization Models.

4 Production planning in sawing units by taking into account the random yield and random demand

In this section, we address the sawmill production planning problem by taking into account the uncertainty in the quality of raw materials and consequently in process yields, as well as uncertainty in products demands. Since the demand and yield possess different uncertain natures, they are modeled separately and then integrated. Demand uncertainty is considered as a dynamic stochastic data process during the planning horizon, which is modeled as a scenario tree. The uncertain yields are then modeled as scenarios with a stationary probability distribution during the planning horizon. Yield scenarios are then integrated into each node of the demand scenario tree, constituting a hybrid scenario tree. Based on the hybrid scenario tree for the uncertain yield and demand, a multi-stage stochastic programming (MSP) model is proposed, which is full recourse for demand scenarios and partial recourse for yield scenarios. In the following, we first present our proposed approach for modeling random yield and demand. Then we provide the multistage stochastic sawmill production planning model in addition to computational results.

4.1 *Modelling the uncertain yield and demand*

We assume that the uncertain demand evolves during the planning horizon as a discrete time stochastic process with finite support. This information structure can be interpreted as a scenario tree (see Figure 10). The nodes at stage t of the tree constitute the states (scenarios) of demand that can be distinguished by information available up to stage t. For each stage a limited number of demand scenarios are taken into account (e.g. *high*, *average*, *low*). In order to define the scenarios for each stage, we can either use the traditional approach of making distributional assumptions, estimating the parameters from historical data, or use scenarios proposed by the experts. In order to keep the resulting multi-stage stochastic model within a manageable size, we assume that the planning horizon is clustered into N stages, where each stage includes a number of periods. In other words, it is supposed that the uncertain demand is stationary during the time periods at each stage. For example, if the demand scenario for the first period at stage n is *high*, it remains the same (*high*) for the remaining periods at stage n; however the demand scenario might change (e.g. to *low*) for the first period in the next stage ($n + 1$).

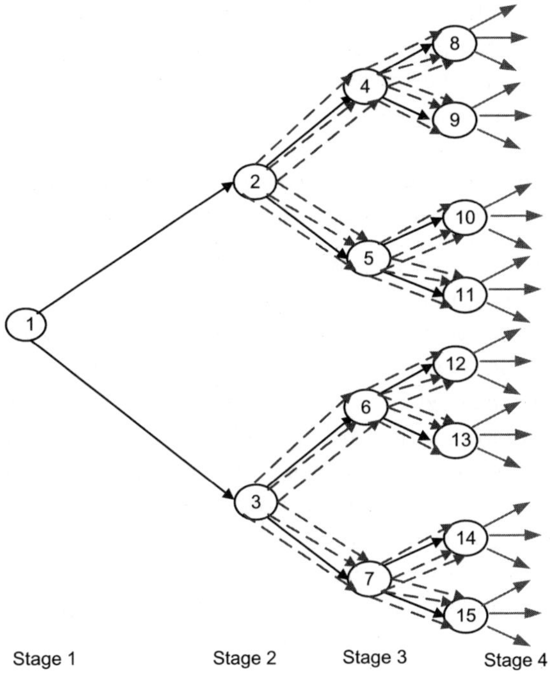

Figure 10 A Hybrid Scenario Tree for Uncertain Demand and Yield.

On the other hand, we assume that raw materials are supplied from the same supply source during the planning horizon. Thus it is supposed that the uncertain yield has a stationary probability distribution. The probability distribution of random yield is estimated based on the approach proposed in Subsection 3.1. A number of scenarios are taken in to account for yields by discretization of the original probability distribution.

In order to have a single stochastic production planning model that considers uncertain yield and demand, yield scenarios are integrated with the demand scenario tree forming a hybrid scenario tree. An example of a four-stage hybrid scenario tree is depicted in Figure 10, where full line branches denote demand scenarios while dashed line branches denote yield scenarios. At each node of the tree, which denotes one demand scenario for the corresponding stage, different yield scenarios can take place (3 scenarios in the example of Figure 10). However, regarding the stationary behavior of uncertain yield, only one of the yield scenarios can be observed during the planning horizon. Thus, the total number of scenarios in the hybrid scenario tree can be computed as the number of leaves in the demand scenario tree multiplied by the number of yield scenarios. (In the example of Figure 10, this number is equal to $8 \times 3 = 24$).

4.2 *Multi-stage stochastic program for sawmill production planning with uncertain yield and demand*

It should be noted that there are several types of formulations for multi-stage stochastic programs including "compact" and "split variable" formulations. In the "compact formulation", the decision variables are indexed by the nodes of the scenario tree, while in the "split variable formulation", the variables are indexed by scenarios. Let us now formulate the problem as a multi-stage stochastic (MSP) model based on the hybrid scenario tree for the uncertain yield and demand. The decision (control) variables of deterministic model (1)–(5) are production plans X_{at}. The inventory and backorder variables I_{pt} and B_{pt} are the consequences (state variables) of the plan. It is worth mentioning that unlike a partial recourse multi-stage stochastic model which considers a fixed control decision variable (e.g. production plan) at the beginning of the planning horizon, a full recourse model updates all decisions as new information is available through the time. In this problem, we assume that the decision maker can adjust the production plan X_{at} for different demand scenarios at each stage of the demand scenario tree. In other words, it is supposed that at the beginning of each stage, enough information on demand is available to the decision maker to select properly among the plans proposed by the MSP model for different scenarios. Thus we have a model with full recourse with respect to the demand scenarios. As we use the compact formulation to represent the problem, the decision variables X_{at} are defined for each node of the demand scenario tree. On the other hand, as the quality of materials is not known before production, the yield scenarios can only be revealed after implementation of the production plan. Thus, the production plan for each node of demand scenario tree should be fixed for all the yield scenarios. In other words, the model becomes partial recourse with respect to the yield scenarios. It is evident that the inventory and backorder of products in each period ($I^i_{pt}(n)$ and $B^i_{pt}(n)$), which are the state variables, depend on the demand scenarios as well as the yield scenarios, thus they are indexed for yield scenarios as well as demand nodes. Regarding the above discussions, the following notation is used in the multi-stage model, in addition to what is provided in 2.2.1. The compact formulation of the multi-stage model follows the notation.

4.2.1 *Notation*

Indices

Tree Scenario tree.
S Number of scenarios for random yields.
i Scenario of random yield.
n, m Node of scenario tree.
$a(n)$ Ancestor of node n in the scenario tree.
t_n Set of time periods corresponding to node n in the scenario tree.

Parameters

$d_{pt}(n)$ Demand of product p by the end of period t at node n of the scenario tree.
$p(n)$ Probability of node n of the scenario tree.
p^i Probability of scenario i for random yield.

Decision variables

$X_{at}(n)$ The number of times each process a should be run in period t at node n of the scenario tree.
$I_{ct}(n)$ Inventory size of raw material of class c by the end of period t at node n of the scenario tree.
$I^i_{pt}(n)$ Inventory size of product p by the end of period t for scenario i of random yield at node n of the scenario tree.
$B^i_{pt}(n)$ Backorder size of product p by the end of period t for scenario i of random yield at node n of the scenario tree.

4.2.2 *Multi-stage stochastic model (compact formulation)*

$$\text{Minimize } Z = \sum_{n\in Tree} p(n)\left(\sum_{t\in t_n}\sum_{c\in \mathrm{C}}\sum_{a\in A} m_{ct}\phi_{ac}X_{at}(n)\right) + \sum_{n\in Tree} p(n)\left(\sum_{i=1}^{S} p^i\left(\sum_{t\in t_n}\sum_{p\in \mathrm{P}}(h_{pt}I^i_{pt}(n)+b_{pt}B^i_{pt}(n))\right)\right)$$

Subject to (17)

$$I_{ct}(n) = I_{ct-1}(m) + s_{ct} - \sum_{a\in A}\phi_{ac}X_{at}(n), \quad n\in Tree, \quad t\in t_n, \quad c\in \mathrm{C},$$

$$m = \begin{cases} a(n), & t-1\notin t_n, \\ n, & t-1\in t_n, \end{cases} \tag{18}$$

$$\sum_{a \in \mathrm{A}} \delta_{ar} X_{at}(n) \le M_{rt}, \quad n \in \mathit{Tree}, \quad t \in t_n, \quad r \in \mathrm{R}, \tag{19}$$

$$I^i_{pt}(n) - B^i_{pt}(n) = I^i_{pt-1}(m) - B^i_{pt-1}(m) + \sum_{a \in \mathrm{A}} \rho^i_{ap} X_{at}(n) - d_{pt}(n),$$
$$n \in \mathit{Tree}, \quad t \in t_n, \quad p \in \mathrm{P}, \quad i = 1, \ldots, S,$$
$$m = \begin{cases} a(n), & t-1 \in t_n, \\ n, & t-1 \notin t_n, \end{cases} \tag{20}$$

$$X_{at}(n) \ge 0, \quad I_{ct}(n) \ge 0, \quad I^i_{pt}(n) \ge 0, \quad B^i_{pt}(n) \ge 0,$$
$$n \in \mathit{Tree}, \quad t \in t_n, \quad c \in C, \quad p \in P, \quad a \in A,$$
$$i = 1, \ldots, S. \tag{21}$$

The first term of the objective function (17) accounts for the expected material cost for demand nodes of the scenario tree. The second term represents the expected inventory and backorder costs for demand nodes and yield scenarios. In model (17)–(21), the decision variables are indexed for each node, as well as for each time period, since the stages do not correspond to time periods. As was mentioned before, each node at a stage includes a set of periods which is denoted by t_n. In this model, there are coupling variables between different stages and these are the ending inventory and backorder variables at the end of each stage. As can be observed in this model, two different node indices (n, m) are used for inventory/backorder variables in the inventory balance constraints ((18) and (20)). More precisely, for the first period at each stage, the inventory or backorder is computed by considering the inventory or backorder of the previous period corresponding to its ancestor node, while for the remaining periods in that stage, the inventory/backorder size of the previous period corresponding to the same node are taken into account.

4.3 *Application of the proposed multi-stage stochastic production planning model for a prototype sawmill*

In this section we report on computational experiments with regard to the proposed multi-stage stochastic programming approach for the prototype sawmill described in Subsection 3.4.1. The objective of our experiment is to investigate the quality of production plans suggested by multi-stage stochastic programming compared to those of the deterministic

LP, and two-stage stochastic programming. We also compute the value of multi-stage stochastic programming (VMSP) for this example. In the following, we first describe our experimental environment and then report on the experimental results.

4.3.1 *Experimental environment*

The hybrid scenario tree for uncertain demand and yield in this example is generated as follows:

Demand Scenario tree

At each stage of the scenario tree, except stage 1, based on the historical data for product demands (per day) in a Quebec sawmill, we estimate a normal distribution for demand. We consider the same probability distribution for all the products. The normal distribution is then approximated by a 3 point discrete distribution by using the Gaussian quadrature method (Miller and Rice, 1983). Since considering each time period as a stage leads to an extremely large number of scenarios, we need to approximate the scenario tree by something more manageable. In our computational experiment, we assumed that the demand for the next 10 days has a stationary behavior, which is a realistic assumption in the lumber market. Thus, we clustered the 30-period planning horizon into 3 stages and hence the multi-stage decision process is approximated by a four-stage one. The first stage consists of time period zero (present time), the second-stage includes periods 1–10, etc. The mentioned approximations result in a scenario tree including 27 demand scenarios and 40 nodes. We consider three different normal distributions for demand with the same mean but different variances (equal to 5% of the mean, 20% of the mean, and 30% of the mean). Thus, three demand trees (DT1, DT2, DT3) and a total of 3 test problems are considered. In all the test problems we consider the same distribution for stages 2 to 4.

Yield Scenarios

As we mentioned in Subsection 4.1, at each node of the demand scenario tree a number of yield scenarios are taken into account. These scenarios are generated based on the approach provided in 3.1. The normal distribution corresponding to the yield of each process was then approximated by three scenarios by Gaussian quadrature. As the randomness of processes yields is the result of non-homogeneity in the quality of logs, we consider three scenarios for the yield of each log class. As we considered three classes of logs in this example, the total number of yield scenarios is equal to $3^3 = 27$.

It should be noted that the same yield scenarios are considered in the three test problems.

The above scenario generation approach for uncertain demand and yield in this sawmill production planning example results in a hybrid scenario tree similar to the one in Figure 10 with 40 nodes, where each node includes 3 branches as demand scenarios and 27 branches as yield scenarios. The total number of scenarios at the end of stage 4 is equal to $27 \times 27 = 729$. The compact multi-stage stochastic model (17)–(21) for this sawmill example is a linear programming (LP) model with nearly 600000 decision variables and 300000 constraints.

CPLEX 10 and OPL 5.1 are used to solve the linear program (17)–(21) and to perform further analysis on the solutions of the test problems. All numerical experiments are conducted on an AMD Athlon^{TM} 64×2 dual core processor 3800+, 2.01 GHz, 3.00 GB of RAM, running Microsoft Windows Server 2003, standard edition.

4.3.2 *Quality of the multi-stage stochastic model solution*

In this section, for the three test problems mentioned in 4.3.1, we compare the solution of the 4-stage stochastic programming model to those of a 3-stage, and 2-stage stochastic programming model as well as the mean-value deterministic model. In the 3-stage model, the 30 periods planning horizon is clustered into 2 stages, each includes 15 periods. In other words, in order to reduce the size of the multi-stage model, it was supposed that the random demand has a stationary behavior during each 15 days. The 2-stage stochastic model corresponds to considering a static probability distribution for the uncertain demand during the planning horizon. Figures 11 and 12

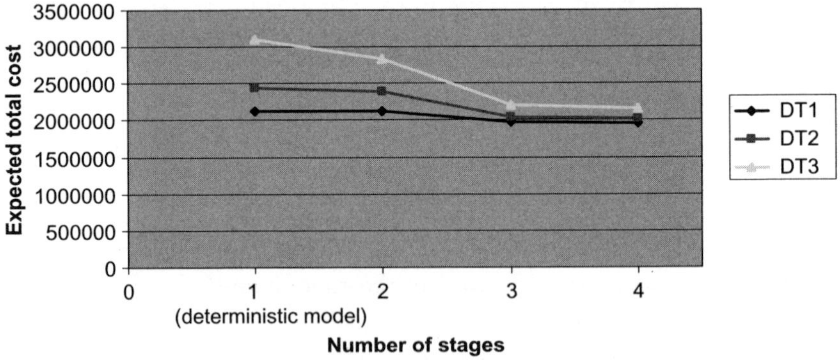

Figure 11 Expected Total Cost Comparison of Different Production Planning Models.

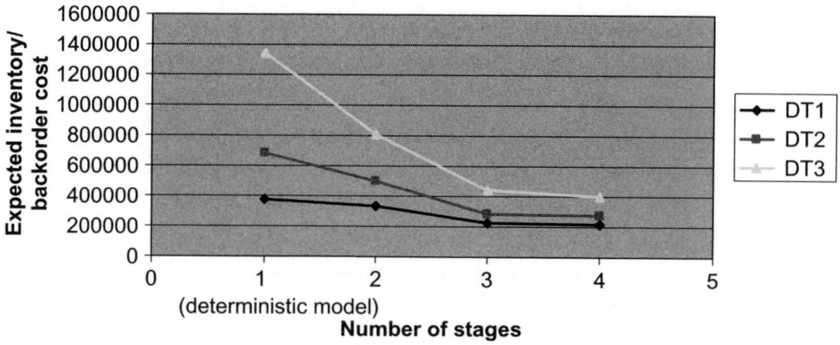

Figure 12 Expected Inventory/Backorder Costs Comparison of Different Production Planning Models.

illustrate the comparison between the total expected cost as well as expected inventory/backorder costs, respectively, of different models for three test problems which are distinguished by the variability of demand at each stage. It should be noted that the expected inventory/backorder costs of 3-stage, 2-stage, and mean-value deterministic models are computed by setting the production plan variables (X_{at}) in the 4-stage stochastic model (17)–(21) as the optimal production plan (X^*_{at}) proposed by the mentioned models. In other words, the expected inventory/backorder costs of production plans proposed by the 3-stage, 2-stage and deterministic model are computed for the hybrid 4-stage scenario tree corresponding to the uncertain yield and demand in each test problem.

As can be observed in Figures 11 and 12, in all the tree test problems the solution of the 4-stage stochastic model is significantly superior to those of the deterministic model. Furthermore, if the uncertain demand is considered as a random variable with a static probability distribution during the planning horizon (as in the two-stage stochastic programming model), the expected material cost as well as the expected inventory/backorder costs of the production plan are considerably higher than those of the multi-stage stochastic model's plan. Finally, by clustering the planning horizon into two stages (as in the 3-stage stochastic programming model) the expected inventory/backorder costs of the plan are higher than those of the 4-stage stochastic model. As the variability of demand increases at each stage, the difference between the expected costs of the various models' plans increases. In other words, the significance of using a multi-stage programming model instead of a two-stage or deterministic model is increased as the variability of demand increases at each stage of the scenario tree.

4.3.3 *Value of multi-stage stochastic programming*

As was mentioned in Subsection 4.2, we considered the production plan (X_{at}) as full recourse with respect to the demand scenarios. In other words, we assumed a flexible production plan that can be adjusted based on the demand scenarios, at different stages. However, in some manufacturing environments the production plan is not flexible and should be fixed at the beginning of the planning horizon. Thus, a partial recourse multi-stage stochastic model should be used to determine the plan. In this section we compare the solutions of multi-stage stochastic programs with full recourse and partial recourse, for the three test problems. Figure 13 compares the VMSP of the three test problems with different variability levels in demand. It can easily be verified in Figure 13 that in all test problems the total cost of the full recourse problem is smaller than that of the partial recourse problem. This should come as no surprise, since the multi-stage model with full recourse offers more flexibility in the production plan decisions with respect to the uncertain states of demand. We denote the optimal objective values corresponding to full recourse and partial recourse multi-stage stochastic programs by v^{FR}, and v^{SR}, respectively. The value of multi-stage stochastic programming (VMSP) is defined as follows (Huang and Shabbir, 2005): $VMSP = v^{SR} - v^{FR}$.

As can be observed in Figure 13, as the variability of demand increases at each stage, considering a full recourse multi-stage stochastic model becomes more significant.

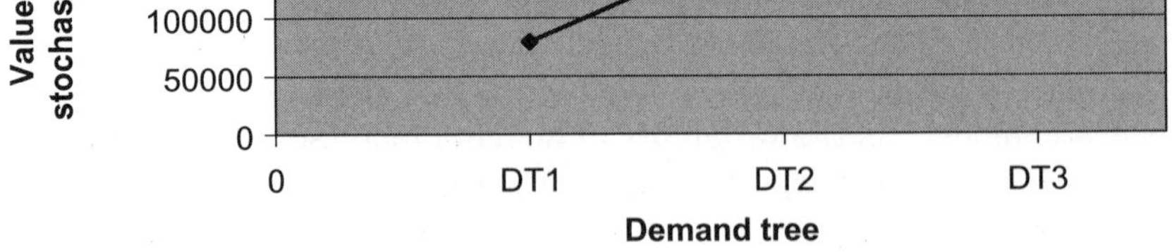

Figure 13 VMSP Comparison of Different Test Problems with Different Demand Variability.

5 Production planning in sawing units under the random yield and demand by taking into account setup constraints

As was mentioned in 2.1, in many sawmills, due to the technological constraints for changing machine setups, only one process family including processes with similar setup requirements can be executed in each period. As a consequence, setup constraints should be taken into account in the production planning model. The resulting sawmill production planning model would be a multi-stage mixed-integer stochastic program (MS-MIP) with full recourse. While the newly developed model is applicable to a larger number of sawmill cases, a challenging problem is faced for solving real-life instances in reasonable time. A mixed-integer solver of a commercial optimization package, namely the CPLEX MIP solver, is not capable of solving the resulting deterministic equivalent model, due to memory shortage. Thus two heuristic solution strategies are proposed to find good approximate solutions. The first algorithm is a scenario decomposition approach which is inspired by the progressive hedging algorithm (PHA) of Rockafellar and Wets (1991), while the second one is an accelerated version of the scenario updating algorithm proposed in Lulli and Sen (2006). In the following, we first provide the MS-MIP sawmill production planning model. Description of the two heuristic algorithms as well as some computational results follows the model.

5.1 *Multi-stage mixed integer (MS-MIP) sawmill production planning model*

By taking into account set-up constraints and modelling the random yield and demand as a hybrid scenario tree similar to the one described in Section 4 the resulting the sawmill production planning model (in its compact formulation) would be similar to model (17)–(21) by adding the following constraints:

$$X_{at}(n) \leq MY_{f_a{}^t}(n) \quad t \in t_n, \quad a \in \mathrm{A}, \quad n \in \mathit{Tree}, \tag{22}$$

$$\sum_{f \in F} Y_{ft}(n) = 1 \quad t \in t_n, \quad n \in \mathit{Tree}, \tag{23}$$

where

$$Y_{ft}(n) = \begin{cases} 1, & \text{if the family} f \text{ of processes is selected at node } n \text{ in period } t \\ 0, & \text{otherwise} \end{cases}.$$

5.2 *Applying the scenario decomposition method to sawmill production planning*

5.2.1 *The general algorithm*

The algorithm is based on the principle of scenario decomposition, as initially proposed in the progressive hedging algorithm (PHA) of Rockafellar and Wets (1991) for multi-stage stochastic linear programs. For each scenario, a deterministic optimization sub-problem is generated and solved. We denote by Z(s) the objective function of the scenario s sub-problem. However, the solution for any particular scenario may not be of any value to us. Since we are not prescient, we have to require solutions that do not necessitate foreknowledge and that will be feasible no matter which scenario is realized. In other words, the solutions should be implementable for all scenarios. The implementability condition indicates that for any pair of scenarios that are indistinguishable at stage t (i.e. scenarios which share the same node at stage t in the scenario tree), the decision variables corresponding to those scenarios and periods at stage t should be identical. The PHA forces implementable solutions at all iterations by adding terms in the objective function to penalize lack of implementability. At each iteration, an estimation of the implementable solution is calculated as the *mean* of the solution values of the scenario sub-problems. The latter are then solved again with adjusted penalties on the difference between the local solution and the average (global) estimation. The general form of the PHA can be represented by the following steps:

Step 1 (initialization)

Set the progressive hedging (PH) iteration counter to $0(k \leftarrow 0)$. Set $w^0(s) \leftarrow 0$.

Step 2

For each scenario s, a deterministic optimization sub-problem is solved. The obtained approximate solutions are determined by $X^0(s)$.

Step 3

Increment the PH iteration counter $(k \leftarrow k+1)$.

For each node n in the scenario tree and for its time index $t = t_n$, evaluate the value of an implementable solution $\bar{X}^{k-1}(t_n, n)$ as the *mean* of all the solution values of scenarios, in the bundle of scenarios that are indistinguishable from s at node n. For all indistinguishable scenarios at node n, set : $\bar{X}^{k-1}(s) = \bar{X}^{k-1}(t_n, n)$.

Step 4

Evaluate with $w^k(s) \leftarrow w^{k-1}(s) + r(X^{k-1}(s) - \bar{X}^{k-1}(s))$ with $r > 0$.

These values correspond to the dual prices associated with the implementability constraints. In other words, they can be interpreted as the value of information.

Step 5

For each scenario s, approximate solutions are obtained for the problem after adding the terms that penalize the lack of implementability in the objective function $Z(s)$. The penalty term is computed as follows:

$$w^k(s)X(s) + \frac{r}{2}\|X(s) - \bar{X}^{k-1}(s)\|^2, \quad r \geq 0.$$

The penalty terms correspond to an augmented Lagrangean strategy.

Step 6

If the termination criteria are not met, then go to step 3. The termination criteria are based on convergence. Nevertheless, we can also terminate based on time, since non-convergence is a possibility (Rockafellae and Wets, 1991). Iterations are counted until the counter k reaches a predetermined value or the algorithm has converged.

5.2.2 *Application to sawmill production planning*

The statement of the above PHA gives rise to two types of optimization sub-problems. The first is a linear MIP problem which is only used at iteration 0 (step 2). The second type, which is used in other iterations (step 5) is the modified form of the first type obtained by adding the quadratic penalty terms in the objective function. Before formulating these sub-problems for the sawmill production planning problem, we introduce the required additional notation. To enforce implementability, we act on two decision variables: X_{at} and Y_{ft}. We use w_{at}^X and w_{at}^Y as the corresponding dual variables corresponding to X_{at} and Y_{ft}, respectively. At each iteration, dual variables are evaluated as described in step 4 of the PHA. Furthermore, in each scenario sub-problem, the decision variables in model (17)–(21) in Subsection 4.2.2 are defined for each demand scenario s instead of each node n in the scenario tree. The demand parameter d_{pt} is also defined for each scenario.

We are now ready to formulate the scenario sub-problem in step 5 of the PHA. Our hybrid scenario tree includes 4 stages, 27 demand scenarios and

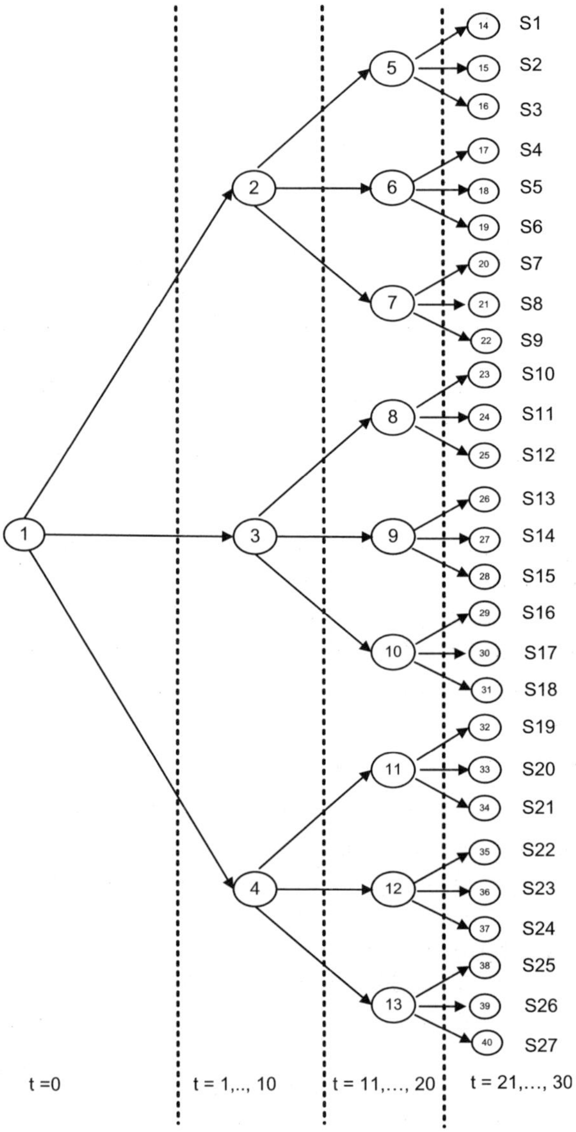

Figure 14 Scenario Tree for the Random Demand.

40 nodes (see Figure 14). The planning horizon is clustered into 3 stages, where each stage includes 10 periods. In the mentioned scenario tree stage 2 includes periods 1–10 and nodes 2–4, stage 3 includes periods 11–20 and nodes 5–13, etc. The total number of yield scenarios at each node of the

demand scenario is equal to S = 27. It is evident that the non-anticipativity constraints should be considered for decision variables corresponding to periods in stages 2 and 3 (periods 1 to 20).

The scenario sub-problem in the PHA (except for iteration 0) can be formulated as follows.

Scenario sub-problem in the PHA (for scenario s)

$$
\begin{aligned}
\text{Minimize } Z(s) = {} & \sum_{t=1}^{T}\sum_{c\in \mathrm{C}}\sum_{a\in \mathrm{A}} m_{ct}\phi_{ac}X_{at}(s) + \sum_{i=1}^{S} p^i \sum_{t=1}^{T}\sum_{p\in \mathrm{P}} (h_{pt}I^i_{pt}(s) \\
& + b_{pt}B^i_{pt}(s)) \\
& + \sum_{a\in A}\sum_{t=1}^{20} w^X_{at}(s)X_{at}(s) + \sum_{f\in F}\sum_{t=1}^{20} w^Y_{ft}(s)Y_{ft}(s) \\
& + (r/2)\sum_{a\in A}\sum_{t=1}^{20}\theta^2_{at}(s) + (r/2)\sum_{f\in F}\sum_{t=1}^{20}\gamma^2_{ft}(s)
\end{aligned}
\tag{24}
$$

Subject to

$$I_{ct}(s) = I_{ct-1}(s) + s_{ct} - \sum_{a\in A}\phi_{ac}X_{at}(s), \quad t = 1, \ldots T, \quad c \in \mathrm{C}, \tag{25}$$

$$\sum_{a\in A}\delta_{ar}X_{at}(s) \le M_{rt}, \quad t = 1, \ldots, T, \quad r \in R, \tag{26}$$

$$
\begin{aligned}
I^i_{pt}(s) - B^i_{pt}(s) = I^i_{pt-1}(s) - B^i_{pt-1}(s) + \sum_{a\in A}\rho^i_{ap}X_{at}(s) - d_{pt}(s), \\
t \in 1, \ldots T, \quad p \in \mathrm{P}, \quad i = 1, \ldots, S,
\end{aligned}
\tag{27}
$$

$$X_{at}(s) \le MY_{f_a{}^t}(s) \quad t = 1, \ldots, T, \quad a \in \mathrm{A}, \tag{28}$$

$$\sum_{f\in F} Y_{ft}(s) = 1 \quad t = 1, \ldots, T, \tag{29}$$

$$\theta_{at}(s) = X_{at}(s) - \bar{X}_{at}(s), \quad t = 1, \ldots, 20, \quad a \in A, \tag{30}$$

$$\gamma_{ft}(s) = Y_{ft}(s) - \bar{Y}_{ft}(s), \quad t = 1, \ldots, 20, \quad f \in F, \tag{31}$$

$$
\begin{aligned}
X_{at}(s) \ge 0, Y_{ft}(s) \in \{0, 1\}, I_{ct}(s) \ge 0, I^i_{pt}(s) \ge 0, B^i_{pt}(s) \ge 0, \\
t = 1, \ldots, T, \quad c \in C, \quad p \in P, \\
a \in A, \quad f \in F, \quad i = 1, \ldots, \ldots, S,
\end{aligned}
\tag{32}
$$

where

$$\bar{X}_{at}(s) = \sum_{s \in B_{s,t}} p^s X_{at}(s)/p(A_{s,t}),$$
$$\bar{Y}_{ft}(s) = \sum_{s \in B_{s,t}} p^s Y_{ft}(s)/p(A_{s,t}), \tag{33}$$

and

- p^s probability of the demand scenario s in the scenario tree
- $B_{s,t}$ bundle of scenarios that are indistinguishable from s in period t
- $A_{s,t}$ the node in the scenario tree corresponding to scenario s in period t
- $p(A_{s,t})$ the probability of node $A_{s,t}$

As some examples of bundles of indistinguishable scenarios $B_{s,t}$ at stage 2 (periods 1 to 10) in the scenario tree of Figure 14, we refer to scenarios 1 to 9 which share node 2, scenarios 10–18 which share node 3, and scenarios 19–27 which share node 4.

Finally, we need to specify the termination criteria for step 7 of the PHA. In this paper, the PHA assumes full convergence for integer variables. In other words, the convergence criterion is satisfied when all integer components of $\bar{X}^k$ are equal to their counterparts in $\bar{X}^{k-1}$. The values of continuous variables are then determined by solving the deterministic equivalent model (17)–(23) when the integer variables are fixed at their converged values.

5.3 *Applying the revised scenario updating procedure to sawmill production planning*

In this section, we first provide a summary of the SU method proposed in Lulli and Sen (2006). Then we present an accelerated SU algorithm which modifies the SU method in two directions: (1) we discuss the validity of proposed bounds on the optimal solution for a full recourse stochastic model, and (2) we propose a new scenario selection rule so as to increase the rate of convergence.

5.3.1 *The scenario updating heuristic*

The scenario updating (SU) procedure proposed by Lulli and Sen (2006) has been motivated by the contamination method (Dupačová, 1995) which has been presented as a tool for post-optimality analysis in scenario-based multi-stage stochastic linear programs (MSLP). The idea behind the method consists of estimating lower and upper bounds for the objective function value of a multi-stage stochastic program when new scenarios are added to the current scenario tree, i.e., it explores the influence of out-of-sample scenarios on the solution of the stochastic program. We describe the SU procedure with the MSLP program given in the following form:

$$\min_{x \in X} f(x, P), \tag{34}$$

with x denoting the vector of first-stage decision variables, f linear in P (probability distribution) and $X \subset R^n$ the set of feasible decision variables. Denoting by $\phi(P)$ the optimal value of the program (34) the contamination method focuses on $\phi(P)$ and its perturbations with respect to the inclusion of additional scenarios. Under appropriate assumptions on the optimal solution set and on the value function, we can bound the objective function value of the perturbed stochastic program. If we consider the perturbation given by a single scenario $\{z\}$ with probability mass p_z and probability distribution P concentrated at scenarios $\xi^1, \ldots, \xi^p$ with probability mass $p_s \forall s \in P$, then the perturbed distribution can be presented as $P' = (1 - p_z) P + p_z z$. Figure 15 illustrates better the perturbation of a current scenario tree P by a scenario $\{z\}$.

The lower bound on the objective function value of the perturbed stochastic program is

$$LB = (1 - p_z)\phi(P) + p_z\phi(z), \tag{35}$$

where $\phi(z)$ is the value of the scenario $\{z\}$ (deterministic) problem. An upper bound on the optimal value of the contaminated program can be

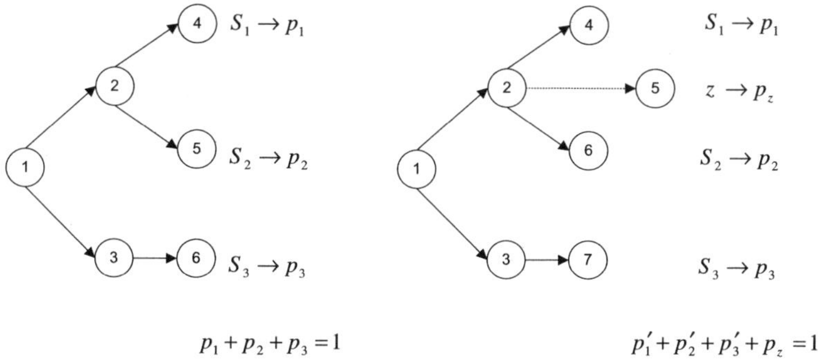

(a) Current scenario tree (P) (b) Perturbed scenario tree (P')

Figure 15 Perturbation of a Scenario Tree by a Single Scenario z.

computed, if the SMIP has complete recourse. We have:

$$UB(1 - p_z)\phi(P) + p_z f(x(P), \delta(z)), \tag{36}$$

where $f(x(P), \delta(z))$ denotes the objective function value of the current solution $x(P)$ under the perturbed scenario tree with a probability distribution $\delta(z)$. The objective function value of the current solution in the production planning model (17)–(23) under the perturbed scenario tree is computed as follows:

(1) Formulate model (17)–(23) for the perturbed scenario tree.
(2) Fix the decision variable values corresponding to nodes in the current (non-perturbed) scenario tree, based on the current solution $x(P)$.
(3) Solve model (17)–(23) with decision variables in the nodes corresponding to the added scenario.

The scenario updating procedure can be summarized as follows.

Initialization

We start with a subset of scenarios $P_0 \subset P$, where P is the set of all scenarios (selection phase). Different decision rules can be used for the selection phase.

Step 1

At iteration k, solve the "reduced" problem, which is comprised of the subset of scenarios P_k, by an appropriate exact/approximate algorithm.

Step 2

Compute an upper and lower bound on the change of the objective value induced by adding a scenario not in the current scenario tree. Those scenarios, which if added to the current scenario tree imply a significant change in the objective function value, are candidates to enter in the next sub-tree (updating phase).

Step 3

If there are no scenario candidates, then stop; otherwise add those scenarios or some of them to the current scenario sub-tree, and go to step 1.

Scenario selection rules

The key issue for an efficient implementation of the scenario updating method concerns the selection of representative scenarios. The goal is to select those scenarios that provide a good approximation of the scenario tree, in terms of quality of the solution. In the following, several decision rules to handle this task are discussed. It should be noted that in Lulli and

Sen (2006) the following scenario selection rules are proposed:

- The *highest probability* rule: Select scenarios in the decreasing order of their probability.
- The *dissimilarity* rule: Solve the deterministic problem for each scenario and compare their first-stage decisions. Select those scenarios which have the most different here and now solutions.
- The *random rule*: select scenarios in a random manner.
- The *mixed* rule: Use an appropriate mix of the other rules.

5.3.2 *The accelerated scenario updating heuristic*

In this subsection we provide the modifications that we propose to adapt the SU method (Lulli and Sen (2006)) for *full recourse* MS models and to improve its rate of convergence.

5.3.2.1 Validity of bounds on the optimal solution in the SU method for full recourse stochastic models

It is worth mentioning that unlike a partial recourse multi-stage stochastic model, which considers a fixed first-stage decision variable (e.g., production plan) at the beginning of the planning horizon, a full recourse model updates the decisions as new information is available through the time. Thus the cost of a multi-stage stochastic model with partial recourse is usually higher than the cost of a full recourse model (as shown in Kazemi Zanjani *et al.* (2010b)). Moreover, the cost of the deterministic model for some scenarios can be quite high, compared to the cost of a multi-stage stochastic model with full recourse. As a consequence, the lower bound defined in (35) might not be appropriate for a full recourse MS-MIP model. More precisely, this lower bound can be much higher than the objective value of the MS-MIP model with full recourse after adding a new scenario to the current scenario tree. On the other hand, both terms in the upper bound defined in (36), $\phi(P)$ and $f(x(\mathrm{P}), \delta(z))$, as well as ϕ_{δ_z} (optimal objective function value of perturbed scenario tree) can be computed for a full recourse MS-MIP model. Thus the proposed upper bound can be considered as a more realistic measure to verify the impact of adding a new scenario to the scenario tree.

5.3.2.2 Increasing the rate of convergence in the SU method through a new scenario selection rule

As was mentioned before, the production planning model in this article is formulated as a full recourse stochastic model. In other words, we do not have any first-stage decision variable in this model. It is possible to use the dissimilarity rule based on the objective function value of each scenario's deterministic model. Moreover, some scenarios with small probabilities might have very large or very small objective values. By integrating them into the scenario tree they might not have considerable impact on the objective function value of the full recourse stochastic model, due to their low probability. Thus, we did not find the dissimilarity rule an efficient rule in a full recourse stochastic model to accelerate the rate of convergence of the algorithm.

In order to increase the convergence rate of the SU algorithm and also to improve the quality of solutions we propose a new scenario selection rule that we refer it as the *stage priority rule* and is described as follows.

— The *stage priority rule*: We define the priority of stages in the scenario tree in increasing order as follows: Stage 1 has the highest priority, while the last stage has the lowest. Thus scenarios should be selected so that the majority of nodes corresponding to high priority stages are taken into account. For example, consider a scenario tree similar to scenario tree (b) in Figure 15, where the current tree includes only scenarios S1 and S2. For the next iteration, we can select among scenario S3 and z. By adding scenario S3, a node corresponding to stage 2 will be considered while by adding scenario z a node corresponding to stage 3 with a lower priority will be added to the scenario tree. So, based on the *stage priority* rule, S3 should be selected as the candidate scenario.

In fact, as the decision variables corresponding to initial stages (high priority ones) are fixed for the following stages, if we consider the majority of nodes in those stages, as we proceed in the algorithm, the impact of adding new scenarios on the objective function value is decreased gradually. This can accelerate the convergence of the algorithm into a good solution. As in many industries (e.g. in sawmills) the production plan is usually updated based on a rolling horizon. The idea is to obtain a production plan for the majority of scenarios (nodes) in the initial stages of the scenario tree (the initial periods). Moreover, mixing the *stage priority* rule with the *highest probability* rule properly, a more balanced approximate scenario tree is resulted, compared to the rules proposed in Lulli and Sen (2006). A more balanced approximate scenario tree can be helpful to avoid overly optimistic or pessimistic approximate solutions.

Finally it is worth mentioning that regarding the large dimensionality of our problem, adding only one scenario considerably increases the complexity of the resulting MS-MIP. Thus, in the updating phase, we add in each iteration of the algorithm one scenario with the largest change in the objective function value. Note that the current solution of the problem should be used as a warm start for the next iteration of the algorithm.

5.4 *Application of the proposed multi-stage MIP stochastic production planning model to a prototype sawmill*

In this subsection, we provide the implementation results of the two solution strategies for the sawmill production planning example described in 3.4.1. In the following, we first provide some details on the implementation of the proposed solution strategies. Then we provide the results of applying them in two numerical examples:

- The first is a large-scale sawmill production planning problem involving simultaneously random yield and demand, for which we cannot find the optimal solution by CPLEX due to its size.
- The second is the same example, but without considering the random yield, for which the optimal solution can be found by CPLEX. It is worth mentioning that the second example is provided in order to evaluate and compare the approximate solution of the proposed approach with the optimal solution.

CPLEX 11 and OPL 6.1 are used to implement the PHA and the RSU method. All numerical experiments are conducted on an AMD Athlon^{TM} 64×2 dual core processor 3800+, 2.01 GHz, 1.00 GB of RAM, running Microsoft Windows Server 2003, standard edition.

5.4.1 *Solution strategy 1 (the scenario decomposition method)*

As was mentioned in 5.2, at iteration 0 of the PHA, the scenario MIP sub-problems are solved to find good integer solutions. Each scenario sub-problem at iteration 0 includes 22,500 constraints and 44,341 decision variables. Good integer solutions that are very close to the optimal one are found by considering a 3000 seconds global time limit for the MIP solver in CPLEX 11. The implementable solutions are computed for the nodes of the demand scenario tree at stages 2 and 3 by averaging the solutions for corresponding scenarios. In the following iterations, the penalty terms are added to the objective function of each scenario sub-problem to hedge against the non-anticipativity. Each scenario MIQP sub-problem includes 23,180 constraints and 45,021 decision variables. The MIQP solver of CPLEX 11 is used to find the optimal solution, which finds the optimum in about 2000 seconds. It was shown in Løkketangen and Woodruff (1996) that the PHA is not sensitive to values of the parameter r. By considering $r = 0.6$, the PHA converges for all integer variables after 3 iterations, which is quite fast. Thus, we did not consider larger values of r. After fixing the binary variables in the deterministic equivalent model at their converged values, this model is solved for continuous decision variables. The optimal solution has an objective function value (cost) equal to 2,172,193. As discussed in Rockafellar and Wets (1991), the converged solution of the PHA in a non-convex case is a local optimal one for the MS-MIP model, thus its objective value can be considered as an upper bound for the optimal solution.

Finally, as the considered case study cannot be solved by CPLEX due to memory problem, we are not able to compare the solution of this solution strategy with the true optimal solution. In the following, we consider the same case study by assuming that the processes yields are deterministic. The latter will result in a 4-stage MIP model with 30,000 constraints and 40,000 decision variables, which can be solved for optimality by CPLEX 11. As the proposed solution strategy (PHA) is applied only for demand scenarios and nodes, the new example still allows us to evaluate its performance for a large scale 4-stage scenario tree. Our computational experiment has led to an optimal objective value equal to 2,147,693 (found by the CPLEX MIP solver). The progressive hedging algorithm, on the other hand, found a local optimum after 3 iterations with an objective value of 2,152,423. The small gap (0.2%) between the objective value of our proposed method and that of the true optimal solution indicates the high quality of this approximate solution.

5.4.2 *The solution strategy 2 (the revised scenario updating method)*

The algorithm is initiated by considering 3 scenarios (among 27) based on the *dissimilarity* rule (i.e., three scenarios with the highest, average and lowest objective function values are selected). At the following iterations, the candidate scenario is selected based on the *stage priority* and the *highest probability* rules by putting more emphasis on the *stage priority* rule. It should be mentioned that for each scenario sub-tree in each iteration of the RSU algorithm the deterministic equivalent model represented by the compact formulation is solved by the MIP solver of CPLEX 11. Table 2 summarizes the results of the scenario updating method for the sawmill example. The third column in Table 2 indicates the size of model (17)–(23) for each scenario sub-tree and column 4 corresponds to its objective

Table 2 Results of the Revised Scenario Updating Method.

Iteration	#Scenarios	#Constraints/ #Decision Variables	Objective Function Value	The Upper Bound	The Relative Gap (%)	Scenario Selection Rule
1	3	60,000/118,240	2,383,544	2,269,416	−5	*dissimilarity*
2	4	82,500/162,580	2,097,401	2,204,807	4.8	*stage priority* + *highest probability*
3	5	97,500/192,140	2,154,496	2,252,988	4.3	*stage priority* + *highest probability*
4	6	112,500/221,700	2,241,741	2,204,618	−1.7	*stage priority* + *highest probability*
5	7	127,500/251,260	2,186,905	2,186,087 ±(0.14%)*	−0.04	*stage priority* + *highest probability*

*The average and standard deviation for 4 candidate scenarios function value. Column 5 indicates the estimated upper bound on the objective function value by adding the candidate scenario selected based on the mentioned rules. The sixth column corresponds to the relative gap between the estimated upper bound and the current objective function value. Positive values in this column indicate a decline in the objective function value while the negative gaps indicate an improvement, after adding the candidate scenario. As can be observed from Table 2, as the algorithm proceeds, the gap between the expected cost of the new sub-tree and the current one decreases. Finally, at iteration 5, the values of this gap computed for 4 candidate scenarios (with the highest stage priority and probability) are negligible. Thus, it can be concluded that adding more scenarios to the current scenario tree would not affect the objective function value significantly and the current solution can be considered as a good approximate one. It would be worth mentioning that at iteration 5, 100%, 78% and 27% of nodes corresponding to stages 2, 3 and 4 in the scenario tree described in section 3 are taken into account. Moreover, at initial iterations, we found an approximate solution for the reduced model in step 1 of the algorithm by limiting the global time of the MIP solver in CPLEX. Only at the last iteration, the problem is solved for the optimal solution. It should be mentioned that by using the scenario selection rules in Lulli and Sen (2006) the algorithm did not convergence until iteration 6.

The obtained results demonstrate the efficiency of the applied scenario selection rules (stage priority and the highest probability) in the RSU method in convergence after a quite small number of iterations. The comparison between the approximate solution of the RSU method with the local optimal solution of the solution strategy 1 (PHA), indicates a small gap (0.4%) between this approximate solution and the upper bound of the optimal objective value. Considering that the approximate solution is obtained by including less than 30% of the scenarios and 50% of the nodes in the original scenario tree, it can be considered as a good solution which can be obtained considerably faster than a local optimal solution (by PHA).

Furthermore, we believe that the solution of the RSU method in this example is a good approximate solution whose objective value is slightly underestimated compared to the optimal objective value. Using the stage priority and the highest probability rules for scenario selection in the RSU method, only 2 nodes are not considered at stage 3 (a low demand after an average demand at stage 2 and a high demand after a low demand at stage 2). Thus we can conclude that the obtained solution is quite realistic until stage 3. However, since at stage 4 only 25% of the scenarios (nodes) are taken into account, we can expect the real expected cost (the true optimal objective value) to be higher than the cost obtained by the RSU method.

Finally, we applied the proposed RSU algorithm to the smaller test problem described in 5.4.1 in order to compare its solution with the optimal solution found by CPLEX. The RSU heuristic found an optimal solution with the objective value equal to 2,144,289 after 5 iterations. The small gap (0.16%) of this approximate solution with the true optimal objective value reveals the high quality of solutions found by this algorithm for large-scale multi-stage stochastic MIP models.

6 Conclusion

In this paper, we developed several models for operational level production planning in the sawing units of sawmills while considering random quality of materials and demand. The proposed models are based on two-stage stochastic programming, robust optimization, and multistage stochastic programming approaches. We also proposed realistic approaches for modeling the random yields and demand. Moreover, we considered setup constraints in this problem in order to widen its application in different types of sawmills. More precisely, the proposed model is also applicable to sawmills that are not equipped with flexible sawing lines and are faced with technological limitations to change machine setups within each production shift. As the resulting large-scale multi-stage stochastic MIP model could not be solved by commercial solvers, we proposed two solution strategies to find good solutions. The first algorithm is based on the progressive hedging algorithm (PHA) which is used for the integer variable convergence. The second algorithm is a scenario updating method (Lulli and Sen, 2006) revised in two directions. First, we used only the proposed upper bound in the algorithm. Moreover, we modified the scenario selection rules and we proposed a new rule to improve the rate of convergence and the quality of the solution. Finally we applied the proposed models and solution methods to a realistic-scale prototype sawmill. The computational results provided enough evidence supporting the effectiveness of the proposed tools for sawmill production planning under uncertainty.

In terms of execution time, the proposed production planning model will be more complex than a deterministic approach when implemented in sawmills. However, generating realistic scenarios that represent the real behavior of demand and yield is essential for finding accurate production plans using the stochastic model. In contrast, the current production planning approach in Quebec sawmills is to apply a deterministic model on a rolling planning horizon. Through planning on a rolling horizon, the uncertainty inherent in demand and yield might be partially addressed, since the plans are updated at predetermined intervals as new information on demand and actual process yields becomes available. In conclusion, before implementing the proposed complex stochastic production planning model in real sawmills, a simulation platform must be developed which makes it possible to implement the deterministic and stochastic approaches on a rolling horizon under realistic circumstances. Such a simulator is essential to better highlight the benefits and/or drawbacks of the stochastic sawmill production planning model.

Acknowledgments

The current study is part of an extensive research project focused on developing robust production planning tools in the presence of uncertain conditions for Quebec sawmills. The latter is one of the projects proposed in the FOR@C research consortium at Laval University. It is financially supported by NSERC (Natural Science and Engineering Research Council of Canada) and the forestry industry in the Quebec province of Canada. The authors would like to acknowledge the financial support provided by the Forest E-business Research Consortium (FOR@C) of Université Laval.

References

Bakir, M. A. and M. D. Byrne. Stochastic linear optimization of an MPMP production planning model. *International Journal of Production Economics*, 55: pp. 87–96, 1998.

Birge, J. R. and F. Louveaux. *Introduction to Stochastic Programming.* New York: Springer, 1997.

Dupačová, J., G. Consigli and S. W. Wallace. Scenarios for multi-stage programs. *Annals of Operations Research*, 100: pp. 25–53, 2000.

Escudero, L. F., P. V. Kamesam, A. J. King and R. J.-B. Wets. Production planning via scenario modeling. *Annals of Operations Research*, 43: pp. 311–335, 1993.

Gaudreault, J., P. Forget, J.-M. Frayret, A. Rousseau and S. D'Amours. Distributed Operations Planning for the Softwood Lumber Supply Chain: Optimization and coordination. Accepted for publication in *International Journal of Industrial Engineering*, 2011.

Huang, K. and A. Shabbir. The value of multi-stage stochastic programming in capacity planning under uncertainty. *Technical Report*, School of Industrial & Systems Engineering, Georgia Tech., 2005.

Kall, P. and S. W. Wallace. *Stochastic Programming.* New York: John Wiley & Sons, 1994.

Kazemi Zanjani, M., D. Ait-Kadi and M. Nourelfath. Robust production planning in a manufacturing environment with random yield: A case in sawmill production planning. *European Journal of Operational Research*, 201: pp. 882–891, 2010a.

Kazemi Zanjani, M., M. Nourelfath and D. Ait-Kadi. A multi-stage stochastic programming approach for production planning with uncertainty in the quality of raw materials and demand. *International Journal of Production Research*, 48(16): pp. 4701–4723, 2010b.

Kazemi Zanjani, M., M. Nourelfath and D. Ait-Kadi. Two heuristic algorithms for a large-scale mixed-integer production planning model with random yield and demand: A case in sawmills. CIRRELT-2010-23 working document, 2010c.

Kazemi Zanjani, M., M. Nourelfath and D. Ait-Kadi. Production planning with uncertainty in the quality of raw materials: A case in sawmills. *Journal of Operations research Society*, 62: pp. 1334–1343, 2011.

Leung, S. C. H., Y. Wu and K. K. Lai. A stochastic programming approach for multi-site aggregate production planning. *Journal of Operations Research Society*, 57: pp. 123–132, 2006.

Leung, S. C. H., S. O. S. Tsang, W. L. Ng and Y. Wu. A robust optimization model for multi-site production planning problem in an uncertain environment. *European Journal of Operational Research*, 181(1): pp. 224–238, 2007.

Løkketangen, A. and D. L. Woodruff. Progressive hedging and tabu search applied to mixed-integer (0,1) multi-stage stochastic programming. *Journal of heuristics*, 2: pp. 111–128, 1996.

Lulli, G. and S. Sen. A heuristic procedure for stochastic integer programs with complete recourse. *European Journal of Operational Research*, 171: pp. 879–890, 2006.

Maness, T. C. and D. M. Adams. The combined optimization of log bucking and sawing strategies. *Wood Fiber Science*, 23: pp. 296–314, 1991.

Maness, T. C. and S. E. Norton. Multiple-period combined optimization approach to forest production planning. *Scandinavian Journal of Forest Research*, 17: pp. 460–471, 2002.

Mendoza, G. A., R. J. Meimban, W. J. Luppold and P. A. Arman. Combined log inventory and process simulation models for the planning and control of sawmill operations. *In Proceedings of the 23rd GIRP International Seminar on Manufacturing Systems*, Nancy (France), 1991.

Miller, A. C. and T. R. Rice. Discrete approximations of probability distributions. *Management Science*, 29(3): pp. 352–362, 1983.

Mulvey, J. M., R. J. Vanderbei and S. A. Zenios. Robust optimization of Large-scale systems. *Operations Research*, 43(2): pp. 264–281, 1995.

Rockafellar, R. T. and R. J.-B. Wets. Scenarios and policy aggregation in optimization under uncertainty. *Mathematics of Operations Research*, 16(1): pp. 119–147, 1991.

Shabbir, A. and N. V. Shahinidis. Robust process planning under uncertainty. *Industrial & Engineering Chemistry Research*, 37: pp. 1883–1892, 1998.

Takriti, S. and A. Shabbir. On robust optimization of two-stage systems. *Mathematical Programming Series A*, 99: pp. 109–126, 2004.

PART III

Papers on Energy

Chapter 14

An Electricity Procurement Model with Energy and Peak Charges

A. B. Philpott* and G. Pritchard†

Summary

We describe a model developed to help minimize the energy procurement costs of a New Zealand process industry that is a high user of electricity. The model accounts for stochastic prices that depend on the hydrological state of the electricity system, as well as transmission charges that are incurred during coincident electricity peaks. We describe how these are modelled and derive a stochastic dynamic programming algorithm that is used to arrange production to meet demand while minimizing the expected costs of electricity procurement.

1 Introduction

This paper deals with a practical problem facing many manufacturing industries with reasonably flexible production: when and how should they procure electricity to minimize the cost of production needed to meet some future demand. This problem is particularly important in process industries that are heavy users of electricity (such as aluminium, food processing and the pulp and paper industry). By shifting production into time periods in which electricity is inexpensive, companies may make considerable savings. This type of behaviour is often called "peak shaving" or "demand response" and there is a substantial literature and consulting activity devoted to doing this efficiently (Borenstein *et al.*, 2002).

The rationale for shifting loads out of peaks comes about because utilities and electricity markets extract higher prices during periods of peak

*Department of Engineering Science, The University of Auckland, Private Bag 92019, Auckland, New Zealand, a.philpott@auckland.ac.nz.

†Department of Statistics, The University of Auckland, Private Bag 92019, Auckland, New Zealand, g.pritchard@auckland.ac.nz.

demand. The economic theory underlying such peak-load pricing has a long history (see e.g. Boiteux, 1960, Crew *et al.*, 1995). In this theory, higher prices should be charged during peak times to reflect the higher utilization of capacity at these times. Pricing policies of this type can be devised to maximize welfare.

In the setting of electricity markets, sellers of electricity do not choose prices with the aim of maximizing total welfare. Rather, the prices emerge from some market-clearing mechanism that meets demand given the supply functions chosen by electricity suppliers. In this environment, prices will vary with time of day, being more expensive in peak hours when the market clears at high-price points on the sellers' supply curves. Consumers who are flexible have incentives to move production out of these peaks. In choosing their supply curves, sellers of electricity should account for this consumer response in seeking to maximize their own profit. A simple model of this phenomenon is discussed in Pettersen *et al.* (2005).

Our focus in this paper is, however, not on the sellers' pricing decisions, but on the purchaser's decision problem given these prices. The prices we consider are market pricing outcomes for energy and reserve, and peak-load tariffs imposed by the grid owner. We assume that the purchaser is not strategic, and so acts as a price taker. The purchaser then uses a mathematical programming model to optimize its production facilities. When prices are known, this becomes a deterministic optimization problem that can be attacked with standard mixed-integer programming software. See for example Henning (1998) and Ashok and Banerjee (2001) for models of this type. When prices are uncertain, but can be modelled with scenarios one might adopt a stochastic programming approach to this problem (see e.g. Philpott and Everett, 2002).

In this paper we focus on the purchaser's decision problem in a particular industrial setting in New Zealand in which a single product is made to meet a known demand occurring at known future points in time. The particular industry we have in mind exports products by ship, and enough stock must be on hand to satisfy a given schedule of ship visits. The production problem is relatively simple when electricity prices are known - one simply produces in sufficiently many low priced periods to generate enough stock to meet demand. When prices are random, but can be modelled as a Markov chain, this problem can be solved as a Markov decision process using dynamic programming (see e.g. Ravn and Rygaard (1994) for such a model for meeting heat and power constraints over time). We follow this line of reasoning in developing our model.

The model we discuss in this paper has two new features. Production facilities that are running may be paid a reserve price l to provide an interruptible load in circumstances where there is some contingency (e.g. a generating unit failure). In other words, the system operator pays l in every trading period for the option to disconnect the power supply to the production units if such an event happens. If the industry has agreed to such a contract then this payout effectively decreases the price p that is paid for electricity. (The decision to offer reserve in this way is exogeneous to our model.) Our stochastic price process is therefore estimated from historical sequences of the net price

$$e = p - l,$$

where both p and l are determined for each trading period by the electricity market clearing mechanism.

The second feature of our model that is novel concerns a payment that depends on the regional peak demand. In New Zealand, the grid owner records the peak demands in each region for each trading period (each having duration half an hour) in the 12 months from September 1 to August 31. These are called *coincident peaks* as they relate to all loads, not just those of the electricity consumer we are modelling. On August 31 each year the grid owner sums each consumer's load in the 100 highest coincident peak periods of the past 12 months. The consumers then pay the grid owner a peak charge M for every megawatt-hour (MWh) purchased in those periods. As regional demand increases towards the daily peak, the purchaser of power is faced with a delicate decision problem: should she shut down her plant in anticipation that the next half hour will be one of the highest 100 periods, even though this fact might not be known until many months in the future?

We show how this problem can be solved by dynamic programming producing a threshold-type policy. This policy requires an estimate at each point in time of the probability that the regional load being observed will exceed the 100th highest regional load observed over the 12 months from September 1 to August 31. This estimate is made using a model for these peak demands.

The paper is laid out as follows. In the next section we formulate the decision problem we wish to solve as a dynamic programming problem, and show that a threshold policy is optimal. We then describe models of random electricity prices and coincident peak demand that have been developed to incorporate some of the serial correlation in these data. In the following section we use these models to extend the dynamic programming model, and illustrate its output on an example with fictitious plant data, but real price and peak demand data from the New Zealand electricity market. The paper concludes with a discussion of the implementation of this model in a practical setting.

2 A dynamic optimization model

In this section we derive a dynamic optimization model for production planning using machines that consume electrical energy. We assume for simplicity that production levels are binary - either machines are all running at full capacity, or they are all turned off. In the last section of the paper we will discuss how this assumption can be relaxed.

The model we describe has two levels of time discretization. We denote by $t = 1, 2, \ldots, T$, the *stages* of a dynamic decision problem, where for each stage we will compute an optimal action. Within each stage there are *trading periods* denoted $k = 1, 2, \ldots, K$. In the New Zealand application, we choose a stage length of one day, divided up into 48 half-hour trading periods.

The decision to be made in any stage t and any trading period k is to determine whether to run the plant given:

1. the current inventory level of our product (z);
2. the current electricity price (p) and price (l) for interruptible load;
3. the current value of total regional demand (x).

We assume that p, l, and x for the trading period are all known at the time the decision is made to shut down. In practice, p, l, and x will be short-term forecasts based on data recorded in previous trading periods. This of course makes them subject to some forecast error, which we assume is neglible in this description. We will discuss the implication of this assumption in the conclusion of the paper.

We consider first the optimal action to be taken at any given stage t, assuming that z is known at the start of stage t. Suppose that the producer requires a given amount y to be produced in stage t. The optimal decision at stage t will depend on realizations of p, l, and x at every $k = 1, 2, \ldots, K$, which we must treat as random variables $P(k)$, $L(k)$, and $X(k)$. We define the random variable $E(k)$ to be the net price per MWh of procuring energy in trading period k, where $E(k) = P(k) - L(k)$ is the difference between the electricity spot price and the price of interruptible load.

First, consider a situation in which the net price and regional demand $(E(k), X(k))$, $k = 1, 2, \ldots, K$, are i.i.d. random variables with joint density $f(e, x)$ over sample space Ω. For any realization (e, x) we wish to decide whether to run the plant or not. Let $\rho(e, x)$ be an indicator function that is 1 if the plant is run (at full capacity) and 0 if it is shut down. The probability that the plant runs in any trading period is

$$\int_{\Omega} f(e, x)\rho(e, x)dedx.$$

Suppose that we observe e and x. Then the expected cost per MWh of running in realization (e, x) is

$$e + M\Pr(x \geq X_{100})$$

where X_{100} is the (random) 100th highest regional demand, and M denotes the peak period charge per MWh. Suppose we know the distribution $G(x) = \Pr(X_{100} \leq x)$. Then the expected cost per MWh of running in realization (e, x) is

$$e + MG(x).$$

In order to produce y tonnes on day t, the producer seeks an indicator function ρ to solve

$$\begin{aligned} P : \text{minimize } & \frac{b}{a}\int_{\Omega}(e + MG(x))f(e, x)\rho(e, x)dedx \\ \text{s.t.} \quad & b\int_{\Omega} f(e, x)\rho(e, x)dedx = y \end{aligned}$$

where b is the capacity of the plant in tonnes per day, and a is the number of tonnes of product from one MWh of electricity consumption.

The Lagrangian for P is

$$\mathcal{L}(\rho, \lambda) = y\lambda + \frac{b}{a}\int_{\Omega}(e + MG(x) - a\lambda)f(e, x)\rho(e, x)dedx.$$

Minimizing $\mathcal{L}$ defines a threshold policy

$$\rho_{\lambda}(e, x) = \begin{cases} 1, \; e + MG(x) \leq a\lambda \\ 0, \; e + MG(x) > a\lambda \end{cases}$$

where λ is chosen so that this policy gives

$$b\int_{\Omega} f(e, x)\rho_{\lambda}(e, x)dedx = y.$$

The optimal value of λ can be interpreted as the marginal value of an extra tonne of production and will be an increasing function of the requirement y.

In a multi-stage setting, we wish to adjust the amount we produce in each stage t depending on how much inventory has been accrued. The optimization problem that we wish to solve is to determine whether to shut down the plant when we observe high values of $e + MG(x)$ on a particular day. The peak values usually occur in the morning and early evening when domestic demand is high. Typically the firm would stop production for these periods for sufficiently high values of $e + MG(x)$. Production typically resumes in the late evening, and the procedure repeats the following morning.

For any given day t in the planning horizon, we seek an optimal threshold value λ. Given this value, the plant will shut down completely in every trading period that has $e + MG(x) > a\lambda$. Given a value of λ, the amount of product produced by policy $\rho_\lambda(e, x)$ is

$$w_t(\lambda) = b \int_\Omega f_t(e, x)\rho_\lambda(e, x)dpdx,$$

and its expected cost is

$$c_t(\lambda) = \frac{b}{a} \int_\Omega (e + MG(x))f_t(e, x)\rho_\lambda(e, x)dedx.$$

Observe that the density $f_t(e, x)$ may now vary with the stage t.

Given this optimization model for each stage t, we can derive a dynamic programming recursion for $t = 1, 2, \ldots, T$. Let $C_t(z)$ be the minimum expected future cost of meeting demand d_τ, $\tau = t + 1, t + 2, \ldots, T$, if there is a stock level of z at the end of day t. Then at the start of day t we seek

$$C_{t-1}(z) = \mathbb{E}\begin{bmatrix} \min_{\rho,y} \frac{b}{a} \int_\Omega (e + MG(x))f_t(e, x)\rho(e, x)dedx \\ \quad + C_t(z + y - d_t) \\ \text{s.t. } y = b \int_\Omega f_t(e, x)\rho(e, x)dedx \end{bmatrix} \tag{1}$$

The stage problem (1) gives Lagrangian

$$\mathcal{L}(\rho, \mu) = \frac{b}{a} \int_\Omega (e + MG(x))f_t(e, x)\rho(e, x)dedx + C_t(z + y - d_t)$$
$$+\lambda y - \lambda b \int_\Omega f_t(e, x)\rho(e, x)dedx$$

and a policy

$$\rho_\lambda(e,x) = \begin{cases} 1, \; e + MG(x) \leq \lambda a \\ 0, \; e + MG(x) > \lambda a \end{cases} \tag{2}$$

where

$$0 \in \lambda + \partial C_t \left(z + b \int_\Omega f_t(e,x)\rho_\lambda(e,x)dedx - d_t \right).$$

Thus λ is the marginal value of storage at $z + b\int_\Omega f_t(e,x)\rho_\lambda(e,x)dedx - d_t$. If we assume that λ is constant over the range of production then this gives a threshold policy defined by (2).

3 Statistical models

The recursion (1) used in the model in the previous section assumes a stagewise independent distribution for (E, X). In reality, (E, X) follows a more complex stochastic process with some stagewise dependence. In the next section, we will derive the dynamic programming recursion for this more complex model, which we now proceed to describe in more detail.

Although we are primarily interested in the electricity price, much of the structure of price series in a hydropower-dominated system (like New Zealand) is derived from the underlying hydrology. This is illustrated in the plot in Figure 1 that shows regional electricity spot prices in New Zealand varying with storage in its largest hydro-electric catchment (the Waitaki system). The first step in modelling prices that depend on hydrology is to develop a process for representing inflows and releases to the hydro catchments. The dynamics of a hydro reservoir are governed by the equation:

$$S_{t+1} = S_t + I_t - R_t$$

where S_t is the stock of water, I_t is inflow and R_t is release of water (for generation and spill). Although releases represent actions of the generators, they are influenced by storage levels to the extent that one can derive reasonable statistical models that represent this (see e.g. Tipping and Read, 2010). In other words high storage levels tend to give large releases and low storage levels give small releases.

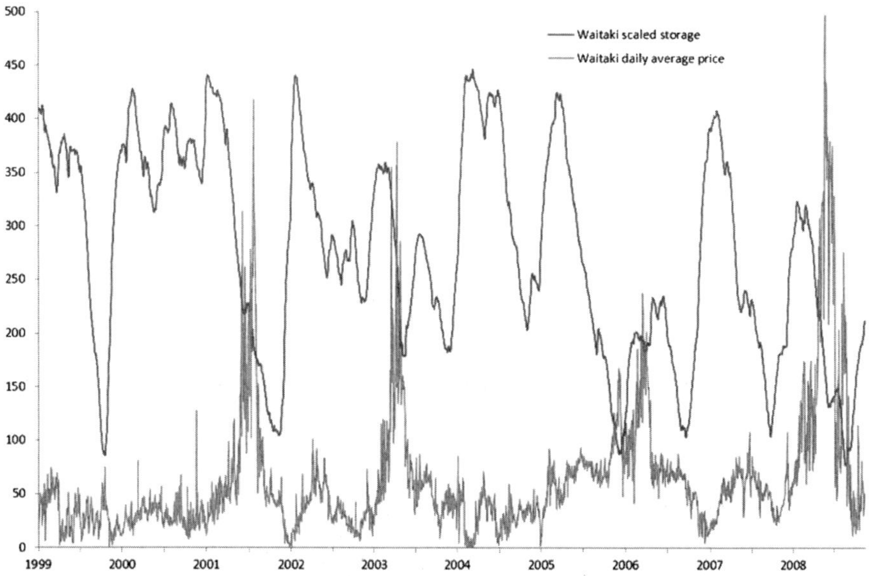

Figure 1 Electricity spot prices and hydro storage levels in the Waitaki hydro system. Prices spike when hydro storage levels drop because of low inflows.

3.1 *Inflows*

We begin our discussion of hydrological processes by discussing inflows. It has been traditional to model hydropower inflows by replaying inflow sequences from past years, starting at the appropriate time of year. This is a generally sound idea, as it reproduces the appropriate distributions, serial correlation structure, etc. But if the sequences are to be regarded as possible scenarios for the immediate future, one should take some account of current conditions. If recent weeks have been dry, one should not use sequences taken from wet years. The following model is an attempt to make a simple adjustment for this.

Our model is:

$$\log I_t = \alpha \log I_{t-1} + T_t^I + \text{error}$$

where I_t is the inflow in week t, and T_t^I is an annual seasonal factor consisting of a second-order trigonometric polynomial with a 1-year period. That is, the log-inflows consist of a fixed seasonal pattern with superimposed red noise.

We can use this in conjunction with historical inflow sequences as follows. The model gives

$$I_t = I_{t-1}^{\alpha} \times (\text{seasonal and random factors}).$$

If the historical sequence is $h_0, h_1, h_2, \ldots$, and we have already observed this year's inflow for the week corresponding to h_0 to be I_0, then a scenario for the following week's inflow is

$$I_1 = \left(\frac{I_0}{h_0}\right)^{\alpha} h_1.$$

For the week after that, we get

$$I_2 = \left(\frac{I_1}{h_1}\right)^{\alpha} h_2 = \left(\frac{I_0}{h_0}\right)^{\alpha^2} h_2,$$

and similarly,

$$I_j = \left(\frac{I_0}{h_0}\right)^{\alpha^j} h_j. \tag{3}$$

Since $\alpha < 1$, we have $\alpha^j \to 0$ and so $I_j \approx h_j$ after the first few weeks.

We use a single model of this type to represent the combined inflows (in energy-equivalent, i.e. gigawatt-hour, terms) for all the large hydro lakes in the New Zealand power system. (The lakes included are named Tekapo, Pukaki, Ohau, Hawea, Te Anau, Manapouri, Taupo, and Waikaremoana.) The fitted value of α is approximately 0.44.

3.2 *Releases*

We now turn attention to the hydroelectric energy released from storage by the electricity industry. The reservoirs included are the same as those in the inflow model.

The model is:

$$R_t = \beta_1 S_t + \beta_2 S_t^2 + \beta_3 S_t^3 + \beta_4 I_t + T_t^R + \text{error} \tag{4}$$

where R_t is the energy released in week t, S_t the total stored energy at the beginning of week t, and T_t^R is an annual seasonal factor consisting of a first-order trigonometric polynomial with a 1-year period.

After fitting this model the residuals are found to be fairly symmetrical about 0, to be approximately normally distributed, and to have little serial correlation.

3.3 *Electricity prices*

Electricity prices are less well-behaved than the hydrology, and so the following models are inevitably somewhat more approximate.

3.3.1 *Price duration curves*

The optimization problem that we wish to solve is to determine whether to shut down the plant when we observe high values of $e + MG(x)$ on a particular day. The key data input in this decision is a *price duration curve*,

$$D(s) = \text{the number of trading periods where } e + MG(x) \leq s.$$

The inverse $D^{-1}(s)$ can be thought of a realization of values of $e + MG(x)$ over a 48-period day, ordered from lowest to highest. Such a curve is illustrated in Figure 2.

If one knows the price duration curve, then it is straightforward to find an optimal threshold policy by conducting a line search for λ that yields the desired production amount of y. Recall that there are K trading periods per day, a is the tonnes produced per MWh, and b is the daily production capacity. Then the optimal λ solves

$$b\frac{D(\lambda)}{K} = y.$$

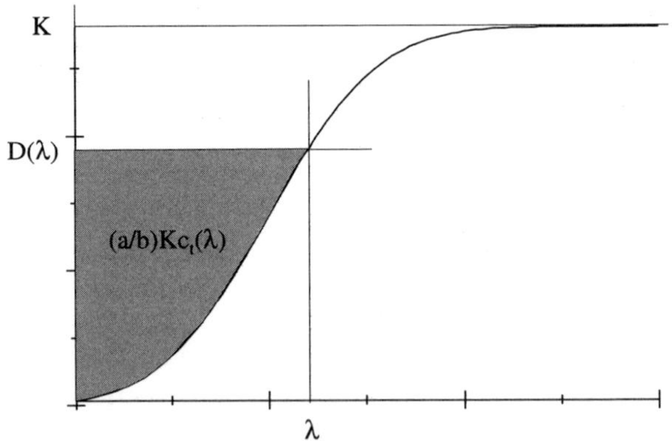

Figure 2 A price duration curve. $D(\lambda)$ is the number of periods with price at most λ. The shaded area when divided by K is the average cost per MWh of policy λ.

The cost of this choice per day is given by the area between the price duration curve and $D(\lambda)$

$$c_t(\lambda) = \frac{b}{aK}\left(\lambda D(\lambda) - \int_0^\lambda D(s)ds\right).$$

In our model, the price duration curve is random, having a finite distribution on each day. This distribution is determined by the weekly hydrology models above, and the random curves on consecutive days are not independent but are correlated in a way that we describe below.

We illustrate the modelling approach we have adopted by applying it to the spot price p; a similar process is used to model $e + MG(x)$. To estimate a model for scenarios of duration curves for p, it helps to transform the data. Some very high prices occur, and we first attempt to rein these in by applying the following transformation to all our price values:

$$P = c\left(\left(1 + \frac{3p}{c}\right)^{1/3} - 1\right);$$

the value $c = \$3/\text{MWh}$ is found to be suitable. Of course, this means that when using models for P, we will eventually have to transform back via

$$p = \frac{(P+c)^3 - c^3}{3c^2}.$$

Let $P_{t,k}$ denote the kth largest (transformed) price occurring on day t ($k = 1, \ldots, 48$). To estimate price-duration curves, we use quantile regression (Koenker and Bassett, 1978). The model equations are:

$$Q_\tau(P_{t,k}) = \sum_{j=0}^{4}\left(\gamma_{1,j}S_t + \gamma_{2,j}\frac{1}{S_t} + \gamma_{3,j}I_t + \gamma_{4,j}H_t\right)f_j(t) + \text{error},$$

where:

1. $Q_\tau(P_{t,k})$ is the τ-quantile of the distribution of $P_{t,k}$;
2. S_t the hydro storage at the beginning of the week that includes t, and I_t the inflow during the week that includes t;
3. H_t a holiday indicator (0 if t is a business day, 1 otherwise);
4. $f_0, \ldots f_4$ are seasonal trigonometric functions with a 1-year period:

$$f_0(t) = 1, \quad f_1(t) = \cos(\omega t), \quad f_2(t) = \sin(\omega t),$$
$$f_3(t) = \cos(2\omega t), \quad f_4(t) = \sin(2\omega t).$$

Fitting these models for $\tau = 0.05, 0.15, \ldots, 0.95$ (and $k = 1, \ldots, 48$) gives a set of ten model price duration curves, each of which is a function of the covariates (S_t, I_t, and H_t, and seasonality) for the day in question.

For some values of the covariates, some of these curves can be slightly non-monotone; if this is important, the prices should be sorted into descending order after they have been computed.

3.3.2 *Transition model*

Each of the ten model scenarios should occur about 1/10th of the time, but this does not happen at random for each day, independently of previous days, as this would ignore the substantial serial dependence. Fitting quantile regressions as above, but for $\tau = 0.1, 0.2, \ldots, 0.9$, allows us to classify each historical price according to which model scenario it most closely resembles (e.g. if a historical price lies between the $\tau = 0.1$ and $\tau = 0.2$ models for the day on which it occurs, it is associated with the $\tau = 0.15$ model scenario). We associate each historical day to a scenario using the classification given by its 6th largest price. This gives a sequence of scenarios which "occurred" historically.

The serial dependence in this sequence of scenarios is much stronger than for a Markov chain with the same one-step transition matrix M. To represent this, we use the following "sticky" random model. Let $X_t \in \{1, \ldots, 10\}$ be the scenario used on day t, and $V_t \in \{1, \ldots, 10\}$ a "background state" pertaining to that day. Then (X_t, V_t) follows a Markov process given by

- X_t is chosen according to the V_{t-1}th row of M;
- $V_t = X_t$ with probability δ, otherwise V_{t-1};

with these random choices being made independently of those on previous days. The value chosen for δ is currently 0.10. This allows for a moderate degree of dependence between X_t and X_{t-r} when r is small (as they are likely to have been picked from the same row of M), which does not disappear too quickly as r gets larger.

3.4 *Peak demand*

The final ingredient in our model is regional peak demand. Figure 3 shows an example of peak demand from September 1, 2006 to August 31, 2007

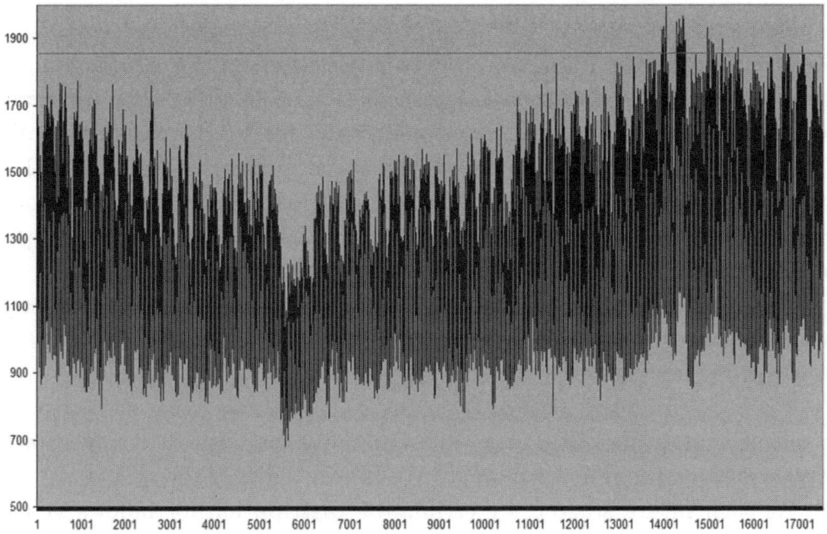

Figure 3 Regional half-hourly demand in 2006–2007. The 100th highest demand realization is shown by the solid horizontal line.

for a region of New Zealand. Observe that most of the peaks occur in the period May 1 to August 31. The horizontal line in the figure is set at the 100th largest demand (1855 MW).

The basis of our peak-load model is a suite of statistical models giving the expected highest, second-highest, etc. load each morning and evening of the winter (April–September). These are simply functions of time (the number of days before or after July 1) and the day of the week (allowing for public holidays). An offset (additive constant) to be added to each model is estimated using recent historical load data beginning on April 1. To this is added a random process chosen so that its serial and cross-correlations between adjacent mornings and evenings are the same as in the data. This allows peak loads to be simulated arbitrarily far ahead. The random process is initialized with a value for the current day predicted from the previous day's loads and air temperatures. The loads are then simulated from the current day until August 31, and the simulation repeated 10,000 times in order to obtain the statistics shown.

Thus for any temperature data provided the model delivers probabilistic estimates of what the 100th largest half-hourly load of the pricing year will be. It gives P5, P50, and P95 estimates (i.e. values with 5%, 50% and 95% probability, respectively, of exceeding the threshold) as well as an estimate of the distribution $G(x)$.

4 Dynamic programming recursion

We now discuss the dynamic programming recursion for determining an optimal policy over the time stages $t = 1, 2, \ldots, T$. The curse of dimensionality precludes a high-dimensional state space. By experimentation we have identified the key states being a national reservoir storage level (represented by $u \in \{1, \ldots, U\}$), electricity price state and background state (represented by $v \in \{1, \ldots, 10\} \times \{1, \ldots, 10\}$) and stock on hand of product ($z \in \{0, 1, \ldots, L\}$).

Based on this structure we have developed for fixed M, a statistical model for duration curves of

$$A = e + M \Pr(X_{100} < x),$$

for all possible values of u and v, as well as probabilities for transitioning between states (u, v) at each stage. We look at each of these in turn.

At each future stage t, we estimate a distribution for inflows at time t, based on the observed inflow at time 0, and simulations using the model represented by (3). This model induces a stochastic process on the storage levels and therefore on the releases according to (4). Thus, at any stage t and reservoir state u, we can estimate a transition probability $q(u, u')$ to state $u' \in \{1, \ldots, U\}$ at the next stage by computing a release decision for each inflow realization using (4) and a sampled error term. The inflows used at each stage in this process are assumed to be statistically independent from those in the preceding stage.

The model for A is very similar to the model for price duration curves described in subsection 3.3.1, except that now we construct duration curves for A rather than p. These curves are estimated a priori and stored for all possible values for t, u and v. We now write

$$D_t(s, u, v) = \text{the number of trading periods in stage } t \text{ and state}(u, v) \text{ where } A \leq s.$$

Given such curves and a choice of threshold λ, we can compute the expected daily production $w_t(\lambda, u, v)$ of the plant, and the expected daily cost of production $c_t(\lambda, u, v)$ that results. Formally

$$w_t(\lambda, u, v) = \frac{b}{K} D_t(\lambda, u, v)$$

and

$$c_t(\lambda, u, v) = \frac{b}{aK} \left(\lambda D_t(\lambda, u, v) - \int_0^{\lambda} D_t(s, u, v) ds \right).$$

We also estimate transition probabilities $r(v, v')$ for moving from state v at time t to state v' at time $t + 1$. (These are independent from the transitions in u.) As described in subsection 3.3.1 we use a sticky random model where the second component of v is a background state that changes in only about 10% of the transitions, and $r(v, v')$ depends only on this second component.

Now we can write down the dynamic programming recursion that we use.

$$\begin{aligned} C_{t-1}(z, u, v) &= h_t(z) + \min_{\lambda}\{c_t(\lambda, u, v)) \\ &\quad + \gamma \sum_{u'} \sum_{v'} q(u, u')r(v, v')C_t(z + w_t(\lambda, u, v) - d_t, u', v')]\} \\ C_T(z, u, v) &= V(z) \end{aligned}$$

where

$$\begin{aligned} C_t(z, u, v) &= \text{future cost at end of stage } t \text{ if in state } (u, v) \\ &\qquad \text{and inventory is } z \\ h_t(z) &= \text{holding cost for inventory } z \text{ over stage } t \\ d_t &= \text{demand for product in stage } t \\ V(z) &= \text{terminal future cost with } z \text{ in stock} \\ \gamma &= \text{daily discount factor} \end{aligned}$$

The dynamic programming recursion gives a threshold value $\lambda = \lambda^*(t, u, v, z)$ at each stage t and state (u, v, z) to determine at what (adjusted) price the company should reduce load. To determine what action to take, the company needs to estimate the current state. This is straightforward for u, but to estimate v (the price state) one requires the current duration curve for A at stage $(t - 1)$ which can be estimated from the previous time period's observations.

To determine whether a shutdown is necessary in trading period k at stage t, the company must estimate the current value of $A(k)$ to compare it with the threshold value. This is computed using the current observation of $x(k)$ and sample values of X_{100} computed by the peak pricing model from subsection 3.4. If the sample values of X_{100} are sorted and compared to $x(k)$ then $\Pr(X_{100} < x(k))$ can be estimated as $x(k)$ is observed. Thus to determine an optimal action in each trading period, the peak pricing model must be solved at the start of the day to provide the appropriate estimate of $\Pr(X_{100} < x(t))$.

We do not allow backlogging of stock, so z is penalized from becoming negative. In fact, given constraints on production capacity, there is a minimum stock level that must be held at each stage to ensure that there will be enough stock on hand to meet demand (which is known a priori). Deviations below this minimum stock level are heavily penalized, as we can guarantee that they will lead to a shortage in demand. Moreover, in our computations we need only consider values of z that lie above this riskzone level. An example of such a riskzone level is shown in Figure 4.

The form of the solution to the dynamic programming recursion is best illustrated by an example. Figure 5 shows a contour plot of the optimal threshold values (with different shadings between contours) for state $(u, v) = (5, 10, 10)$ and varying z levels over the same period covered by the ship visits shown in Figure 4. In this example we have used 10 reservoir states, so $(5, 10, 10)$ corresponds to an average reservoir storage state (5), but high electricity price and background states (10). The shaded region at the top of the figure corresponds to the range $\lambda^*(t, u, v, z) \in [0, 20]$.

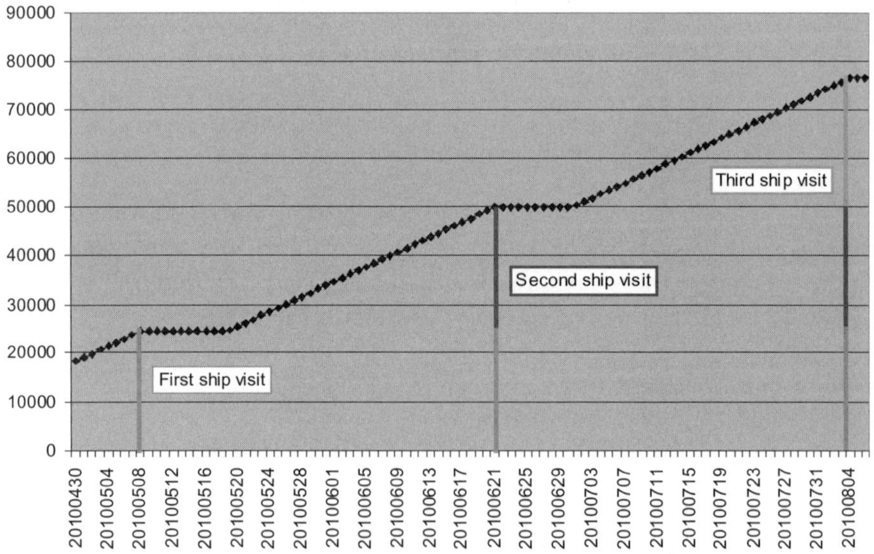

Figure 4 A typical riskzone that includes demand. Ship visits occur on May 8, June 21 and August 4. The model must maintain a cumulative production above the blue line to ensure that the demand from these ship visits can be met. The sloping sections grow at the maximum production rate.

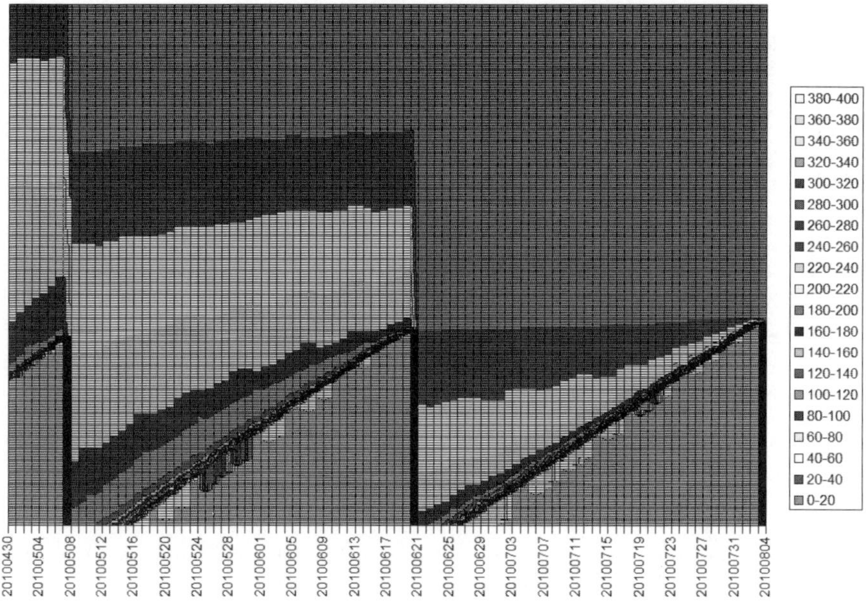

Figure 5 Contours of threshold values $\lambda^*(t, u, v, z)$ for state $(u, v) = (5, 10, 10)$. Shadings of (t, z) indicate different ranges of threshold values. The contours are vertical at times where demand deliveries must be made.

One can see from the plot that for fixed t, $\lambda^*(t, 5, 10, 10, z)$ decreases as the storage z increases. In the top region we have more than enough stock to meet the ship visits, and to avoid inventory costs a threshold value of close to zero is optimal. In this region the plant should shut whenever A gets above 20. The triangular regions where (t, z) lies in the riskzone are easy to see, and the contours of $\lambda^*(t, 5, 10, 10, z)$ become close together as (t, z) approaches the boundaries of these, since, to avoid a stockout, the optimal value of λ must increase at a higher rate to ensure production is not interrupted.

The most valuable information from our model is not so much the threshold policy that it computes, but the marginal value of storage λ that is obtained. This gives the slope of the expected future cost $C_t(z, u, v)$ of meeting the demand from time t to T, with z in stock, given an optimal threshold policy is followed. The future cost is defined for all possible values of the national storage state (u) and market state (v). It is also defined for all possible levels of storage (z) above the riskzone. An example of such a marginal cost curve for June 5, 2010 is shown in Figure 6.

This information can be used to guide a daily optimization model that incorprates the details of the company's plant operations. To do this a plan

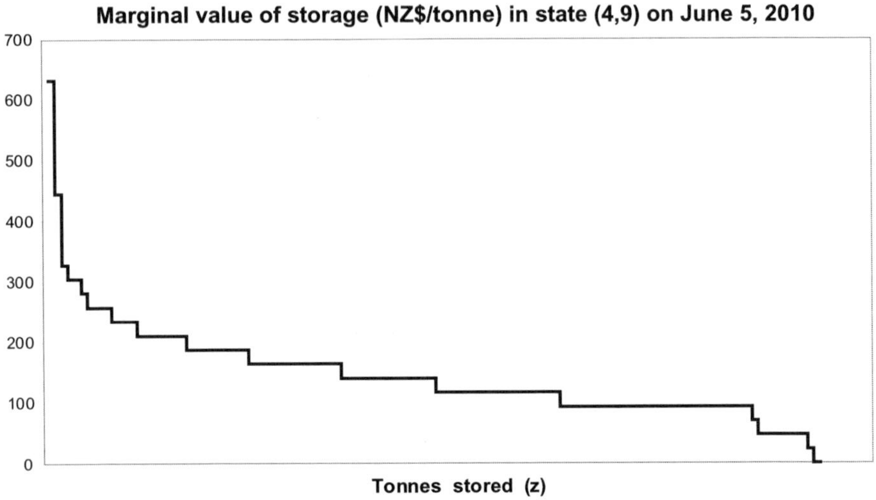

Figure 6 Marginal value of product stored as a function of inventory level (z), for market state (5,10,10) at $t = 2010605$ (as annotated on Figure 5).

is computed for the current day that uses forecasts (or scenarios) for $A(t)$ in the current day, and minimizes today's cost plus the expected future cost as defined by $C_t(z, u, v)$.

The daily optimization can be of higher fidelity than the optimizations used to compute the marginal value of storage, and include several operating units, ramping constraints, startup costs and shutdown costs. The marginal cost curve ensures that a close to optimal trade off is made between running and increasing the amount of inventory stored, and shutting down and avoiding energy and possible peak charges.

An example of such a model is the following mixed integer programming model of the unit commitment type. To see how this would work, assume that we have predicted $A(t)$ over the course of the coming day. This gives the expected cost of purchasing a MWh of electricity (accounting for maximum–demand charges, electricity price and reserve.) Let

$$x(t) = \begin{cases} 1 & \text{if plant runs in period } t \\ 0 & \text{otherwise} \end{cases}$$

Let N be the cost of switching the plant off, and let y denote the total amount of product produced in a day. Suppose the amount of product in inventory is z_0 and we increase this to $z_0 + z$. Let $a_j + b_j(z_0 + z)$, $j = 1, 2, \ldots, J$, be cutting planes representing the future expected savings from producing z over the day. The savings function is then given by the function

$$Q(z) = \min_{j=1,\dots,J} \{a_j + b_j(z_0 + z)\}.$$

The mixed integer program we solve is:

$$\begin{aligned}
\min \ & \sum\nolimits_{t=1}^{48} Nx(t) + A(t)y(t) - \theta \\
\text{s.t.} \ & x(t) \geq y(t) - y(t+1), \\
& \frac{k}{48} \sum\nolimits_{t=1}^{48} y(t) = z, \\
& \theta \leq a_j + b_j(z_0 + z), \qquad\qquad j = 1, 2, \dots, J, \\
& x(t), y(t) \in \{0, 1\}.
\end{aligned}$$

This gives a sequence $x(t)$ of ones and zeros that can be used to determine when to shut down the plant. The switching cost N will prevent too many shutdowns in a day. The optimal value z^* obtained will be the total production over the day. The optimal value θ^* from this model will satisfy

$$\theta^* \leq a_j + b_j(z_0 + z^*), \ j = 1, 2, \dots, J,$$

and be as large as possible to minimize the objective of MIP and so

$$\begin{aligned}
\theta^* &= \min_{j=1,\dots,J} \{a_j + b_j(z_0 + z^*)\} \\
&= Q(z_0 + z^*)
\end{aligned}$$

the future savings from producing z^*.

5 Conclusions

This paper has described a simple but effective model for peak-shaving industrial electricity demand. The model relies on a "top-down" statistical model of electricity prices and regional peak demand. Peak shaving models are well understood in a deterministic framework, but have received little attention in an uncertain environment, and none when peak charges are incurred in hindsight. Our model is a first attempt at constructing a model for problems of this type. Although the model is specifically adapted for use in the New Zealand context, the approach should be readily applicable to other settings, albeit with some structural changes to the statistical models.

The current version of our model assumes that the national inflows are independent from week to week. This assumption is made for computational convenience, and it does have implications for the policies that are computed. In essence, if a week is dry then the model will not assume that the next week is more likely to be dry and so it will not be as conservative in maintaining production as it should be. Similarly if the week is wet, then the model might recommend producing more than needed because it does not account for an increased probability of high inflows in the following week.

The effect of stagewise dependence in national inflows can be investigated by increasing the state space to include the previous day's inflows, with an increase in computational complexity. However the extent of this effect is not as bad as it might seem at first because the dynamic programming model is in some sense "self correcting". This happens because in applying the solution from the model, a sequence of dry inflow weeks and a less conservative policy will eventually give a low national storage level, and lower than needed inventory levels. As this occurs, the optimization model will increase the threshold value of λ until the user starts to shut down less and act more conservatively.

In our model we have assumed that the plant operator can predict e and x immediately prior to the trading period to which these pertain (so that they can shut the plant before incurring the charge $e+MG(x)$). In practice, large firms have data feeds to these parameters, and so the forecasts are relatively precise (in contrast to a prediction that the next period's regional demand will be one of the 100 highest). The model could incorporate some uncertainty in these forecasts with some increase in complexity by estimating probability distributions of e, and x about forecast values $\hat{e}$, and $\hat{x}$. The threshold policy then becomes

$$\rho_\lambda(\hat{e},\hat{x}) = \begin{cases} 1, \mathbb{E}[e + MG(x) \mid \hat{e},\hat{x}] \le a\lambda \\ 0, \mathbb{E}[e + MG(x) \mid \hat{e},\hat{x}] > a\lambda \end{cases}$$

requiring some additional computation in the dynamic programming recursion.

A futher restrictive assumption in our model is that a single commodity is being produced. Even the process industries that we have in mind in this work admit different product types (e.g. corresponding to purity of aluminium or basis weight of paper). Unfortunately, the curse of dimensionality means that increasing the number of commodities increases the complexity of the dynamic programming problem considerably, unless the production processes can be decoupled so that they do not share costs of electricity procurement. With this proviso, the complexity of the production scheduling process on the day of operation can be considerably more complex than the one we have modelled, as illustrated in the previous section. The model of this paper gives some end conditions on a daily scheduling model that will ensure that the production schedule makes appropriate tradeoffs in electricity procurement.

References

Ashok, S. and R. Banerjee. An optimization model for industrial load management. *IEEE Transactions on Power Systems*, 16(4): pp. 879 –884, 2001.

Boiteux, M. Peak-load pricing. *The Journal of Business*, 33(2): pp. 157–179, 1960.

Borenstein, S., M. Jaske, and A. Rosenfeld. Dynamic pricing, advanced metering and demand response in electricity markets. Technical report, Center for the Study of Energy Markets, University of California Berkeley, 2002.

Crew, M. A., C. S. Fernando, and P. R. Kleindorfer. The theory of peak-load pricing: a survey. *Journal of Regulatory Economics*, 8: pp. 215–248, 1995.

Henning, D. Cost minimization for a local utility through CHP, heat storage and load management. *International Journal of Energy Research*, 22: pp. 691–713, 1998.

Koenker, R. and G. Bassett Jr. Regression quantiles. *Econometrica*, 46(1): pp. 33–50, 1978.

Pettersen, E., A. B. Philpott, and S. W. Wallace. An electricity market game between consumers, retailers and network operators. *Decision Support Systems*, 40 : pp. 427–438, 2005.

Philpott, A. B. and G. R. Everett. Pulp electricity demand management. In *Proceedings of the 37th Annual Conference of ORSNZ*, pp. 55–63, 2002.

Ravn, H. F. and J. M. Rygaard. Optimal scheduling of coproduction with a storage. *Engineering Optimization*, 22(4): pp. 267–281, 1994.

Tipping, J. and E. G. Read. Hybrid bottom-up/top-down modeling of prices in deregulated wholesale power markets. In S. Rebennack, P. M. Pardalos, M. V. F. Pereira, and N. A. Iliadis, eds., *Handbook of Power Systems II*, Energy Systems, pp. 213–238. Berlin Heidelberg: Springer, 2010.

Chapter 15

A Stochastic Game Model Applied to the Nordic Electricity Market

Stein-Erik Fleten* and Tek Tjing Lie†

Summary

The paper presents an investigation of market power potential in a deregulated hydro and thermal based electricity market through a two-stage stochastic Cournot model. The optimal hydro and forward strategies are examined when generation companies take into account the commitment effect of these decisions, i.e. the influence on the future play of opponents. The generation capacity of each company is regarded as given. In the first stage, firms decide on forward contracting, hydro and thermal generation (including nuclear) for the current spot market. In the second stage, they learn about water inflows to reservoirs before deciding on generation for the second stage spot market. Using cost and capacity data and historical inflow series from the Scandinavian market, large pure hydro producers are found to have little or no market power on the seasonal level, and that the largest producer, Vattenfall, has some incentive to withdraw capacity from the market. The simulated price in an oligopoly case is less than 7% higher than under perfect competition, thus the market is expected to be reasonably competitive. Still, competition authorities should monitor and possibly prevent market concentration to increase.

List of Symbols

a_t, b_t — constant coefficients of the inverse demand function, for stage t.

C_p — cost of thermal generation of player p.

c_p — coefficient of the quadratic term of the thermal cost function of player p.

*Department of Industrial Economics and Technology Management, Norwegian University of Science and Technology, Trondheim, Norway, stein-erik.fleten@iot.ntnu.no.
†Department of Electrical & Electronic Engineering, Auckland University of Technology, New Zealand, tek.lie@aut.ac.nz.

d_p — coefficient of the linear term of the thermal cost function of player p.
$E()$ — expected value.
f_p — energy sold forward by player p.
g_p^t — thermal generation of player p.
g_p^h — hydro generation of player p.
I_p — residual inflow of water of player p.
$\mathcal{N}$ — set of pure hydropower producers.
p^f — forward price.
P_t — the electricity price in stage t.
$\mathcal{P}$ — set of all players.
Q_t — the total electricity demand/generation in stage t.
U_p — objective function of player p.
$Var()$ — variance.
λ — parameter of risk aversion.
Π_p — profit function of player p.

1 Introduction

In recent years, the traditional vertically integrated electric utility structure has been subject to change due to the introduction of a competitive market scheme in many countries worldwide. The objective of this restructuring is to create a competitive environment for trade in electricity to improve market efficiency and social welfare. However, due to high barriers caused by long construction periods of power plants and large capital investments needed to enter the electricity market, and also due to isolation between consumers and producers caused by transmission constraints and transmission losses, the new electricity markets are more akin to oligopoly markets in which only a few producers serve a given geographic area than perfectly competitive markets. In a perfectly competitive market, a large number of firms compete with one another and each of them is a price taker. Social welfare is maximized since the market price is equal to the marginal cost of supply and also equal to the marginal willingness to pay for consumption. In oligopoly markets, there are only a few firms that compete with one another. Companies owning generation assets can increase profits through exercising market power, by using their ability to raise market price by their bidding behavior. Such exercise of market power, raising prices above marginal cost, lead to productive and allocative inefficiencies and inefficient signals for new investment. Therefore, it is very important for a system regulator to analyze the competitiveness of a set of generators and to recognize and curb market power.

Common policy measures against market power include price caps and "windfall" taxes on excess profits. However, those approaches may mask the real market trading situation and could have long-run disadvantages. The best way is for the regulator to identify which particular generators are exercising market power and deal with them individually. The use of the proposed two stage oligopoly model is suitable for such an analysis.

This paper analyzes the Nordic electricity market using a two-stage numerical oligopoly model. The focus is on the largest firms' market power potential in a mixed market, i.e. one where the generation has both hydro and thermal sources. The purpose is to provide insight on how a market with hydropower capacity constraints and forward contracting possibilities functions.

In the Nordic electricity market the electricity systems are closely connected, and there has been extensive trade of electricity between countries in the region since the 1960s. This is mainly to take advantage of the differences in prices of electricity due to different generation technologies that each country in the region has. For example, Norway has only hydro, the Swedish system has hydro, nuclear and a small share of conventional thermal, the Finnish system is similar to the Swedish except it has a smaller hydro and larger conventional thermal share, and Denmark for the most part relies on fossil-based thermal plants, in addition to a growing share of wind power.

The demand for electricity in the Nordic region can be met mostly by hydro and nuclear sources, which are the lowest in terms of production costs. See Figure 1, which displays a market cross of the Nordic electricity

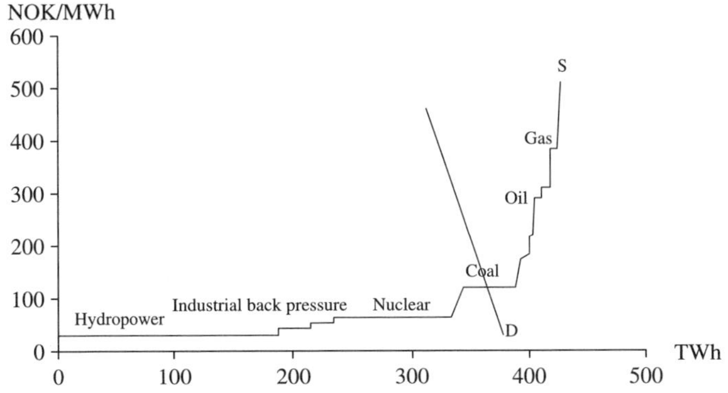

Figure 1 Short Run (one year) Supply and Demand in the Nordic Electricity Market.

market. Coal-fired thermal plants are the marginal technology in the normal situation. In a dry season oil and gas fired plants are the marginal source. On the other hand, in a wet season the nuclear plants may be marginal. Due to space limitations, a detailed description of the Scandinavian market is omitted here but it is given by, e.g. Halseth (1998); Johnsen *et al.* (1999).

Studies on the Nordic electricity market system are conducted to analyze the potential market power of large producers such as Statkraft, Norway and Vattenfall, Sweden. We investigate whether there are incentives to use the reservoirs strategically, either by withholding water now and releasing more later, or vice versa.

Many researchers have used simulations of oligopolistic electricity markets (Bushnell, 2003; Kelman *et al.*, 2001; Ramos *et al.*, 1998; Scott and Read, 1996; Villar and Rudnick, 2003). They have studied how oligopolists in the electricity market balance gaming in the spot market as well as in the future market. However, some studies have adopted open-loop analysis, i.e. the players are not allowed to condition their strategies on past moves by opponents, so they cannot influence the future play of opponents by committing to future physical or financial delivery of electricity.

An aspect that is abstracted from in this analysis is transmission constraints. There are significant capacity between regions in Scandinavia, and about 70% of the time there are no transmission bottlenecks out of or into an average price zone. However, in practice transmission congestion can interact with market power, see e.g. de la Torre *et al.* (2003); Ellersdorfer (2005); Hobbs (2001); Jing-Yuan and Smeers (1999); Wang *et al.* (2004) for suggestions on how to handle this.

Based on the work by Klemperer and Meyer (1989), Green and Newbery (1992) study the UK electricity market using a model of equilibrium in supply functions. Producers optimize over a set of smooth cost-quantity functions to select a bid curve giving maximum expected profit. Newbery (1998) and Green (1999) extend the model to include the effect of contracts and potential entry. Halseth (1998) studies the Nordic electricity market using the supply function framework incorporated in a comprehensive model that includes transmission constraints, several demand categories and monthly time steps split into three load segments. However, electricity contracts are not modeled, and he is unable to capture the dynamic market power aspect of hydro generation.

Another approach is followed by von der Fehr and Harbord (1993), who model the market as a sealed bid auction, introducing more realistic functions, including discrete steps, to represent the generators' pool bids. However, both the supply function equilibrium approach and the auction approach do not lend themselves well to electricity markets where there is an aspect of a quantity game. Hydro scheduling seems best regarded as a quantity game, thus the hydro aspect of our model and of the market makes the case for a Cournot analysis. Note however, that the Cournot approach will give a higher estimate of the degree of market power than both the supply function equilibrium and the auction approach (Klemperer and Meyer, 1989).

Scott and Read (1996) introduce gaming concepts into hydrothermal systems management. A multistage stochastic programming algorithm is utilized to solve the optimization problem with a Cournot market equilibrium superimposed on it at each stage. The work focuses on developing a practical method to optimize hydro reservoirs in a deregulated environment, and the contract position is regarded as exogenously given.

Borenstein and Bushnell (1999) model the California electricity market using a static Cournot framework, and find that the potential for market power is particularly large in seasons when hydro generation is low. Amundsen *et al.* (1998) employ a Cournot oligopoly model of the Scandinavian electricity market to study the market power and price effects of free trade between the Nordic countries. The model takes transmission constraints into account. They find that Cournot equilibrium prices are close to perfectly competitive prices when inter-country trade is allowed, as it is today. None of these analyses take into account the effect of contracts or potential dynamic hydro gaming.

Based on Allaz (1992) and Allaz and Vila (1993), Powell (1993) studies the effect of forward trading between risk averse distributors and producers in the British electricity market. In contrast to the Scandinavian forward market, which we believe is very competitive, the British contract market is non-transparent and difficult to enter. Consequently, Powell models the forward market without external speculators. In any case, Murphy and Smeers (2010) show that the results of Allaz and Vila (1993) do not easily generalize to dynamic electricity markets.

In an empirical analysis of the Norwegian market, Johnsen *et al.* (1999) find evidence of use of market power in the southern part of the country. The largest Norwegian producer, Statkraft, is not among the four largest producers in that region. In addition, the paper points to a number of "incidents", such as the 1992 Statkraft annual report admitting wasting water to avoid further price decreases, Statkraft announcing they would not bid below 100 NOK/MWh (October 1992), price changes around 27 October 1998 when the day ahead prices rose by 300% compared to the day before in the Bergen electricity price zone, and the early December 1998 price increase of nearly 400%. The survey by Fridolfsson and Tangerås (2009) indicates that no studies have found evidence of systematic exploitation of system level market power on Nord Pool.

In an extension of Scott and Read (1996), Bushnell (2003) attempts to quantify the potential for use of market power by hydro-based producers in the western US electricity market. He finds that hydro may profitably be scheduled to off-peak periods from peak periods. Our work adds the contract dimension to the analysis so that the tradeoff between strategic incentives and risk hedging can be captured, an important feature in electricity markets. In addition, the work in this paper in more oriented towards energy, whereas Bushnell's analysis is more capacity oriented due to the share of thermal generation technology in California and adjacent states.

The main contribution of this work lies in the analysis of the commitment aspects of hydropower decisions and forward trading, and its influence on the potential for use of market power. A two-stage game model where the two stages are interlinked through two mechanisms is proposed. First, the quantities of hydropower available in each period are linked through the constrained water reservoirs, and second, the firms can use forward contracts in the first period to commit to second period trades. The model is calibrated to the Nordic electricity market, and we analyze numerically the properties of the Nash equilibrium.

The applied model is based on the assumption that no water will be spilled, and that the aggregate hydro capacity of the numerous smaller hydropower producers bar the market power potential of large hydro producers. Although we find that thermal generation capacity can profitably be withheld from the market, the high number of independent producers still makes the market reasonably competitive.

The paper is organized as follows. Section 2 explains the main modeling considerations, and Section 3 explains the structure of the model. Section 4 presents the main data for the case studied and Section 5 analyzes the results of the studies. Section 6 concludes.

2 Electricity market modeling considerations

The Nordic power system is characterized by its large hydro share, about 50%. Due to the large variability of water inflow in the region, the hydropower dependency means very volatile prices and also high variability in firm profits. In this environment it is hard to judge whether market power is used, e.g. Amundsen and Bergman (2003) write that "no clear conclusion can be drawn with respect to the question of whether price manipulation has taken place or not". The cost of hydro generation depend on the producers' subjective expectations regarding future spot prices and inflow conditions. It is not possible for outsiders to observe these expectations, making it difficult to assess whether observed prices significantly exceed marginal costs. Therefore we set out to build a model of the market to see how profitable market power strategies are, i.e. scheduling that involves holding back thermal generation capacity from the market or scheduling hydro from high-demand periods to lower demand periods in order to increase prices and profit.

We assume that the companies condition their strategies not only on the history of water inflow and their own decisions, but also on the history of moves by opponent players. These are known as "closed loop strategies".

Although the Nordic electricity market has a large number of independent producers, it has been alleged that the largest firms use their market power to manipulate spot prices. In a game theoretic model of the market, we investigate the potential for use of such market power. Since this market has a 50% hydro share, it is necessary to include hydro-specific aspects in the model. Many authors have pointed out that the existence of a contract market in which producers and consumers commit to future delivery and purchase of product increases the degree of spot competition (Allaz, 1992; Allaz and Vila, 1993; Green, 1999; Thille, 2003). Recognizing that firms will use both hydro and contracting decisions as strategic precommitments to affect future competition, the model is non-cooperative, dynamic and of closed-loop type.

We consider the capacity of all players given; there is no capacity being scrapped or new capacity being installed. Phasing in new capacity takes roughly two years or more (depending on the size and location), and ignoring entry/exit dynamics is justified by our choice of a short (one year) study horizon. However, scrapping old capacity can be done on relatively short notice, and the Swedish producers Vattenfall and Sydkraft report having scrapped old high-cost thermal capacity in 1998. This is the kind of thermal capacity that typically would be withheld to keep prices up in a dry year. This capacity had a low utilization factor, and the decision means avoiding maintenance costs, but can also be seen as an attempt to avoid the political turmoil resulting from hesitating to use that capacity in cases when it is obviously necessary. We are unable to capture such political dynamics in our model.

3 A two-stage model

3.1 *Introduction*

The producers are assumed to be risk averse and can trade on a forward market. In addition, we assume that the companies' decisions on the spot and forward market are made simultaneously in every stage. The stochastic variables are electricity prices (endogenous) and water inflow to reservoirs (exogenous).

The uncertainty of the inflows and prices has significant effects on varying the financial performances of the firms. The number of stages must be restricted in order to achieve computational tractability; thus we use two stages with a continuous action space.

In the first stage, the generation companies can buy or sell forward contracts that call for delivery of electricity in the second stage. It is assumed that these forward contracts are binding and observable commitments. In the second stage, the Cournot game is played in quantity. The firms' payoff functions are modified by the positions that they have already taken on the forward market.

The model has two or more producers with mixed hydrothermal generation technology, a competitive fringe and speculators. The forward market is perfectly competitive. We assume that it is optimal not to spill any water in any state, and check the validity of this assumption after calculating the equilibrium.

The complete game form of the model is as follows: There are two stages. In stage 1 the firms decide on hydro release, thermal generation and contracting. The contracts are of the forward type, having delivery and settlement in the second stage. Contracts are priced at their expected payoff. In both stages, the demand side is modeled as by price-taking behavior. In the second stage, the producers learn about inflow and then decide on thermal generation. Hydro release is just what is left in storage. All decision makers are risk averse with a mean-variance objective function.[1]

We solve the second stage first, taking the first stage decisions as given, then solve the reduced form of the first stage (i.e. we substitute in the optimal second stage solutions). The first stage is now a Cournot game where the payoff functions include recourse. This is solved by

[1]The competitive fringe is assumed to be risk neutral in aggregate. Most of the fringe are vertically integrated firms having both short and long term contracts with their customers, providing a hedge.

taking derivatives of the first stage variables and solving the corresponding nonlinear complementarity problem. We also have a perfectly competitive model with the same structure for comparison purposes.

The game unfolds as follows:

- The two or more producers determine their forward sales and their generation outputs (hydro and thermal) optimal positions simultaneously by considering the price expectation, the production possibilities, and the number of forward markets available. The spot price is given by the inverse demand function.
- The speculators announce their bids. Since it is assumed that there are risk neutral speculators with correct expectations operating in the forward market, the forward price is the expected spot price at equilibrium. Thus, the speculators' profits will be zero.
- The two or more firms determine their thermal generation output (the hydro generation output is determined as a residual). The spot price is given by the inverse demand function.

The formal model is explained next. Detailed derivations can be found in Fleten (2000).

3.2 *The second stage*

When making second stage decisions, all uncertainty has been revealed. The relevant decision is the thermal generation, because hydro generation is determined as the residual of inflow I and hydro generation in the first stage g^h: $I - g^h$. First stage hydro generation and forward trading f_p are regarded as given. The inverse demand function is given as:

$$P = a - bQ, \tag{1}$$

where P is price and Q is consumption, equal to the generation. Subscripts indicating the second stage has been left out.

Let f_p be the energy sold forward in stage 1, having delivery in stage 2, $C_p()$ the thermal cost function with c_p as the coefficient for the quadratic term and d_p for the linear term. The profit function for player p is

$$\begin{aligned}\Pi_p &= P(I_p - g_p^h + g_p^t - f_p) - C_p(g_p^t) \\ &= (a - bQ)(I_p - g_p^h + g_p^t - f_p) - (c_p g_p^{t2} + d_p g_p^t) \end{aligned} \tag{2}$$

The profit from selling forward is $(p^f - P)f_p$, where p^f is the forward price. However, the forward price is a constant in the second stage, so the term $p^f f_p$ can be ignored in the profit expression to be optimized. When the producer has *sold* forward, i.e. $f_p > 0$, the higher the second stage price, the lower the forward profit. This means that companies have an incentive component to *decrease* price (e.g. by increasing thermal generation), due to the forward sale. If the firm has *bought* forward, the forward profit will increase when price increases, so in this case the market power incentives are exacerbated.

There are some constraints that are ignored, such as $g_p^t \geq 0$. However, we assume that these constraints are satisfied, and check them in the calculations.

For players without thermal plants, belonging to the set $\mathcal{N}$, $g_p^t = 0$. For a player having thermal plants, the first order condition under the Cournot assumption gives

$$g_p^t = n_p[a - bQ - d_p - b(I_p - g_p^h - f_p)]$$

where $n_p = (2c_p + b)^{-1}$. The thermal generation, by substituting for total demand/supply Q, becomes

$$g_p^t = n_p[au - d_p - b(uS + I_p - g_p^h - f_p)] \tag{3}$$

where $u = (1 + bR)^{-1}$, $R = \sum_{p\in\mathcal{P}\setminus\mathcal{N}} n_p$ and

$$S = \sum_{p\in\mathcal{P}} (I_p - g_p^h) - \sum_{p\in\mathcal{P}\setminus\mathcal{N}} n_p[d_p + b(I_p - g_p^h - f_p)].$$

3.3 *The first stage*

We model the first stage as a complementarity problem. For example,

$$\frac{\partial U}{\partial g_p^t} \leq 0 \quad \text{complements} \quad g_p^t \geq 0.$$

This means that at least one of the above inequalities is binding. In the first stage the players decide on thermal and hydro generation and forward contracting. Hydro and contracting decisions carry over to the next stage, i.e. they become "capacity" commitments. We assume that the players can be risk averse, so we use an objective function for each player as follows:

$$U_p = \mathrm{E}\Pi_p - \lambda \mathrm{Var}(\Pi_p)$$

Let P_t be the spot price in stage t, Q_t the aggregated generation, p_f the forward price, and g_p^* the optimal second stage thermal generation.

The overall risky profit is

$$\begin{aligned}\Pi_p &= P_1(Q_1)(g_p^h + g_p^t) + p_f f_p - C_p(g_p^t)\\ &\quad + P_2(Q_2)(I_p - g_p^h + g^*(g_p^h, f_p) - f_p) - C_p(g^*(g_p^h, f_p))\end{aligned} \tag{4}$$

We assume there are enough well-informed speculators in the forward market to drive the forward price to the expected value of the second stage spot price[2]: $p^f = \mathrm{E}P_2(Q_2)$.

The resulting set of complementarity conditions are:

$$g_p^t \geq 0 \quad \text{complements} \quad \frac{\partial \Pi_p}{\partial g_p^t} \leq 0 \tag{5}$$

$$0 \leq g_p^h \leq \min_{\omega \in \Omega} I_p^\omega \quad \text{complements} \quad -\frac{\partial U_p}{\partial g_p^h}. \tag{6}$$

for $p \in \mathcal{P}$, where Ω is the set of outcomes for the random inflow.

Equation (6) is the first order condition of maximization with respect to hydro release and can equivalently be expressed as

$$\frac{\partial U_p}{\partial g_p^h} - \mu_p \leq 0, \quad g_p^h \left(\frac{\partial U_p}{\partial g_p^h} - \mu_p\right) = 0$$

$$\mu_p \left(g_p^h - \min_{\omega \in \Omega} I_p^\omega\right) = 0$$

$$\mu_p \geq 0, \quad 0 \leq g_p^h \leq \min_{\omega \in \Omega} I_p^\omega$$

where μ_p is a Lagrange multiplier.

In addition, the first order condition regarding contracting can be expressed as

$$\frac{\partial U_p}{\partial f_p} = 0 \tag{7}$$

In the implementation we also model the fringe and pure thermal players and separate between run-of-the-river inflow, which must be used for production immediately or be wasted, and reservoir inflow, which can be stored. The PATH solver with AMPL is used to solve the system (5)–(7) (Dirkse and Ferris, 1995; Ferris *et al.*, 1999a,b; Fourer *et al.*, 1993). This amounts to finding a solution to the nonlinear inequality system made up by the first order conditions of all players, where the inverse demand functions are substituted in for spot market price.

[2]This reflects the fact that the economic value of a forward contract is zero.

3.4 *Comments*

There is a certain asymmetry in the model: In the first period the firms make both output decisions and contracting decisions regarding the next period, whereas in the second period the firms make only output decisions. This means that in the first period the firms make output decisions in the absence of forward contracts made for that period, while in the second period the firms make output decisions taking into account forward contracts made earlier for that period. A possible remedy for this asymmetry could be to include an exogenous forward position that come to delivery in the first stage. The consequence would be that the firms produce more (less) in the first period had the firms sold (bought) forward, compared to the current situation, where there are no forward contracts with delivery in the first stage. Instead of speculating on what could be reasonable forward positions that firms take with them into the game, we use a zero prior forward position and leave this issue for possible future work.

Another remedy for the minor asymmetry could be to include more stages into the model. However, this would increase the complexity of the model greatly. Speculating on the qualitative results of such a model, regarding the hydro release, we believe such a model would predict that there would be at least one period of fierce competition in which much hydro is released. Put another way, the producer can postpone the bulk of the release one, maybe two periods, but eventually it has to produce. We believe that aspect is taken care of in a qualitative sense in our two-stage model.

Regarding contracts, it is a little more complex. For example, in a three-stage model, there would be two contracts in the first stage having delivery in the second and third period respectively, and in the second stage there would be one contract and in the third stage there would be none. We still believe that contracts will be used as before, mainly to increase or decrease the degree of competition in conjunction with hydro release and for mitigating risk.

We believe the proposed model can provide a starting point in learning and analyzing the large firms' strategic decisions regarding power scheduling, contracting and coordination between those two activities. In addition, it can provide information regarding trade-offs between risk and expected return for large producers.

4 Case study

We consider the largest power producerss in Nordic region (as the situation was per 1997[3]): Statkraft (Norway), Sydkraft and Vattenfall (Sweden), Fortum (Finland), and Elsam and Elkraft (Denmark). The rest of the producers are considered as price takers and are grouped into a competitive fringe. The different gencos have different technologies and capacities as shown in Figure 2.[4]

The producers are assumed to be risk averse. Table 1 shows the risk aversion parameter, λ, for each company. The parameters are chosen such that the ratio of λ·variance and expected profit is 0.15 in the oligopoly case, i.e. there is a reasonable balance between risk and expected profit for each

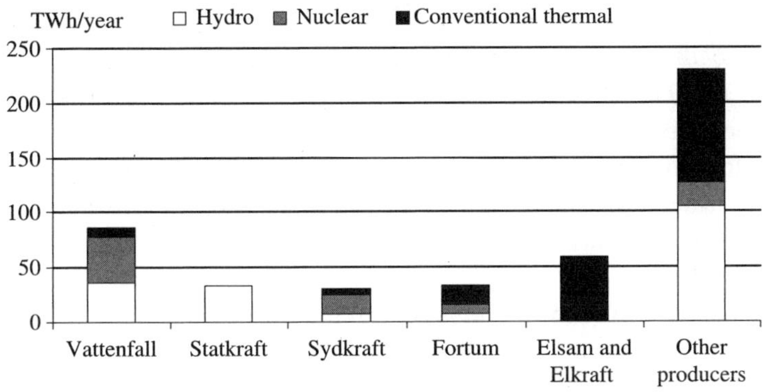

Figure 2 The Main Firms' Capacity, TWh/Year.

Table 1 Risk Aversion Parameter λ in Per Million Norwegian Kroner ($[\text{MNOK}]^{-1}$), for Use in the Objective Function $U = \text{E}(\Pi) - \lambda \text{Var}(\Pi)$, where Π is Profit.

Vattenf.	**Statkr.**	**Sydkr.**	**Fortum**	**Elsam**	**Elkr.**	**Fringe**
0.006	0.007	0.06	0.11	0.3	0.35	0.0

[3]Since then, the two Danish producers have merged, but the main results still apply.
[4]Fortum (1998); Sydkraft (1998) report having larger capacities than displayed in the table; however, much of that extra capacity is operated by relatively independent firms partly owned by the larger firms. See Sørgard (1997) for an analysis of cross-ownership in this industry.

producer. These parameters will be varied to see the impact of a change in risk aversion.

We use historical inflow data from 1931–1990 to estimate the inflows used in the model; it is 60 scenarios. In the second stage of our model it is assumed that producers release an amount of water giving an end of period reservoir level equal to the one at the beginning of the first stage. In reality, there are multiyear reservoirs that buffer the large inflow variations. To mimic that effect, the variance of the inflows were reduced by 25–50% for each firm. Since individual plant data is unavailable, we model one reservoir for each company.

The direct production cost for the hydro-based power is assumed to be negligible but for the thermal is represented as follows:

$$C_p(g_p^t) = c_p(g_p^t)^2 + d_p g_p^t \tag{8}$$

The coefficients for the cost function are in the ranges $c_p \in [1.1, 3.2]$ and $d_p \in [60, 73]$. The fit of the linear marginal cost to the cost data is good for most producers; however for the two Swedish firms, who have constant low cost nuclear generation for the greater part of their total thermal energy capacity, other cost functions would be more accurate. We believe that using other cost functions would not affect our results significantly, at least not qualitatively.

The demand in each period is represented by linear functions (Eqn. (1)) and is shown in Figure 3. The demand parameters were found by calibrating the model to yield a realistic price and quantity level. The resulting average price elasticity of demand is −0.6, which is within a range of plausible elasticities (Sørgard, 1997). The two stages together could constitute a length of any period, but in this paper we use a six month summer season

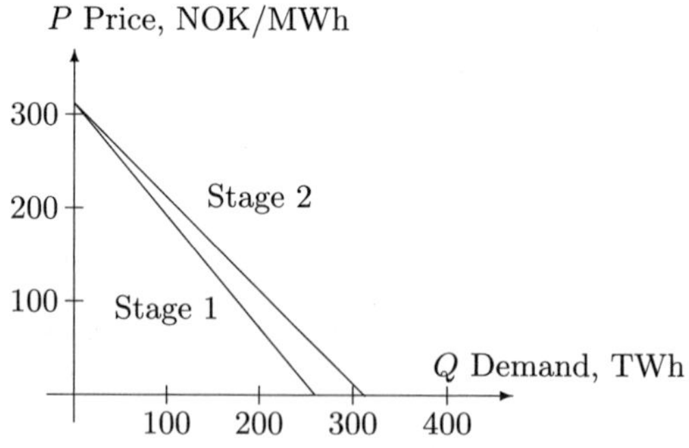

Figure 3 Demand Functions; One Per Season.

for the first stage, and a six month winter season for the second stage.[5] Due to this short horizon, no discounting is used.

5 Analysis of the results

Figure 4 shows the generation of the producers under perfect competition (PC) and two cases of Cournot oligopoly. For second stage generation the average value over the 60 scenarios is displayed. The oligopoly case marked O is the base case oligopoly, where the firms are risk averse with coefficients as shown in Table 1 and forward contracting is allowed. The second oligopoly case is for risk neutral companies (O+), so that the only incentive for entering into contracts is strategic, i.e. increased profit due to the influence on the second stage game.

From the results, one can see that the largest producer, Vattenfall, wants to reduce thermal production in order to force the spot price up. The oligopolistic equilibrium spot price is less than 7% higher than under perfect competition, as shown in Table 2. This is low compared to empirical results in the industrial organization literature (Bresnahan, 1989). Due to the hydro capacity of the fringe, the prices are equal for both stages. From Figure 4, one can see that Vattenfall, due to its low thermal cost, has to pull the load in increasing prices. The other firms including the fringe benefit

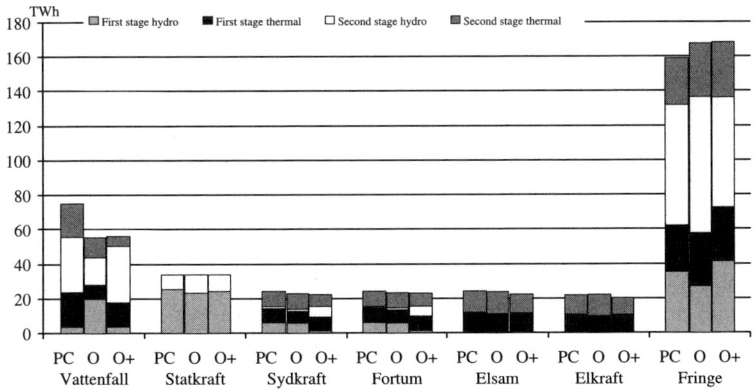

Figure 4 The Expected Total Production in TWh for Perfect Competition (PC), Base Case Oligopoly (O) and an Oligopoly Where All Producers are Risk Neutral (O+).

[5]The demand curves in Figure 3 is per season, while the demand curve in Figure 1 is for a year.

Table 2 Equilibrium sPot Price in NOK/MWh, for Perfect Competition (PC), Base Case Oligopoly (O), Oligopoly without a Forward Market (O*), an Oligopoly Case where the Strategic Incentives of Vattenfall Regarding Contracting has been Removed (O**), Higher Risk Aversion Oligopoly (O***), Risk Neutral Oligopoly (O+), and an Oligopoly Case Where Statkraft is Considered as a Price-Taker (O++).

	Equilibrium Spot Price (NOK/MWh)						
Stage	**O**	**O***	**O****	**O*****	**O+**	**O++**	**PC**
1 and 2	124.8	125.8	122.6	124.5	126.3	124.7	117.0

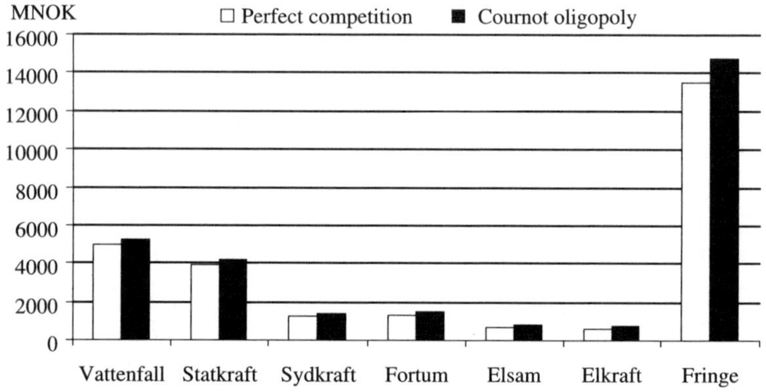

Figure 5 Expected Profit in MNOK for each Producer Under Perfect Competition and Cournot Competition Respectively.

from Vattenfall's efforts (see Figure 5). Our findings are consistent with the results of Halseth (1998).

In the base case oligopoly (O) Vattenfall is committing to a low second stage hydro generation level by choosing a high first stage hydro output. This is for a large part offset by the hydro scheduling of the fringe firms.

When all the decision makers are risk neutral (no hedging incentives), the spot price tends to go higher. This is consistent with the results of Allaz (1992). In this case all thermal generators commit to forward purchases (see Figure 6), consistent with the low level of generation in the second stage. This is due to the fact that the producers try to prevent each other from taking a position as a Stackelberg leader on the spot market. As a result, the producers trade forward which end up generating less. Offsetting this is the shifted hydro output from the first period to the last.

Running the model without the opportunity to trade forward, we find that, compared to when forward contracting is allowed, the total output is marginally decreased. This is consistent with other studies.

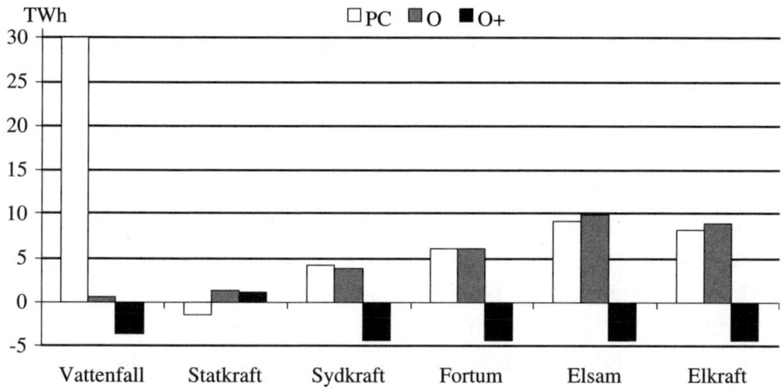

Figure 6 Forward Sale in TWh for Perfect Competition (PC), Base Case Oligopoly (O), and an Oligopoly Case where the Firms are Assumed to be Risk Neutral (O+).

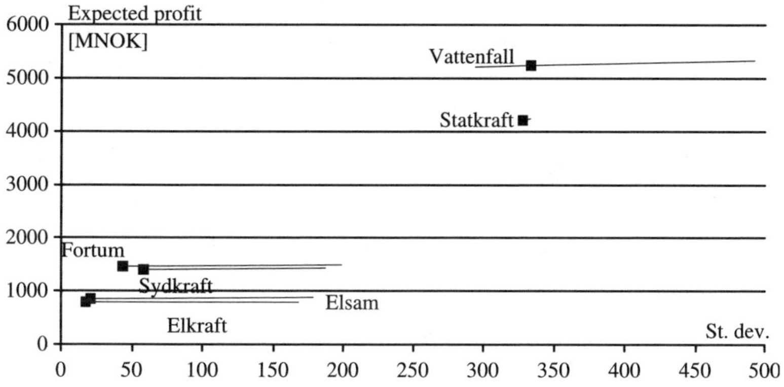

Figure 7 Efficient Frontiers Under Oligopoly.

If we reduce risk aversion and continue to allow forward contracting, the total output goes down and the prices go up. Producers having thermal capacity start *buying* forward, i.e. f becomes negative. This causes second stage thermal generation to decrease. Thus, risk aversion on the part of thermal producers is good for competition. Statkraft, which is a pure hydro producer, has no strategic incentives regarding forwards; it only trades forward to hedge risk. This is due to the fact that marginal revenue is unaffected by forward trading for pure hydro producers.

Figure 7 depicts the profit and standard deviation of profit for each generation company under Cournot competition. The efficient frontiers are indicated with solid lines. Statkraft has little opportunity to change neither its risk nor its expected profit. The other firms can reduce the risk from the risk neural point on the right end of the frontiers, without sacrificing significantly in terms of reduced expected profit. The extra profit from buying more/selling less forward is small compared to total profit.

We investigate more closely the severity of the tradeoff between reduced risk and increased expected profit with regard to the contract position for the largest producer. The strategic incentives for Vattenfall regarding the forward contracts were removed. This amounts to a certain irrational behavior of the firm, as it no longer recognizes the price influence of forward commitments. The results show that the optimal forward position for Vattenfall from a risk hedging point of view is to *sell* about two thirds of the expected second stage generation (see Figure 6). In addition, the total output generation is also greater. These results indicate that Vattenfall's contract position is very sensitive to its perception of its own market power.

Finally, we investigate the level of potential of a large hydro producer like Statkraft. We change the behavior of Statkraft into that of a price-taking firm. The total output increases only marginally, from 347 TWh to 348 TWh. One can conclude that Statkraft has very little potential market power on the seasonal level, when there are no transmission constraints.

Although Vattenfall may have incentives to use market power, no empirical evidence has been put forward to support a claim that this power has been used. If that happens, Vattenfall faces the risk of being split up in a pure hydro and a pure thermal part. Furthermore, Vattenfall is a vertically integrated utility, and has probably signed long term contracts for a significant share of their expected generation, i.e. more than indicated by our model. This weakens Vattenfall's incentives to use market power.

Remark: The study done in this paper is using the historical data up to 1997, and thus the data may not be completely accurate reflections for the current situation in the Nordic market. However, we believe that the proposed technique can be applied to the current situation to get an approximate reflection of the Nordic market.

6 Conclusion

This paper presents a two-stage stochastic Cournot model of the Scandinavian electricity market. We have investigated the strategies of large producers by taking into account the influence on the future play of the opponents.

The results indicate that a large pure hydro producer like Statkraft does not have significant long term (seasonal) market power. Vattenfall, however, which is the largest producer in the region, has incentives to reduce its total thermal production in order to increase the market spot price.

In addition, the simulation studies lean toward concluding that the electricity market is reasonably competitive and there seems to be little market power being exercised by the dominant firms. This is shown in the simulated price in an oligopoly case which is less than 7% higher than under perfect competition. In the future years however, demand is expected to increase more than supply, making strategic bidding and withholding capacity potentially more profitable. For these reasons, the market regulator should monitor[6] the development of the market, and possibly prevent the market shares of the dominant companies to increase.

Acknowledgements

The authors acknowledge support from the Norwegian Research Council and Energy Norway. The authors would like to thank SINTEF Energy Research, Norway, for providing inflow data and some of the capacity/cost data. We are grateful for comments from Olav Fagerlid, Daniel Christiansen and Stein W. Wallace.

References

Allaz, B. Oligopoly, uncertainty and strategic forward transactions. *International Journal of Industrial Organization*, 10: pp. 297–308, 1992.

Allaz, B. and J.-L. Vila. Cournot competition, forward markets and efficiency. *Journal of Economic Theory*, 59: pp. 1–16, 1993.

Amundsen, E. S., L. Bergman, and B. Andersson. Competition and prices on the emerging Nordic electricity market. Working Paper Series in Economics and Finance No. 217, Stockholm School of Economics, 1998.

Amundsen, E. S. and L. Bergman. The deregulated electricity markets in Norway and Sweden: A tentative assessment. In J.-M. Glachant and D. Finon, editors, *Competition in European Electricity Markets: A Cross-Country Comparison*, pp. 110–132. Edward Elgar Publishing, 2003.

Borenstein, S. and J. Bushnell. An empirical analysis of the potential for market power in California's electricity industry. *Journal of Industrial Economics*, 47(3): pp. 285–323, 1999.

Bresnahan, T. K. Empirical studies of industries with market power. In R. D. Willig and R. Schmalensee, eds., *Handbook of Industrial Organization. Vol. II*, chapter 17, pp. 1011–1057. Amsterdam: North-Holland, 1989.

[6]See e.g. Sandsmark and Tennbakk (2010) for a monitoring procedure adapted to the case of a hydropower dominated system.

Bushnell, J. A mixed complementarity model of hydro-thermal electricity competition in the Western US. *Operations Research*, 51(1): pp. 80–93, 2003.

de la Torre, S., A. J. Conejo, and J. Contreras. Simulating oligopolistic pool-based electricity markets: A multiperiod approach. *IEEE Transactions on Power Systems*, 18(4): pp. 1547–1554, 2003.

Dirkse, S. P. and M. C. Ferris. The PATH solver: A non-monotone stabilization scheme for mixed complementarity problems. *Optimization Methods and Software*, 5: pp. 123–156, 1995.

Ellersdorfer, I. A multi-regional two-stage Cournot model for analyzing competition in the German electricity market. *Proceedings of the 7th European Energy Conference 2005 "European Energy Markets in Transition", Bergen*, 2005.

Ferris, M. C., R. Fourer, and D. M. Gay. Expressing complementarity problems and communicating them to solvers. *SIAM Journal on Opimization*, 9 : pp. 991–1009, 1999a.

Ferris, M. C., C. Kanzow, and T. S. Munson. Feasible descent algorithms for mixed complementarity problems. *Mathematical Programming*, 86 : pp. 475–497, 1999b.

Fleten, S.-E. *Portfolio management emphasizing electricity market applications. A stochastic programming approach.* PhD thesis, Norwegian University of Science and Technology, Trondheim, 2000. Doktor Ingeniøravhandling 2000: pp. 16.

Fortum, 1998. Annual Report.

Fourer, R., D. M. Gay, and B. W. Kernighan. *AMPL: A Modeling Language for Mathematical Programming.* San Francisco: The Scientific Press, 1993.

Fridolfsson, S.-O. and T. Tangerås. Market power in the Nordic electricity wholesale market: A survey of the empirical evidence. *Energy Policy*, 37(9): pp. 3681–3692, 2009.

Green, R. The electricity contract market in England and Wales. *The Journal of Industrial Economics*, 47(1): pp. 107–124, 1999.

Green, R. and D. M. Newbery. Competition in the British electricity spot market. *Journal of Political Economy*, 100(5): pp. 929–953, 1992.

Halseth, A. Market power in the Nordic electricity market. *Utilities Policy*, 7(4): pp. 259–268, 1998.

Hobbs, B. F. Linear complementarity models of Nash-Cournot competition in bilateral and POOLCO power markets. *IEEE Transactions on Power Systems*, 16(2): pp. 194–202, 2001.

Jing-Yuan, W. and Y. Smeers. Spatial oligopolistic electricity models with Cournot generators and regulated transmission prices. *Operations Research*, 47(1): pp. 102–112, 1999.

Johnsen, T. A., S. K. Verma, and C. D. Wolfram. Zonal pricing and demand-side bidding in the Norwegian electricity market. University of California Energy Institute, POWER Working Paper-063, 1999.

Kelman, R., L.A.N. Barroso, and M.V.F Pereira. Market power assessment and mitigation in hydrothermal systems. *IEEE Transactions on Power Systems*, 16(3): pp. 354–359, 2001.

Klemperer, P. D. and M. A. Meyer. Supply function equilibria in oligopoly under uncertainty. *Econometrica*, 57(6): pp. 1243–1277, 1989.

Murphy, F. and Y. Smeers. On the impact of forward markets on investments in oligopolistic markets with reference to electricity. *Operations research*, 58(3): pp. 515–528, 2010.

Newbery, D. M. Competition, contracts, and entry in the electricity spot market. *Rand Journal of Economics*, 29(4): pp. 726–749, 1998.

Powell, A. Trading forward in an imperfect market: The case of electricity in Britain. *The Economic Journal*, 103: pp. 444–453, 1993.

Ramos, A., M. Ventosa, and M. Rivier. Modeling competition in electric energy markets by equilibrium constraints. *Utilities Policy*, 7(4): pp. 233–242, 1998.

Sandsmark, M. and B. Tennbakk. Ex post monitoring of market power in hydro dominated electricity markets. *Energy Policy*, 38(3): pp. 1500–1509, 2010.

Scott, T. J. and E. G. Read. Modelling hydro reservoir operation in a deregulated electricity sector. *International Transactions in Operations Research*, 3(3–4): pp. 209–221, 1996.

Sørgard, L. Domestic merger policy in an international oligopoly: The Nordic market for electricity. *Energy Economics*, 19(2): pp. 239–253, 1997.

Sydkraft, 1998. Annual Report.

Thille, H. Forward trading and storage in a Cournot duopoly. *Journal of Economic Dynamics and Control*, 27(4): pp. 651–665, 2003.

Villar, J. and H. Rudnick. Hydrothermal market simulator using game theory: Assessment of market power. *IEEE Transactions on Power Systems*, 18(1): pp. 91–98, 2003.

von der Fehr, N.-H. M. and D. C. Harbord. Spot market competition in the UK electricity industry. *Economic Journal*, 103(418): pp. 531–546, 1993.

Wang, Z., Y. Li, and S. Zhang. Oligopolistic equilibrium analysis for electricity markets: A nonlinear complementarity approach. *IEEE Transactions on Power Systems*, 19(3): pp. 1348–1355, 2004.

Chapter 16

Multi-Lag Benders Decomposition for Power Generation Planning with Nonanticipativity Constraints on the Dispatch of LNG Thermal Plants

Andre L. Diniz* and Maria E. P. Maceira†

Summary

Multi-stage Benders decomposition has long been applied to mid and long term power generation planning, which is usually conceived as a stochastic problem, where the dispatch of the plants at each node of the scenario tree is determined based on current system conditions and on a number of possible future realizations of the random variables. However, recent logistic issues have forced the dispatch of Liquefied Natural Gas (LNG)-based thermal plants to be determined some months in advance to its actual generation, which leads to nonanticipativity constraints on the dispatch of these plants. This paper proposes a multi-lag variant of Benders decomposition, where additional state variables corresponding to the pre-order dispatch of LNG plants are introduced to the recourse function of each stage. Artificial variables are defined so that the number of state variables does not depend directly on the number of LNG plants in the system. Numerical results show the efficiency of the application of the proposed approach in the stochastic dual dynamic programming and L-shaped solving methodologies used to solve the long and mid-term generation planning problems for the large-scale Brazilian system.

1 Introduction

The power generation planning problem aims to determine optimal policies for the dispatch of hydro and thermal generation plants, so as to meet the demand requirements along a multi-year planning horizon (Wood and

*CEPEL, Electric Energy Research Center, Energy Optimization and Environment Department, Av Horácio Macedo, 354 – P.O.Box 21941 911, Ilha Do Fundão, Rio de Janeiro, RJ, Brazil, diniz@cepel.br.
†CEPEL, Electric Energy Research Center, Energy Optimization and Environment Department, Av Horácio Macedo, 354 – P.O.Box 21941 911, Ilha Do Fundão, Rio de Janeiro, RJ, Brazil, elvira@cepel.br.

Wollemberg, 1996; Sachdeva, 1982; Yeh, 1985; Rotting and Gjelsvik, 1999; Maceira *et al.*, 2002). The power dispatch of all hydro and thermal plants must be coordinated in time and space, for several reasons, as for example:

(i) hydro generation uses water stored in the reservoirs. Therefore, the use of water today affects the availability of water for power generation in the future;
(ii) hydro plants are spread throughout several river basins. For this reason, operation of downstream plants is strongly dependent on the operation of upstream plants in the same cascade;
(iii) the existence of a transmission grid allows a given load to be met by any power plant in the system, given that transmission limits are not exceeded.

Another key point in generation planning is the representation of uncertainties, especially when longer time horizons are considered. For predominantly hydro systems, it is of particular importance to take into account the stochastic inflows to the reservoirs (Maceira and Bezerra, 1997; Faber and Stedinger, 2001). Finally, the nonlinear relation between the turbined outflow and the power output of the hydro plants (Hicks *et al.*, 1974) and thermal unit commitment constraints in the short term horizon (Hobbs *et al.*, 2001) make the generation planning problem a large-scale, multi-stage stochastic mixed-integer nonlinear optimization problem.

In practice, this problem is handled in a hierarchical way by defining long, medium and short-term models (Tufegdzic *et al.*, 1996; Rotting and Gjelsvik, 1999; Maceira *et al.*, 2002), with proper coordination strategies among them. In this paper we consider the mid /long term generation planning in vertically integrated systems, which is done in a centralized way by a so-called Independent System Operator (ISO) in a cost-minimization basis. This problem is usually formulated as a stochastic linear program and solved by multi-stage Benders decomposition approaches (Pereira and Pinto, 1991; Rotting and Gjelsvik, 1999; Maceira *et al.*, 2002).

In the usual formulation of the hydrothermal planning problem, the generation dispatch for each hydro and thermal plant is determined at each node of the scenario tree, based on the current system conditions when the node is reached and taking into account a sample of possible future inflows scenarios for the hydroelectric plants. However, recent logistic issues have forced the dispatch of some thermal plants fueled by Liquefied Natural Gas

(LNG) to be determined K months in advance to its actual generation. As a consequence, the "wait and see" procedure for the dispatch of these plants at each node of the scenario tree in the stochastic program becomes a "here and now" decision under uncertainty to be taken at previous time steps. In this sense, nonanticipativity constraints on the dispatch of these plants must be included.

This paper proposes a so called multi-lag variant of the multi-stage Benders decomposition (MSBD) strategy to consider this pre-order requirement in the dispatch of LNG plants in the long and mid term stochastic hydrothermal generation problems (Rotting and Gjelsvik, 1999; Maceira *et al.*, 2002). The main contribution as compared to previous works in the literature is the consideration of higher order time-linking dependencies among decision variables, which increases the complexity of these methods both from the theoretical and from the practical point of view. In our approach, such higher order lags are properly considered in the recourse function at each node of the scenario tree by including additional state variables related to the dispatch of LNG plants up to K months ahead. Artificial variables are created in order to limit the dimension of the state-space, so that it does not depend directly on the number of LNG plants in the system.

This paper is organized as follows. In section 2 we present a general discussion of the applications of stochastic programming to generation planning problems. In section 3 we present the traditional formulation of the specific problem considered in this paper and briefly describe the single-lag MSBD applied to solve the problem. In section 4 we formulate the problem with the inclusion of LNG thermal plants and present the multi-lag MSBD approach proposed in this paper. In section 5 we present the numerical results of the application of this approach to the real large scale Brazilian system, as well as sensitivity analysis on the value of the lag K and the number of LNG plants. Finally, section 6 states the conclusions of this work.

2 Stochastic optimization applied to generation planning of electrical systems

In this section we present an overview of stochastic programming applications to the generation planning of electrical energy systems. We consider a general stochastic program as follows:

$$\begin{aligned}
&\min c(x_1) + \underset{\xi_2}{\mathrm{E}}\left[\varphi_2(x_1,\xi_2)\right] \\
&s.t.\ g_1(x_1) = b_1 \\
&\qquad x_1 \in X \\
&\text{where, for } t = 2,\ldots,T, \\
&\varphi_t(x_{t-1},\xi_t) = \min_{x_t}\left\{c(x_t) + \underset{\xi_{t+1}|\xi_t,\ldots,\xi_{t-p+1}}{\mathrm{E}}\left[\varphi_{t+1}(x_t,\xi_{t+1})\right]\right\} \\
&s.t.\ g_t(x_t) = b_t(x_{t-k,k=1,\ldots,K},\xi_t) \\
&\qquad x_t \in X
\end{aligned} \tag{2.1}$$

where x_t is the vector of decision variables at time step t, typically the vector of hydro and thermal generations for all plants in the system; $c(.)$ is the objective function that in most of the applications consists in minimizing system costs (Wood and Wollemberg, 1996). Risk-averse formulations may also be considered (Guigues and Römisch, 2010).

The set of constraints for each time step t is denoted by $g_t(.)$ and comprises the system demand equation, water conservation equations in the reservoirs, hydro production function of the hydro plants (Diniz and Maceira, 2008), and possibly a set of additional operation constraints for the generating units and transmission lines. We consider random variables ξ in the right hand side b_t of some constraints. In the literature, such variables typically consist of the water inflows to the reservoirs in predominantly hydro systems (Maceira and Bezerra, 1997; Rotting and Gjelsvik, 1999) or the power demand in predominantly thermal systems (Nowak and Römisch, 2001). The recursive term $\varphi_{t+1}(.)$ is the recourse function for the subproblem of time step t, which can be obtained iteratively by applying decomposition-based approaches to solve the problem.

Expression (2.1) is quite general, since it comprises: two stage ($T = 2$) (Slyke and Wets, 1969) or multi-stage ($T > 2$) (Birge, 1985) stochastic programs; linear, nonlinear and mixed-integer stochastic optimization problems (Ermoliev and Wets, 1988), depending on the formulation for $c(.)$ and $g(.)$ and the feasible set X; independent (Higle and Sen, 1996; Linderoth *et al.*, 2006) or dependent (Frauendorfer, 1988; Infanger and Morton, 1996; Maceira *et al.*, 2002) random variables, whether density functions $f(\xi_{t+1}|\xi_t,\ldots,\xi_{t-p+1})$ and $f(\xi_{t+1})$ are equal or not, where p is the maximum order of an autoregressive model for the dependent case; a single lag ($K = 1$) (Pereira and Pinto, 1991; Rotting and Gjelsvik, 1999; Philpott and Guan, 2008) or a multi-lag ($K > 1$, as in this paper) dependence among decision variables.

2.1 *Output of the model*

For the mid-term planning problem, the main output of the model is the solution x_1 (e.g. hydro and thermal generation) as well as the Lagrange multipliers of the demand equation for the first stage subproblem, which are used to set the energy spot prices in the market (Rotting and Gjelsvik, 1999, Maceira *et al.*, 2002). In the long term planning models, the main output is the set of recourse functions for each stage throughout the planning horizon, which define an optimal operation policy for the system as a function of the system state at the beginning of each time step. Such policy may be used to simulate the operation of the system in the near future for a set of candidate scenarios (e.g., to assess the probability of load shedding) and to give a boundary condition to the mid-term planning model.

2.2 *Representation of the stochastic process*

In most of the works in the literature of stochastic programming the random variables are considered to be independent (Higle and Sen, 1996; Linderoth *et al.*, 2006). However, since this hypothesis is hardly applicable for the practical operation of hydroelectric systems, temporal or spatial correlation should be represented at least approximately.

Temporal correlation. Interstage dependence has been addressed since the early works in hydrothermal planning (Little, 1955), where Markov models were considered (i.e., water inflows in the previous time step were included in the recourse function φ). Later on, stochastic models derived from time series analysis were employed in a number of works, including periodic autoregressive (Par-P) stochastic models (Maceira and Bezerra, 1997). Some of the theoretical results obtained by assuming independent variables may be applied to the random noises of such models (Frauendorfer, 1999; Infanger and Morton, 1996; Chiralaksanakul and Morton, 2003).

Spatial correlation. Cross correlation among water inflows to different plants may also be considered in order to represent hydrological diversity within large power systems. However, few works have addressed this subject (Birge and Wallace, 1986; Edirisinghe and Ziemba, 1994; Maceira and Bezerra, 1997).

2.3 *Solution techniques*

We now discuss the application of some of the most important optimization techniques in the literature of stochastic programming (Ermoliev and

Wets, 1988; Shapiro *et al.*, 2009) to the generation planning problem of hydrothermal systems. Two general procedures can be considered to tackle this problem: *deterministic sampling* or *stochastic sampling* approaches, as discussed below.

Deterministic sampling approaches. In this first type of approach a single tree is supposed to represent completely the stochastic process and the deterministic equivalent problem (Wets, 1974) is solved by an optimization technique — which may or not lead to a decomposition of the problem — with a deterministic-like stopping criterion. One of the leading techniques is the L-shaped method for two-stage (Slyke and Wets, 1969) or multi-stage problems (Birge, 1985), where a time decomposition is employed (in the latter case it is also referred to as nested decomposition). Many improvements of this method have been proposed in the literature, based on the management of the cuts (Birge and Louveaux, 1988, Zakeri *et al.*, 2000), the ways to transverse the scenario tree (Gassmann, 1990, Dempster and Thompson, 1999), or by proposing regularizing terms to the nondifferentiable objective function (Ruszczyński, 1993). Scenario decomposition methods that relax nonanticipativity constraints such as progressive hedging (Santos *et al.*, 2009) have also been proposed.

Hydrothermal generation planning is a multi-stage problem in nature, since the regularization period for large reservoirs (i.e., from totally full to totally empty) may take several months or years. Therefore, those deterministic sampling-based approaches are very difficult to implement in practice, because it is impossible to build a scenario tree to represent the uncertainty on the water inflows to the reservoirs in the future, even if a very small number of scenarios is considered in each time step. Therefore, methods based on stochastic sampling should probably be applied.

Stochastic sampling approaches. In this second type of approach it is assumed that the complete tree of the problem cannot be enumerated and should be approximated by successive sampling, which may be done by two ways, labeled in the literature as *interior* and *exterior* sampling approaches (Chiralaksanakul and Morton, 2003; Linderoth *et al.*, 2006).

Exterior sampling approaches. By this procedure the problem is solved successively for different scenario trees, and in each run a deterministic-based technique may be applied. One of the leading methods of this type is the so-called *Sample Average Approximation* (SAA) (Kleywegt *et al.*, 2001; Xu and Meng, 2007; Pagnoncelli *et al.*, 2009). However, the main drawback of the application of such methods for multi-stage problems is the large computational burden involved in having to solve many deterministic equivalent subproblems, one for each scenario tree. Indeed, most of the applications of SAA are limited to two-stage problems, and application of sampling approaches of this type for multistage problems are rare (Chiralaksanakul and Morton, 2003). As a consequence, for long term planning in a multi-year time horizon, interior sampling approaches (i.e., sampling while the optimization problems is being solved) should probably be employed.

Interior sampling approaches. By this procedure the scenarios for the random variables are sampled as the optimization problem is being solved iteratively by a decomposition-like method. Lower and upper statistical bounds for the optimal solution are obtained during the forward and backward passes of such methods, and the process stops when a given stopping criteria is reached (Morton, 1998). The leading solving techniques of this type are stochastic dual dynamic programming (SDDP) (Pereira and Pinto, 1991) and stochastic decomposition (SD) (Higle and Sen, 1996). However, we note that the SD approach has only recently been extended to a problem with more than two stages (Higle *et al.*, 2010), while the SDDP approach has long been applied for practical applications in energy systems (Rotting and Gjelsvik, 1999; Maceira *et al.*, 2002). The convergence properties and sampling techniques of the SDDP approach have been subjects of intensive research (Hindsberger, 2002; Philpott and Guan, 2008; Shapiro, 2010).

Besides the main approaches discussed above, additional techniques for multi-stage problems can also be found in the literature (Chen and Powell, 1999; Küchler and Vigerske, 2007; Beltran *et al.*, 2008).

3 Mid/long term hydrothermal coordination (HTC) problem

We briefly describe in this section the mid term (MT) and long term (LT) generation planning models for hydrothermal systems — namely NEWAVE and DECOMP, respectively (Maceira *et al.*, 2002), which were used to implement the multi-lag MSBD approach proposed in this paper. Such models have been used for more than 10 years by the Brazilian system operator (ISO) for the generation planning of the large-scale Brazilian electrical system (Maceira *et al.*, 2008). The objective function of both models — whose main characteristics are summarized in Table 1, is the minimization of total thermal generation costs in the system.

Table 1 Main Characteristics of the MT and LT Generation Planning Models Considered in this Paper.

	Horizon	Time Step	Sampling Approach	Solving strategy	Model for the Hydro Plants
NEWAVE (LT)	5 years	monthly	Stochastic; exterior	SDDP	Equivalent energy reservoirs
DECOMP (MT)	1 year	Weekly/ monthly	Deterministic	L-shaped	Hydro plants in cascade

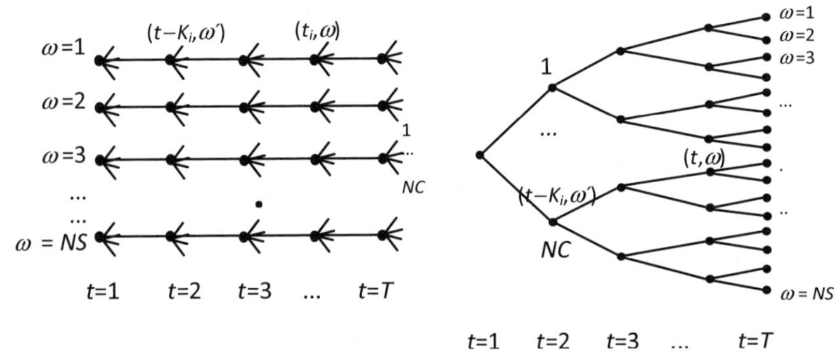

Figure 1 Schematic Illustration of the Scenario Tree for both the LT Model (left) and the MT Model (right) Considered in this Paper. Typical Values are $NC = 20$ and $NS = 200$.

Uncertainty in these models is related to the water inflows to the reservoirs in the system, which are modeled by a periodic autoregressive model (Maceira and Bezerra, 1997) (Par-P). In the MT model a deterministic (fixed) scenario tree is considered, while in the LT model the "complete tree" is composed of NC scenarios for each time step. The SDDP algorithm applied to solve the LT model uses NS scenario samples during the forward pass and all NC scenarios for the corresponding time step at each node during the backward pass. Figure 1 illustrates the scenario tree in both models. In the sequel, a coordinate pair (t, ω) indicates a scenario ω at time step t, and for a given node (t', ω), $\Omega^t_{(t',\omega)}$ is the set of all its descendent nodes at time step t.

Coordination among these models is performed by including a future cost function (FCF) at the end of the planning horizon of the MT model. This FCF is a multivariate piecewise linear function that gives expected system costs beyond the planning horizon of the MT model as a function of the vector of final storages in the reservoirs and the P last energy inflows to the system, where P is the order of the Par-P model used to generate the scenarios. Actually, the FCF is the recourse function for the corresponding month which was obtained previously by solving the LT problem applying the SDDP approach, with the same initial system conditions.

3.1 *Modeling assumptions*

All nonlinear constraints of the problem are approximated by piecewise linear functions, as for example the quadratic thermal generation costs and the hydro generation function (Diniz and Maceira, 2008). As a consequence, the problem may be formulated as a linear program (LP), with a convex objective function and a polyhedral feasible set. Since all constraints are modeled with slack variables (e.g., load shedding variables) to ensure primal feasibility, the stochastic problem has at least relatively complete recourse. Finally, due to the discretization of the random variable with finite support at the right hand side of the hydro balance equation, the recourse function for each stage is piecewise linear and convex (Pereira and Pinto, 1991) and the value of the optimal solution is bounded. The length of the planning horizon for the LT model has shown to be long enough to mitigate end-horizon effects (Grinold, 1983).

3.2 *Subproblem formulation*

The decision vector for the overall problems is $x = \{x_{gt}, x_{gh}, x_q, x_v\}$, where x_{gt} and x_{gh} are the thermal and hydro generations and x_q, x_v are the turbined outflow and storage vectors for the operation of the NH hydro plants of the system, along all scenarios and time steps of the planning horizon. Denoting by superscripts $x^{t,\omega}$ the components of x related to a given time step t and scenario ω, the linear program for each node in the scenario tree for both models is shown in Eq. (3.1).

$$
\begin{aligned}
&\min \left\langle c, x_{gt_i}^{t,\omega} \right\rangle + \left(\tfrac{1}{1+\beta}\right) \varphi^{t,\omega}\left(x_v^{t,\omega}, \xi^{t-p,\,\omega, p=0,\ldots P-1}\right) \\
&s.t. \\
&\ell x_{gt}^{t,\omega} + \ell x_{gh}^{t,\omega} = d^t \\
&x_v^{t,\omega} - x_v^{t-1,\omega} + D x_q^{t,\omega} = \xi^{t,\omega} \\
&x_{gh_i}^{t,\omega} = h_i(x_{q_i}^{t,\omega}, x_{v_i}^{t,\omega}),\ i = 1, \ldots, NH \\
&\underline{x^{t,\omega}} \le x^{t,\omega} \le \overline{x^{t,\omega}}
\end{aligned}
\tag{3.1}
$$

The objective function is the sum of thermal generation costs (given by linear incremental costs c) and expected system future costs given by the recourse function $\varphi(.)$ for the current system conditions, taking into account the discount rate β. The first constraint is the power balance to meet the system demand d, where ℓ indicates a vector of 1's with appropriate dimensions; the second group of constraints are the water conservation equation for all reservoirs, where ξ is the random vector of water inflows and D is the incidence matrix that links hydro plants in a same cascade. The third set of constraints are the piecewise linear hydro generation functions (Diniz and Maceira, 2008) h_i for each plant i. Lower and upper bounds for all variables are denoted by $\underline{x}$ and $\overline{x}$. There are also several additional operating constraints that are omitted here for simplicity (Maceira *et al.*, 2002; Maceira *et al.*, 2008).

The overall HTC problem comprises the set of subproblems (3.1) for all time steps t and scenarios ω, according to the recursive expression (2.1) and the scenario tree structure shown in Figure 1. The problem structure for this traditional formulation of the HTC problem is depicted in the left part of Figure 2, where it can be noted that decision variables of time step t are only linked directly to decision variables of time $t-1$. We call such structure a single-lag problem.

3.3 *Traditional single-lag MSBD solving strategy*

The methods applied to solve the HTC problem for the MT and LT models are the multi-stage L-shaped (Birge, 1985) and the SDDP (Pereira and

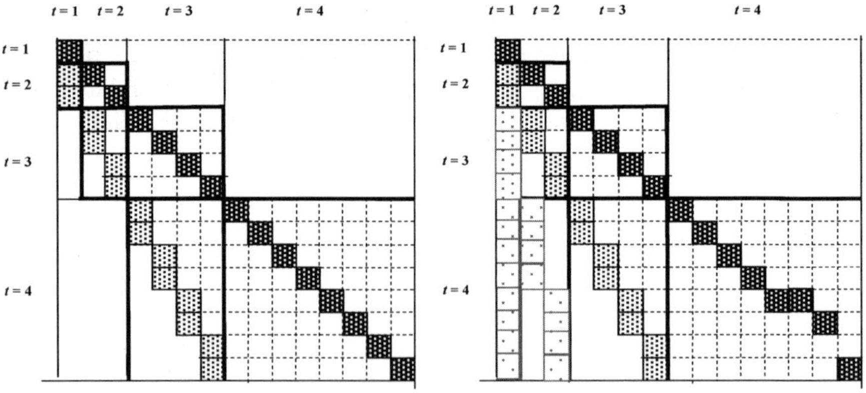

Figure 2 Structure of the Constraints Matrix of the HTC Problem in the Traditional Single-Lag Case (left) and in the Multi-Lag Case (right).

Pinto, 1991) techniques, respectively, which are both multi-stage Benders Decomposition (MSBD) approaches. In both methods the recourse function $\varphi^{t,\omega}(.)$ for each node is approximated by applying Benders cuts, in an iterative procedure composed of two phases:

- a forward pass, from stages 1 to T, where values of state variables for each node are obtained from the solution $\hat{x}^{t-1,\omega}$ of the ancestor node in the tree;
- a backward pass, from stage $T-1$ to 1, using the same state variables $\hat{x}^{t-1,\omega}$ of the forward pass and building Benders cut to the ancestor nodes.

During this process, a new Benders cut to refine the future cost function for each node ω' at stage $t-1$ is obtained from the solution of the subproblems for the corresponding set $\Omega^t_{(t-1,\omega')}$ of descendent scenarios ω at time step t, as follows:

$$\varphi^{t-1,\omega'} \geq \sum_{\omega \in \Omega^t_{(t-1,\omega')}} p_{\omega|\omega'} \left[z^{t,\omega *} + \left\langle \frac{\partial z^{t,\omega}}{\partial x^{t-1,\omega}}(\hat{x}^{t-1,\omega}), x^{t-1,\omega} - \hat{x}^{t-1,\omega} \right\rangle \right] \tag{3.2}$$

where $z^{t,\omega *}$ is the optimal value for the subproblem of node (t, ω) and $\partial z^{t,\omega}/\partial x^{t-1,\omega}(.)$ is the vector of partial derivatives of $z^{t,\omega *}$ with respect to the state variables of this node; $p_{\omega|\omega'}$ is the conditional probability of scenario $\omega$with respect to scenario ω'.

Equation (3.2) defines a multivariate linear cut to the recourse function of node $(t-1,\omega)$, which is a function of decision variables of this node.

Remark. Due to the sharing cut property of the SDDP approach (Infanger and Morton, 1996), in this method the recourse function is unique for all nodes in a same time step and can be denoted as $\varphi^t(.)$.

At the end of each forward pass, an upper bound $\overline{Z}$ for the optimal solution Z^* of the problem is obtained as the average value among all scenarios. Such bound is deterministic for the L-shaped method and statistic for the SDDP approach. A lower bound $\underline{Z}$ is also obtained at the end of each backward pass as the optimal value for the subproblem for t =1. In our implementation, the solving procedure stops for the MT model when the relative difference $(\overline{Z} - \underline{Z})/\underline{Z}$ is sufficiently small, and for the LT model when $\underline{Z}$ is within a $(1-\alpha)$ confidence interval for the value of $\overline{Z}$ (Pereira and Pinto, 1991; Maceira *et al.*, 2002). Alternative stopping criteria for the SDDP approaches have been proposed in the literature (Morton, 1998; Hindsberger, 2002).

4 HTC coordination with nonanticipativity constraints on the dispatch of LNG thermal plants

Recently, Liquefied Natural Gas (LNG)-based thermal plants have been considered as an alternative to provide additional generation capacity to electric power systems. However, contrary to conventional thermal plants, their dispatch must be determined K months prior to its actual dispatch, due to logistic issues. From the mathematical point of view, such requirements turns the "wait and see" procedure for the dispatch of thermal plants at a node (t, ω) into a "here and now" decision under uncertainty to be taken at an ancestor node ($t - K$, ω'), as illustrated in Figure 1.

4.1 *Subproblem formulation*

By denoting x_{LNG} the generation of the LNG thermal plants and K_i the required "anticipation lag" for each plant i, the formulation of each subproblem of the HTC problem becomes as shown in (4.1), where nc and ng are the number of conventional and LNG-based thermal plants in the system. The discount rate should be applied to the generation cost of the LNG plant at $t + K_i$since it is being evaluated at time step t.

$$\begin{aligned}
&\min \sum_{i=1}^{nc} c_i x_{gt_i}^{t,\omega} + \sum_{i=1}^{ng} \frac{c_i x_{LNG_i}^{t+K_i,\omega''}}{(1+\beta)^{K_i}} \\
&\qquad + \left(\frac{1}{1+\beta}\right) \varphi^{t,\omega}\left(x_v^{t,\omega}, \xi^{t-p,\omega,\, p=0,\dots P-1}\right) \\
&s.t. \\
&\ell x_{gt}^{t,\omega} + \ell x_{gh}^{t,\omega} = d^t - \sum_{i=1}^{ng} x_{LNG_i}^{t,\omega} \\
&x_v^{t,\omega} - x_v^{t-1,\omega} + D x_q^{t,\omega} = \xi^{t,\omega} \\
&x_{gh_i}^{t,\omega} = h_i(x_{q_i}^{t,\omega}, x_{v_i}^{t,\omega}),\ i = 1,\dots, NH \\
&\underline{x^{t,\omega}} \le x^{t,\omega} \le \overline{x^{t,\omega}}
\end{aligned} \tag{4.1}$$

We note that, for each LNG plant i, a single value $x_{LNG_i}^{t+K_i,\omega''}$ must be set for its generation at all future scenarios ω'' in the set $\Omega_{(t,\omega)}^{t+K_i}$. In the other side, the generation $x_{LNG_i}^{t,\omega}$ at this node (t, ω) is placed at the right hand side of the power demand equation since it has already been decided at a previous node ($t - K_i$, ω'), where $\omega \in \Omega_{(t,\omega')}^{t+K_i}$.

If we consider the overall deterministic equivalent problem, the pre-order requirement in the dispatch of LNG plants leads to the addition of the nonanticipativity constrains (4.2) in the problem, which are illustrated

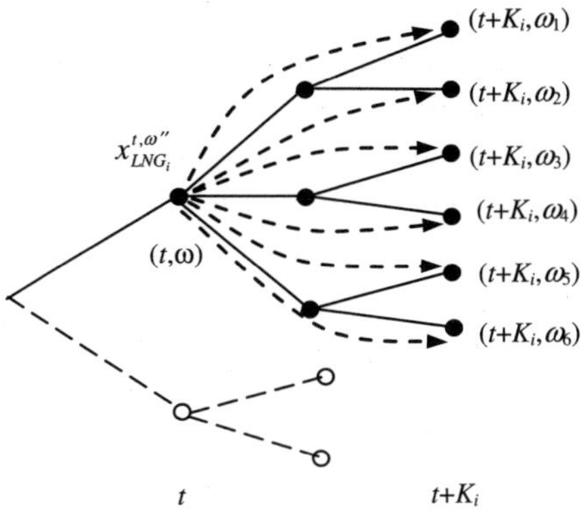

Figure 3 Example of a Set of Nonanticipativity Constraints for the Dispatch of LNG Plants, for the Case Where $K_i = 2$.

in Figure 3.

$$\begin{cases} x_{LNG_i}^{t+K_i,\omega''} = x_{LNG_i}^{t+K_i,\omega_1} \\ x_{LNG_i}^{t+K_i,\omega''} = x_{LNG_i}^{t+K_i,\omega_2} \\ \cdots \\ x_{LNG_i}^{t+K_i,\omega''} = x_{LNG_i}^{t+K_i,\omega_6} \end{cases} \tag{4.2}$$

The constraints matrix of the deterministic equivalent problem with such time-linking constraints on the dispatch of LNG plants has the lower block diagonal structure shown in the right part of Figure 2, which reveals a multi-lag dependence among decision variables. Applications of Benders decomposition to generation planning problems are restricted to the single lag case (Morton, 1996; Rotting and Gjelsvik, 1999; Philpott and Guan, 2008). Higher lags have only been considered for the stochastic process that generates water inflows scenarios to the reservoirs, which are data and not decision variables of the optimization problem (Pereira and Pinto, 1991; Infanger and Morton, 1996; Maceira *et al.*, 2008).

In order to properly take into account nonanticipativity constraints (4.2) in the HTC problem, we proceed to solve a multi-lag version of the MSBD approach as described in the next section.

4.2 *Proposed multi-lag MSBD approach*

At a given node (t, ω), the generation of each LNG plant $i$at time step $t+K_i$ should be decided, but the dispatch from time step $(t+1)$ to time step

$(t + K_i - 1)$ has already been defined at previous nodes. These generation values impact future system operation costs, since they decrease the demand to be satisfied by the remaining energy sources. Therefore, the recourse function φ at node (t, ω) should be augmented to include all variables related to future thermal generation in the LNG plants that have already been decided at this point.

However, in order to avoid a large increase in the state-space of the problem, we define additional artificial variables $s_{LNG}^{t+1}, s_{LNG}^{t+2}, \ldots, s_{LNG}^{t+K}$ that represent the sum of LNG-based thermal generations from time steps $(t+1)$ to $(t + K)$, where $K = \max_i\{K_i\}$. Therefore, the number of additional state variables does not depend on the number of LNG plants, but on the maximum lag size K.

In a general iteration of the MSBD algorithm, the subproblem at a given node (t, ω) is formulated as:

$$\min \sum_{i=1}^{nc} c_i x_{gt}^{t,\omega} + \sum_{i=1}^{ng} \frac{c_i x_{LNG_i}^{t+K_i,\omega''}}{(1+\beta)^{K_i}} + \left(\frac{1}{1+\beta}\right) \varphi^{t,\omega}$$

$$s.t.$$

$$\ell x_{gt}^{t,\omega} + \ell\, x_{gh}^{t,\omega} = d^t - \hat{s}_{LNG}^{t,\omega} \qquad \leftarrow \lambda_d^{t,\omega^*} \tag{4.3}$$

$$\sum_{\substack{i=1,\\ |k=K_i}}^{ng} x_{LNG_i}^{t+k,\omega''} - s_{LNG}^{t+k,\omega''} = 0, \; k = 1, \ldots, K$$

$$(\ldots)$$

$$\varphi^{t,\omega} - \gamma_{v_j} x^v - \sum_{i=1}^{ng} \left[\gamma_{LNG_j}^{t,\omega,K_i} x_{LNG_i}^{t+K_i,\omega''}\right]$$

$$\geq \gamma_{0_j} + \sum_{i=1}^{ng} \left[\sum_{k=1}^{K_i-1} \gamma_{LNG_j}^{t,\omega,k} \hat{x}_{LNG_i}^{t+k,\omega''}\right] j = 1, \ldots J^{t,\omega}. \qquad \leftarrow \lambda_{\phi_j}^{t,\omega*}$$

The last set of constraints comprise all $J^{t,\omega}$ Benders cuts for node (t, ω) that were already built in previous iterations, where $\gamma_0, \gamma_v, \gamma_{LNG}$are respectively the independent term and coefficients related to hydro storage and generation of LNG plants (coefficients related to past inflows to reservoirs are omitted for simplicity). For each LNG plant i, the terms related to coefficients γ_{LNG} are split into two parts:

- coefficients for variables $\{\hat{x}_{LNG_i}^{t+k,\omega''}, k = 1, \ldots K_i - 1\}$ of the current iteration, which are already known and becomes part of the right hand side of the cut;

- a coefficient for $x_{LNG_i}^{t+K_i,\omega''}$, which is a decision variable of this node and is placed on the left hand side of the cut.

The presence of the first terms above is the main difference between the traditional single-lag MSBD and the multi-lag MSBD proposed in this paper. Such terms must be "carried on" to predecessor nodes until the state variables become decision variables at previous nodes in the tree.

After the subproblems for all nodes $(t,\,\omega) \in \Omega^t_{(t-1,\omega')}$ are solved in the backward pass and optimal multipliers λ_d^{t,ω^*},$\lambda_\varphi^{t,\omega^*}$ of the demand equation and the Benders cuts are obtained, a new Benders cut (4.4) is built to the predecessor node $(t-1,\omega')$:

$$\varphi^{t-1,\omega} = \sum_{\omega \in \Omega^t_{(t-1,\omega')}} p_{\omega|\omega'} \begin{bmatrix} z^{t,\omega^*} + \sum\limits_{i=1}^{NH} \lambda_{h_i}^{t,\omega^*} (x_{v_i}^{t-1} - \hat{x}_{v_i}^{t-1}) \\ + \lambda_d^{t,\omega^*} (s_{LNG}^{t,\omega} - \hat{s}_{LNG}^{t,\omega}) \\ + \sum\limits_{k=1}^{K-1} \left[\sum\limits_{j=1}^{J^{t,\omega}} \lambda_{\varphi_j}^{t,\omega*} \gamma_{LNG_j}^{t,\omega,k} \right] (s_{LNG}^{t+k,\omega} - \hat{s}_{LNG}^{t+k,\omega}) \end{bmatrix} \tag{4.4}$$

where $\lambda_h^{t^*}$ are the Lagrange multipliers of the water conservation equations and $\hat{x}_v^{t-1}$ are the state variables related to hydro storages.

Each coefficient (as shown in (4.3)) of this new Benders cut to the subproblem of node $(t-1,\omega')$ is given by:

$$\begin{cases} \gamma_{0_J}^{t-1,\omega} = \sum\limits_{\omega \epsilon \Omega^t_{(t-1,\omega')}} p_{\omega|\omega'} \begin{bmatrix} z^{t,\omega^*} - \sum\limits_{i=1}^{NH} \lambda_{h_i}^{t,\omega^*} \hat{x}_{v_i}^{t-1} - \lambda_d^{t,\omega^*} \hat{s}_{LNG}^{t,\omega} \\ - \sum\limits_{k=1}^{K} \left(\sum\limits_{j=1}^{J^{t,\omega}} \lambda_{\phi_j}^{t,\omega*} \gamma_{LNG_j}^{t,\omega,k} \right) \hat{s}_{LNG}^{t+k,\omega} \end{bmatrix} \\ \gamma_{v_{ji}}^{t-1,\omega} = \sum\limits_{\omega \epsilon \Omega^t_{(t-1,\omega')}} p_{\omega|\omega'} \lambda_{h_i}^{t,\omega^*}, \quad i = 1, \ldots, NH \\ \gamma_{LNG_j}^{t-1,\omega,k} = \begin{cases} \sum\limits_{\omega \epsilon \Omega^t_{(t-1,\omega')}} p_{\omega|\omega'} \lambda_d^{t,\omega^*}, & k = 1 \\ \sum\limits_{\omega \epsilon \Omega^t_{(t-1,\omega')}} p_{\omega|\omega'} \left[\sum\limits_{j=1}^{J^{t,\omega}} \lambda_{\phi_j}^{t,\omega*} \gamma_{LNG_j}^{t,\omega,k-1} \right], & k = 2, \ldots K \end{cases} \end{cases} \tag{4.5}$$

Now we look in detail the structure of coefficient $\gamma_{LNG_{ji}}^{t,\omega,K_i}$, which multiplies the variable $x_{LNG_i}^{t+K_i,\omega''}$ for a given plant i and time step t. By applying recursively the Benders cut (4.4)–(4.5) from time steps $t + K_i$ to $t+1$ during the backward pass of the multi-lag L-shaped algorithm, it can be shown that the following expression is obtained:

$$\begin{cases} \gamma_{LNG_j}^{t,\omega,K_i} = \sum_{\omega \in \Omega_{(t,\omega')}^{t+K_i}} \left(p_{\omega|\omega'} \lambda_d^{t+K_i,\omega^*}\right) \prod_{k=1}^{K_i-1} \sum_{j=1}^{J^{t,\omega}} \left(\lambda_{\phi_j}^{t+k,\omega*}\right) \\ = \sum_{\omega \in \Omega_{(t,\omega')}^{t+K_i}} p_{\omega|\omega'} \lambda_d^{t+K_i,\omega^*} \left(\frac{1}{1+\beta}\right)^{K_i} \end{cases} \tag{4.6}$$

since the dual formulation of (4.2) ensures that $\sum_{j=1}^{J^{t,\omega}} \lambda_{\varphi_j}^{t+k,\omega*} = \frac{1}{1+\beta}$ at all nodes (t,ω). Expression (4.6) is quite intuitive because it states explicitly that the incremental future "benefit" of thermal generation at $t+K_i$ is the weighted average of the system marginal costs at all descendent nodes of (t,ω) at time step $t+K_i$. The decision about the value of each thermal unit $x_{LNG_i}^{t+K_i}$ is taken by comparing the incremental cost c_i of the plant with its future benefit $\gamma_{LNG_j}^{t,\omega,K_i}$, both discounted at time step t.

Remark. In the practical implemental of the LT and MT models, the system is split into several subsystems, interconnected among each other. As a consequence, there is a power demand equation for each subsystem s and the multiplier λ_d to be considered in the Benders cut is related to the subsystem to which the LNG plant is located. Although these details were not considered in the problem formulation for the sake of simplicity, they were carefully taken into account in our implementation. The number of additional state variables related to future generation of LNG plant becomes $\sum_{s=1}^{S} \max_{i \in \psi_S} K_i$, where ψ_s is the set of LNG plants belong to subsystem i and S is the number of subsystems.

5 Numerical results

The proposed multi-lag MSBD approach was applied to set the real generation dispatch of the Brazilian system taking into account the pre-order dispatch of LNG plants. The system comprises 120 hydro plants, 40 conventional thermal plants and 5 subsystems.

We first present the results for the real case performed by the Brazilian ISO with the first LNG plant (Linhares) that was put in operation in Brazil from February 2011, whose summarized data is shown in Table 2. The planning horizon in the LT model was 56 months with 200 scenarios in

Table 2 Main Operation Data for the LNG Plant Considered in the Real Brazilian System.

	Minimum generation	Maximum generation	Generation cost (c_i)	Anticipation lag (K_i)
DECOMP (MT)	0 MW	200 MW	62.4 \$/MWh	2 months

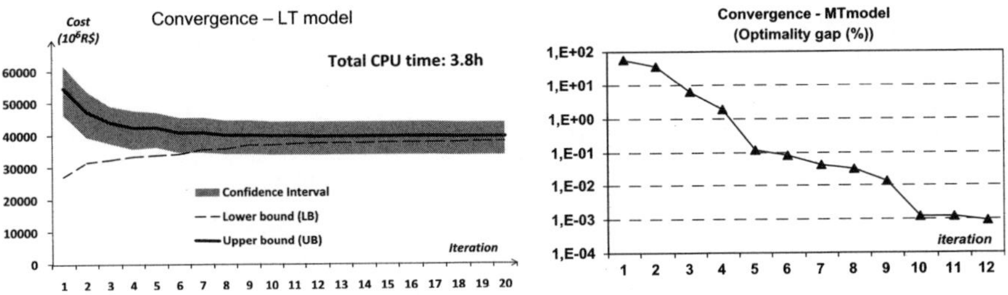

Figure 4 Convergence Behaviour of the Proposed MSBD Approach for the SDDP (left) and L-Shaped Methods (right).

the forward pass, and 20 scenarios in the backward pass at each node. The planning horizon for the MT model was 2 months with 353 scenarios in the second month.

The convergence behavior of the SDDP approach applied to the LT model and the L-shaped method applied to the MT model are shown in Figure 4. Total CPU time was 3.8 hours for LT model and 37.8 minutes for the MT model.

In the next sections we consider a study case with 40 LNG thermal plants, with anticipation lags ranging from 1 to 3 months and a time horizon of 3 months, with a scenario tree composed of $1 \times 50 \times 150$ nodes in the MT model.

5.1 *Dispatch of LNG thermal plants*

The dispatch of the LNG plant that is decided at a given node (t, ω) is supposed to be used at all descendent nodes at time step $t + K_i$, regardless of which scenario will be revealed. Therefore, it is impossible to set at each of these descendent nodes the ideal "wait and see" dispatch for the thermal plants that would be obtained in the traditional formulation of the HTC problem. Such dispatch is determined based on a comparison of

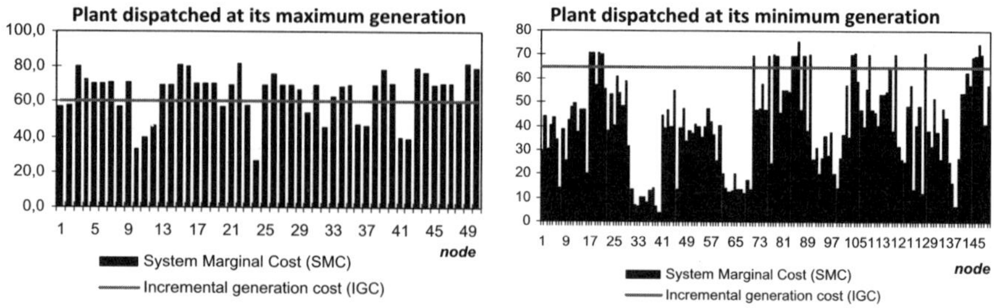

Figure 5 Comparison Between Incremental Cost of the Unit and System Marginal Costs in Future Nodes to which the Dispatch of the Plant was Set.

the incremental cost c_i of the thermal plant and the system marginal cost $\lambda_d^{t+K_i}$ (the plant would be fully dispatched whenever $c_i < \lambda_d^{t+K_i}$).

We analyze the dispatch decided at node (1,1) for two thermal plants in the MT model. The first plant has an anticipation lag of $K_i = 1$ and was set to dispatch at its maximum output at all 50 nodes of stage 2. The second plant has $K_i = 2$ and was set to dispatch at its minimum output at all 150 nodes of stage 3. The graphs in Figure 5 compare the incremental cost of the plant (flat line) with the distribution of $\lambda_d^{t+K_i}$ at all future nodes at time step $t + K_i$. The dispatch proved to be "correct" in 34 of the 50 future scenarios for the first plant (in the left) and in 126 of the 150 scenarios for the second plant (in the right).

5.2 *Sensitivity analysis*

Finally, we present in Table 3 a sensitivity analysis of the CPU time to solve the MT problem with the number of LNG plants (in the left) and the size of the anticipation lag K_i (in the right). In this second case, an entry "X/Y/Z" indicates that the number of plants with $K_i = 1, 2$ and 3 are X, Y and Z respectively.

Table 3 Sensitivity Analysis on the Number of LNG Plants (left) and on the Size of the Anticipation Lag Ki-MT Model.

# LNG plants	**CPU time (min)**	**# plants with K_i=1/2/3**	**CPU time (min)**	**# plants with K_i=1/2/3**	**CPU time (min)**
—	19.5	—	19.5	10/20/10	35.7
10	25.4	40/00/00	21.2	10/10/20	36.2
20	28.1	30/10/00	25.4	00/20/20	38.1
30	42.1	20/20/00	26.6	00/10/30	39.6
40	37.7	20/10/10	29.5	00/00/40	29.9

We note that CPU times tend to increase reasonably when higher order lags K_i are considered. Even though this issue is not relevant to our problem, since the anticipation lag for LNG plants will rarely be higher than 3 months, a future work may consist in strategies to mitigate such effects, as for example the use of multi-period/multi-scenarios stages in the time decomposition of the problem (Dempster and Thompson, 1998; Santos and Diniz, 2009).

6 Conclusions

This paper proposes a multi-lag variant of multi-stage Benders decomposition, where the recourse function of each node may include decision variables of several previous time steps. The approach was applied in the L-shaped and stochastic dual dynamic programming solving strategies employed to solve the stochastic long and mid term generation planning problem for hydrothermal systems, where the dispatch of some liquefied natural gas-based thermal plants must be decided K months prior to its actual dispatch. Numerical results illustrate the effectiveness of the proposed approach for a real application in the large-scale Brazilian electrical system.

Acknowledgments

The authors would like to acknowledge the contributions given by Vitor Duarte, Tiago Santos, Michel Pompeu, Débora Penna, Fernanda Costa and Ana Sabóia during the development of this work.

References

Beltran, C., L. F. Escudero, R. E. Rodrigues-Ravines, Multi-stage stochastic linear programming: an approach by events, *Optimization Online*, 2008.

Birge, J. R., Decomposition and partitioning methods for multistage stochastic linear programs, *Operations Research*, 33(5): pp. 989–1007, 1985.

Birge, J. R. and F. V. Louveaux, A multicut algorithm for two-stage stochastic linear programs, *European Journal of Operational Research*, 34(3): pp. 384–392, 1988.

Birge, J. R. and S. W. Wallace, Refining bounds for stochastic linear programs with linearly transformed independent random variables, *Operations Research Letters*, 5(2): pp. 73–77, 1986.

Chen, Z. L. and W. B. Powell, Convergent cutting-plane and partial-sampling algorithm for multistage stochastic linear programs with recourse, *Journal of Optimization Theory and Applications*, 102(3): pp. 497–524, 1999.

Chiralaksanakul, A. and D. P. Morton, Assessing policy quality in multi-stage stochastic programming, Technical Report, The University of Texas at Austin, 2003.

Dempster, M. and R. T. Thompson, Parallelization and aggregation of nested Benders decomposition, *Annals of Operations Research*, 81: pp. 163–187, 1998.

Dempster, M. and R. T. Thompson, EVPI-based importance sampling solution procedures for multistage stochastic linear programmes on parallel MIMD archiectures, *Annals of Operations Research*, 90: pp. 161–184, 1999.

Diniz, A. L. and M. E. P. Maceira, A four-dimensional model of hydro generation for the short-term hydrothermal dispatch problem considering head and spillage effects, *IEEE Transactions on Power Systems*, 23(3): pp. 1298–1308, 2008.

Edirisinghe, N. C. P. and W. T. Ziemba, Bounds for two-stage stochastic programs with fixed recourse, *Mathematics of Operations Research*, 19(2): pp. 292–313, 1994.

Ermoliev, Y. and R. J.-B. Wets, Numerical techniques for stochastic optimization, Springer Series in Computational Mathematics, v.10, Springer-Verlak, 1988.

Faber, B. A. and J. R. Stedinger, Reservoir optimization using sampling SDP with ensemble streamflow predicion (ESP) forecasts, *Journal of Hydrology*, 249: pp. 113–133, 2001.

Frauendorfer, K., Solving SLP, Recourse problems with arbitrary multivariate distributions — the dependent case, *Mathematics of Operations Research*, 13(3): pp. 377–394, 1988.

Gassmann, H. I., MSLiP: A computer code for multistage stochastic linear programming problem, *Mathematical Programming*, 47: pp. 407–423, 1990.

Grinold, R. C., Model building techniques for the correction of end effects in multistage convex programs, *Operations Research*, 31(3): pp. 407–431, 1983.

Guigues, V. and Römisch, W., Sampling-based decomposition methods for risk-averse multistage stochastic programs, Optimization Online, 2010.

Hicks, R. H., C. R. Gagnon, S. L. S. Jacoby and J. S. Kowalik, Large scale, nonlinear optimization of energy capability for the pacific northwest hydroelectric system, *IEEE Transactions on Power Apparatus and Systems*, 93(5): pp. 1604–1612, 1974.

Higle, J. L. and S. Sen, Stochastic decomposition: a statistical method for large scale stochastic linear programming, *Nonconvex Optimization and its Applications*, 8: 1996.

Higle, J. L., B. Rayco and S. Sen, Stochastic scenario decomposition for multistage stochastic programs, *IMA Journal of Management Mathematics*, 21: pp. 39–66, 2010.

Hindsberger, M., Interconnected hydro-thermal systems — Models, methods, and applications, Phd Thesis, Tech. Univ. Denmark, 2002.

Hobbs, B. F., M. H. Rothkopf, R. B. O'Neil and H. Chao (eds.), *The Next Generation of Electric Power Unit Commitment Models*, USA: Kluwer Academic Publisher, 2001.

Infanger, G. and D. P. Morton, Cut sharing for multistage stochastic linear programs with interstage dependency, *Mathematical Programming*, 75(2): pp. 241–256, 1996.

Kleywegt, A., A. Shapiro and T. Homem-de-Mello, The sample average approximation method for stochastic discrete optimization, *Siam Journal on Optimization*, 12: pp. 479–502, 2001.

Küchler, C. and S. Vigerske, Decomposition of multistage stochastic programs with recombining scenario trees, *Technical Report*, Humboldt University, 2007.

Linderoth, J., A. Shapiro and S. Wright, The empirical behavior of sampling methods for stochastic programming, *Annals of Operations Research*, 142: pp. 215–241, 2006.

Little, J. D., The use of storage water in a hydroelectric system, *Operations Research*, 3(2): pp. 187–197, 1955.

Maceira, M. E. P. and C. V. Bezerra, Stochastic Streamflow model for Hydroelectric Systems, *5th PMAPS Conference Proceeding*, Vancouver, Canada, 1997.

Maceira, M. E. P., L. A. Terry, F. S. Costa, J. M. Damazio and A. C. G. Melo, Chain of optimization models for setting the energy dispatch and spot price in the Brazilian system, *PSCC Conference Proceedings*, Sevilla, Spain, 2002.

Maceira, M. E. P., V. S. Duarte, D. D. J. Penna, L. A. M. Moraes and A. C. G. Melo, Ten years of application of stochastic dual dynamic Programming in official and agent studies in Brazil — Description of the Newave program, *16th PSCC Proceedings*, Glasgow, 2008.

Morton, D. P., An enhanced decomposition algorithm for multistage stochastic hydroelectric scheduling, *Annals of Operations Research*, 64: pp. 211–235, 1996.

Morton, D. P., Stopping rules for a class of sampling-based stochastic programming algorithms, *Operations Research*, 46(5): pp. 710–718, 1998.

Nowak, M. P. and W. Römisch, Stochastic Lagrangian relaxation applied to power scheduling in a hydro-thermal system under uncertainty, *Annals of Operations Research*, 100(1): pp. 251–271, 2001.

Pagnoncelli, B. K., S. Ahmed and A. Shapiro, Sample average approximation method for chance constrained programming: theory and applications, *Journal of Optimization Theory and Applications*, 142(2): pp. 399–416, 2009.

Pereira, M. V. F. and L. M. V. G. Pinto, Multi-stage stochastic optimization applied to energy planning, *Mathematical Programming*, 52: pp. 359–375, 1991.

Philpott, A. B. and Z. Guan, On the convergence of stochastic dual dynamic programming and related methods, *Operations Research Letters*, 36(4): pp. 450–455, 2008.

Rotting, T. A. and A. Gjelsvik, Stochastic dual dynamic programming for seasonal scheduling in the Norwegian power system, *IEEE Transactions on Power Systems*, 7(1): pp. 273–279, 1999.

Ruszczńsk, A., Regularized decomposition of stochastic programs: algorithmic techniques and numerical results, *Tech. Rep.*, IIASA, 1993.

Sachdeva, S. S., Bibliography on optimum reservoir drawdown for the hydroelectric-thermal power system operation, *IEEE Transactions on Power Apparatus and Systems*, 101(6): pp. 1487–1496, 1982.

Santos, T. N. and A. L. Diniz, A New Multiperiod Stage Definition for the Multistage Benders Decomposition Approach Applied to Hydrothermal

Scheduling, *IEEE Transactions on Power Systems*, 24(3): pp. 1383–1392, 2009.

Santos, M. L. L., E. L. Silva, E. C. Finardi and R. Gonçalves, Practical aspects in solving the medium-term operation planning problem of hydrothermal power systems by using the progressive hedging method, *International Journal of Electrical Power and Energy Systems*, 31(9): pp. 546–552, 2009.

Shapiro, A., Analysis of stochastic dual dynamic programming method, Optimization Online, 2010.

Shapiro, A., D. Dentcheva and A. Ruszczńsk, Lectures on Stochastic Programming: Modelling and Theory, MPS-SIAM Series on Optimization, 2009.

Tufegdzic, N., R. J. Frowd and W. O. Stadlin, A coordinated approach for real-time short term hydro scheduling, *IEEE Transactions on Power Systems*, 11(4): pp. 1698–1704, 1996.

Van Slyke, R. and R. J.-B. Wets, L-shaped linear programs with application to optimal control and stochastic programming, *SIAM Journal on Applied Mathematics*, 17: pp. 638–663, 1969.

Wets, R. J.-B., Stochastic programs with fixed recourse: The equivalent deterministic program, *SIAM Review*, 16(3): pp. 309–339, 1974.

Wood, A. and B. Wollemberg, Power Generation Operation and Control, John Wiley and Sons, 2nd Edition, New York, 1996.

Xu, H. and F. Meng, Convergence analysis of sample average approximation methods for a class of stochastic mathematical programs with equality constraints, *Mathematics of Operations Research*, 32(3): pp. 648–668, 2007.

Yeh, W., Reservoir management and operations models: a state-of-the-art review, *Water Resources Research*, 21(12): pp. 1797–1818, 1985.

Zakeri, G., A. B. Philpott and D. M. Ryan, Inexact cuts in Benders decomposition, *Siam Journal on Optimization*, 10(3): pp. 643–657, 2000.

PART IV

Papers on Telecommunications

Chapter 17

Stochastic Second-Order Cone Programming in Mobile Ad-Hoc Networks: Sensitivity to Input Parameters

Francesca Maggioni*, Marida Bertocchi†, Elisabetta Allevi‡, Florian A. Potra§, and Stein W. Wallace**

Summary

In this paper sensitivity analysis is adopted in order to reveal the role of randomness of a stochastic second-order cone program (Maggioni *et al.*, 2009) for mobile ad-hoc networks starting from the semidefinite stochastic location-aided routing (SLAR) model, described in Ariyawansa and Zhu (2006) and Zhu *et al.* (2011). The algorithm looks for a destination node and sets up a route by means of the *expected zone*, the region where the sender node expects to find the destination node and the *requested zone* defined by the sender node for spreading the route request to the destination node. The movements of the destination node are represented by ellipses scenarios, randomly generated by uniform and normal distributions in a neighborhood of the initial position of the destination node. Sensitivity analysis is performed by considering an increasing number of scenarios, different costs of flooding and latency penalty. Evaluation of *Expected Value of Perfect Information EVPI* and *Value of Stochastic Solution VSS*

*Department of Mathematics, Statistics, Computer Science and Applications, University of Bergamo, Via dei Caniana 2, Bergamo 24127, Italy, francesca.maggioni@unibg.it.
†Department of Mathematics, Statistics, Computer Science and Applications, University of Bergamo, Via dei Caniana 2, Bergamo 24127, Italy, marida.bertocchi@unibg.it.
‡Department of Quantitative Methods, University of Brescia, Contrada S. Chiara, 50, Brescia 25122, Italy, allevi@eco.unibs.it.
§Department of Mathematics & Statistics, University of Maryland, Baltimore County, U.S.A. The work of this author was supported in part by the National Science Foundation under Grant No. 0139701, potra@umbc.edu.
**Department of Finance and Management Science, Norwegian School of Economics, NO-5045 Bergen, Norway, stein.wallace@nhh.no.

(Maggioni and Wallace, 2010; Birge, 1970) allows us to find the range of values in which it is convenient the deterministic versus the stochastic approach.

Keywords: Sensitivity analysis, mobile ad-hoc networks, second-order cone programming, stochastic programming, scenarios generation, evaluation of the deterministic solution.

1 Introduction

Sensitivity analysis is a key issue when one wishes to study the robustness of solutions of problems, that is, to see how they might change if the input data were different or subject to errors. A change in the solution or its structure would indicate the need for further investigation. In the case neither changes, the proposed solution can be considered as an appropriate guide for decision making. When the decision has to be made under uncertainty, the solution given by sensitivity analysis could no longer be appropriate: it may not being able to capture the response to information at times at which decisions are made (see Wallace, 2000 and Higle and Wallace, 2003, the latter with a worked-out example). Moreover, there could be stochastic programs (Ruszczyński and Shapiro, 2003; Shapiro, 2008) where, even the uncertainty is crucial for the problem, the solutions are the same along the different scenarios and do not reveal properly the stochastic structure of the considered problem. In such a case, sensitivity analysis techniques are useful to find the parameters range within which the solutions are scenarios dependent. This is the case considered in this paper where sensitivity analysis is applied to a stochastic second order cone program for mobile ad hoc networks (Maggioni *et al.*, 2009).

Wireless mobile hosts, able of communicating with each other in the absence of a fixed infrastructure, are now a standard in our world economy. The Mobile Ad hoc NETworks (MANET), consisting of wireless mobile nodes, and the related routing protocols have been extensively studied in the last 15 years (see Ko and Vaidya, 2000; Vyas, 2000 and the references therein).

Ko and Vaidya 2000 suggest a special approach based on the use of local information for decreasing the overhead of route discovery. Their algorithm, known as the Location-Aided Routing (LAR) protocol, tries to reduce the number of nodes to whom the requested route is propagated by using local information given, for example, by the Global Positioning System (GPS). The building blocks behind the algorithm are the *expected zone*, and the *requested zone*. The former is the region where the sender node S expects to find the destination node D in an elapsed time t_1, while the latter is the region defined by the sender node for spreading the route request to the destination node. An interesting approach, suggested by Ariyawansa and Zhu (2006), proposes to structure the problem of the identification of the expected zone as a two-stage stochastic semidefinite programming problem, called stochastic location-aided routing (SLAR). In Zhu *et al.* (2011), a similar formulation is used in conjunction with a greedy process of directional search, and numerical results are given for a limited number of scenarios. The results from Zhu *et al.* (2011) reveal that, in general, SLAR can significantly reduce the overall overhead than existing deterministic algorithms, because the location uncertainty in the routing problem is better captured.

In Maggioni *et al.* (2009), SLAR has been formulated as a stochastic two-stage second order cone programming with a significant lower computational complexity than the corresponding semidefinite program. This allows to solve problems with a much larger number of scenarios (20250) than it is possible with the semidefinite model (500). The solution of the stochastic two-stage second-order cone programming with many scenarios allows us to compute a new expected zone that is very appropriate for practical applications, and to validate our model.

The paper is organized as follows: Section 2 contains the problem description and formulation of SLAR as a stochastic two-stage second order cone programming problem. Section 3 describes scenario generation for the stochastic model. Section 4 describes the results obtained by the sensitivity analysis and Section 5 concludes the paper.

2 Problem description and model formulation

2.1 *Stochastic location-aided routing (SLAR) in mobile ad-hoc networks*

We consider a location-aided routing problem in a wireless ad-hoc network (Ko and Vaidya, 2000). A wireless ad-hoc network consists of a group of mobile nodes that communicate with each other in the absence of a fixed infrastructure. Therefore, the design of routing protocols becomes a crucial issue and a number of routing protocols have been proposed with the goal of searching for a route when hosts move. One of the most successfully algorithm is the Stochastic Location-Aided Routing (SLAR) (Ko and Vaidya, 2000), based on the use of local information (given for example by the Global Positioning System, GPS). When a *sender node*, say S, needs to find a route to a *destination node*, say D, S broadcasts a route

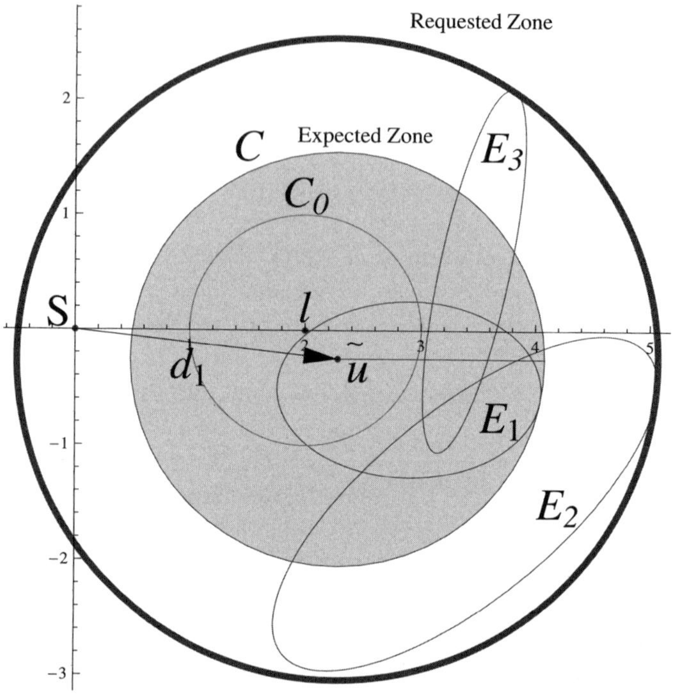

Figure 1 Expected Zone (grey circle) with SLAR Algorithm in the Case of the Ellipses Scenarios E_1, E_2 and E_3.

request to all its neighbors. On receiving the route request, D responds by sending a route reply message to S following the reverse path of the route request received by D. If the route request message does not get to D, the protocol allows S to initiate a new route discovery with an expanded requested zone.

The main concepts behind the algorithm are the *expected zone* and *requested zone*. The former is the region in the shape of a circle C (see Figure 1), where S (for simplicity fixed at the origin $0 \in \mathrm{R}^n$) expects to find D after an elapsed time t_1, based on the knowledge that node D was located at l at time t_0 and its lowest velocity is v. The latter is the region defined by S which includes the expected zone, for spreading the route request to reach D in case it does not belong to the expected zone.

One of the main characteristics of a mobile ad hoc network is the mobility of the nodes: the movements of D are then represented by ellipses scenarios E_k, $k = 1, \ldots, K$ (see Figure 1), randomly generated by uniform and normal distributions in a neighborhood of the starting position l of

the destination node. This choice corresponds to a typical real situation in which people are moving along preferred directions (for example different motorways) identified by the length of the main semiaxis $\sigma_{1,k}$ and angle φ_k of the ellipse E_k with the possibility to exit from the motorway for short distances (length of the second semiaxis $\sigma_{2,k}$).
Our proposal for implementing SLAR algorithm uses the following three-stage procedure (Maggioni *et al.*, 2009):

(1) calculate the initial expected zone (circle C) where the destination node is expected to be at time t_1; the disk C is required to contain the smallest disk C_0 centred in l and radius $v(t_1 - t_0)$ corresponding to the minimum speed v at which D is supposed to move (assuming a radial direction). The route request is then sent from the source node to cover this circle. Notice that the main decisions at this stage are the center $\tilde{u} \in \mathrm{R}^n$ and radius $r = \sqrt{\tilde{u}^T \tilde{u} - \gamma}$ of the circle:

$$C = \{u \in \mathrm{R}^n : u^T u - 2\tilde{u}^T u + \gamma \leq 0\}. \tag{1}$$

(2) The route request is sent out to look for D by flooding inside the expected zone C. If D is in C, no further action is needed (see the ellipse E_1 in Figure 1); the route request reaches the destination and the reply message is sent back to the source. Then a route is established between the source and the destination node.

(3) In case the destination node D is not found in stage 2, D should be in an ellipse E_k, $k = 1, \ldots, K$, not covered by C (see ellipses E_2 and E_3 in Figure 1). The disk C is then enlarged in order to cover the ellipse E_k and to get a new circle C_k^*

$$C_k^* = \{u \in \mathrm{R}^n : u^T u - 2\tilde{u}^T u + \gamma - \zeta_k \leq 0\}, \tag{2}$$

with the same center $\tilde{u} \in \mathrm{R}^n$ of C and radius $\sqrt{\tilde{u}^T \tilde{u} - \gamma + \zeta_k}$ enlarged by the quantity $\zeta_k \in \mathrm{R}^+ \cup \{0\}$.
We propose as a natural candidate for the requested zone the circle

$$\tilde{C} = \{u \in \mathrm{R}^n : u^T u - 2\tilde{u}^T u + \tilde{\gamma} \leq 0\}, \tag{3}$$

where $\tilde{\gamma} = \gamma - \sum_{k=1}^{K} p_k \zeta_k$.
Note that in Ariyawanza and Zhu (2006) the second stage decision consists in choosing a unique disk $\hat{C}$, the smallest disk containing all the ellipses E_k, $k = 1, \ldots, K$ and the circle C_0:

$$\hat{C} = \{u \in \mathrm{R}^n : u^T u - 2\hat{u}^T u + \hat{\gamma} \leq 0\}, \tag{4}$$

with the same center $\hat{u} \in \mathrm{R}^n$ of the expected zone C and radius $\sqrt{\hat{u}^T\hat{u} - \gamma + \hat{\zeta}}$ enlarged by the quantity $\hat{\zeta} \in \mathrm{R}^+ \cup \{0\}$.

2.2 *Stochastic second-order cone model (SSOCP) for SLAR*

In Ariyawansa and Zhu (2006) the problem described in the previous section has been modelled with a *stochastic semidefinite formulation* SSDP.
We remind that the general formulation of a stochastic semidefinite problem with recourse (SSDP) in standard primal form (see Ariyawansa and Zhu, 2006), is given by:

$$\begin{aligned} \min\ & c^T x + E[Q(x,\omega)] \\ \text{s.t.}\ & \sum_{i=1}^{n} x_i A_i \preceq B, \end{aligned} \tag{5}$$

where $Q(x,\omega)$ is the minimum of the problem

$$\begin{aligned} \min\ & q(\omega)^T y \\ \text{s.t.}\ & \sum_{i=1}^{n} x_i S_i(\omega) + \sum_{i=1}^{p} y_i T_i(\omega) \preceq C(\omega), \end{aligned} \tag{6}$$

and

$$E[Q(x,\omega)] = \int_{\Omega} Q(X,\omega) P(\mathrm{d}\omega), \tag{7}$$

is the expected value of $Q(x,\omega)$. The first stage data $\mathcal{D} = (c, A_1, \ldots, A_n, B) \in \mathrm{R}^n \times (\mathrm{R}_s^{m\times m})^{n+1}$, appearing in (5), are deterministic, while the second stage data

$$\begin{aligned} \mathcal{R}(\omega) = (q(\omega), S_1(\omega), \ldots, S_n(\omega), T_1(\omega), \ldots, T_p(\omega), C(\omega)) \in \mathrm{R}^p \\ \times (\mathrm{R}_s^{l\times l})^{n+p+1}, \end{aligned} \tag{8}$$

are stochastic and depend on a random event $\omega \in \Omega$, where (Ω, P) is an underlying probability space.

On the other hand, the *deterministic second-order cone programming* (DSOCP) problem consists in minimizing a linear function over the intersection of an affine set and a Cartesian product of second-order (Lorentz) cones:

$$\mathcal{K}_n := \{x = (x_0; \bar{x}) \in \mathrm{R}^n : x_0 \in \mathrm{R},\ x_0 \geq \|\bar{x}\|\}, \tag{9}$$

where $\|\cdot\|$ refers to the standard Euclidean norm, and n is the dimension of $\mathcal{K}_n$ (see Alizadeh and Goldfarb, 2003).

Since

$$x_0 \geq \|\bar{x}\| \Leftrightarrow Arw(x) := \begin{pmatrix} x_0 & -\bar{x}^T \\ -\bar{x} & x_0 I \end{pmatrix} \succeq 0, \tag{10}$$

it follows that DSOCP is a particular case of DSDP and by means of (10) its general formulation is as follows:

$$\begin{aligned} &\min && f^T\mathbf{x} \\ &\text{subject to} && \|A_i\mathbf{x} + \mathbf{b}_i\| \leq \mathbf{c}_i^T\mathbf{x} + d_i, \quad i = 1, \ldots, N. \end{aligned} \tag{11}$$

However, the computational effort per iteration required by an interior point methods to solve the DSOCP problem is substantially less than that required to solve the corresponding DSDP problem (Alizadeh and Goldfarb, 2003).

A *stochastic second order cone formulation* (SSOCP) of the SLAR problem has been then introduced in Maggioni *et al.* (2009) allowing to solve the problem with a much larger number of scenarios (20250) than what is possible with stochastic semidefinite program (SSDP). The stochastic second-order cone SSOCP model for the SLAR problem presented in Maggioni *et al.* (2009) is composed by two (n + 1)-dimensional second-order cone constraints, $n(K+1)$ 3-dimensional second-order cone constraints and with all the other constraints linear as follows:

$$\min \alpha d_1 + \beta d_2 + \sum_{k=1}^{K} p_k \beta_1 \zeta_k \tag{12}$$

$$\text{s.t. } r \geq 0,\ \tau \geq 1,\ \gamma \leq \tau\|l\|^2 - \tau(t_1 - t_0)^2 v^2 - 1^T r, \tag{13}$$

$$\begin{pmatrix} r_j + \tau - 1 \\ 2\,(\tau l_j - \tilde{u}_j) \\ r_j - \tau + 1 \end{pmatrix} \in \mathcal{K}_3, \quad j = 1, \ldots, n, \tag{14}$$

$$\begin{pmatrix} d_1 \\ \tilde{u} \end{pmatrix} \in \mathcal{K}_{n+1}, \quad \begin{pmatrix} \sqrt{d_2 + \gamma} \\ \tilde{u} \end{pmatrix} \in \mathcal{K}_{n+1} \tag{15}$$

$$h_k = Q_k^T(\delta_k g_k + \tilde{u}),\ \zeta_k \geq 0,\ k = 1, \ldots, K, \tag{16}$$

$$s_k \geq 0,\ \delta_k \geq \frac{1}{\lambda_{\min}(\Lambda_k)},\ \zeta_k \geq \gamma + 1^T s_k - \delta_k \nu_k,\ k = 1, \ldots, K, \tag{17}$$

$$\begin{pmatrix} s_{kj} + \delta_k \lambda_{kj} - 1 \\ 2h_{kj} \\ s_{kj} - \delta_k \lambda_{kj} + 1 \end{pmatrix} \in \mathcal{K}_3, \quad j = 1, \ldots, n,\ k = 1, \ldots, K. \tag{18}$$

The decision variables in model (12)-(18) are

$$d_1, d_2, \tilde{u}, \gamma, \tau, r, \zeta_1, \cdots, \zeta_K, \delta_1, \cdots, \delta_K, s_1, \cdots, s_K, h_1, \cdots, h_K, \tag{19}$$

all the other quantities being considered constant.

A key step of the model is to determine a cost-effective initial expected zone so as to balance the message flooding cost with latency to reach the destination node D. The objective function (12) shows that the cost of choosing the expected region C is proportional to the distance $d_1 = \sqrt{\tilde{u}^T \tilde{u}}$ of the centre $\tilde{u}$ from the source node S and to the radius r.

The constraints (13) and (14) represent the condition of inclusion of disk C_0 in the first stage disk C; (15) and (16) represent respectively an upper bound of the distance of center of C from the origin 0 (position of S) and an upper bound on the radius of disk C. Finally (17) and (18) represent the condition of inclusion of the ellipse E_k into the disk C_k^* at scenario $k = 1, \ldots, K$.

3 Scenario Generation

In our numerical experiments we have generated the ellipsoids of the form

$$E = \{u \in \mathrm{R}^n : u^T H u + 2g^T u + \nu \leq 0\}, \tag{20}$$

where $H \in \mathrm{R}_s^{n \times n}$ is a given positive definite matrix, $g \in \mathrm{R}^n$ is a given vector, and $\nu \in \mathrm{R}$ is a given real number such that

$$\nu < g^T H^{-1} g. \tag{21}$$

Equation (20) can also be written as

$$E = \{u \in \mathrm{R}^n : \left\| H^{1/2}(u - u^0) \right\| \leq 1\}, \tag{22}$$

with $H = Q \Lambda Q^T$ the spectral decomposition of H, where $Q = (q^1 q^2 \ldots q^n)$ is the matrix whose columns are the eigenvectors of H and $\Lambda = Diag(\lambda_1; \ldots; \lambda_n)$ is the diagonal matrix of the corresponding eigenvalues with

$$u^0 = -H^{-1} g, \quad \rho = \sqrt{g^T H^{-1} g - \nu}.$$

In the two-dimensional case ($n = 2$) an ellipsoid (now an ellipse) is completely defined by its center $u^0 = (u_1^0, u_2^0)$, the angle φ between the first axis of the ellipse and the u_1-axis of the coordinate system, and the lengths σ_1, σ_2 of the two semi-axes.

For given φ, $u^0 = (u_1^0, u_2^0)$, σ_1, σ_2, we can represent the ellipsoid in the form (20) by setting

$$H = Q \begin{pmatrix} \sigma_1^{-2} & 0 \\ 0 & \sigma_2^{-2} \end{pmatrix} Q^T, \quad g = -Hu^0. \quad \nu = u^{0T} H u^0 - 1. \tag{23}$$

In our numerical experiments we have randomly generated these quantities, and then we have obtained ellipses of the form (20) by using (23). The quantities

$$u_{1,k}^0, u_{2,k}^0, \varphi_k, \sigma_{1,k}, \sigma_{2,k}, \quad k = 1, 2, \ldots, K, \tag{24}$$

corresponding to the ellipses (20) are obtained by using:

- the uniform distribution in the interval $(\sqrt{8} - 1, \sqrt{8} + 1)$ for generating $u_{1,k}^0$;
- the normal distribution $\mathcal{N}(0, 0.5)$ for generating $u_{2,k}^0$;
- the uniform distribution in the interval $[0, \frac{\pi}{2}]$ for generating φ_k;
- the normal distribution $\mathcal{N}(2, 1)$ for generating $s_{u_1,k}$;
- the normal distribution $\mathcal{N}(1, 0.5)$ for generating $s_{u_2,k}$.

In order to have realistic scenarios, we have imposed the following upper and lower bounds:

$$\begin{cases} u_2^{0,min} \leq u_{2,k}^0 \leq u_2^{0,max}, & k = 1, \ldots, K, \\ \sigma_1^{min} < \sigma_{1,k} \leq \sigma_1^{max}, & k = 1, \ldots, K, \\ \sigma_2^{min} < \sigma_{2,k} \leq \sigma_2^{max}, & k = 1, \ldots, K. \end{cases}$$

Ellipses scenarios are randomly generated in MATLAB 7.4.0, according to the method described above, where we have set

$$u_2^{0,min} = -1, \; u_2^{0,max} = 1, \; \sigma_1^{min} = \sigma_2^{min} = 0.1, \; \sigma_1^{max} = \sigma_2^{max} = 3.$$

The SSOCP model was implemented in GAMS 22.5, by using the second order cone programming solver from the software package Mosek (http://www.mosek.com/).

In our computational experiments we have supposed that the scenarios are equiprobable; furthermore we have fixed the location $l = (2, 0)$ of the node D at initial time $t_0 = 0$, its lowest speed $v = 1$, and the final time $t_1 = 1$, so that the disk C_0 has center $l = (2, 0)$, and unitary radius,

$$C_0 = \{u = (u_1, u_2) \in \mathrm{R}^2 : u_1^2 + u_2^2 - 4u_1 + 3 \leq 0\}.$$

4 Sensitivity analysis and numerical results

In this section we first compare the SSOCP approach (12)–(18) and the SSDP model proposed by Ariyawansa and Zhu (2006) and then we analyze the sensitivity with respect to input parameters.

4.1 *Comparison of SSOCP versus SSDP (Ariyawansa and Zhu)*

We set the first stage costs $\alpha = \beta = 1$, meaning that the cost of choosing the radius of the first stage circle is the same than the cost of choosing the location of its center $\tilde{u}$, and the second stage cost $\beta_1 = 1.5$ paying more for a recourse decision.

Figure 2 refers the expected and requested zone solutions C and $\tilde{C}$ for the SSOCP model and Figure 3 shows C and $\hat{C}$ for the SDSP by Ariyawansa and Zhu (2006). In the SSDP model we can observe that the higher second stage cost forces the first stage circle C to contain all the scenarios ellipses

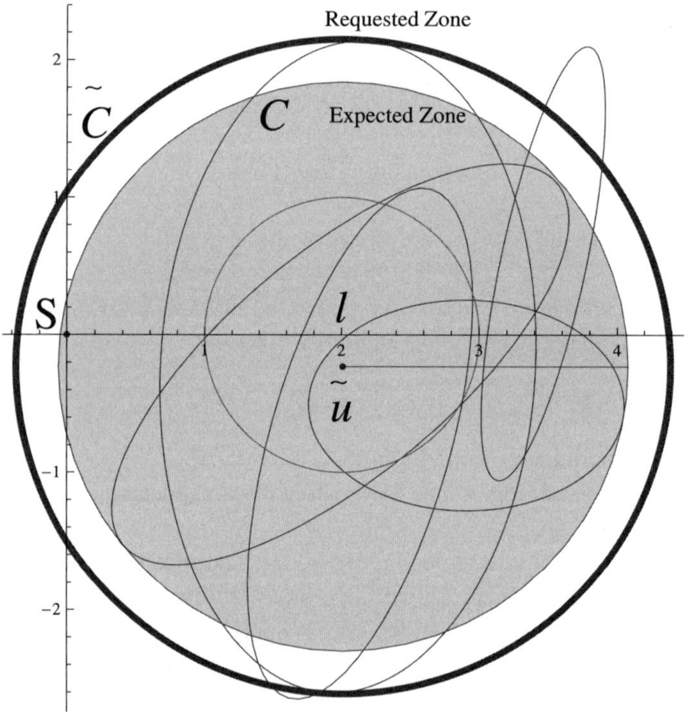

Figure 2 SSOCP Model Solution.

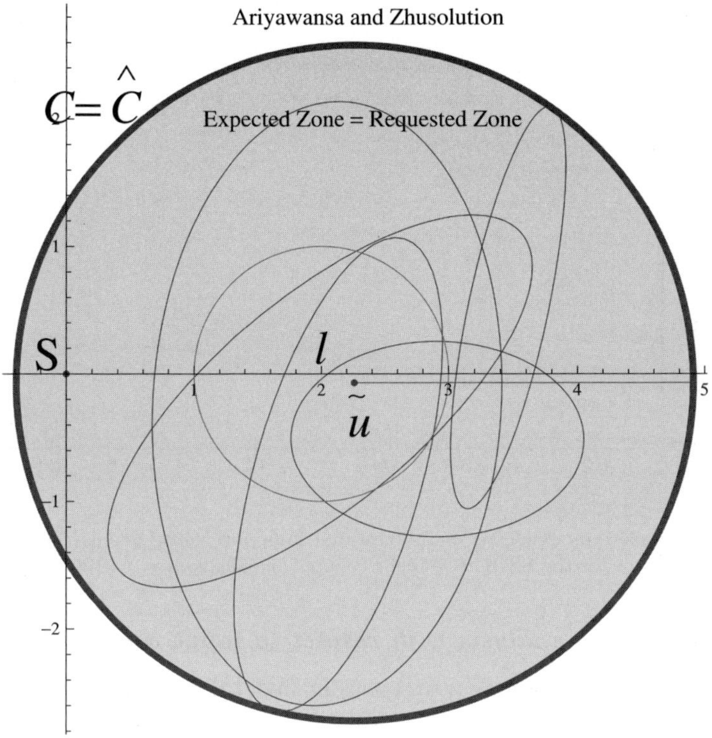

Figure 3 SSDP Model Solution Proposed in Ariyawansa and Zhu (2006).

Table 1 Case of Five Scenarios: Comparison between the SSOCP Solution and the SSDP Solution Proposed in Ariyawansa and Zhu (2006).

	d_1	d_2	$\tilde{u}_1$	$\tilde{u}_2$	γ	τ	r	**Obj. value**
SSOCP	2.02	4.27	2.01	−0.23	−0.17	2.49	2.07	8.52
SSDP (see Ariyawansa and Zhu (2006))	2.26	7.04	2.26	−0.07	−1.91	2.65	2.65	9.31

E_k, $k = 1, \ldots, 5$ and consequently $C = \hat{C}$. This does not happen for the SSOCP model where the requested zone is the average over all second stage circles C_k^* with a consequent lower cost than SSDP as reported in Table 1 (8.52 instead of 9.31). As we will discuss in the next section, in SSOCP model, the expected zone will coincide with the requested zone $C = \tilde{C}$ for $\beta \geq 2.7$.

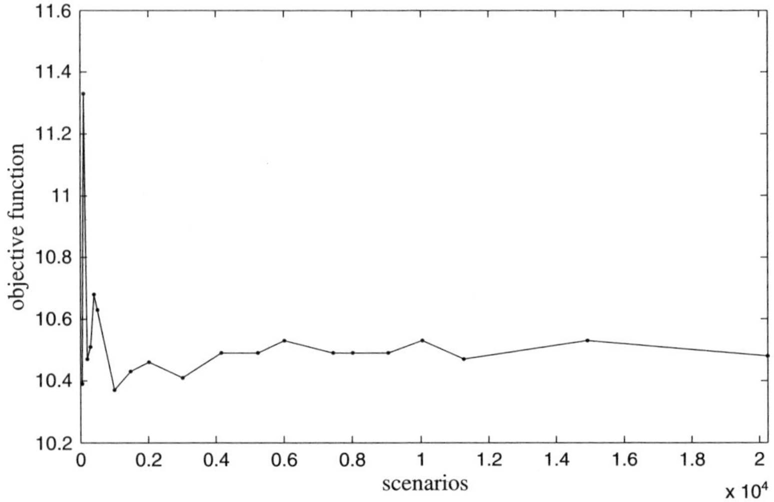

Figure 4 Sensitivity Analysis of the Optimal Function Value Versus the Number of Ellipses Scenarios for the SSOCP Model Proposed in Maggioni *et al.* (2009).

4.2 *Sensitivity analysis with respect to input parameters*

Among the input parameters we consider first (a) the number of scenarios in the scenario tree and then (b) different costs of flooding and making corrections on the radius of the second stage circles C_k^*.

(a) In order to validate the model, we have first analyzed the sensitivity of the solutions to the number of scenarios up to 20250. Figures 4 and 5 show the stabilization of the objective function as the number of scenarios increases respectively in the SSOCP (12)-(18) and the SSDP model proposed in Ariyawansa and Zhu (2006). From the results we can deduce an *in-sample stability,* i.e. the optimal objective values are approximately the same when different numbers of scenarios are considered (for a definition of in-sample stability see Kaut and Wallace 2007). In particular, irrespectively of the scenario tree considered, the SSDP approach (Ariyawansa and Zhu, 2006) has higher optimal objective values than the SSOCP model.

(b) We keep the values of first stage costs fixed to $\alpha = \beta = 1$ and we let the second stage cost β_1 varying from 0 to 4.

Sensitivity analysis versus the second stage flooding cost β_1 is reported in Figures 6 and 7. From Figure 6 we note that for a low value of $\beta_1 \in [0, 0.4]$ the expected zone C coincides with the initial

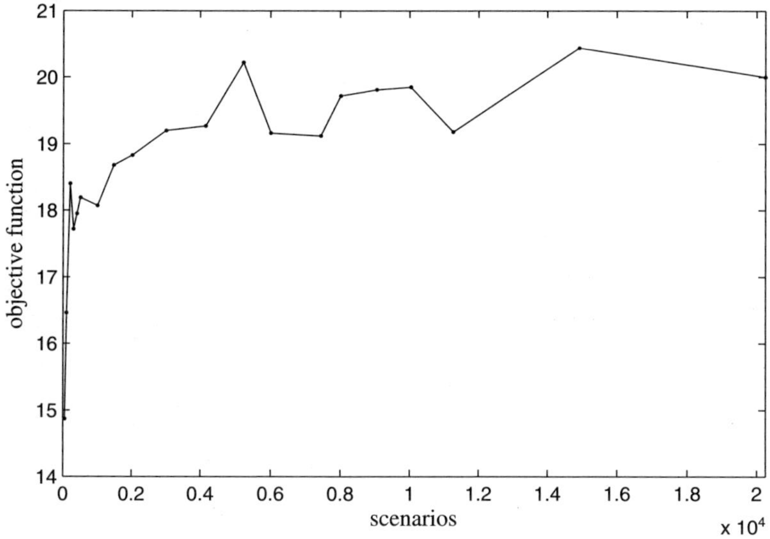

Figure 5 Sensitivity Analysis of the Optimal Function Value Versus the Number of Ellipsoid Scenarios for the SSDP Model Proposed in Ariyawansa and Zhu (2006).

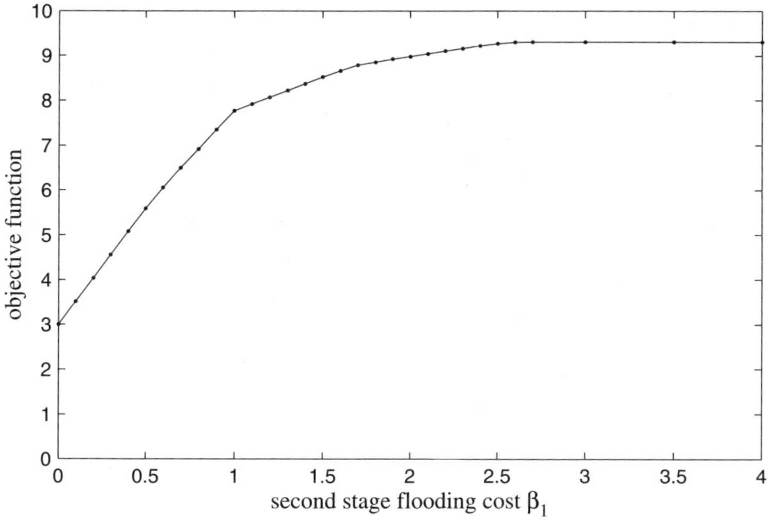

Figure 6 Sensitivity Analysis of the Objective Function Versus the Second Stage Flooding Cost $\beta_1 \in [0, 4]$ for the SSOCP Model Proposed in Maggioni *et al.* (2009).

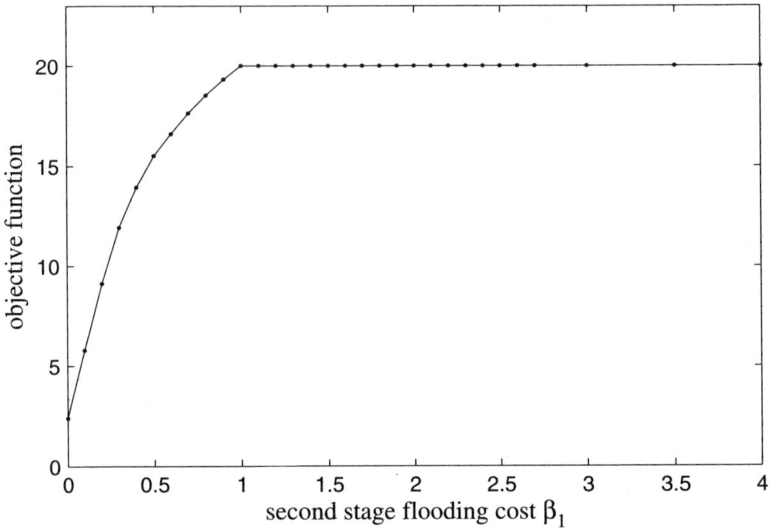

Figure 7 Sensitivity Analysis of the Objective Function Versus the Second Stage Flooding Cost $\beta_1 \in [0, 4]$ for the SSDP Model Proposed by Ariyawansa and Zhu (2006).

smallest disk C_0 with an associated total cost equal to 3; this means that it is more convenient to enlarge the radius of each second stage disk C_k^* instead of moving the centre in the first stage. Conversely, for $\beta_1 \geq 2.7$, it is too expensive to enlarge the radius at the second stage and consequently the expected zone C already contains all the K ellipses E_k, $(k = 1, \ldots, K)$ and the disk C_0; the associated total cost stabilizes on the value of 9.31 (see Figure 6).

From Figure 7 we observe that in the SSDP model (Ariyawansa and Zhu, 2006) for $\beta_1 \geq 1$ the expected zone C already contains all the scenarios ellipses E_k, $k = 1, \ldots, K$ (against $\beta_1 \geq 2.7$ for SSOCP). For a detailed sensitivity analysis of the SSDP model (see Zhu *et al.*, 2011).

To check the importance of modeling the randomness of the parameters, we compare the optimal solutions and objective value of the stochastic model with those obtained from the corresponding *deterministic* model (or *expected value problem*), where the unique scenario is represented by an ellipse with the center $(u_1^{mean}, u_2^{mean}) = (2.5056, -0.2461)$, the angle $\varphi^{mean} = 1.2866$, the semi-axes $\sigma_1^{mean} = 1.7448$, $\sigma_2^{mean} = 0.8586$, given respectively by the mean of the centers, of the angles and of the semi-axes of the ellipses E_k, $k = 1, \ldots, 5$ (for more details on the ellipses scenarios see Maggioni *et al.*, 2009).

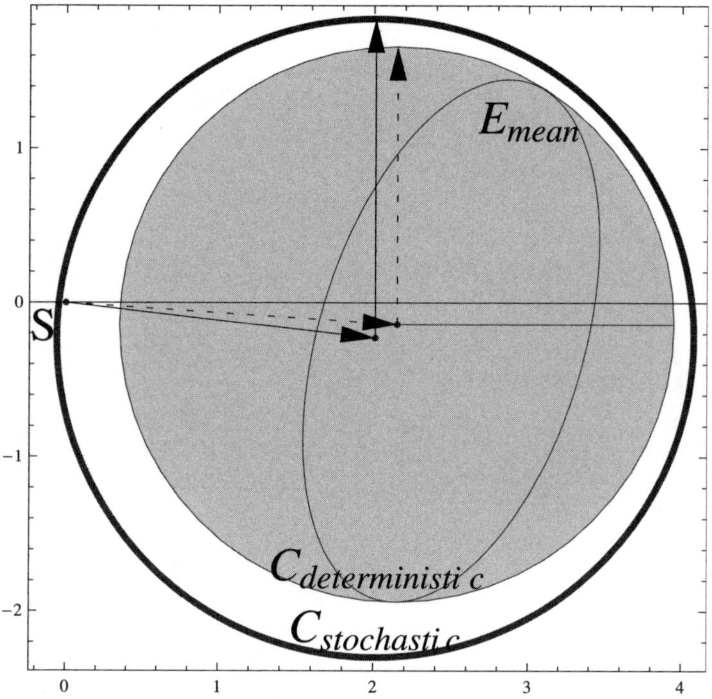

Figure 8 Comparison between the First Stage Stochastic Solution $C_{stochastic}$ and the *Mean Value Solution* (grey circle $C_{deterministic}$) for the Mean Scenario E_{mean}, Mean of the Ellipses Scenarios $E_1, \ldots, E_5$ Plotted in Figure 9.

Because in a deterministic problem the future is completely known, a recourse action is not required and the consequent total cost is lower (5.38 instead of 8.52 of the stochastic case). The resulting expected region (see Figures 8, 9 and Table 2) appears to be too small (the radius r is 1.8 instead of 2.07 of the stochastic case) and the centre is located furthermost from the sender node S (the distance $d_1 = 2.16$ instead of 2.02). For an analysis of the quality of the Expected Value Solution, see Maggioni and Wallace (2010).

Figure 10 refers to the *Value of Stochastic Solution VSS*, defined as

$$VSS = EEV - RP \tag{25}$$

where EEV is the optimal value of the stochastic model using as first stage variable the mean value solution and RP is the optimal value of the stochastic model (see Birge, 1970 and Birge and Louveaux, 1997). Figure 10 shows that for low β_1 we save time by solving the

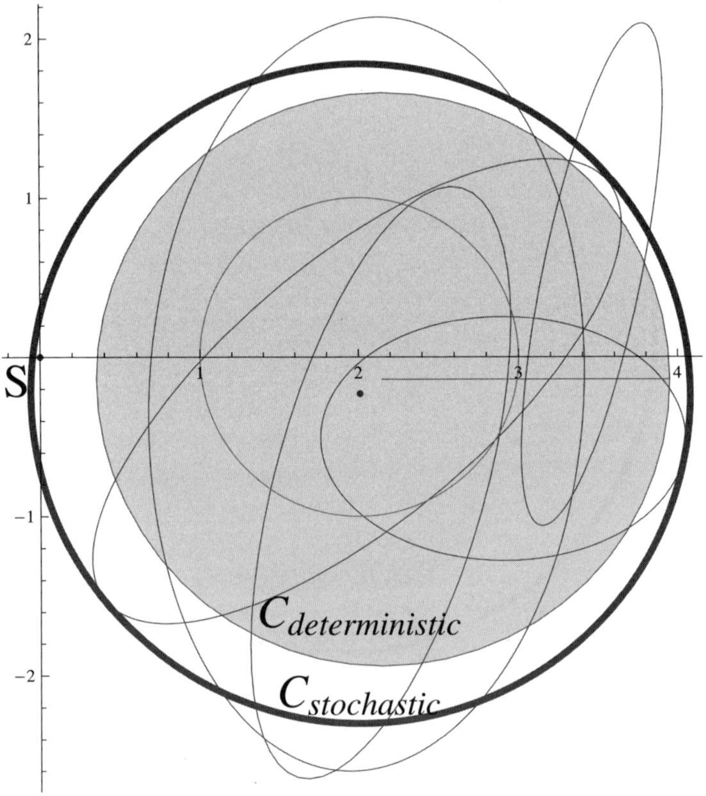

Figure 9 Comparison between the First Stage Stochastic Solution $C_{stochastic}$ and the *Mean Value Solution* (grey circle $C_{deterministic}$) for the Mean Scenario E_{mean} (Plotted in 8), Mean of the Ellipses Scenarios $E_1, \ldots, E_5$.

Table 2 Case of Five Scenarios: Comparison between the Stochastic and Deterministic Disk C.

	d_1	d_2	$\tilde{u}_1$	$\tilde{u}_2$	γ	τ	r	**Obj. value**
deterministic	2.16	3.23	2.15	−0.14	1.42	1.75	1.80	5.38
stochastic SSOCP	2.02	4.27	2.01	−0.23	−0.17	2.49	2.07	8.52

deterministic mean value problem instead of the complex stochastic one. Moreover, the monotonically increasing behavior of the VSS means that the the deterministic first stage radius is not enough to contain all the scenarios ellipses. As a consequence, in order to establish a successful communication between the nodes S and D, the radius of the circle has to be enlarged with an higher second stage price.

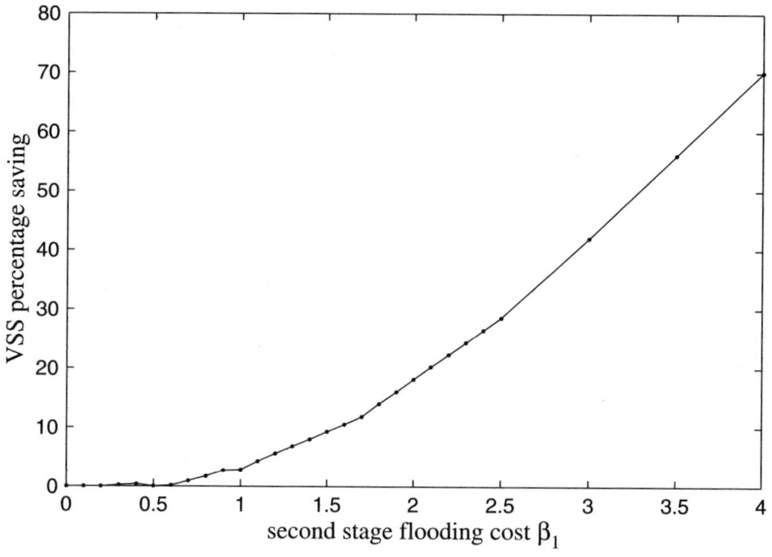

Figure 10 Sensitivity Analysis of the *Value of Stochastic Solution VSS* (% saving) Versus the Second Stage Flooding Cost $\beta_1 \in [0, 4]$.

Notice that the monotonically increasing behavior of the VSS with respect to second stage costs, is a general feature of cost minimization problems where the recourse action is necessary to fulfill all the second stage constraints.

We finally evaluate the *Expected Value of Perfect Information* $EVPI$ (Birge and Louveaux, 1997) defined as

$$EVPI = RP - WS \tag{26}$$

where WS is the optimal solution when the uncertainty is revealed. $EVPI$ represents how much we should be ready to pay in return for complete information in advance, about the direction and velocity of the destination node D. We refer to Figure 11: when the second stage flooding cost is low ($\beta_1 \in [0, 0.2]$) $WS = RP$, that is, the first stage circle is chosen in order to coincide with the initial smallest disk C_0 and at each scenario k the ellipse E_k is recovered by enlarging the second stage circle C_k^*. Notice that, for a second stage flooding cost $\beta_1 < 1$, the deterministic problem solved at each scenario is still a two stage problem because of the updating of the circle C_k^*.

From Figure 11 we can see that the percentage saving by postponing the decision, monotonically increases with β_1, until it stabilizes on the

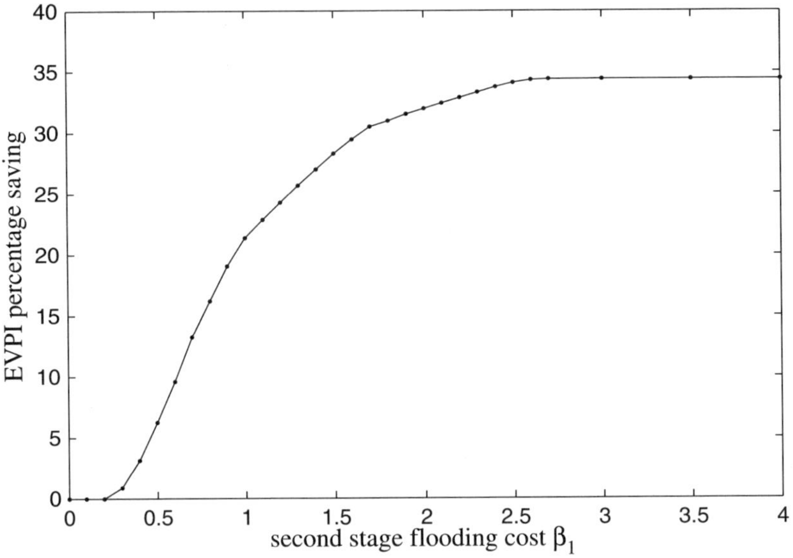

Figure 11 Sensitivity Analysis of the *Expected Value of Perfect Information EVPI* (% saving) Versus the Second Stage Flooding Cost $\beta_1 \in [0, 4]$.

value 34.37%. The stabilizing behavior means that for a second stage flooding cost greater than 1 the first stage circle already includes all the scenarios ellipses and disk C_0. A summary of the results discussed in this section can be found in Maggioni *et al.* (2010).

5 Conclusion

In this paper sensitivity analysis is applied to a stochastic second order cone program for mobile ad hoc networks, by considering the influence of different number of scenarios in the solution and different costs of flooding and making corrections on the radius of the requested zone. The results show a stabilization of the solution beyond 4000 scenarios. Further, a small flooding cost coefficient implies a large expected zone and in the opposite case, in order to minimize the number of nodes involved in the flooding, a small circle is picked up. Moreover, a higher cost on the radius of the requested zone puts more penalty on the route discovery latency indicating that it is preferable to find node D at the first stage. As a consequence the expected zone corresponds to the largest area already containing all the ellipses scenarios. This fact is observed already at lower flooding costs in the SSDP model proposed by Ariyawanza and Zhu (2006). We have also considered classical measures in stochastic programming such as the Expected Value of Perfect Information *EVPI* and Value of Stochastic Solution *VSS*: *EVPI* stabilizes to a plateau region because of the rejection of the upgraded circle at the second stage. The convex behavior of the *VSS* versus the latency penalty shows the range of values in which we save time by solving the deterministic mean value problem instead of using a complex stochastic structure. The results obtained show that also in a stochastic context sensitivity analysis represents a useful tool for a deeper understanding of the considered problem and to reveal the randomness of the model hidden in the wrong choice of parameters.

References

Alizadeh, F. and D. Goldfarb. Second-order cone programming, *Mathematical Programming* Ser. B, 95(1): pp. 3–51, 2003.

Ariyawansa, K. A. and Y. Zhu. A preliminary set of applications leading to stochastic semidefinite programs and chance-constrained semidefinite programs, *Applied Mathematical Modelling*, 35(5): pp. 2425–2442, 2006.

Ariyawansa, K. A. and Y. Zhu. Stochastic semidefinite programming: a new paradigm for stochastic optimization, *4OR, A Quarterly Journal of Operations Research*, 4(3): pp. 239–253, 2006.

Birge, J. R. The Value of Stochastic Solution in Stochastic Linear Programs with Fixed Recourse, *Mathematical Programming*, 24: pp. 314–325, 1970.

Birge, J. R. and F. Louveaux. *Introduction to stochastic programming*, New York: Springer Verlag, 1997.

Higle, J. L. and S. W. Wallace. Sensitivity Analysis and Uncertaintly in Linear Programming, *Interfaces*, 33(4): pp. 53–60, 2003.

Kaut, M. and S. W. Wallace. Evaluation of scenario generation methods for stochastic programming, *Pacific Journal of Optimization*, 3(2): pp. 257–271, 2007.

Ko, Y.-B. and N. H. Vaidya. Location-Aided Routing (LAR) in mobile ad hoc networks, *Wireless Network*, 6(4): pp. 307–321, 2000.

Maggioni, F., F. Potra, M. Bertocchi and E. Allevi. Stochastic second order cone programming in mobile ad hoc networks, *Journal of Optimization Theory Applications*, 143: pp. 309–328, 2009.

Maggioni, F. and S. W. Wallace. Analyzing the quality of the expected value solution in stochastic programming, *Annals of Operations Research*, 1–18, ISSN: 0254-5330, (2010) doi: 10.1007/s10479-010-0807-x.

Maggioni, F., S. W. Wallace, M. Bertocchi and E. Allevi. Sensitivity Analysis in Stochastic Second Order Cone Programming for Mobile Ad Hoc Networks, *Sixth International Conference on Sensitivity Analysis of Model Output, Procedia Social and behavioral Sciences*, 2(5): pp. 7704–7705, 2010.

Ruszczyński, A. and A. Shapiro. *Stochastic programming.* Series Handbooks in Operations Research and Management Science. Amsterdam: Elsevier Science B.V., 2003).

Shapiro, A. Stochastic programming approach to optimization under uncertainty, *Mathematical Programming Series B*, 112(1): pp. 183–220, 2008.

Vyas, N. Mobility pattern based routing algorithm for mobile ad hoc wireless networks, Ms Thesis, CSE Department, Florida Atlantic University, August 2000.

Wallace, S. W. Decision Making Under Uncertainty: is sensitivity analysis of any use?, *Operation Research*, 48(1): pp. 20–25, 2000.

Zhu, J., J. Zhang and K. Patel. Location-Aided Routing with Uncertainty in Mobile Ad-hoc Networks: A Stochastic Semidefinite Programming Approach, *Mathematical and Computer Modelling*, 53(11): pp. 2192–2203, 2011.

Chapter 18

Stochastic Frequency Assignment Problem

Wadie Benajam*, A. Gaivoronski† and Abdel Lisser‡

Summary

We consider the Frequency Assignment Problem (FAP) in the case when the important components of problem formulation are uncertain and can be modelled by random variables. In particular, we look at the case when the interference between sources and/or demand for frequences have random nature. Several optimization models which belong to stochastic programming approach are considered. Solution schemes based on semidefinite relaxations are developed. Presented results including numerical experiments confirm that stochastic programming modeling tools coupled with semidefinite relaxations constitute a promising approach for solving frequency assignment problems under uncertainty.

1 Introduction

Frequency Assignment Problem (FAP) is a whole class of combinatorial optimization problems which deal with assignment of frequencies in wireless telecommunication networks to different transmitters. The growing interest in this problem is related to increasing importance of wireless networks and related services both in the telecommunication industry and society as a whole. The demand for mobile telecommunication services grows with impressive rates, but the basic resource used for transmission of communication flows (radio spectrum) remains virtually unchanged. Therefore, in order to be functional the wireless networks must reuse this resource in

*LRI, Université Paris-Sud. PCRI, Bat. 650, Rue Noetzlin, 91190 Gif-sur-Yvette, France, wadie.benajam@lri.fr.
†Department of Industrial Economy and Technology Management, Norwegian University of Science and Technology, Trondheim, Norway, alexei.gaivoronski@iot.ntnu.no.
‡LRI, Université Paris-Sud. PCRI, Bat. 650, Rue Noetzlin, 91190 Gif-sur-Yvette, France, abdel.lisser@lri.fr.

different geographical locations without jeopardizing the quality of service. This can be a challenging task because if the same radio frequencies are used in neighboring locations, this may lead to severe interference problems. The objective of FAP is to provide optimization tools for wireless network planners which will allow efficient and even optimal use of limited resource of radio spectrum while maintaining the required quality of service. Usually this results in challenging combinatorial optimization problems, see (Aardal *et al.*, 2003; Eisenblätter *et al.*, 2002; Murphey *et al.*, 1999) for a survey of different FAP formulations, the history of research and further references.

One of important features in many applications of FAP to design of mobile networks is the uncertainty about the data which describe the problem. There are multiple sources of this uncertainty. The propagation conditions for radio signals can change depending on time of the day or season. The demand for mobile services experience daily and seasonal variations. Even more importantly, the demand changes due to the introduction of new services and subscription of new users. Unpredictable failures may occur in the network itself. This inherent uncertainty should be taken into account in order to design mobile networks which can be resilient to its effects. One way to do this is to come up with a probabilistic description of uncertainty and to use appropriate modeling and optimization tools specifically designed for solving the optimization problems under uncertainty, and in particular stochastic programming (Birge and Louveaux, 1997; Ermoliev and Wets, 1988; Kall and Wallace, 1994; Wallace and Ziemba, 2005; Wets, 1982), see also (Andrade *et al.*, 2004, 2005; Bonatti *et al.*, 1994; Gaivoronski, 2005; Sen *et al.*, 1994; Tomasgard *et al.*, 1998) for different aspects of stochastic programming models in the network design and planning.

For this reason extensions of FAP to stochastic case have considerable application value. Despite this, practically all research on FAP was dedicated to deterministic case. The reason for this are serious computational difficulties which are associated even with the deterministic FAP. The stochasticity adds another layer of complexity to already difficult problem which could lead to computationally intractable models. Presumably, this deterred researches from addressing the stochastic FAP. The objective of this paper is to show that despite these difficulties it is possible to develop tractable stochastic FAP models based on judicious combination of stochastic programming techniques with the recent advances in combinatorial optimization, and in particular with semidefinite programming (Wolkowicz *et al.*, 2000). In this direction we describe several stochastic formulations of FAP, develop the algorithm based on combination of Benders decomposition or L-shaped method (Van Slyke and Wets, 1969; Wets, 1982) with the semidefinite relaxation and provide results of numerical experiments. The models developed in this paper belong to the general class of stochastic integer programming models which is an active field of research (Ahmed *et al.*, 2004; Caroe and Schultz, 1999; Klein Haneveld and Van der Vlerk, 1999; Laporte and Louveaux, 1993; Schultz *et al.*, 1998; Sen and Sherali, 2004; Gaivoronski *et al.*, 2011).

The rest of the paper is organized as follows. We start with the definition of deterministic FAP in Section 2 which is used as the starting point for developing stochastic FAP models. Semidefinite relaxation of this problem is described in Section 3, it serves as a building block for developing the solution algorithm for stochastic FAP. Section 4 is dedicated to the stochastic FAP problems. We start with the simplest one which can be classified as an expectation stochastic programming problem for frequency allocation. This is followed by more involved formulation which takes consideration of possible reallocations of frequencies at a later dates in response to changing demand. This problem belongs to the family of stochastic programs with recourse. After that the numerical scheme based on combination of Benders decomposition and semidefinite relaxation is described in Section 5. Encouraging results of initial numerical experiments with these problems are reported in Section 6. The paper concludes with summary in Section 7.

2 Deterministic frequency assignment problem

In this section we present one specific and yet typical formulation of deterministic FAP problem which will be utilized for development of stochastic variants of this problem.

It is necessary to assign n frequencies to m sites which in GSM terminology are called base stations in such a way that demand for frequencies is satisfied and the interference between different frequencies is minimal. The frequencies are represented as a set of positive integers $i = 1, \ldots, n$. For every pair (i, j) of frequencies the distance ρ_{ij} is defined:

$$\rho_{ij} = |i - j|$$

Let us introduce further the following notations:

d_i — demand for frequencies at site i, this is a nonnegative integer number;

w_{ij}^{kl} — interference attained if frequency k is assigned to site i and frequency l is assigned to site j;

x_i^k — decision binary variable which equals 1 if frequency k is assigned to site i and zero otherwise;

In these notations the problem is to find assignment which minimizes the total interference, namely

$$\min_{x_i^k} \sum_{i,j,k,l} w_{ij}^{kl} x_i^k x_j^l \tag{1}$$

under the following constraints:

1. *Demand satisfaction*

$$\sum_k x_i^k = d_i, \quad \forall i \tag{2}$$

2. Close frequencies are not assigned to adjacent sites due to high interference

$$x_i^k + x_j^l \leq 1, \quad \forall i, j, k, l : i, j \text{ adjacent}, \quad |k - l| \leq c_1 \tag{3}$$

3. Close frequencies are not assigned to the same site. This is again due to high interference

$$x_i^k + x_i^l \leq 1, \quad \forall i, k, l : |k - l| \leq c_2 \tag{4}$$

One way to deal with constraints (3), (4) is to assign high interference numbers to frequencies and site combinations which violate these constraints. Then the problem is reduced to the problem of minimization of objective (1) under the demand satisfaction constraints (2).

3 Semidefinite relaxation of deterministic FAP

The problem (1)–(4) belongs to the class of quadratic assignment problems (Cela, 1998) which is one of the most challenging classes of combinatorial optimization problems. Even in the deterministic case the exact solution of such problems is beyond reach in large majority of nontrivial cases. For this reason different approximate approaches were developed. We do not pursue here the aim of giving a comprehensive comparison of these methods. Our objective is different: to select a promising method which can be incorporated into a larger computational scheme for solution of the stochastic FAP problem.

Our choice for such approximation scheme is semidefinite relaxation. The reason for this choice is that the quadratic form of the objective function lends itself naturally to semidefinite programming (Goemans and Rendl, 2000). In addition, resulting linear programming problems can be incorporated relatively easily into iterative decomposition techniques of stochastic programming (Birge and Louveaux, 1997).

Let us derive a semidefinite relaxation of the problem (1)–(4). The objective is to lift the original problem into the space of symmetric real matrices where the relaxed problem will assume the form

$$\begin{gathered}\min_Z C \cdot Z \\ A_i \cdot Z = b_i, \quad i = 1, \ldots, m_1 \\ Z \succeq 0\end{gathered} \tag{5}$$

where $Z, C, A_i \in S^{n_1}$, the space of real, symmetric $n_1 \times n_1$ matrices. The inequality $Z \succeq 0$ means Z is positive semidefinite. The notation $C \cdot Z$ stands for the inner product of the symmetric matrices C and Z:

$$C \cdot Z = \sum_{i=1}^{n_1} \sum_{j=1}^{n_1} C_{ij} Z_{ij}$$

The relaxed problem (3) can be solved by linear programming methods after transforming the semidefinite constraint $Z \succeq 0$ into the equivalent set of linear constraints (Wolkowicz *et al.*, 2000).

In order to give a SDP relaxation (3) for the FAP model given by (1), (2), (3) and (4) we assume $c_1 = 2$ and $c_2 = 3$. This covers the most important practical cases, the general values of c_1, c_2 can be treated along the same lines. Let N be the set of sites and M the set of frequencies. We also introduce a new binary variable $y_{i,j}^{l,k} = x_i^l x_j^k$. The FAP can be written as:

1. Demand satisfaction

$$\sum_{l=1}^{m} y_{i,i}^{l,l} = d_i, \quad \forall\, i \in N$$

2. Close frequencies are not assigned to the same site again due to high interference

$$y_{i,i}^{l,k} = 0, \quad \forall\, i \in N, \quad \text{and} \quad |l - k| \leq 3.$$

or

$$\begin{cases} y_{i,i}^{l,l+1} = 0 \\ y_{i,i}^{l,l+2} = 0 \quad \forall i \in N, \ \forall l \in M \\ y_{i,i}^{l,l+3} = 0 \end{cases}$$

3. Close frequencies are not assigned to adjacent sites due to high interference

$$\begin{cases} y_{i,j}^{l,l} = 0 \\ y_{i,j}^{l,l+1} = 0 \quad 1 \leq i,j \leq n, \ j \text{ co-site of } i \text{ and } l \in M \\ y_{i,j}^{l,l+2} = 0 \end{cases}$$

These constraints can be also written by the mean of the matrix trace by choosing a matrix accordingly. Let

$$Y = \begin{bmatrix} Y_{1,1} & Y_{1,2} & \cdots & Y_{1,n} \\ Y_{2,1} & Y_{2,2} & \cdots & Y_{2,n} \\ \vdots & \vdots & \ddots & \vdots \\ Y_{n,1} & Y_{n,2} & \cdots & Y_{n,n} \end{bmatrix}$$

where

$$Y_{i,j} = \begin{bmatrix} y_{i,j}^{1,1} & y_{i,j}^{1,2} & \cdots & y_{i,j}^{1,m} \\ y_{i,j}^{2,1} & y_{i,j}^{2,2} & \cdots & y_{i,j}^{2,m} \\ \vdots & \vdots & \ddots & \vdots \\ y_{i,j}^{m,1} & y_{i,j}^{m,2} & \cdots & y_{i,j}^{m,m} \end{bmatrix}$$

Let W be the interference matrix such that:

$$W = \begin{bmatrix} W_{1,1} & W_{1,2} & \cdots & W_{1,n} \\ W_{2,1} & W_{2,2} & \cdots & W_{2,n} \\ \vdots & \vdots & \ddots & \vdots \\ W_{n,1} & W_{n,2} & \cdots & W_{n,n} \end{bmatrix}$$

where

$$W_{i,j} = \begin{bmatrix} w_{i,j}^{1,1} & w_{i,j}^{1,2} & \cdots & w_{i,j}^{1,m} \\ w_{i,j}^{2,1} & w_{i,j}^{2,2} & \cdots & w_{i,j}^{2,m} \\ \vdots & \vdots & \ddots & \vdots \\ w_{i,j}^{m,1} & w_{i,j}^{m,2} & \cdots & w_{i,j}^{m,m} \end{bmatrix}$$

Then the SDP relaxed FAP can be written as:

$$(SDPFAP)\begin{cases} Min \quad Trace(W*Y) \\ s.c \\ \qquad Trace(Y_{ii}) = d_i \quad i \in N \\ \qquad \begin{cases} y_{i,i}^{l,l+1} = 0 \\ y_{i,i}^{l,l+2} = 0 \qquad i \in N,\ l \in M \\ y_{i,i}^{l,l+3} = 0 \end{cases} \\ \qquad \begin{cases} y_{i,j}^{l,l} = 0 \\ y_{i,j}^{l,l+1} = 0 \qquad 1 \le i,j \le n, j \text{ co-site of } i \text{ and } l \in M \\ y_{i,j}^{l,l+2} = 0 \end{cases} \\ \qquad diag(Y) = y \\ \qquad Y - yy^t \succeq 0 \end{cases} \tag{6}$$

where $y = (x_i^l)_{1 \le l \le m, 1 \le i \le n}$ is the decision variable vector.

Let Z_{il} and Z_{ijl} be matrices which define each co-site and co-station constraint respectively. The SDPFAP can be also written as:

$$(SDPFAP1)\begin{cases} Min \quad Trace(W*Y) \\ s.c \\ \qquad Trace(Y_{ii}) = d_i \qquad i \in N \\ \qquad Trace(Z_{il}Y_{il}) = 0 \qquad \forall i,l \\ \qquad Trace(Z_{ijl}Y_{ijl}) = 0 \quad \forall i,j,l,\ i \text{ and } j \text{ co-site} \\ \qquad diag(Y) = y \\ \qquad Y - yy^t \succeq 0 \end{cases} \tag{7}$$

which is a specific case of the formulation (3).

4 Stochastic problem

Several sources of uncertainty are present in the frequency assignment which potentially can make solution of deterministic problem (1)–(2) inadequate. The main such sources are the following:

1. The interference between sources in reality changes considerably with changing atmospheric conditions, time of the day and of the year, etc. For this reason it is more realistic to assume that the interference between the frequency k assigned to site i and and the frequency l assigned to

site j is described by the joint probabilistic distribution $H_{ij}^{kl}(w)$ instead of numbers w_{ij}^{kl}.
2. The demand for frequencies have daily, weakly and seasonal patterns and on top of such patterns there is a considerable random variation. Any treatment of demand as a known constant is bound to bring about inadequate results. We assume that the demand at sites $i = 1, \ldots, m$ is a random vector with joint distribution $P(d)$ which is concentrated in a finite number of discrete points.

Another source of uncertainty may come in the form of equipment failures but we are not going to consider it here.

In the rest of this section we are going to formulate several optimization problems which has as solutions the robust frequency assignment policies taking into account this uncertainty. Two cases are distinguished:

— *Non-adaptive case* when the frequency assignment is done on the basis of the probabilistic description of demand and the interference patterns before any specific realization of these random variables becomes known. In stochastic programming terminology this is the expectation model which in this particular case can be reformulated as a stochastic problem with simple recourse (Birge and Louveaux, 1997).
— *Adaptive case.* In this case the initial assignment is done before a more specific information about demand and interference patterns becomes available. However, after specific realization of these random variables becomes known one can update the frequency assignment taking into account this new information. The assignment can change again after the new change of the demand and/or interference patterns. In stochastic programming terminology this will be a stochastic FAP problem with recourse (Birge and Louveaux, 1997).

4.1 *Non-adaptive case*

The frequency assignment decision is made in this case on the basis of probability distributions of demand and interference patterns before any specific demand or interference realization becomes known. This assignment is not changed afterwards. For this reason some part of the demand will not be satisfied. Therefore our objective function will contain two components: the first component will describe the average interference while the second component defines the average non-satisfied demand. In order to put such disparate things in the same objective function we have to make them commeasurable. One way of doing so is to translate them into costs. Indeed, the interference negatively influence the quality of service (QOS) and may lead to some calls being lost and to user dissatisfaction with the quality of service. We assume that this deterioration of QoS can be expressed as the opportunity costs for the service provider. Similarly, the underassignment of frequencies to a particular site will result in the deterioration of QoS and the lost demand not dissimilar to the deterioration of QoS due to interference. We assume again that this can be expressed by the opportunity costs. In the modeling terms this results in the weighting coefficients to be assigned to the different components of the objective function. The resulting problem is the following

$$\min_{x_i^k} \mathbb{E}_w \sum_{i,j,k,l} w_{ij}^{kl} x_i^k x_j^l + g\mathbb{E}_d \sum_i \max\left\{0, d_i - \sum_k x_i^k\right\} \tag{8}$$

with no other constraints except possible constraints (3)–(4). Here g is the coefficient which makes commeasurable the two components of the objective function as outlined in the above discussion, $\mathbb{E}_w$ denotes the expectation with respect to random interference patterns w_{ij}^{kl} and $\mathbb{E}_d$ denotes the expectation with respect to random demand d_i. The first term describes the contribution of the interference into the deterioration of QoS, while the second term reflects the penalties for not meeting demand. It is assumed that the overassignment of frequencies does not bring any penalty, although any such penalties can be easily accounted in the model. This expression can be further transformed as follows. First of all observe that

$$\mathbb{E}_w \sum_{i,j,k,l} w_{ij}^{kl} x_i^k x_j^l = \sum_{i,j,k,l} (\mathbb{E}_w w_{ij}^{kl}) x_i^k x_j^l = \sum_{i,j,k,l} \bar{w}_{ij}^{kl} x_i^k x_j^l$$

where we denoted

$$\bar{w}_{ij}^{kl} = \mathbb{E}_w w_{ij}^{kl}$$

In other words, the detailed information about the distribution of the interference coefficients is not exploited in this model, all that is necessary to know is their average values $\bar{w}_{ij}^{kl}$. For this reason we do not consider this type of uncertainty in our subsequent exposition.

Moreover, since each given demand d_i take a value from a finite set of integer numbers, the joint demand distribution $P(d)$ is concentrated in a finite number of points $d^r = (d_1^r, \ldots, d_m^r)$ with weights p_r, $r = 1, \ldots, R$. These points will be called *demand scenarios*. Note that in the case when the total number R of demand scenarios is too high, one can approximate the original distribution with distribution $\hat{P}(d)$ concentrated in a smaller number of points. Thus, the problem (8) is equivalent to the following:

$$\min_{x_i^k} \sum_{i,j,k,l} \bar{w}_{ij}^{kl} x_i^k x_j^l + g \sum_r \sum_i p_r \max \left\{ 0, d_i - \sum_k x_i^k \right\} \tag{9}$$

which, after introducing auxiliary variables v_i^r for each of the sites $i = 1, \ldots, m$ and for each scenario $r = 1, \ldots, R$ brings the problem back to familiar quadratic form:

$$\min_{x_i^k, v_i^r} \sum_{i,j,k,l} \bar{w}_{ij}^{kl} x_i^k x_j^l + g \sum_r \sum_i p_r v_i^r \tag{10}$$

$$\sum_k x_i^k + v_i^r = d_i^r, \quad \forall\, i, r \tag{11}$$

where v_i^r is integer and nonnegative. Constraints (3)–(4) can be added to this formulation. In the stochastic programming terminology the problem (10)–(11) is the *deterministic equivalent* of the problem (8).

4.2 *Stochastic FAP with recourse (adaptive case)*

Here we assume that there exist a technical possibility to change the frequency assignment on a semi-permanent basis. The obvious use of such capability is to react flexibly to changing demand patterns updating assignment after the current demand pattern becomes known. Two cases can be distinguished.

— There are no costs or other inconveniences involved in the changing of the current frequency assignment. In this case the problem is decomposed into a collection of deterministic frequency assignment problems of the type (1)–(4) with each problem corresponding to a given demand pattern. These problems are solved when the specific demand pattern emerges and their solution is implemented. In the case when there is a fixed collection of alternating demand scenarios these problems can be solved once and for all at the stage of the system design.

— Reassignment costs are substantial enough to be taken into account. This is the problem which is going to be discussed in this section.

In this case we have the current demand pattern $d = (d_1, \ldots, d_m)$ and the collection of future demand scenarios $d^r = (d_1^r, \ldots, d_m^r)$ which will materialize with probabilities p_r. The full description of the future uncertainty may include also the interference patterns, then each scenario will be $(d_1^r, \{w_{ij}^{klr}\})$ where $\{w_{ij}^{klr}\}$ is the interference attained under scenario r if frequency k is assigned to site i and frequency l is assigned to site j. One way to simplify the description of the future uncertainty is to consider the interference patterns to be the same for all scenarios and equal to their average values $\bar{w}_{ij}^{kl}$.

The problem is to select the current assignment $X = \{x_i^k\}_{i=1:n}^{k=1:m}$ minimizing the total costs which consist of the following components.

1. The QoS degradation cost due to the interference.
2. The QoS degradation cost due to partial satisfaction of current demand.
3. The averaged anticipated costs associated with the period of the future when the frequency reassignment will be performed. These costs are defined for each scenario and include:
   - the QoS degradation cost due to the interference for new assignment;
   - the anticipated cost of QoS degradation due to the partial satisfaction of demand;
   - the cost for updating the frequency assignment, this cost is assumed to be proportional to the number of frequency reassignments.

Repeating the argument of the previous section which led to the problem (10)–(11) from the problem (8) we obtain the following problem:

$$\min_{x_i^k, v_i} \sum_{i,j,k,l} \bar{w}_{ij}^{kl} x_i^k x_j^l + g_1 \sum_i v_i + g_2 \sum_r p_r Q_r(X) \tag{12}$$

$$\sum_k x_i^k + v_i = d_i, \quad \forall i \tag{13}$$

where g_2 is the coefficient which makes commeasurable average combined future costs with the current ones. This problem belongs to the class of *stochastic problems with recourse* which is an important problem class in stochastic programming (Birge and Louveaux, 1997). $Q_r(X)$ is the combined anticipated reassignment and QoS deterioration cost for a given demand scenario d^r. This cost depends on the current frequency assignment X. The cost $Q_r(X)$ is obtained from the solution of the following *recourse* problem solved for each demand scenario $r = 1, \dots, R$

$$\min_{x_i^{kr}, y_i^{kr}, z_i^{kr}, v_i^r} \sum_{i,j,k,l} w_{ij}^{klr} x_i^{kr} x_j^{lr} + g_1 \sum_i v_i^r + g_3 \sum_i \sum_k (y_i^{kr} + z_i^{kr}) \tag{14}$$

$$x_i^{kr} + z_i^{kr} - y_i^{kr} = x_i^k, \quad \forall i, k \tag{15}$$

$$\sum_k x_i^{kr} + v_i^r = d_i^r, \quad \forall i \tag{16}$$

$$x_i^{kr}, y_i^{kr}, z_i^{kr} \in \{0,1\}, \quad v_i^r \geq 0$$

The following new notations are introduced in the description of problem (14)–(16):

x_i^{kr} — binary variable which equals 1 if frequency k is assigned to site i under scenario r;
y_i^{kr} — binary variable which equals 1 if frequency k is newly assigned to site i under scenario r;
z_i^{kr} — binary variable which equals 1 if frequency k is released from site i under scenario r;
g_3 — cost of releasing or assigning a frequency.

The first and the second terms in the expression (14) represent respectively the QoS degradation cost due to interference and the partial demand satisfaction. The last term represents reassignment costs. Variables y_i^{kr}, z_i^{kr} and constraints (15) are introduced to take account of the reassignment costs while (16) is the demand satisfaction constraint. Substituting (14)–(16) into (12)–(13) we obtain the following *deterministic equivalent* formulation:

$$\min_{\substack{x_i^k, v_i, \\ x_i^{kr}, y_i^{kr}, z_i^{kr}, v_i^r}} \sum_{i,j,k,l} \bar{w}_{ij}^{kl} x_i^k x_j^l + g_1 \sum_i v_i + g_2 \sum_r p_r$$

$$\times \left(\sum_{i,j,k,l} w_{ij}^{klr} x_i^{kr} x_j^{lr} + g_1 \sum_i v_i^r + g_3 \sum_i \sum_k (y_i^{kr} + z_i^{kr}) \right) \tag{17}$$

$$\sum_k x_i^k + v_i = d_i, \quad \forall i \tag{18}$$

$$\sum_k x_i^{kr} + v_i^r = d_i^r, \quad \forall i, r \tag{19}$$

$$x_i^{kr} + z_i^{kr} - y_i^{kr} = x_i^k, \quad \forall i, k, r \tag{20}$$

$$x_i^k, x_i^{kr}, y_i^{kr}, z_i^{kr} \in \{0,1\}, \quad v_i, v_i^r \geq 0$$

This problem is becoming simpler if the full demand satisfaction is assumed under all demand scenarios. In this case we obtain:

$$\min_{\substack{x_i^k,\\ x_i^{kr}, y_i^{kr}, z_i^{kr}}} \sum_{i,j,k,l} \bar{w}_{ij}^{kl} x_i^k x_j^l + g_2 \sum_r p_r \left(\sum_{i,j,k,l} w_{ij}^{klr} x_i^{kr} x_j^{lr} + g_3 \sum_i \sum_k (y_i^k + z_i^k) \right) \tag{21}$$

$$\sum_k x_i^k = d_i, \quad \forall i \tag{22}$$

$$\sum_k x_i^{kr} = d_i^r, \quad \forall i, r \tag{23}$$

$$x_i^{kr} + z_i^{kr} - y_i^{kr} = x_i^k, \quad \forall i, k, r \tag{24}$$

$$x_i^k, x_i^{kr}, y_i^{kr}, z_i^{kr} \in \{0, 1\}$$

Additional constraints (3)–(4) can be added to these formulations. Solutions of these problems will give the frequency assignment which balances between the current QoS degradation due to interference and the future reassignment and interference costs. Note that only the current assignment X will be implemented. In the future, when the demand pattern changes, the problem will be solved again taking into account the new information about the demand scenarios and the solution of this new problem will be implemented instead of assignments x_i^{kr}. The process will continue with moving window of two periods.

5 Semidefinite relaxations for stochastic FAP

In this section we are going to propose several semidefinite relaxations of stochastic FAP models described in the previous section. First we concentrate on the non-adaptive case and the expectation model (10)–(11). This allows us to develop the conceptual approach for utilization of semidefinite relaxations within the stochastic programming framework and perform a full fledged feasibility study including numerical experiments. After this we provide SDP relaxation for more demanding adaptive case with stochastic FAP recourse problem (12)–(16).

We are going to exploit two ways to solve the stochastic programming problems with the finite number of scenarios (Birge and Louveaux, 1997). One possibility is to construct deterministic equivalent of the problem and apply to it the general solution procedures. Another way is to exploit the specific structure of the stochastic program in order to develop the iterative decomposition procedure like in Van Slyke and Wets (1969). These two approaches correspond to two different types of semidefinite relaxation of the problem (10)–(11) which we develop in this paper. First we are going to apply the semidefinite relaxation to the deterministic equivalent (10)–(11). Another way is to use SDP relaxation inside a decomposition solution procedure to obtain upper and lower bounds for the solution of the original problem.

5.1 *SDP relaxation of the deterministic equivalent problem (nonadaptive case)*

Let us reformulate the problem (10)–(11) a little bit relaxing the equality in the constraint (11).

$$\min_{x_i^k, v_i^r} \sum_{i,j,k,l} \bar{w}_{ij}^{kl} x_i^k x_j^l + g \sum_r \sum_i p_r v_i^r \tag{25}$$

$$\sum_k x_i^k + v_i^r \geq d_i^r, \quad \forall i, r \tag{26}$$

and suppose that $v_i^r \in \{0,1\}$. This assumption covers the largest part of practical situations and in the case when it is too restrictive we can introduce several 0–1 variables which will model the single integer variable v_i^r. Similarly to the section 3 we introduce $y_{i,j}^{l,k} = x_i^l x_j^k$ together with the following notations:

$$Y = \begin{bmatrix} Y_{1,1} & \cdots & Y_{1,n} \\ \vdots & \ddots & \vdots \\ Y_{n,1} & \cdots & Y_{n,n} \end{bmatrix}, \quad Y_{i,j} = \begin{bmatrix} y_{i,j}^{1,1} & \cdots & y_{i,j}^{1,m} \\ \vdots & \ddots & \vdots \\ y_{i,j}^{m,1} & \cdots & y_{i,j}^{m,m} \end{bmatrix},$$

$$\tilde{W} = \begin{bmatrix} \tilde{W}_{1,1} & \cdots & \tilde{W}_{1,n} \\ \vdots & \ddots & \vdots \\ \tilde{W}_{n,1} & \cdots & \tilde{W}_{n,n} \end{bmatrix}, \quad \tilde{W}_{i,j} = \begin{bmatrix} \tilde{w}_{i,j}^{1,1} & \cdots & \tilde{w}_{i,j}^{1,m} \\ \vdots & \ddots & \vdots \\ \tilde{w}_{i,j}^{m,1} & \cdots & \tilde{w}_{i,j}^{m,m} \end{bmatrix}$$

$$V_r^T = (v_1^r, \ldots, v_N^r), \quad U = \begin{bmatrix} Y & 0 & \cdots & 0 \\ 0 & V_1 V_1^T & \cdots & 0 \\ \vdots & \vdots & \ddots & \vdots \\ 0 & 0 & \cdots & V_R V_R^T \end{bmatrix},$$

$$D = \begin{bmatrix} \tilde{W} & 0 & \cdots & 0 \\ 0 & gp_1 I_N & \cdots & 0 \\ \vdots & \vdots & \ddots & \vdots \\ 0 & 0 & \cdots & gp_N I_N \end{bmatrix}$$

where I_N is $N \times N$ unit matrix. Then SDP lifting of (25) is the following:

$$\min_U trace(D * U) \tag{27}$$

subject to constraint

$$U - diag(U) diag(U)^T \succeq 0 \tag{28}$$

In order to represent constraints (26) let us consider a submatrix of matrix U which is denoted by U_{ir}

$$U_{ir} = \begin{bmatrix} Y_{ii} & 0 \\ 0 & (v_i^r)^2 \end{bmatrix}$$

Then the corresponding constraint is

$$trace(U_{ir}) \geq d_i^r \tag{29}$$

Thus, positive semidefinite relaxation of problem (25)–(26) is the problem (27)–(29) where co-cite and co-station constraints should be added similarly to (7).

5.2 *SDP relaxation within decomposition algorithm (nonadaptive case)*

Let us first describe the Benders decomposition algorithm for this problem. After that SDP relaxation will be used to provide an approximation to the solution of the master problem which is necessary to solve on each iteration of the decomposition algorithm. The advantage of this approach is that it allows to substitute the solution of large deterministic equivalent problem by solution of a succession of considerably smaller problems. This approach proved to be very productive in stochastic programming (Birge and Louveaux, 1997; Ermoliev and Wets, 1988; Kall and Wallace, 1994).

1. *Initialization.* Select a feasible frequency assignment x_i^{k0} and take $u^{+0} = +\infty$, $u^{-0} = -\infty$, these are the current upper and lower bounds to the optimal value of the problem.

2. *Generic step.* Suppose that by the beginning of step s we have the current frequency assignment x_i^{ks} and bounds u^{+s}, u^{-s}. Then on step s we do the following:

2a. Solve subproblem

$$\min_{v_i^r} \sum_r \sum_i gp_r v_i^r \tag{30}$$

$$\sum_k x_i^{ks} + v_i^r \geq d_i^r, \quad \forall i, r \tag{31}$$

which has the dual

$$\max_{\mu_i^r \geq 0} \sum_r \sum_i \mu_i^r \left(d_i^r - \sum_k x_i^{ks} \right)$$

$$0 \leq \mu_i^r \leq gp_r, \quad \forall i, r$$

with the explicit solution

$$\mu_i^{rs} = \begin{cases} gp_r & \text{if} \qquad d_i^r \geq \sum_k x_i^{ks} \\ 0 & \text{otherwise} \end{cases} \tag{32}$$

2b. Add cut

$$z_0 \geq \sum_{i,j,k,l} \bar{w}_{ij}^{kl} x_i^k x_j^l + \sum_i \sum_r \left(d_i^r - \sum_k x_i^k \right) \mu_i^{rs}$$

to the master problem and take

$$u^{+0} = \min \left\{ u^{+0}, \sum_{i,j,k,l} \bar{w}_{ij}^{kl} x_i^{ks} x_j^{ls} + \sum_i \sum_r \left(d_i^r - \sum_k x_i^{ks} \right) \mu_i^{rs} \right\}$$

for the new upper bound for the optimal value. Solve the master problem

$$\bar{z}_0 = \min_{z_0, x_i^k} z_0 \tag{33}$$

$$z_0 - \sum_{i,j,k,l} \bar{w}_{ij}^{kl} x_i^k x_j^l + \sum_i \sum_k x_i^k \sum_r \mu_i^{rq} \geq \sum_i \sum_r d_i^r \mu_i^{rq}, \quad q = 1, \ldots, s \tag{34}$$

and take $u^{-0} = \bar{z}_0$. Stop if $u^{-0} - u^{-0} < \epsilon$ where ϵ is some prespecified tolerance. Otherwise take $x_i^{k,s+1}$ to be the solution of (33)–(34) and go to the iteration $s+1$.

In this algorithm subproblems (30)–(31) has explicit solution, but the master problem (33)–(34) is very difficult. We use SDP relaxation to obtain the lower bound for it. We have to lift the inequality (34) into the space of positive semidefinite matrices. Observe that this inequality can be expressed as follows:

$$trace\left(\begin{bmatrix} 1 & b_q^T \\ b_q & -\tilde{W} \end{bmatrix} * \begin{bmatrix} z_0 & x^T \\ x & Y \end{bmatrix}\right) \geq \sum_i \sum_r d_i^r \mu_i^{rq}$$

where

$$x = (x_1^1, \ldots, x_1^m, \ldots, x_n^1, \ldots, x_n^m),$$

$$b_q = \frac{1}{2}\left(\overbrace{b_{1q}, \ldots, b_{1q}}^{n \text{ times}}, \ldots, \overbrace{b_{nq}, \ldots, b_{nq}}^{n \text{ times}}\right), \quad b_{iq} = \sum_r \mu_i^{rq}$$

Denoting now

$$U = \begin{bmatrix} z_0 & x^T \\ x & Y \end{bmatrix}, \quad A = \begin{bmatrix} 1 & 0_{nm}^T \\ 0_{nm} & 0_{nmxnm} \end{bmatrix},$$

$$B_q = \begin{bmatrix} 1 & b_q^T \\ b_q & -\tilde{W} \end{bmatrix}, \quad C = \begin{bmatrix} 0 & -\frac{1}{2}1_{nm}^T \\ -\frac{1}{2}1_{nm} & I_{nm} \end{bmatrix}, \quad D = \begin{bmatrix} 0 & 0_{nm} \\ 0_{nm} & I_{nm} \end{bmatrix}$$

we obtain the following relaxation of the master problem:

$$\min_U trace(AU) \tag{35}$$

$$trace(B_q U) \geq \sum_i \sum_r d_i^r \mu_i^{rq}, \quad q = 1, \ldots, s \tag{36}$$

$$DU - diag(DU) diag(DU)^T \succeq 0 \tag{37}$$

Compared to SDP relaxation of the whole deterministic equivalent developed in the previous subsection, this approach requires repeated solution of SDP relaxation which dimension is much smaller and grows by one cut with each iteration. This SDP relaxation problem (35)–(37) should be solved on each iteration of the decomposition algorithm and its solution should be mapped to the space of 0–1 variables.

5.3 SDP relaxation of the deterministic equivalent for stochastic FAP with recourse (the adaptive case)

We consider here the problem (21)–(24) where the demand is fully satisfied. SDP relaxation of the more general problem (17)–(20) can be developed in the similar manner. To get a semidefinite relaxation of this problem, we extend the notations used above for the deterministic equivalent of non adaptive case. Let us introduce $y_{i,j}^{l,k,r} = x_i^{lr} x_j^{kr}$ together with the following notations:

$$\bar{Y}^r = \begin{bmatrix} Y_{1,1}^r & \cdots & Y_{1,n}^r \\ \vdots & \ddots & \vdots \\ Y_{n,1}^r & \cdots & Y_{n,n}^r \end{bmatrix}, \quad Y_{i,j}^r = \begin{bmatrix} y_{i,j}^{1,1,r} & \cdots & y_{i,j}^{1,m,r} \\ \vdots & \ddots & \vdots \\ y_{i,j}^{m,1,r} & \cdots & y_{i,j}^{m,m,r} \end{bmatrix}$$

$$\bar{Y} = \begin{bmatrix} \bar{Y}^1 & & \\ & \ddots & \\ & & \bar{Y}^r \end{bmatrix}$$

$$V_r = (y_1^{1r}, \ldots, y_1^{Mr}, y_N^{1r}, \ldots, y_N^{Mr}), \quad Z_r = (z_1^{1r}, \ldots, z_1^{Mr}, z_N^{1r}, \ldots, z_N^{Mr})$$

$$U = \begin{bmatrix} Y & 0 & 0 & \cdots & 0 & \cdots & \cdots & 0 \\ 0 & \bar{Y} & 0 & \cdots & 0 & \cdots & \cdots & 0 \\ 0 & 0 & V_1 V_1^T & \cdots & 0 & \cdots & \cdots & 0 \\ \vdots & \vdots & \vdots & \ddots & \vdots & & & 0 \\ 0 & 0 & 0 & \cdots & V_R V_R^T & 0 & \cdots & 0 \\ 0 & 0 & 0 & \cdots & 0 & Z_1 Z_1^T & \cdots & 0 \\ \vdots & \vdots & \vdots & \vdots & \vdots & \vdots & \ddots & \vdots \\ 0 & 0 & 0 & \cdots & 0 & 0 & \cdots & Z_R Z_R^T \end{bmatrix}$$

$$\tilde{W} = \begin{bmatrix} \tilde{W}_{1,1} & \cdots & \tilde{W}_{1,n} \\ \vdots & \ddots & \vdots \\ \tilde{W}_{n,1} & \cdots & \tilde{W}_{n,n} \end{bmatrix}, \quad \tilde{W}_{i,j} = \begin{bmatrix} \tilde{w}_{i,j}^{1,1} & \cdots & \tilde{w}_{i,j}^{1,m} \\ \vdots & \ddots & \vdots \\ \tilde{w}_{i,j}^{m,1} & \cdots & \tilde{w}_{i,j}^{m,m} \end{bmatrix}$$

$$\bar{W}^r = \begin{bmatrix} W_{1,1}^r & \cdots & W_{1,n}^r \\ \vdots & \ddots & \vdots \\ W_{n,1}^r & \cdots & W_{n,n}^r \end{bmatrix}, \quad W_{i,j}^r = \begin{bmatrix} w_{i,j}^{1,1,r} & \cdots & w_{i,j}^{1,m,r} \\ \vdots & \ddots & \vdots \\ w_{i,j}^{m,1,r} & \cdots & w_{i,j}^{m,m,r} \end{bmatrix}$$

$$\bar{W} = \begin{bmatrix} \bar{W}^1 & & \\ & \ddots & \\ & & \bar{W}^r \end{bmatrix}$$

$$D = \begin{bmatrix} \tilde{W} & 0 & 0 & \cdots & 0 & 0 & \cdots & 0 & \cdots & 0 \\ 0 & g_2p_1\bar{W}^1 & 0 & \cdots & 0 & 0 & \cdots & 0 & \cdots & 0 \\ \vdots & \vdots & \ddots & \vdots & 0 & 0 & \cdots & 0 & \cdots & 0 \\ 0 & 0 & \cdots & g_2p_r\bar{W}^r & 0 & 0 & \cdots & 0 & \cdots & 0 \\ 0 & 0 & \cdots & 0 & g_2p_1g_3I_{NM} & \cdots & 0 & 0 & \cdots & 0 \\ \vdots & \vdots & \cdots & \vdots & \vdots & \ddots & \vdots & \vdots & \cdots & \vdots \\ 0 & 0 & \cdots & 0 & 0 & \cdots & g_2p_rg_3I_{NM} & 0 & \cdots & 0 \\ 0 & 0 & 0 & \cdots & 0 & 0 & \cdots & g_2p_1g_3I_{NM} & \cdots & 0 \\ \vdots & \vdots & \vdots & \vdots & \vdots & \vdots & \vdots & \vdots & \ddots & \vdots \\ 0 & 0 & 0 & \cdots & 0 & 0 & \cdots & 0 & \cdots & g_2p_rg_3I_{NM} \end{bmatrix}$$

where I_{NM} is $NM \times NM$ unit matrix.

We reformulate the matrix U by using the following sub-matrices $U_1 = Y$, $U_2 = \bar{Y}$,

$$U_3 = \begin{bmatrix} V_1V_1^T & \cdots & 0 \\ \vdots & \ddots & \vdots \\ 0 & \cdots & V_RV_R^T \end{bmatrix}, \quad U_4 = \begin{bmatrix} Z_1Z_1^T & \cdots & 0 \\ \vdots & \ddots & \vdots \\ 0 & \cdots & Z_RZ_R^T \end{bmatrix}$$

then

$$U = \begin{bmatrix} U_1 & 0 & 0 & 0 \\ 0 & U_2 & 0 & 0 \\ 0 & 0 & U_3 & 0 \\ 0 & 0 & 0 & U_4 \end{bmatrix}$$

With the help of these notations the objective function can be written as:

$$\min_U trace(D * U)$$

and the constraints (22) and (23) are given by:

$$trace((U_1)_{ii}) = d_i, \quad \forall i$$

$$trace((U_2)_{ii}^r) = d_i^r, \quad \forall i, r$$

The lifting of constraint (24) requires some additional notations. We define the following matrices:

- $(nm \times nm)$–matrix $(A_1)_i^k$ where the diagonal component $im + k$ is equal to -1.
- $(nmR \times nmR)$–matrix $(A_2)_i^{k,r}$ where the component with row number $(r-1)nm + im + k$ and column number $(r-1)nm + im + k$ is equal to 1

All other components of these matrices are equal to zero.

Taking finally

$$A_i^{k,r} = \begin{bmatrix} (A_1)_i^k & 0 & 0 & 0 \\ 0 & (A_2)_i^{k,r} & 0 & 0 \\ 0 & 0 & -(A_2)_i^{k,r} & 0 \\ 0 & 0 & 0 & (A_2)_i^{k,r} \end{bmatrix}$$

we can write constraints (24) as follows:

$$trace(A_i^{k,r} * U) = 0, \quad \forall i, k, r$$

Putting together all the components derived so far we obtain the following SDP relaxation of the stochastic FAP problem with recource:

$$\min_U trace(D * U) \tag{38}$$

$$trace((U_1)_{ii}) = d_i, \quad \forall i, \tag{39}$$

$$trace((U_2)_{ii}^r) = d_i^r, \quad \forall i, r, \tag{40}$$

$$trace(A_i^{k,r} * U) = 0, \quad \forall i, k, r, \tag{41}$$

$$U - diag(U) diag(U)^T \succeq 0. \tag{42}$$

6 Numerical experiments

In this section we test the SDP relaxation of Benders decomposition and deterministic equivalent problem for the nonadaptive stochastic FAP. The objective of these experiments is to validate the feasibility of our approach and create a toolbox for further development of this algorithmic scheme for full fledged recourse problem (14)–(16). The data set is based on a generic data set characterizing the mobile cellular networks. All experiments are carried out on a PC using Linux operating system. The algorithms are coded in C++. We use CPLEX mip solver (IBM, 1990–2011) to solve linear programming relaxations and the bundle spectral code (Helmberg and Rendl, 1999) to solve the SDP relaxation problems.

Table 1 gives details on dimensions of the test problems. The column *sites* contains the number of sites and *frequencies* contains the radio resources expressed in terms of frequencies. We solved four variants of each instance. Each variant is related to the number of scenarios generated, i.e. 5, 10, 15 and 30 scenarios respectively.

As what can be expected from a feasibility study, the size of the instances is relatively small due to our emphasis on stochastic features of

Table 1 Problem Instances.

Instances	**Sites**	**Frequencies**
Fap1	3	12
Fap2	5	14
Fap3	6	16
Fap4	8	20

stochastic FAP. In fact, the difficulty of stochastic FAP can be considerable. This can be shown by linearizing the quadratic terms of the deterministic equivalent FAP in order to reduce the problem to a linear one. The resulting LP sizes are given in Table 2 for each instance and each variant. The number of 0–1 variables ranges from 792 to 13280. This exceeds the capabilities of the current state of the art LP commercial packages. For instance, CPLEX mip solver solves to optimality only $FAP1$ and $FAP2$ and their variants. Moreover, it is well known from combinatorial optimization that linear relaxation of such problems provide weak lower bounds, i.e. $z_{0_{lb}} = 0$ for all tested instances.

Table 3 gives details of SDP master program. The column *Variables* contains the number of variables and the column *Constraints* gives the number of constraints. Columns under 'MP Iterations Number' correspond to the number of iterations of Benders decomposition which coincide with the number of solutions of SDP relaxation of the master program. Observe that the master program iterations are less than 20 for all instances. The number of constraints corresponds to the last master program i.e. once all Benders cuts are added.

Tables 4 and 5 give details of Benders decomposition results. The column marked by LB represents the lower bound. The column marked by OPT is the integer optimal solution given by solving deterministic LP relaxations by CPLEX. Notice that only the best feasible solutions are reported for *Fap*3 and *Fap*4. The column marked by UB contains the upper bound given by Benders Decomposition. The column under GAP is the duality gap, i.e. $\frac{UB-LB}{LB}$.

CPLEX did not solve instances with more than 5 scenarios for *Fap*3 and *Fap*4 and it has not found a feasible solution after hours of computing. Benders decomposition solves all instances with a small duality gap i.e. 10.6% for the worst case and around 5% in average.

Table 6 details the balance between costs and penalties in the objective function (25). The column COST gives the cost part in the objective function which corresponds to interferences whilst the column PENALTY

Table 2 The Size of LP Equivalents.

Instances	S = 5		S = 10		S = 15		S = 30	
	Variables	Constraints	Variables	Constraints	Variables	Constraints	Variables	Constraints
Fap1	717	2076	732	2091	747	2106	792	2151
Fap2	2580	7765	2605	7790	2630	7815	2705	7890
Fap3	4782	14172	4812	14202	4842	14232	4932	14322
Fap4	13080	38976	13120	39016	13160	39056	13280	39176

Table 3 The Sizes of SDP Relaxations.

	Master SDP Program					
			MP Iterations Number			
Instances	**Variables**	**Constraints**	**S = 5**	**S = 10**	**S = 15**	**S = 30**
Fap1	666	100	4	3	7	4
Fap2	2484	356	7	7	10	9
Fap3	4656	271	7	9	6	7
Fap4	12880	457	15	19	16	17

Table 4 Lower and Upper Bounds for 5 and 10 Scenarios.

	S = 5				S = 10			
Data	**LB**	**OPT**	**UB**	**GAP**	**LB**	**OPT**	**UB**	**GAP**
Fap1	312	320	324	3.8	292	301	310	6.2
Fap2	766	776	803	4.8	741	759	807	8.9
Fap3	915	940†	955	4.4	820	863†	907	10.6
Fap4	9772	11157†	10186	4.2	9850	‡	10250	4.1

†: Best feasible solution found by CPLEX.
‡: No feasible solution found by CPLEX.

Table 5 Lower and Upper Bounds for 5 and 10 Scenarios.

	S = 15				S = 30			
Data	**LB**	**OPT**	**UB**	**GAP**	**LB**	**OPT**	**UB**	**GAP**
Fap1	296	302	304	2.7	299	299	299	0
Fap2	932	965	1002	7.5	970	983	1004	3.5
Fap3	875	903†	910	4.0	996	1017†	1029	3.3
Fap4	10336	‡	11036	6.8	9428	‡	9837	4.3

†: Best feasible solution found by CPLEX.
‡: No feasible solution found by CPLEX.

corresponds to the part given by unsatisfied demands. Observe that penalty part grows with the size of instances and number of scenarios. Our approach gives advantage to penalty rather than to interferences, i.e. the model prefer reject demand and minimize interferences. This strategy seems to be more realistic in practice.

Summarizing, these numerical experiments show the advantage of the solution approach based on combination of SDP relaxation with Benders decomposition over brute force application of standard commercial software to deterministic equivalent of stochastic SDP problem. They confirm that Stochastic SDP can be solved by exploiting its structure and

Table 6 Costs Versus Penalties in the SDP Formulation.

	S = 5			S = 10			S = 15			S = 30		
Data	**UB**	**Cost**	**Penalty**	**UB**	**Cost**	**Penalty**	**UB**	**Cost**	**Penalty**	**UB**	**Cost**	**Penalty**
Fap1	324	164	160	310	145	165	304	122	182	299	80	219
Fap2	803	223	620	807	267	540	1002	214	978	832	200	632
Fap3	955	279	676	907	237	570	910	250	660	1029	199	830
Fap4	10186	2286	7900	10250	3050	7200	11036	1122	9914	9837	1016	8821

provide the motivation for development of decomposition schemes for more sophisticated stochastic SDP models.

7 Summary

This paper presents a stochastic programming approach to modeling and solving of the Frequency Assignment Problem under uncertainty. From the beginning it was not clear whether it is feasible to add uncertainty modeling level to this problem which is already very challenging in deterministic setting. This paper gives a positive answer to this question which is its main contribution. We developed the modeling framework and numerical approaches based on Semidefinite Programming. Numerical experiments confirm that our numerical approach based on exploitation of the problem structure enables solution of problems not solvable with direct application of the benchmark commercial software. The further research will be using this framework to extend these results to more sophisticated problems with higher dimensions.

References

Aardal, K., S. Van Hoesel, A. Koster, C. Mannino, and A. Sassano. Models and solution techniques for frequency assignment problems. *4OR*, 1: pp. 261–317, 2003.

Ahmed, S., M. Tawarmalani, and N. V. Sahinidis. A finite branch and bound algorithm for two-stage stochastic integer programs. *Mathematical Programming*, 100: pp. 355–377, 2004.

Andrade, R., A. Lisser, N. Maculan, and G. Plateau. Planning network design under uncertainty with fixed charge. *Computational Optimization and Applications*, 29(2): pp. 127, 2004.

Andrade, R., A. Lisser, N. Maculan, and G. Plateau. B&B frameworks for the capacity expansion of high speed telecommunication networks under uncertainty. *Annals of Operations Research*, 140(1): pp. 49–65, 2005.

Birge, J. R. and F. Louveaux. *Introduction to Stochastic Programming.* New York: Springer, 1997.

Bonatti, M., A. Gaivoronski, P. Lemonche, and P. Polese. Summary of some traffic engineering studies carried out within RACE project R1044. *European Transactions on Telecommunications*, 5: pp. 207–218, 1994.

Caroe, C. C. and R. Schultz. Dual decomposition in stochastic integer programming. *Operations Research Letters*, 24(1–2): pp. 37–45, 1999.

Cela, F. *The Quadratic Assignment Problem: Theory and Algorithms.* Massachussets, USA: Kluwer, 1998.

Eisenblätter Andreas, Martin Grötschel, and Arie M. C. A. Koster. Frequency planning and ramifications of coloring. *Discussions Mathematicae, Graph Theory*, (22): pp. 51–88, 2002.

Ermoliev, Yu. and R. J.-B. Wets, eds., *Numerical Techniques for Stochastic Optimization.* Berlin: Springer Verlag, 1988.

Gaivoronski, A. A. Stochastic optimization problems in telecommunications. In S. W. Wallace and W. T. Ziemba, eds., *Applications of Stochastic Programming*, pp. 669–704. SIAM & MPS, 2005.

Gaivoronski, A. A., A. Lisser, R. Lopez and H. Xu. Knapsack problem with probability contraints. *Journal of Global Optimization*, 49(3): pp. 397–493, 2011.

Goemans, M. and F. Rendl. Combinatorial optimization. In H. Wolkowicz, R. Saigal, and L. Vandenberghe, eds., *Handbook of Semidefinite Programming: Theory, Algorithms and Applications*, pp. 343–360. Kluwer, 2000.

Klein Haneveld, W. K. and M. H. Van der Vlerk. Stochastic integer programming: General models and algorithms. *Annals of Operations Research*, 85: pp. 39–57, 1999.

Helmberg, C. and F. Rendl. A spectral bundle method for semidefinite programming. *SIAM Journal on Optimization*, 10(3): pp. 673–696, 1999.

IBM. *IBM ILOG CPLEX Optimizer*, 1990–2011.

Kall, P. and S. Wallace. *Stochastic Programming.* New York: John Wiley and Sons, 1994.

Laporte, G. and F. V. Louveaux. The integer L-shaped method for stochastic integer programs with complete recourse. *Operations Research Letters*, 13: pp. 133–142, 1993.

Murphey, R. A., P. M. Pardalos, and M. G. C. Resende. Frequency assignment problems. In Z. Du D and P.M. Pardalos, eds., *Handbook of Combinatorial Optimization, Supplement Volume A.* Kluwer Academic Publishers, 1999.

Schultz, R., L. Stougie, and M. H. Van der Vlerk. Solving stochastic programs with complete integer recourse: A framework using Groebner Bases. *Mathematical Programming*, 83(2): pp. 229–252, 1998.

Sen, S., R. D. Doverspike, and S. Cosares. Network planning with random demand. *Journal of Telecommunications Systems*, 3: pp. 11–30, 1994.

Sen, S. and H. D. Sherali. Decomposition with branch-and-cut approaches for two stage stochastic mixed-integer programming. Technical report, University of AZ, 2004.

Tomasgard, A., J. A. Audestad, S. Dye, L. Stougie, M. H. Van der Vlerk, and S. W. Wallace. Modelling aspects of distributed processing in telecommunications networks. *Annals of Operations Research*, 82: pp. 161–184, 1998.

Van Slyke, R. and R. J.-B. Wets. L-shaped linear programs with application to optimal control and stochastic programming. *SIAM Journal on Applied Mathematics*, 17: pp. 638–663, 1969.

Wallace, S. W. and W. T. Ziemba, eds., *Applications of Stochastic Programming.* MPS-SIAM Series in Optimization. SIAM & MPS, 2005.

Wets, R.J.-B. Stochastic programming: Solution techniques and approximation schemes. In A. Bachem, M. Groetshel, and B. Korte, eds., *Mathematical Programming: The State of the Art.* Berlin: Springer Verlag, 1982.

Wolkowicz, H., R. Saigal, and L. Vandenberghe, eds., *Handbook on Semidefinite Programming. Theory, Algorithms and Applications.* Kluwer Academic Publishers, 2000.

INDEX

632196

A TIARA FOR THE TWENTIETH CENTURY

A Tiara for the Twentieth Century

The Collected Poems of

Suzanne Richardson Harvey

FITHIAN PRESS, MCKINLEYVILLE, CALIFORNIA, 2009

Printed in the United States of America

Published by Fithian Press
A division of Daniel and Daniel, Publishers, Inc.
Post Office Box 2790
McKinleyville, CA 95519
www.danielpublishing.com

Distributed by SCB Distributors (800) 729-6423

LIBRARY OF CONGRESS CATALOGING-IN-PUBLICATION DATA
Harvey, Suzanne R., (date)
A tiara for the twentieth century : the collected poems / by Suzanne Richardson Harvey.
p. cm.
ISBN-13: 978-1-56474-489-0 (pbk. : alk. paper)
ISBN-10: 1-56474-489-2 (pbk. : alk. paper)
I. Title.
PS3608.A78924T53 2009
811'.6--dc22

2009017170